Instrument / Commercial Manual

JEPPESEN
Sanderson Training Products

Second Edition 1988
Third Edition 1989
Fourth Edition 1990
Fifth Edition 1990
Sixth Edition 1992
Seventh Edition 1993
Eighth Edition 1994
Ninth Edition 1994
Tenth Edition 1995
Eleventh Edition 1996

55 Inverness Drive East, Englewood, CO 80112-5498
International Standard Book Number 0-88487-231-9

JS314704L

PREFACE

Congratulations on your decision to continue your pilot training. The Jeppesen Sanderson *Instrument/Commercial Manual* includes the material you need to complete your instrument training, as well as that needed for your commercial pilot certificate. It is based on the "study/review" concept of learning. This means that detailed material is presented in an uncomplicated way, then important points are summarized through use of bold type and color. The manual incorporates many design features that will help you get the most out of your study and review efforts. These include:

- **Margin Notes** — The margin notes, which are printed in color, summarize key points from the text. They emphasize material contained in the FAA computer exams, as well as other important data. You are encouraged to add your own study notes in the wide margins provided on each page.
- **Illustrations** — Illustrations are carefully planned to complement and expand upon concepts introduced in the text. Color in the illustrations and the accompanying captions flag them as items that warrant your attention during both initial study and review.
- **Bold Type** — Important new terms in the text are printed in bold type, then defined.
- **Checklist** — A checklist appears at the end of each section to help you verify your understanding of principal concepts.
- **Abbreviations** — Abbreviations explain many of the most frequently used aviation acronyms and abbreviations.
- **Federal Aviation Regulations** — Appropriate FARs are presented in a separate FAR Reference Book, which includes both instrument and commercial exercises designed to test your understanding of pertinent regulations.

The manual is the key element in the Jeppesen Sanderson training materials. Although it can be studied alone, there are several other components which we recommend to make your flight training as complete as possible. These include the Exercise Book, Syllabus, and Jeppesen FAA Instrument Rating and Commercial Pilot Airmen Knowledge Study Guides. The Study Guides are published in both printed and electronic format. You may also note that the table of contents for your manual contains cross-references to video presentations. These video programs are available for your use at participating schools and are designed to enhance and complement your study. When used together, these various elements provide an ideal framework for you and your instructor as you prepare for the FAA airmen knowledge and practical tests.

TABLE OF CONTENTS

CHAPTER 1

PRINCIPLES OF INSTRUMENT FLIGHT

INTRODUCTION

This chapter provides the foundation for your entire instrument training course. You already have some experience in controlling an aircraft by instrument reference for limited periods of time. Now, you will acquire the knowledge and skills needed to fly instruments on a continuing basis. This technique, called attitude instrument flying, is based on your understanding of the instruments and your skill in interpreting the information they display. You must use this information to achieve precise aircraft control. The first section contains a description of the flight instruments which highlights their unique operating characteristics. In the next section, you learn how to apply that information as you develop your basic attitude flying skills — instrument interpretation, cross-check, and aircraft control. The concluding section covers instrument navigation, completing the final element of basic instrument flight. This section provides important information you will need as you begin flying instrument approaches.

SECTION A

FLIGHT INSTRUMENT SYSTEMS

In the IFR environment, proper instrument interpretation is the basis for aircraft control. In part, this skill depends on an understanding of how a particular instrument or system functions. This section reviews the instrument systems and concentrates on helping you understand and interpret instrument indications. It also covers the limitations of each instrument, as well as the appropriate IFR checks. This knowledge enables you to determine quickly what an instrument is telling you and translate it into a control response.

FARs specify the instruments that must be installed and operating when you fly in IFR weather.

An important part of your aircraft selection and preflight inspection is determining whether it is legal to operate in IFR conditions. Besides the basic instruments required by FARs for VFR flight, if you fly in IFR conditions, your aircraft must have an operating gyroscopic rate-of-turn indicator, slip-skid indicator, gyroscopic bank and pitch indicator (artificial horizon or attitude indicator), and a gyroscopic direction indicator (directional gyro or heading indicator). Your aircraft must also be equipped with a sensitive altimeter which is adjustable for barometric pressure, and a clock that has a sweep-second pointer or digital presentation that displays hours, minutes, and seconds. If any of these instruments are missing or not functioning, you cannot legally depart and fly in IFR weather.

As the pilot in command, you must ensure the required checks and inspections have been done prior to IFR flight in controlled airspace.

FARs also specify the systems that must be inspected before IFR operations. For example, before you can operate an aircraft in controlled airspace under IFR conditions, you must ensure that the static system has been inspected within the preceding 24 calendar months. Other systems, such as the transponder and VOR, have IFR inspection requirements that will be covered in following sections. It is your responsibility as pilot in command to determine that each system has been checked and found to meet FAR requirements for instrument flight.

GYROSCOPIC FLIGHT INSTRUMENTS

Electrical and vacuum-powered gyros are used as backups to each other in case one system fails.

The three gyroscopic instruments found in your aircraft are the attitude indicator, heading indicator, and turn coordinator. A combination of vacuum (suction) and electrical power is typically used to operate these instruments. On most small airplanes, the vacuum system supplies power to the attitude and heading indicators, while the electrical system powers the turn coordinator. This configuration provides a backup in case one system fails.

Gyros are affected by two principles — rigidity in space and precession. **Rigidity in space** means that, once a gyro is spinning, it tends to remain in a fixed position and resists external forces applied to it. This principle allows a gyroscope to be used to measure changes in the attitude or direction of an airplane.

A spinning gyroscope tends to remain in a fixed position and will resist any deflection.

Precession is the tilting or turning of a gyro in response to pressure. The reaction to this pressure does not occur at the point where it was applied; rather, it occurs at a point, in the direction of rotation, 90° from where the pressure was applied. This principle allows the gyro to determine a rate of turn by sensing the amount of pressure created by a change in direction. Precession can also cause minor errors in some instruments.

When force is applied to a gyroscope, the reaction occurs at a point 90° from where the force was actually applied.

ATTITUDE INDICATOR

The attitude indicator provides a substitute for the natural horizon. It is the only instrument that provides an immediate and direct indication of the airplane's pitch and bank attitude. You make pitch changes, in bar widths or fractions of bar widths, with reference to the movement of the horizon bar in relation to the miniature airplane. You make bank changes by adjusting the bank angle of the aircraft in relation to the pointer and bank index of the instrument. [Figure 1-1]

The attitude indicator is the only instrument that simultaneously displays pitch and bank.

ERRORS

Centrifugal force in a turn can cause some attitude indicators to precess, creating errors in both pitch and bank. In vacuum-driven attitude indicators, the pendulous vanes are moved by centrifugal force, and the gyro precesses toward the inside of the turn. These errors are usually minor and result in deviations of no more than five degrees of bank and one bar-width of pitch. The effect is greatest in a 180° steep turn. For example, when you roll out of a 180° steep turn to straight-and-level flight by

The precession caused by centrifugal force can result in a number of small errors in the attitude indicator during a 180° turn.

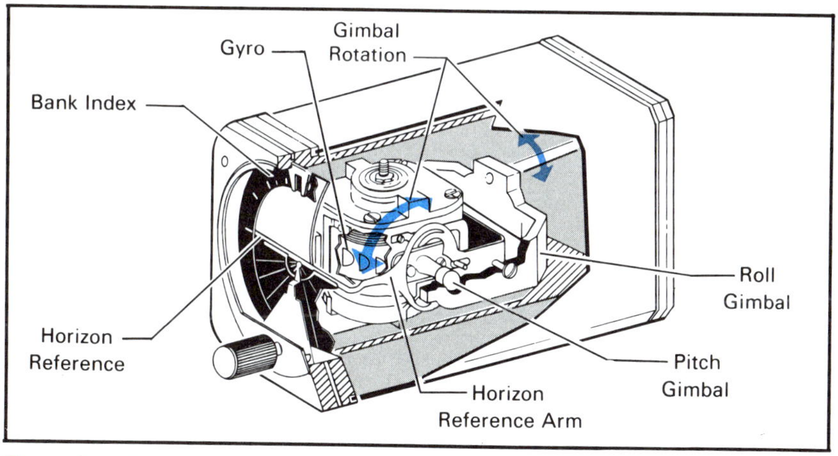

Figure 1-1. The gyro in the attitude indicator spins in the horizontal plane. Two mountings, or gimbals, are used so that both pitch and roll can be sensed simultaneously. Due to rigidity in space, the gyro remains in a fixed position relative to the horizon as the case and airplane rotate around it.

visual reference, the attitude indicator will show a slight climb and turn in the opposite direction. At the end of a 360° turn, the precession you induced during the first 180° is cancelled out by precession in the opposite direction during the second 180° of turn. A skidding turn precesses the gyro toward the inside of the turn. After the aircraft returns to straight-and-level, coordinated flight, the miniature aircraft shows a turn in the direction opposite the skid.

You may also induce precession errors during a rapid acceleration or deceleration.

Acceleration and deceleration also may induce precession errors, depending upon the amount and extent of the force applied. During acceleration, the horizon bar moves down, indicating a climb. If you apply control pressure to correct this indication, it will result in a pitch attitude lower than the instrument shows. The opposite error results from deceleration.

As long as the correct vacuum pressure is maintained, modern attitude indicators are usually very reliable instruments. Some function properly during 360° of roll or 85° of pitch. Gyros in older indicators, however, will "tumble" from their plane of rotation beyond approximately 100° of bank or beyond 60° of pitch. Older indicators often employ caging devices so you can stabilize the spin axis after the gyro has tumbled.

HEADING INDICATOR

The heading indicator displays movement about the vertical axis.

The heading indicator is usually vacuum powered and senses movement about the vertical axis. In most light airplanes, it has no automatic north-seeking capability built into it. [Figure 1-2]

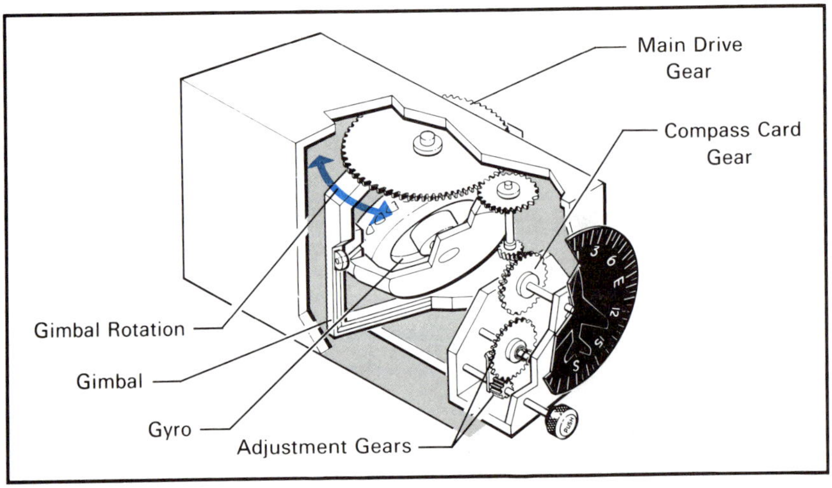

Figure 1-2. The gyro in the heading indicator spins in the vertical plane. Rigidity in space enables it to maintain this alignment. A single gimbal permits the gyro to sense changes in direction.

ERRORS

Banking the airplane induces precession, which can cause the heading to creep or drift from the proper setting. Therefore, you must align the heading indicator with the magnetic compass before flight and check it at approximately 15-minute intervals during flight. When you do an in-flight alignment, make sure you are in straight-and-level, unaccelerated flight to ensure an accurate magnetic compass indication.

Due to precession, you must periodically align the heading indicator with the magnetic compass.

Like attitude indicators, the vacuum-powered heading indicator has operating limits in both pitch and roll. If you exceed these limits, the gyro may tumble, which may damage the instrument. If the indicator has tumbled, you must reset it by aligning it with a known heading or with a stabilized indication from the magnetic compass.

TURN INDICATORS

One of the main functions of the turn coordinator and turn-and-slip indicator is to allow you to establish and maintain constant rate turns. A **half-standard rate turn** takes four minutes for 360° of turn. At this rate, you turn one and one-half degrees per second. A **standard-rate turn** is a turn at a rate of three degrees per second. At this rate, you will complete a 360° turn in two minutes. The bank required is directly related to your true airspeed (TAS). As TAS decreases, you must decrease the angle of bank to maintain a standard-rate turn. Conversely, an increase in true airspeed requires an increase in bank to maintain standard rate. [Figure 1-3]

To maintain a standard-rate turn (three degrees per second), you must increase bank as TAS increases.

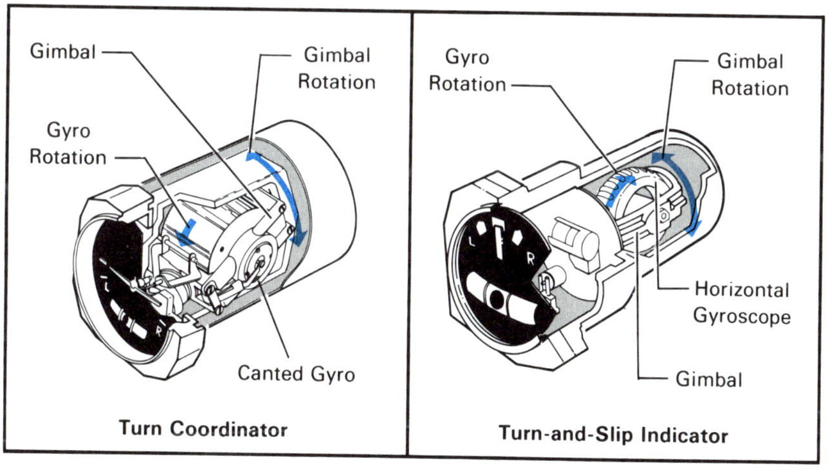

Figure 1-3. The gyros in both the turn coordinator and turn-and-slip indicator are mounted so that they rotate in the vertical plane. The gimbal in the turn coordinator is set at an angle, or canted, which means precession allows the gyro to sense both rate of roll and rate of turn. The gimbal in the turn-and-slip indicator is horizontal. In this case, precession allows the gyro to sense only rate of turn. When the miniature airplane or needle is aligned with the turn index, you are in a standard-rate turn. Since a coordinated turn requires you to bank the airplane, the turn coordinator provides an indirect indication of bank.

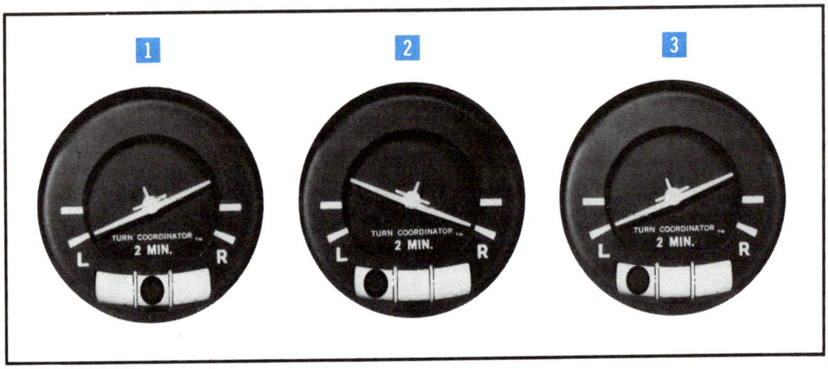

Figure 1-4. In a coordinated turn (instrument 1), the ball is centered. In a skid (instrument 2), the rate of turn is too great for the angle of bank, and the ball moves to the outside of the turn. Conversely, in a slip (instrument 3), the rate of turn is too small for the angle of bank, and the ball moves to the inside of the turn. In both cases, you need to adjust either the angle of bank or the rate of turn.

The ball in the turn coordinator defines the quality of a turn.

Another part of the turn coordinator is the inclinometer. The position of the ball defines the quality of the turn, or whether you have used the correct angle of bank for the rate of turn. [Figure 1-4]

The inclinometer shows the relationship between the opposing horizontal forces in a turn.

Another way to analyze the coordination of a turn is by comparing opposing forces. The horizontal component of lift causes an airplane to turn and opposes centrifugal force. The inclinometer reflects the state of these opposing forces. For example, during a coordinated turn, the horizontal component of lift is balanced by centrifugal force, and the ball in the inclinometer is centered. During a skid, centrifugal force exceeds the horizontal component of lift. This makes the rate of turn too great for the angle of bank, and the ball moves to the outside of the turn. In a slip, the aircraft is banked too much for the rate of turn. Since the horizontal component of lift exceeds centrifugal force, the ball moves to the inside of the turn.

Slipping or skidding also alters the normal load factor you experience in turns. This happens because the wings must generate enough lift to support the weight of the airplane and overcome centrifugal force. Since a skid generates a higher than normal centrifugal force, load factor is increased. In a slip, load factor decreases because centrifugal force is lower than normal.

INSTRUMENT CHECKS

The preflight check of the gyro instruments ensures they are operating properly.

The preflight check of the gyro-operated instruments and their power sources takes on additional importance when you are planning a flight under instrument conditions. It is far better to find an inoperative instrument or system during the preflight than it is to cope with it after you are airborne. Before you turn on the master switch or start the engine, make sure the instruments that have power indicators are displaying off indications. These indicators are your first sign that an instrument has failed. The inclinometer should be full of fluid, with the ball resting at its lowest point. When you turn on the master switch, listen to the electrically driven gyros. There should not be any abnormal noises, such as grinding sounds, that would indicate an impending failure. When you start the engine, listen to the vacuum-driven gyros. The gyros normally reach full operating speed in approximately five minutes. During this time, it is common to see some vibration in the instruments. When the gyros have stabilized, the miniature airplane in the turn coordinator and the horizon bar in the attitude indicator should be level while the airplane is stopped or taxiing straight ahead. During turns, the turn coordinator (or turn-and-slip indicator) and heading indicator should display a turn in the correct direction. The ball in the inclinometer should swing to the outside of the turn. The attitude indicator should not tilt more than five degrees during normal turns. Align the heading indicator with the magnetic compass. Then, recheck it prior to takeoff to ensure it has not precessed significantly. A precession error of no more than 3° in 15 minutes is acceptable for normal operations.

You should include the suction gauge and ammeter in your pretakeoff check to ensure the gyro instruments are receiving adequate power. If the vacuum pressure is outside its normal range, the vacuum-driven indicators become unreliable. This normally affects the attitude and heading indicators. Some airplanes are equipped with a vacuum warning light, which illuminates when the vacuum pressure drops below the acceptable level. Low- and high-voltage warning lights are available on some airplanes to detect electrical problems.

Low vacuum pressure usually causes the attitude and heading indications to become unreliable.

MAGNETIC COMPASS

The magnetic compass is the only direction-seeking instrument in most light airplanes. Although the compass appears to move, it is actually mounted in such a way that the aircraft turns about the compass card as the card maintains its alignment with magnetic north.

ERRORS

The magnetic compass can only give you reliable directional information if you understand its limitations and inherent errors. These include magnetic variation, compass deviation, and magnetic dip.

VARIATION

Magnetic variation is not as significant to IFR operations as it is during VFR flight, since most of your flying is along airways which are aligned with magnetic north. However, you must still be able to convert between true and magnetic north, so let's take just a moment to review. Variation is the angular difference between the true and magnetic poles. The amount of variation depends on where you are located in relation to the earth's magnetic and true north poles. Navigation charts connect points of equal variation by use of isogonic lines. To convert directions expressed in terms of true north to magnetic north, subtract easterly variation from and add westerly variation to the true direction. Reverse the process to convert magnetic to true.

DEVIATION

Deviation is produced by the magnetic fields of the airplane and its electronic equipment.

Magnetic fields are produced by the metal and electrical accessories within the airplane. These distort the lines of magnetic force produced by the earth and cause the compass to swing away from the correct heading. Manufacturers often install compensating magnets within the compass housing to reduce the effects of deviation. These magnets are usually adjusted with the engine running and all electrical equipment operating. Deviation error, however, cannot be completely eliminated. Therefore, you must correct for deviation by using a compass correction card, which is mounted near the compass. Deviation varies from one heading to the next because the lines of force interact at different angles.

MAGNETIC DIP

Acceleration, deceleration, and turning errors are caused by magnetic dip.

Magnetic dip is the result of the vertical component of the earth's magnetic field. This dip is virtually nonexistent at the magnetic equator, since the lines of force are parallel to the earth's surface and the vertical component is minimal. As you move a compass toward the poles, the vertical component increases, and magnetic dip becomes more apparent at higher latitudes. Magnetic dip is responsible for compass errors during acceleration, deceleration, and turns.

Acceleration and deceleration errors occur during speed changes and are most apparent on headings of east and west.

Magnetic dip causes **acceleration and deceleration errors,** which are fluctuations in the compass during changes in speed. In the northern hemisphere, the compass swings toward the north during acceleration and toward the south during deceleration. When the speed stabilizes, the compass returns to an accurate indication. This error is most pronounced when you are flying on headings of east or west and decreases gradually as you fly closer to a north or south heading. The error doesn't occur when you are flying directly north or south. The memory aid, ANDS (Accelerate North, Decelerate South), may help you recall this error. In the southern hemisphere, the error occurs in the opposite direction (accelerate south, decelerate north).

The vertical component of the earth's magnetic field causes a turning error in the compass.

Turning error is most apparent when you are turning to or from a heading of north or south. This error, referred to as northerly turning error, increases as you near the poles, due to magnetic dip and the vertical component of the earth's magnetic field. There is no turning error when you are flying near the equator.

The magnetic compass indication lags during turns from north and leads during turns from south.

When you make a turn from a northerly heading in the northern hemisphere, the compass gives an initial indication of a turn in the opposite direction. It then begins to show the turn in the proper direction, but it lags behind the actual heading. The amount of lag decreases as the turn continues, then disappears as the airplane reaches a heading of east or west. When you are making a turn from a southerly heading, the compass gives an indication of a turn in the correct direction, but it leads the actual heading. This error also disappears as the airplane approaches a heading of east or west.

Accurate compass-only turns require bank limits of 15° to 18° because of the way the compass is constructed. You must lead the roll-out for turns to north and lag the roll-out for turns to south. To compute lead or lag, use the earth's latitude in your local area plus or minus the normal roll-out lead for a 15° banked turn (usually one-half the bank angle). At 35°N latitude, as an example, a right turn to north requires a roll-out point of 318° (360-35-7). A right turn to south requires a roll-out point of 208° (180 + 35 - 7). At the same latitude, the lead points for left turns are 42° when turning to north and 152° when turning to south. Turning errors are reversed in the southern hemisphere.

Latitude must be considered when you are computing the appropriate roll-out point from magnetic compass turns.

OSUN
over shooT South
under Shoot North

INSTRUMENT CHECK

Prior to flight, make sure the compass is full of fluid. During the taxi, the compass should swing freely and indicate known headings. Since the magnetic compass is required for all flight operations, the aircraft should never be flown with a faulty compass.

The compass should be full of fluid and rotate freely.

PITOT-STATIC INSTRUMENTS

The pitot-static instruments (airspeed indicator, altimeter, and vertical speed indicator) operate on the principle of differential air pressure. **Pitot pressure**, also called impact, ram, or dynamic pressure, is directed only to the airspeed indicator, while **static pressure,** or ambient pressure, is directed to all three instruments. An alternate static source, which is included in many airplanes, allows you to select a secondary static source in the event the main port becomes blocked. In nonpressurized aircraft, the alternate static source usually is located in the cabin. [Figure 1-5]

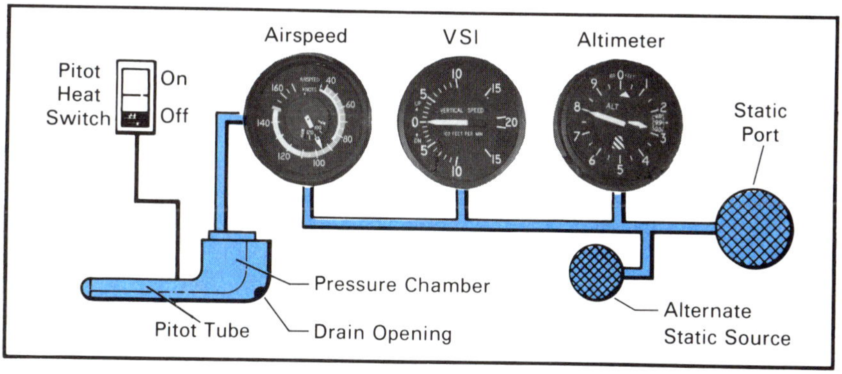

Figure 1-5. Ram air pressure is supplied only to the airspeed indicator, while static pressure is used by all three instruments. Electrical heating elements may be installed to prevent ice from forming on the pitot head. A drain opening to remove moisture is normally included.

AIRSPEED INDICATOR

The recommended airspeed for a given situation is not affected by elevation or temperature.

The airspeed indicator displays the speed of your airplane through the air by comparing ram air pressure with static air pressure — the greater the differential, the greater the speed. The instrument displays the result of the pressure differential as **indicated airspeed** (IAS). Manufacturers use this speed as the basis for determining aircraft performance. When an indicated airspeed is specified for a given situation, such as takeoff or landing, you normally should use that speed, regardless of the elevation or temperature. [Figure 1-6]

When discussing the airspeed indicator, it is helpful to understand the different types of airspeed as well as the various V-speeds associated with the instrument. The following is a brief review of these speeds.

AIRSPEEDS

CAS is IAS corrected for installation and instrument errors.

Calibrated airspeed (CAS) is indicated airspeed corrected for installation and instrument errors. Although these errors are minimized by the manufacturer, they cannot be totally eliminated throughout the full range of operating speeds, weights, and flap settings. Generally, the difference between indicated and calibrated is greatest at slow speeds and smallest at cruising speeds. Calibrated airspeed corrections are found in the pilot's operating handbook. CAS equals TAS in standard atmospheric conditions at sea level.

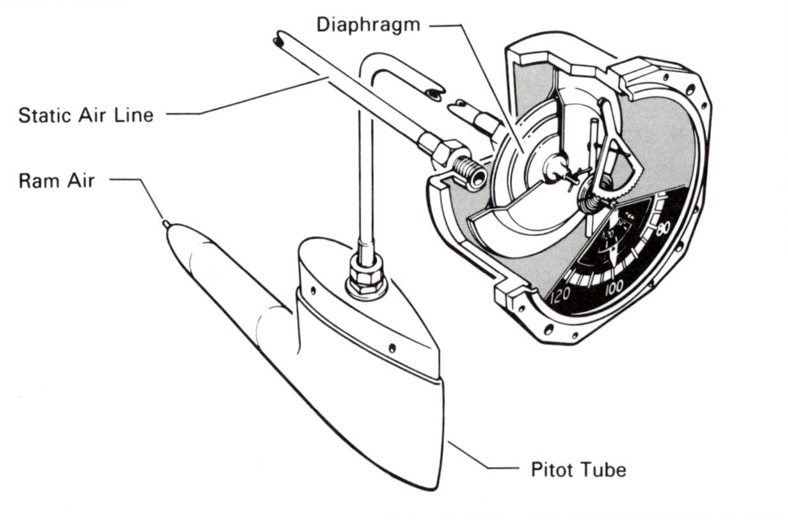

Figure 1-6. Ram air pressure from the pitot tube is directed to a diaphragm inside the airspeed indicator. The airtight case is vented to the static port. As the diaphragm expands or contracts, a mechanical linkage moves the needle on the face of the indicator. This process helps to explain how the instrument reacts when one or both of the air sources become blocked.

Equivalent airspeed (EAS) is calibrated airspeed corrected for the compression of air at a particular altitude. In the higher speed ranges, an aircraft passes through the atmosphere so rapidly that the air is compressed in front of it. This error, often called compressibility, causes EAS to be lower than CAS. Many electronic and mechanical flight computers incorporate provisions to compensate for this error. However, it is generally considered insignificant when you are operating below 200 KIAS and below 20,000 feet.

EAS is CAS corrected for compressibility.

True airspeed (TAS) compensates for air pressure and temperature and represents the true speed of your airplane through the air. At sea level on a standard day, CAS (or EAS, as appropriate) is equal to TAS. As atmospheric pressure decreases or air temperature increases, the density of the air decreases. As the air density decreases at a given indicated airspeed, such as during a climb, the true airspeed increases. At a constant power setting and altitude, an increase in outside air temperature will also result in an increase in TAS. The reverse is true with a decrease in temperature or an increase in pressure. You can estimate true airspeed by adding two percent of the indicated airspeed value for each 1,000-foot increase in altitude. However, you should use your flight computer to correct CAS (or EAS) for pressure altitude and temperature (density altitude) to obtain an accurate TAS.

TAS is EAS corrected for nonstandard pressure and temperature.

In high performance aircraft, some limiting airspeeds are based on the speed of sound. These aircraft usually have a Mach indicator or Mach meter in addition to the airspeed indicator. **Mach** simply shows the ratio of the aircraft's true airspeed to the speed of sound. For example, a speed of Mach .8 means the aircraft is flying at 80% of the speed of sound.

Mach is the ratio of the aircraft's true airspeed to the speed of sound.

V-SPEEDS AND COLOR CODES

Several airspeed limitations, called **V-speeds**, are shown on the airspeed indicator. If you know the significance of each color code, you can easily identify the safe speed ranges for the various phases of flight. [Figure 1-7]

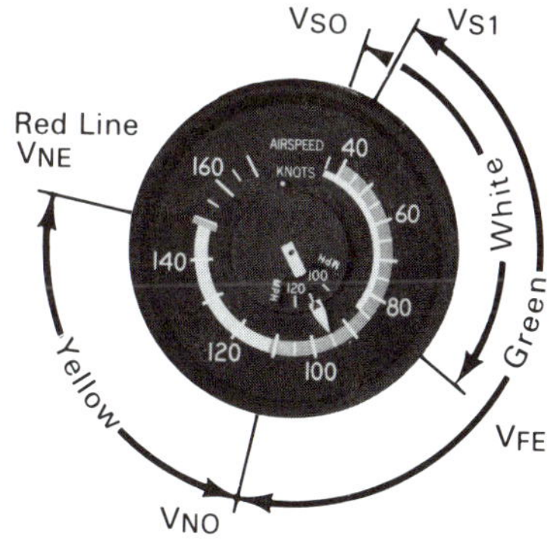

Color	Meaning
White	Flap Operating Range
Green	Normal Operating Range
Yellow	Caution Range
Red	Never-Exceed Speed

Figure 1-7. The airspeed indicator is divided into color-coded arcs that define speed ranges. The boundaries of these arcs also identify airspeed limitations.

V_{S0} is the stalling speed, or minimum steady flight speed, in the landing configuration. In light airplanes, this is the power-off stall speed at the maximum landing weight in the landing configuration (gear and flaps down). **V_{S1}** is the stalling speed, or minimum steady flight speed, obtained in a specified configuration. For light airplanes, this is the power-off stall speed at the maximum takeoff weight in the clean configuration (gear up, if retractable, and flaps up). You should check the POH for specific information on your airplane. **V_{FE}** is the maximum speed with the flaps fully extended, while **V_{NO}** is the maximum structural cruising speed. Finally, **V_{NE}** denotes the never-exceed speed.

Several important airspeeds, such as V_A, V_{LE}, and V_{LO}, are not marked on the airspeed indicator.

Although the color-coding on the face of the airspeed indicator defines many speed limits and operating ranges, other limitations are not shown, but are equally important. During gusty or turbulent conditions, the aircraft should be flown at or below the design maneuvering speed, **V_A**. The load factor that can be imposed on the wing below this speed is within safe limits. At or below V_A, the airplane will stall before excessive G-forces can build up. This important limit is listed in your pilot's operating handbook or on a placard. Airplanes with retractable landing gear also list a **V_{LE}**, or the maximum speed with the landing gear extended. **V_{LO}** is the maximum speed for extending or retracting the landing gear.

INSTRUMENT CHECK

Before you taxi, make sure the airspeed indicator reads zero. If a strong wind is blowing directly at the airplane in the tiedown area, the indicator may not read zero until you start to taxi and turn the airplane away from the wind. As you accelerate during the initial takeoff roll, make sure the airspeed indicator is increasing at an appropriate rate. If it is not, discontinue the takeoff.

ALTIMETER

The altimeter is an essential instrument when you are operating in IFR conditions. It provides vital information that helps you maintain aircraft control, terrain clearance, and separation from other aircraft. An understanding of how the altimeter operates and its limitations will enable you to read and interpret the indications accurately. [Figure 1-8]

ALTITUDES

Altimetry involves more than a simple measurement of height. To an aircraft designer, altitude is significant for determining performance which is based on air density. To aeronautical chart designers, the surveyed altitude of ground obstructions is critically important. As a pilot, you depend on the accuracy of this charted information for obstacle clearance. The measurement of altitude shown on your altimeter is especially important, since it is your only source of height information. The following is a review of the various altitudes used in aviation.

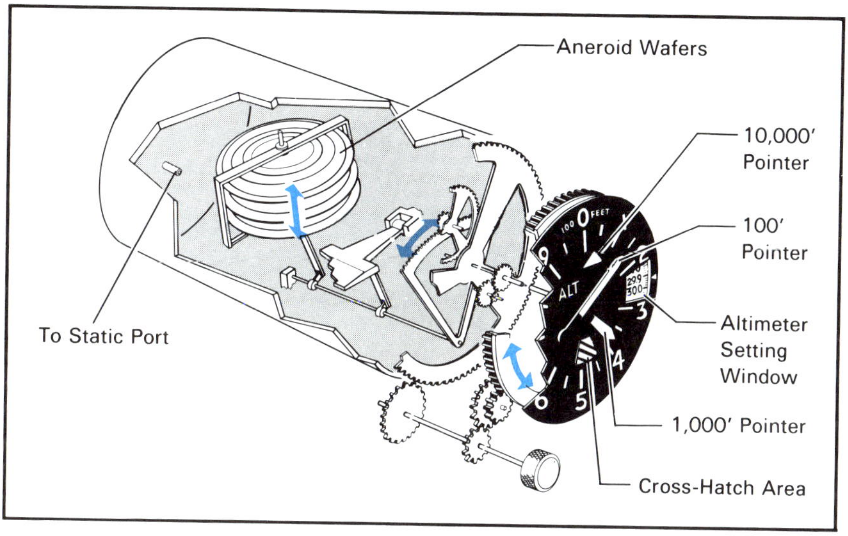

Figure 1-8. The main component of the altimeter is a stack of sealed aneroid wafers. They expand and contract as atmospheric pressure from the static source changes. The mechanical linkage translates these changes into pointer movements on the indicator. In addition to pointers, some indicators include a cross-hatched area which is visible when the altimeter is reading below 10,000 feet.

Indicated altitude is read directly from the altimeter when it is correctly adjusted to the local altimeter setting. This altitude is referred to as mean sea level (MSL). Altitudes assigned to aircraft in controlled airspace under instrument flight rules are indicated altitudes, except for flights operating in the high altitude route structure.

Pressure altitude is displayed on the altimeter when it is set to the standard sea level atmospheric pressure of 29.92 in. Hg. Regulations require that you set the altimeter to 29.92 when you are operating at and above 18,000 MSL. Altitudes at and above 18,000 feet MSL are called flight levels (FL).

To read pressure altitude, set your altimeter to 29.92 in. Hg.

Density altitude is pressure altitude corrected for nonstandard temperature. It is a theoretical value used to determine airplane performance. Performance charts for many older aircraft are based on density altitude. When density altitude is high (temperatures above standard), aircraft performance is reduced. Pressure altitude and density altitude are equal when the temperature is standard for that pressure altitude.

Density altitude is equal to pressure altitude when standard temperature conditions exist.

True altitude is the actual height of an object above mean sea level. On aeronautical charts, the elevation of such objects as airports, towers, and TV antennas are true altitudes. Unfortunately, your altimeter displays true altitude in flight only under standard conditions. Nonstandard temperature and pressure causes your indicated altitude to differ from true altitude. You can calculate approximate true altitude with a flight

True altitude and pressure altitude are equal only when standard atmospheric conditions exist.

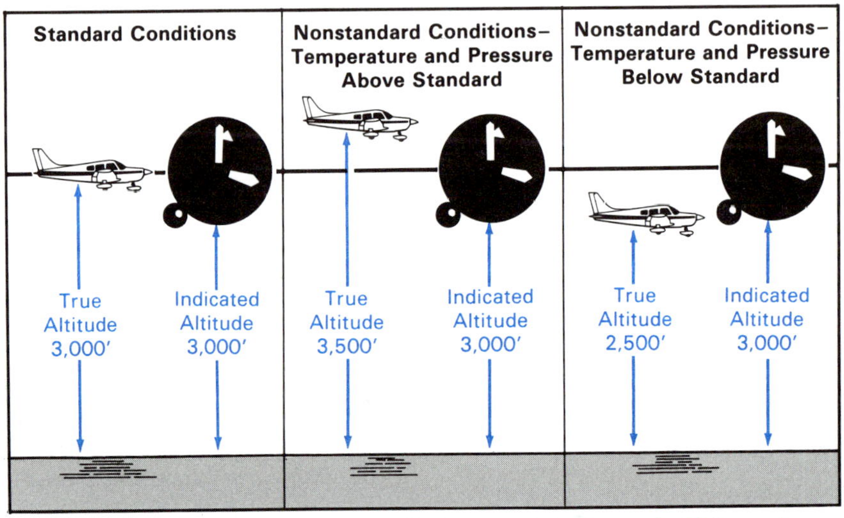

Figure 1-9. When nonstandard conditions exist, true altitude may be higher or lower than indicated altitude. When atmospheric pressure or temperature is higher than standard, your true altitude is higher than your indicated altitude. When pressure is lower or temperature is colder than standard, your true altitude is lower than your indicated altitude.

computer, but the computation is based on the assumption that pressure and temperature lapse rates match standard values. Normally, these lapse rates vary from those in a perfectly standard atmosphere. In two cases, however, true and indicated altitude are equal. One occurs during flight, when you have the correct altimeter setting and atmospheric conditions match International Standard Atmospheric (ISA) values. The other situation occurs on the ground. When you set the altimeter to the local pressure setting, it indicates the field elevation, which is a true altitude. [Figure 1-9]

Absolute altitude is the actual height of the airplane above the earth's surface. It is commonly referred to as height above ground level (AGL). During instrument approaches, absolute altitude is used to define the height above the airport (HAA), height above the touchdown zone (HAT), and the threshold crossing height (TCH).

ALTIMETER SETTING

The altimeter indicates your altitude in relation to the pressure level set in the altimeter.

The altimeter senses the current atmospheric pressure, but it indicates height in feet above the barometric pressure level set in the altimeter window. This is a very important concept to keep in mind when using the altimeter. For example, if the current altimeter setting, 29.82, is set in the window, the altimeter would indicate the height of the airplane above the surface pressure level of 29.82.

Since local barometric pressure rarely remains constant, you must adjust the altimeter setting in the window for changes in pressure. If you do not,

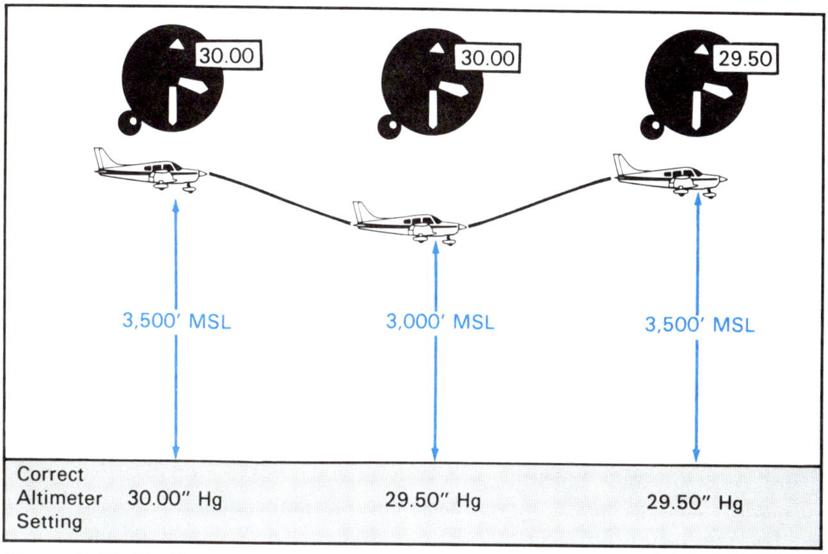

Figure 1-10. In this example, the altimeter setting decreased from 30.00 to 29.50, a change of 0.50 in. Hg. Since one inch equals 1,000 feet, 0.50 inches equals 500 feet (1,000 x .50 = 500). You are flying from a high to a low pressure area, so your indicated altitude is higher than the actual altitude. Resetting the altimeter to 29.50 will cause the indicated altitude to decrease 500 feet, and you would then climb back to your desired altitude.

the indicated altitude will be incorrect. For example, assume your altimeter is set to 30.00 in. Hg, and you are flying at 3,500 feet indicated. If you fly into an area where the pressure is 29.50 in. Hg, the altimeter will sense this decrease in pressure as an increase in altitude and will display a higher reading. Your response will be to lower the nose of the airplane and descend to maintain your "desired" altitude. A good memory aid is, "When flying from high to low or hot to cold, look out below." For abnormally high (above 31.00 in. Hg) or abnormally low (below 28.00 in. Hg) altimeter setting procedures, refer to the AIM. [Figure 1-10]

When flying from high to low or hot to cold, look out below.

In the event you are departing an airport where you cannot obtain a current altimeter setting, you should set the altimeter to the airport elevation. After departure, obtain the current altimeter setting as soon as possible from the appropriate ATC facility. During IFR flight, ATC periodically provides the altimeter setting to use. By using the current barometric pressure setting, you can minimize altimeter errors.

A current altimeter setting for the area you are flying over is essential for reliable altitude indications.

INSTRUMENT CHECK

In addition to the required static system check, you should make sure the altimeter is reading accurately during the IFR preflight check. To accomplish this, set the altimeter to the current altimeter setting. If it is within 75 feet of the actual elevation of that location, the altimeter is generally considered acceptable for use.

Altimeter errors greater than 75 feet are considered unacceptable.

VERTICAL SPEED INDICATOR

The VSI displays changes in ambient air pressure as a rate of climb or descent.

The vertical speed indicator (VSI), sometimes called a vertical velocity indicator (VVI), measures how fast the ambient air pressure increases or decreases as the aircraft climbs or descends. It then displays this change as a rate of climb or descent in feet per minute. Since the VSI measures only the rate at which air pressure changes, air temperature has no effect on this instrument. [Figure 1-11]

Rough control use can cause erratic VSI indications.

Since the VSI gives an immediate indication of a change in vertical direction, it is used as a trend indicator. After a period of six to nine seconds, the VSI stabilizes and displays the actual rate of climb. Rough control techniques or turbulence can further extend the lag period and cause erratic or unusable rate indications. Because the VSI is supplied with static air pressure, a clogged static port makes this instrument unusable.

Some aircraft may be equipped with an instantaneous vertical speed indicator (IVSI). This device incorporates accelerometers to compensate for the lag found in the typical VSI.

INSTRUMENT CHECK

The VSI is not required for instrument flight; however, it is an extremely useful instrument. Before starting the aircraft engine, check to see that the VSI reads zero. When you start the engine, you may see the VSI fluctuating up and down due to the propeller slipstream. In straight-and-level flight, the VSI should indicate zero. If it indicates a value other than zero, simply apply the amount of error to the VSI for in-flight indications.

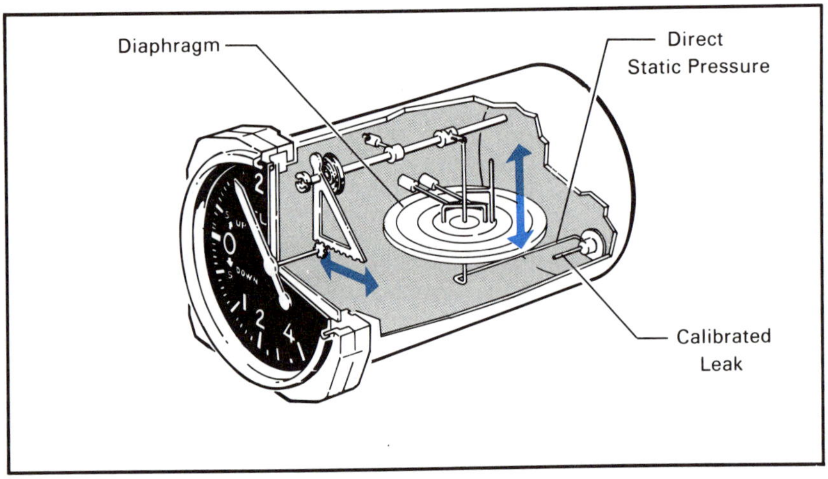

Figure 1-11. Although the sealed case and diaphragm are both connected to the static port, the air inside the case is restricted through a calibrated leak. When the pressures are equal, the needle reads zero. As you enter a climb or descent, the static pressure inside the diaphragm instantly changes, and the needle registers an immediate change in vertical direction. When the pressure differential stabilizes at a definite ratio, the needle registers the rate of altitude change.

SYSTEM ERRORS

The pitot-static instruments are usually very reliable. Gross errors almost always indicate blockage of the pitot tube, the static port, or both. Blockage may be caused by moisture (including ice), dirt, or even insects.

During preflight, you should make sure the pitot tube cover is removed. Then, check the pitot and static port openings. If they are clogged, the openings should be cleaned by a certificated mechanic. It is also possible for the pitot tube to become blocked by visible moisture during flight when temperatures are near the freezing level. If you are flying in visible moisture and your airplane is equipped with pitot heat, it should be on to prevent pitot tube icing.

PITOT BLOCKAGE

The airspeed indicator is the only instrument affected by a pitot tube blockage. The system can become clogged in one of two ways. First, the ram air inlet can clog, while the drain hole can remain open. In this situation, the pressure in the line to the airspeed indicator will vent out the drain hole, causing the airspeed indicator to drop to zero. Often, this occurs when ice forms over the ram air inlet.

A blockage of the pitot tube affects only the airspeed indicator.

The second situation occurs when both the ram air inlet and drain hole become clogged. When this occurs, the air pressure in the line is trapped. The airspeed indicator will no longer indicate changes in airspeed. If the static port is open, the indicator will react the same as an altimeter, showing an increase in airspeed as altitude increases and a decrease in speed as altitude decreases.

A blockage of the drain hole and ram air input causes the airspeed indicator to react like an altimeter.

STATIC BLOCKAGE

If the static system is clogged, the airspeed indicator will continue to react to changes in airspeed, since ram air pressure is still being supplied by the pitot tube. However, the readings will not be correct. When you are operating above the altitude where the static port became clogged, the airspeed will read lower than it should. Conversely, when you are operating below that altitude, the indicator will read higher than the correct value. The amount of error is proportional to the distance from the altitude where the static system became clogged. The greater the difference, the greater the error.

A static port blockage affects the airspeed indicator, altimeter, and VSI.

Since the altimeter determines altitude by measuring ambient air pressure, any blockage of the static port will "freeze" the altimeter in place and make it unusable. The VSI freezes at zero, since its only source of pressure is from the static port. After verifying a blockage of the static system by cross-checking the other flight instruments, you should attempt to provide an alternate source of static pressure.

In many aircraft, an alternate static source is provided as a backup for the main static source. In nonpressurized aircraft, the alternate source

When you have selected the alternate static source, the altimeter will read a little high, the airspeed a little fast, and the VSI will show a momentary climb.

usually is located inside the aircraft cabin. Due to the slipstream, the pressure inside the cabin is less than that of outside air. Therefore, when you select the alternate static source, the altimeter will read a little higher and the airspeed a little faster than they should. Also, the vertical speed will show a momentary climb. In the case of a pressurized aircraft with a static line leak inside the pressurized compartment, the altimeter will read lower than the actual flight altitude, due to the increased static pressure. The airspeed may also read lower than it should, and the vertical speed indicator may indicate a momentary descent.

If the aircraft is not equipped with an alternate static source, you can break the glass of the vertical speed indicator to allow ambient air pressure to enter the static system. Of course, this makes the VSI unreliable for instrument reference. Do not use this technique on pressurized aircraft until you depressurize the cabin.

CHECKLIST

After studying this section, you should have a basic understanding of:

✓ **Gyroscopic principles** — What rigidity in space and precession are and how these principles affect gyroscopic instruments.

✓ **Attitude indicator** — What its function is and how it reflects the pitch and bank attitude of an aircraft.

✓ **Heading indicator** — What its function is, why it precesses, and how to keep it properly aligned.

✓ **Turn coordinator** — What the instrument displays and how to maintain a standard-rate turn and coordinated flight.

✓ **Magnetic compass** — What its basic operating principles are and how to compensate for its inherent errors.

✓ **Pitot-static system** — How the pitot-static system operates and what instruments are affected.

✓ **Airspeed indicator** — How the airspeed indicator functions; what the differences are between IAS, CAS, EAS, and TAS; what V-speeds are; and what the airspeed indicator color codings mean.

✓ **Altimeter** — How the altimeter functions and what the differences are between indicated, pressure, density, true, and absolute altitudes.

✓ **Vertical speed indicator** — How the VSI functions as a trend and rate indicator.

✓ **System errors** — How various blockages of the pitot-static system affect the airspeed indicator, altimeter, and vertical speed indicator.

SECTION B

ATTITUDE INSTRUMENT FLYING

Attitude instrument flying is the basic technique of controlling an airplane by reference to flight instruments, rather than outside visual reference. Once you have acquired the basic knowledge and developed the fundamental skills, you will find that you have the same precise control by instrument reference as you do with outside visual reference.

CONCEPTS AND SKILLS

Before discussing individual flight maneuvers, you should be aware that there are two generally accepted methods in use for teaching attitude instrument flying. Both methods use the same flight instruments, and both require the same responses for attitude control. However, they differ in the degree of reliance on the attitude indicator. One method is based on the concept that airplane performance depends on how you control the attitude and power relationships of the airplane. This is referred to as the **control and performance concept** of attitude instrument flying, which designates certain flight instruments as control and others as performance instruments. In general, this method relies heavily on the attitude indicator during most maneuvers and is particularly well suited for high performance turbojet aircraft.

The primary/support concept of attitude instrument flying groups the flight instruments according to pitch, bank, and power control functions.

The second method of attitude instrument flying regards the attitude of the airplane as a function of pitch, bank, and power control. For a given maneuver, there are specific pitch, bank, and power instruments that you should use to control the airplane and obtain the desired performance. Those instruments which provide the most pertinent and essential information during a given condition of flight are termed **primary** instruments. Those which back up and supplement the primary instruments are termed **supporting** instruments. As a result, this method is termed the **primary/support concept** of attitude instrument flying and is recommended by the FAA for instrument training conducted in light aircraft with low operating speeds.

This method of attitude instrument flying does not lessen the value of any individual instrument. The attitude indicator, for example, is still the instrument which provides basic attitude reference. Since it is the only instrument that provides instant and direct aircraft attitude information, it should be considered primary during any change in pitch or bank attitude. After the new attitude is established, other instruments become primary, and the attitude indicator usually becomes a supporting instrument. As you will see, this discussion follows the FAA guidelines for primary and supporting instruments.

INSTRUMENT FLYING SKILLS

The three skills used in instrument flying are cross-check, instrument interpretation, and aircraft control.

To achieve smooth, positive control of the aircraft during instrument flight maneuvers, you will need to develop three fundamental skills. They are cross-check, interpretation, and aircraft control. The function of instrument training is to develop and integrate these skills to a high level of proficiency in order to maintain any prescribed flight path.

Common errors in cross-check include fixation, omission, and emphasis.

Instrument scan, or **cross-check,** is the first fundamental skill. It requires logical and systematic observation of the instrument panel. A methodical and meaningful instrument scan is necessary to make appropriate changes in aircraft attitude and performance. During initial training, you will form definite habit patterns regarding your instrument cross-check. With practice, you will soon learn which instruments to look at and what to look for when maintaining a particular flight attitude. Some of the common scanning errors include fixation, omission, and emphasis. **Fixation** occurs when you stop your scan and stare at a single instrument. Fixation may not always be a result of a faulty cross-check; it may be caused by problems with instrument interpretation or even aircraft control. Another problem in cross-checking is **omission**. This occurs when you fail to include pertinent instruments at the right time. Errors of omission often occur following attitude changes. **Emphasis** on a single instrument instead of a combination of instruments is another cross-check error which is common during your initial training. Even though you are maintaining an active scan, you are relying too heavily on one instrument. For example, you can maintain level flight reasonably well using the attitude indicator, but you must include the altimeter in your scan to maintain precise altitude control.

To interpret instruments properly, you must first learn how each one of them operates.

The second fundamental skill is **instrument interpretation**. This begins with a knowledge of how each instrument operates and an awareness of the instrument indications that represent given pitch and bank attitudes for the aircraft you are flying. For example, if an existing level pitch attitude is to be confirmed, you should refer to the altimeter, vertical speed indicator (VSI), airspeed indicator, and attitude indicator. If you need to confirm bank attitude, the instruments for interpretation are the heading indicator, turn coordinator, and attitude indicator. You must learn what performance to expect and what instruments to interpret in order to control airplane attitude during a given maneuver.

The third and last skill is **aircraft control**. This skill is actually the result of cross-check and accurate interpretation of the instruments. Aircraft control involves adjustments to pitch, bank, and power in order to achieve a desired flight path. Another important aspect of aircraft control is trim. Trimming to relieve control pressures after the desired attitude is established is even more important during instrument flight than it is during visual, or contact, flying. An improperly trimmed airplane requires continuous control pressures, increases tension, interrupts your cross-check, and may result in abrupt or erratic control. When properly trimmed, aircraft control is simply a matter of applying control pressures to maintain or change the desired airplane attitude as dictated by cross-check and instrument interpretation.

The most common sensory illusion during instrument flight is **spatial disorientation**. This sensation is caused when the brain receives conflicting messages from your sensory organs. You are more subject to spatial disorientation when you rely on body signals to interpret flight attitudes. This can create an incorrect mental image of your position, attitude, or movement in relation to what is actually occurring to your airplane. For example, decelerating while turning in one direction can create the sensation of turning in the opposite direction. Another example is found during a level turn that produces a load factor of 1.5 positive Gs. This maneuver creates the illusion of a climb rather than level flight. Inversion illusion can occur where an abrupt change from a climb to straight-and-level flight can create the feeling of tumbling backward. Somatographic illusion occurs during rapid acceleration on takeoff and makes you feel like the aircraft is in a nose-up attitude. When you do not have outside visual references, your kinesthetic, or postural, sense often interprets the centrifugal force of turns as a sensation of rising or falling. To avoid or overcome spatial disorientation, you must learn to disregard false body sensations and rely on the flight instruments. Other factors can also help you avoid spatial disorientation during instrument flight. For example, avoid sudden head movements, particularly during takeoffs, prolonged constant-rate turns, and approaches to landings. Also, remember that illness, medication, alcohol, fatigue, sleep loss, and mild hypoxia are likely to increase susceptibility to spatial disorientation.

During instrument flight, the sensations that lead to spatial disorientation can be overcome by reading and interpreting the flight instruments, then acting accordingly.

BASIC FLIGHT MANEUVERS

Precise aircraft control requires smooth coordination of pitch, bank, power, and trim. With instrument references you need to develop a high degree of proficiency in scanning, as well as interpretation. Then you must translate the readings on the instruments into correct control responses. With practice, your level of proficiency will increase to the point where aircraft control is almost automatic.

STRAIGHT-AND-LEVEL FLIGHT

Be sure to align the miniature aircraft with the horizon bar when you are still on the ground. This way, the attitude indicator will reflect approximate level flight at normal cruise speed. In a given configuration, the factors that determine the pitch attitude required to maintain level flight are airspeed, air density, and aircraft weight. During straight-and-level, unaccelerated flight, you can align the miniature airplane more closely with the horizon bar. Use this reference as a basis for estimating changes in pitch or bank when you are performing any of the other instrument flight maneuvers. Other instruments also reflect attitude changes immedi-

Airspeed, air density, and aircraft weight determine the level flight pitch attitude for a given configuration.

ately. That is, a bank change is apparent on the heading indicator and turn coordinator, while a pitch change is reflected on the altimeter, VSI, and airspeed indicator. However, always make attitude changes on the attitude indicator, not the supporting instruments.

ALTITUDE CONTROL

You can maintain the desired altitude by establishing a specific pitch attitude on the attitude indicator and trimming the aircraft. When you have established the attitude, scan the altimeter, VSI, and airspeed indicator to determine if any change is occurring. The altimeter normally is considered the primary pitch instrument during level flight, since it provides the most pertinent altitude information. The supporting pitch instruments include the VSI, airspeed indicator, and attitude indicator. The VSI provides both trend and rate information and immediately reflects initial vertical movement of the aircraft. If a departure from the desired altitude occurs, it is reflected first on the VSI and next on the altimeter. By evaluating the initial rate of movement of these instruments, you can estimate the amount of pitch change you need to restore level flight. The amount of pitch change usually is small and requires only a fraction of a bar width of change on the attitude indicator. [Figure 1-12]

If you deviate from an assigned altitude, use the attitude indicator, altimeter, and VSI to make pitch corrections.

When a deviation from the desired altitude occurs, your judgment and experience in a particular aircraft dictate the rate of correction. As a guide, adjust the pitch attitude to produce a rate of change which is double the amount of altitude deviation and use power as necessary. If your aircraft is descending at 300 f.p.m. and is 100 feet below the desired altitude, you need to make a correction using a climb rate of 200 f.p.m. First, make an initial pitch adjustment to stop the descent and then initiate the approximate climb rate. Maintain this pitch attitude on the attitude indicator until the vertical speed stabilizes. A further pitch adjustment may be necessary to produce the desired climb rate. When you approach the altitude, select a leadpoint on the altimeter for initiating a pitch attitude change for leveloff. [Figure 1-13]

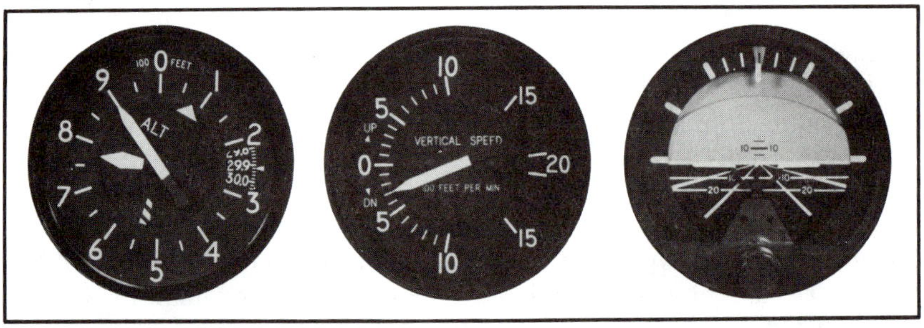

Figure 1-12. These instruments represent a pitch deviation from level flight. The altimeter shows that you are 100 feet below your desired altitude, and the VSI reflects a 300 f.p.m. rate of descent. As a general rule, use one-half bar width corrections on the attitude indicator for pitch deviations of 100 feet or less and use initial full-bar width corrections for larger deviations.

Figure 1-13. Estimate your leadpoint for leveloff by using 10% of the vertical velocity rate. If your rate of correction to the desired altitude is 200 f.p.m., begin the leveloff 20 feet before you reach the desired altitude.

HEADING CONTROL

You can maintain the desired heading by establishing a zero bank attitude on the attitude indicator and detecting deviations on the heading indicator. The heading indicator is considered the primary bank instrument, since it provides the most pertinent heading information. Supporting bank instruments include the turn coordinator and attitude indicator. When a deviation occurs, establish a definite angle of bank on the attitude indicator to produce a suitable rate of turn. For small variations, use an angle of bank equal to the degrees of heading deviation. [Figure 1-14]

Figure 1-14. In this example, the aircraft has drifted 10° off the desired heading of 340°. To correct this, you should establish a bank of 10° to the left. For larger corrections, limit the bank angle so you don't exceed a standard-rate turn on the turn coordinator. In addition, don't exceed a maximum bank angle of 30° on the attitude indicator.

AIRSPEED CONTROL

You can maintain airspeed by referring to the airspeed indicator and making adjustments to either pitch or power. If a change in the power setting is indicated, select the approximate setting and check the airspeed indicator to determine whether you will need to make further corrections to power and/or pitch.

To make suitable power adjustments, you need to know the approximate power required to establish a desired airspeed. If you do not know this information or do not have it readily available, you should acquire it by actual flight experience in various configurations. The airspeed indicator is considered the primary power instrument for straight-and-level flight. The supporting power instrument is the tachometer or manifold pressure gauge.

Pitch and power adjustments are closely related. Adjustment of one usually requires appropriate adjustment of the other. For example, if you adjust pitch during flight, you also must adjust power to prevent a change in airspeed. During an airspeed change in straight-and-level flight, the altimeter is primary for pitch, heading indicator is primary for bank, and tachometer or manifold pressure gauge is primary for power. To increase the airspeed any appreciable amount while maintaining straight-and-level flight, advance power beyond the setting required to maintain the new desired airspeed. As the airspeed increases, adjust the pitch attitude downward to maintain altitude. When the airspeed approaches the desired indication, reduce the power to an estimated setting that maintains the new airspeed. When you reduce airspeed, reverse this procedure.

LEVEL TURNS

The altimeter is the primary pitch instrument during roll-in to a turn. Once a standard-rate turn has been established, the turn coordinator is the primary bank instrument.

During instrument flight, you normally make turns at a standard rate of three degrees per second. To enter a level turn, establish the desired angle of bank by reference to the bank index on the attitude indicator (primary bank). Scan the turn coordinator (supporting bank) to confirm control coordination and proper bank angle for the desired rate of turn. During the roll-in to the turn, the altimeter is the primary pitch instrument. After the level turn is established, the turn coordinator is primary for bank, and the altimeter is primary for pitch control. The attitude indicator is a supporting instrument for both pitch and bank. The VSI is a supporting pitch instrument, and the airspeed indicator is primary power.

BANK CONTROL

Rate of turn varies with changes in true airspeed and angle of bank. You can quickly estimate the approximate angle of bank required for a standard-rate turn by dividing the true airspeed in knots by a factor of 10 and adding 5° to the resultant figure. For example, the angle of bank required for a standard-rate turn at a true airspeed of 110 knots is 11 plus 5°, or a 16° angle of bank.

The rate of turn at any given airspeed actually depends on the horizontal component of lift, which is directly proportional to the angle of bank. You will find that, for a given airspeed, the rate of turn increases as the angle of bank increases. The radius of turn also is affected by the horizontal component of lift. Therefore, airspeed and angle of bank also affect radius of turn. [Figure 1-15]

Remember to lead your roll-out to stop the turn on the desired heading. You can estimate the leadpoint by using one-half the angle of bank. For example, if the bank angle is 10°, initiate a roll-out 5° before you reach the desired heading. As you gain experience, you may vary the leadpoint to meet your own piloting techniques.

ALTITUDE CONTROL

When you initiate a turn and increase the bank, the aircraft tends to descend, due to loss of vertical lift. Therefore, adjust the pitch attitude by reference to the attitude indicator (increase angle of attack). Then, scan the altimeter and VSI to confirm that pitch adjustments are adequate. If you decrease airspeed in a level turn by reducing power, you must either decrease the angle of bank and/or increase the angle of attack to maintain altitude. An increase in airspeed during a level turn requires an increase in bank angle and/or a decrease in angle of attack. During airspeed changes, the altimeter is primary for pitch control, supported by the attitude indicator and VSI.

The supporting pitch instruments for an airspeed change in a level turn are the attitude indicator and the VSI.

When rolling out of a turn, you must reduce the back pressure or trim used to maintain altitude during the turn. This is necessary to prevent

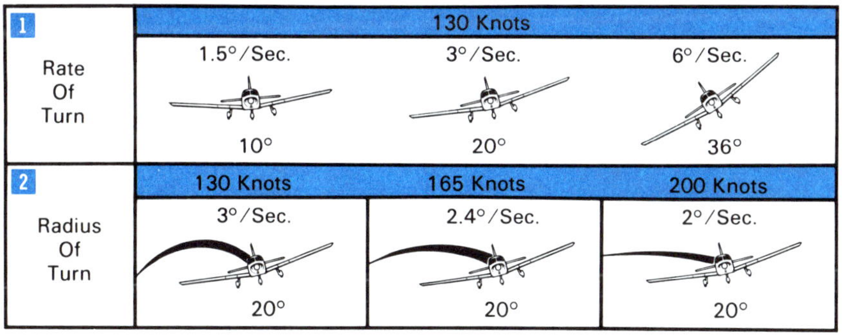

Figure 1-15. A specific angle of bank and true airspeed will always produce the same rate and radius of turn, regardless of aircraft type. If you increase only angle of bank, the rate of turn will increase (item 1). If you increase only the true airspeed, the radius of turn will increase (item 2). You can see that, as you increase airspeed during a level turn, the turn rate gets slower and the turn radius gets larger. On the other hand, a reduction in airspeed and/or an increase in bank angle will result in a faster rate of turn with a smaller radius.

the aircraft from climbing because of the increased vertical lift available. Therefore, scan the pitch attitude shown on the attitude indicator in the same manner as during roll-in.

AIRSPEED CONTROL

An aircraft tends to lose airspeed in a level turn because the increase in angle of attack causes induced drag to increase. This requires an increase in power to maintain airspeed. The amount of power required varies with airspeed; the slower the airspeed, the greater the power requirement. At slow speeds, it may be desirable to add an estimated amount of power as you establish the turn, rather than wait until an airspeed loss becomes apparent.

STEEP TURNS

You normally perform steep turns by instrument reference at a bank angle of approximately 45°. Roll in slowly, using the attitude indicator for pitch and bank references. As the bank angle steepens, you will need to increase back pressure and power to maintain altitude and airspeed. After you have established the turn, retrim the aircraft as necessary. [Figure 1-16]

Figure 1-16. These are the instrument indications of a steep turn to the left. During steep turns, scan the airspeed indicator, altimeter, and VSI to determine whether you need to make adjustments to the aircraft attitude or power. Make adjustments to pitch and bank immediately and precisely, before you reach the point where back elevator pressure tightens the turn without raising the nose. If too steep a bank occurs, you must first reduce the bank angle before you increase the back pressure.

You should expect a tendency to climb when you are rolling out of a steep turn and make adjustments to the trim, power, and back pressure. Steep turns are a confidence maneuver and are not recommended for normal instrument flight.

CLIMBS AND DESCENTS

Climbing and descending maneuvers are divided into two general categories — constant airspeed or constant rate. Do the constant airspeed maneuver by maintaining a constant power indication and varying the pitch attitude as required to maintain a specific airspeed. Do the constant rate maneuver by varying both power and pitch as required to maintain a constant airspeed and vertical velocity. You may perform either type of climb or descent while maintaining a constant heading or while turning. It is best to practice these maneuvers using airspeeds, configurations, and altitudes corresponding to those which you will use in actual instrument flight.

CONSTANT AIRSPEED CLIMBS

When entering a climb from straight-and-level flight at cruise airspeed, simultaneously increase the power to the climb power setting and smoothly apply back pressure. Adjust the miniature airplane on the attitude indicator to approximately two bar widths above the horizon. This pitch adjustment is an average and varies with the aircraft, the desired airspeed, and the rate of climb you select. A smooth, slow power application increases the pitch attitude so that you will need only slight control pressures to make the desired pitch change. During the transition, the attitude indicator is the primary pitch instrument and a supporting bank instrument. The tachometer or manifold pressure gauge is the primary power instrument. You also should scan the heading indicator (primary bank) and turn coordinator (supporting bank) to confirm coordinated straight flight. Since the combination of climb power, torque, and P-factor causes a left-turning tendency, you must apply right rudder pressure to maintain a constant heading. [Figure 1-17]

When transitioning from straight-and-level flight to a constant airspeed climb, the attitude indicator, heading indicator, and tachometer or manifold pressure gauge are the primary instruments for pitch, bank, and power, respectively.

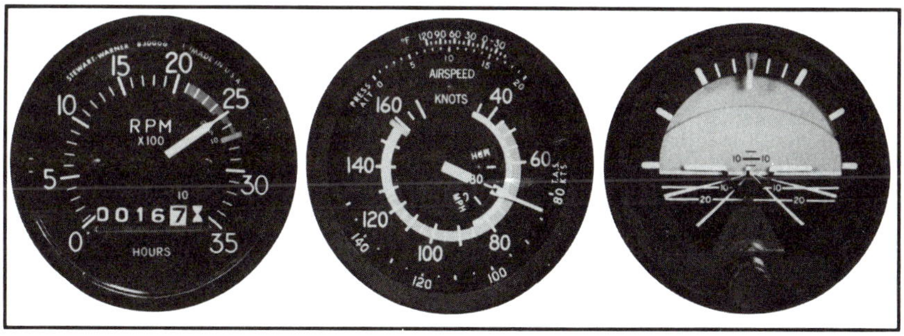

Figure 1-17. This illustration shows the instrument indications for a constant airspeed climb. Since power is constant, the airspeed indicator is primary for pitch. Any corrections to airspeed must be made with pitch adjustments on the attitude indicator.

CONSTANT AIRSPEED DESCENTS

You perform constant airspeed descents with a constant power setting using the pitch attitude of the airplane to control airspeed. For example, to enter a descent from cruise without changing airspeed, simultaneously reduce the power smoothly to the desired setting and reduce the pitch attitude slightly, using the attitude indicator, so the airspeed remains constant. The degree of pitch change and power reduction will vary according to the particular training airplane. Once you have established the power and pitch attitude, the airspeed indicator becomes the primary pitch instrument supported by the attitude indicator. Variations in airspeed dictate pitch changes, and small corrections can be made using the attitude indicator.

Any change in power also requires a corresponding change in pitch attitude in order to maintain a constant airspeed during the descent. When you make a power change, you must cross-check the airspeed indicator and attitude indicator to ensure the speed remains constant. As you adjust power, trim off control pressures.

CONSTANT RATE CLIMBS

During transition to a constant rate climb, the airspeed indicator is primary for pitch until the rate stabilizes.

You can perform constant rate climbs by maintaining the desired constant vertical velocity and airspeed. During this maneuver, use pitch attitude control to establish and maintain the vertical velocity, and use power to control the airspeed. For example, if the airplane is in level flight at cruise power and airspeed, establish the desired rate of climb as the first step in entering the maneuver. Do this by increasing the control back pressure and cross-checking the VSI and the attitude indicator. Apply power simultaneously in anticipation of the decrease in airspeed. As power is increased, the airspeed indicator is primary for pitch control until the VSI approaches the desired value. The heading indicator is the primary bank instrument for establishing a straight climb, and the tachometer or manifold pressure gauge is primary for power during the transition.

Once you have established the pitch attitude to produce the rate of climb, adjust the power to maintain the appropriate airspeed. You can determine the need for subsequent pitch adjustments by cross-checking the VSI, which is the primary pitch instrument. You can see the need for power adjustments on the airspeed indicator, which is the primary power instrument. The primary bank instrument in a straight climb is the heading indicator, while the supporting bank instruments include the attitude indicator and turn coordinator.

Due to the performance capabilities of most training airplanes, a constant airspeed may not be possible if you maintain a constant vertical velocity.

For example, in nonturbocharged airplanes, the performance capabilities decrease as the altitude increases. Therefore, to maintain a constant rate of climb, increase the control back pressure, thereby decreasing the airspeed. If you increase power gradually to maintain airspeed, you will reach a point where the engine is developing full power and the airspeed will begin to dissipate. Nevertheless, it's important to understand the principles of the maneuver so you may use the technique when you begin to fly airplanes with higher performance capabilities.

CONSTANT RATE DESCENTS

Whether you make a constant rate descent at cruise speed or approach speed, the control procedures are identical. As in a constant rate climb, power controls airspeed and pitch attitude controls rate. Enter the descent by simultaneously adjusting the nose of the miniature airplane just below the horizon bar on the attitude indicator and reducing power to a predetermined setting. When you reduce power, the aircraft may have a tendency to turn right, so you may need to use slight left rudder pressure.

Techniques for scanning the other instruments are the same as those you use in a climb. Since you desire a constant rate in addition to a constant airspeed, it is important to scan the VSI. [Figure 1-18]

LEVELOFF LEADPOINT

Determine the leveloff leadpoint for both climbs and descents by using 10% of the vertical velocity rate when you approach the desired altitude.

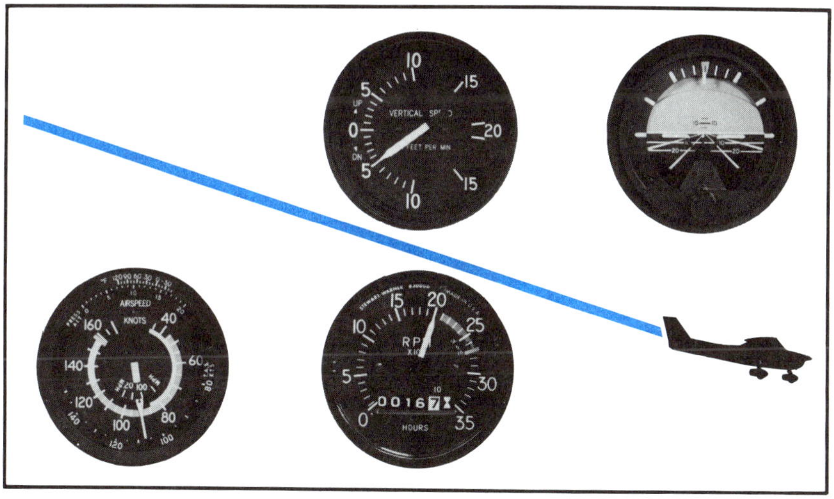

Figure 1-18. Once the airplane stabilizes in a constant rate descent, the VSI is the primary pitch instrument and the attitude indicator provides supporting pitch information. The airspeed indicator is primary for power control; it is supported by the tachometer or manifold pressure gauge.

Approximately 10% of the vertical velocity should be used to determine the number of feet to lead the leveloff from a climb to a specified altitude.

For example, if the climb or descent rate is 500 f.p.m., initiate the appropriate leveloff procedure when the airplane is 50 feet from the desired altitude. Then, position the nose to a level flight attitude and monitor the altimeter and vertical velocity to maintain level flight. Adjust the power until you reach the desired cruise airspeed and adjust the trim to relieve control pressure.

In some situations, you may need to level off at an airspeed higher than your descent airspeed. If so, assuming a 500 f.p.m. descent rate, you should begin adding power when you are 100 to 150 feet above the desired altitude.

CLIMBING AND DESCENDING TURNS

After changing airspeed in a level turn, the primary power instrument is the airspeed indicator.

Climbing or descending turns require you to use various combinations of the procedures previously discussed. Since the vertical component of lift decreases in turns, you should place more emphasis on pitch control. When you are performing a rate climb or descent in a turn, adjust pitch to maintain the desired rate and adjust power to maintain the desired airspeed. During these maneuvers, the VSI is primary for pitch, while the turn coordinator is primary for bank. The airspeed indicator is primary for power.

UNUSUAL ATTITUDE RECOVERY

Failure of the attitude indicator, spatial disorientation, wake turbulence, lapse of attention, or abnormal trim conditions can cause you to enter an unusual flight attitude. Although such cases are rare, you must know how to recover from an unusual attitude during instrument flight either with or without the attitude indicator.

To enter an unusual attitude, you will be asked to remove your hands and feet from the controls and close your eyes, while your instructor flies the aircraft through various maneuvers calculated to induce spatial disorientation. At some point, your instructor will indicate that you should take control of the airplane to make a proper recovery.

NOSE-HIGH ATTITUDE

The indications of a nose-high unusual attitude are a nose-high attitude, decreasing airspeed, rapid gain in altitude, and high rate of climb. Therefore, the initial objective for recovery is to prevent a stall. [Figure 1-19]

NOSE-LOW ATTITUDE

The indications of a nose-low unusual attitude are a nose-low attitude, increasing airspeed, rapid loss of altitude, and a high rate of descent. The primary objective of a nose-low unusual attitude recovery is to avoid a critically high airspeed and load factor. [Figure 1-20]

Figure 1-19. These are the indications of a typical nose-high unusual attitude. Before you initiate a correction, cross-check the instruments to confirm the reliability of the attitude indicator. To recover, you should almost simultaneously increase power, decrease pitch (reducing angle of attack), and roll the wings level.

Figure 1-20. To recover from a nose-low unusual attitude, simultaneously reduce power, roll wings level, then increase pitch attitude to stop acceleration, and gently raise the nose to the level flight attitude. If you raise the nose before rolling the wings level, the increased load factor may result in an accelerated stall, a spin, or a force exceeding the aircraft design load factor.

Recovery from unusual attitudes with partial panel is basically the same as that with a full panel, except that you use the turn coordinator to stop the turn. In addition, you must use the pitot-static instruments for pitch attitude reference. When the needles of the altimeter and airspeed indicators stop and reverse direction, you are passing through a level flight attitude. At this point, neutralize elevator pressure to maintain a level pitch attitude.

STALLS

You may practice stalls during instrument training to demonstrate that the recognition and recovery procedures under instrument conditions are exactly the same as under visual conditions. Recover from a stall by immediately reducing the angle of attack and positively regaining normal flight attitude by coordinated use of flight and power controls.

PARTIAL PANEL FLYING

The altimeter, airspeed indicator, and turn coordinator normally are the instruments to use if your gyroscopic instruments have failed.

Demonstration of basic attitude instrument flying skills on partial panel is required for the instrument rating practical test. The term **partial panel** usually means you have lost one or both of your vacuum-powered gyroscopic instruments, and you are controlling the aircraft primarily with the altimeter, airspeed indicator, and turn coordinator. The VSI and magnetic compass also are available during partial panel maneuvers.

ATTITUDE INDICATOR

Airliners, most military aircraft, and some light aircraft have attitude indicators which are powered by the aircraft electrical system. This type of indicator is equipped with an OFF flag which appears when the electrical power fails or the gyro is not operating at the proper speed. If the attitude indicator fails on an aircraft equipped this way, check the circuit breakers to see whether power can be restored to the instrument.

Many light aircraft are not equipped with dual, independent, gyroscopic heading or attitude indicators. In many cases, these instruments are powered with only a single vacuum source. A vacuum pressure loss or inconsistencies with associated flight instruments often indicate failure of a vacuum-operated attitude indicator. For example, if the attitude indicator shows that the aircraft is sharply nose-up, but the airspeed is not decreasing and the altitude is not increasing, you may assume that the attitude indicator is inaccurate. In this case, you have no choice but to use other sources for aircraft attitude information.

The turn coordinator provides a reliable indication of wing position when you use it in conjunction with the heading indicator. If the heading indicator has failed, the miniature aircraft of the turn coordinator becomes the primary bank instrument. The turn coordinator functions much like the attitude indicator with respect to bank attitude, as long as the ball is centered. However, when you bank the aircraft, but do not actually turn because of cross-control pressures, the miniature airplane in the turn coordinator indicates a normal wings-level flight attitude with the ball in the inclinometer off center.

If the aircraft is equipped with a turn-and-slip indicator, you may use it for heading change information. However, the vertical needle of the turn-and-slip indicator deflects only when the aircraft actually is turning. In addition, the vertical turn needle is somewhat sensitive, so you must use an average reading in turbulent air.

The VSI, together with the altimeter and the airspeed indicator, gives a very accurate indication of pitch attitude. If you raise the nose and the aircraft begins to climb while the power setting is constant, the VSI shows a positive indication, the altimeter shows an increase in altitude, and the airspeed indicator shows a decrease.

When the attitude indicator has failed, you must modify your scan pattern and omit the failed instrument from your cross-check. Because you must use several instruments to get the same information you usually obtain from the attitude indicator, you must increase your scan rate significantly. You may find it easier to fly partial panel by covering the inoperative attitude indicator with a piece of paper, or anything else available, to prevent subconscious use of the false indications displayed.

To reduce confusion and prevent the use of failed instruments, you may find it helpful to cover an inoperative instrument.

When you apply a flight control movement, the aircraft must overcome a small amount of inertia while becoming established on the new flight path. When the attitude indicator is inoperative, this lag becomes very noticeable, especially if you apply large or rapid corrective control movements. Therefore, you should keep the aircraft trimmed properly at all times and make small, well-controlled corrections.

HEADING INDICATOR

While flying IFR, you are most likely to detect a failure of the heading indicator if you experience difficulty remaining on your intended course with reference to the navigation indicator. When this type of deviation occurs, check the heading indicator against the magnetic compass.

If the attitude indicator is operating, it can assist you in maintaining your course. You can also maintain a heading simply by keeping the miniature airplane level and the ball centered in the turn coordinator. In addition, you use the magnetic compass for heading information, provided the aircraft is in wings-level, unaccelerated flight. Keep in mind that failure of the vacuum source on many light aircraft usually results in failure of both the attitude indicator and heading indicator.

TIMED TURNS

You can turn 180° at standard rate in one minute.

You make timed turns by initiating a standard-rate turn with reference to the turn coordinator. This type of turn results in a heading change at the rate of three degrees per second. For example, if you wish to make a 180° turn, begin timing as you roll into the turn, maintain the wingtip on the index for 60 seconds, then start recovery to straight-and-level flight. You also can make turns at one-half standard rate (1.5° per second) by keeping the wingtip midway between the index and level flight. This requires a very shallow bank angle, since a 360° turn requires four minutes.

For increased accuracy in timed turns, start the turn when the second hand on the aircraft clock passes one of the cardinal numbers. At the end of the predetermined number of seconds required to make the desired heading change, begin to roll out of the turn. Try to make the rate of roll constant for both rolling in and rolling out of the turn.

After the aircraft has stabilized in level flight, check the magnetic compass to determine whether you completed the turn on the desired heading. Make any necessary corrections by using the same procedure, then recheck the magnetic compass after it has stabilized.

IFR flight with gyro-stabilized instruments inoperative is considered a semiemergency situation, because you may not be able to comply immediately and accurately with all ATC clearances. You should notify ATC of the situation as soon as it occurs. When you must make an approach without the use of the heading indicator, request a no-gyro radar approach if the destination airport has this capability. However, it is usually a better course of action to divert to an alternate airport which has more favorable weather conditions.

LOGGING INSTRUMENT FLIGHT TIME

FAR Part 91 requires a safety pilot to occupy the other control seat during simulated instrument flight. The safety pilot must be appropriately rated in the aircraft and have at least a private pilot certificate with category and class ratings appropriate to the aircraft being flown.

According to FAR Part 61, a pilot with at least a private pilot certificate may log as instrument flight time only that flight time during which the aircraft is operated solely by reference to instruments in either actual or simulated instrument flight conditions. Each logbook entry must include the place and type of each instrument approach completed, and the name of the safety pilot for each simulated instrument flight. Instrument flight instructors may log as instrument time all flight time during which they act as a flight instructor in actual instrument weather conditions.

CHECKLIST

After studying this section, you should have a basic understanding of:

✓ **Primary/support concept** — What the primary and supporting instruments are for different flight attitudes and maneuvers.

✓ **Instrument flying skills** — Why cross-check, instrument interpretation, and aircraft control are important.

✓ **Cross-check** — What the importance is of developing an effective scan and avoiding the errors of fixation, omission, and emphasis.

✓ **Spatial disorientation** — What the false illusions are that can be experienced during instrument flight and how best to overcome them.

✓ **Straight-and-level flight** — Which techniques apply for altitude, heading, and airspeed control.

✓ **Level turns** — What the primary and supporting instruments are for altitude, bank, and airspeed control.

✓ **Climbs and descents** — What the different techniques for constant airspeed and constant rate climbs and descents are, including climbing and descending turns.

✓ **Unusual attitudes** — What the proper methods are for recognition and recovery from nose-high and nose-low unusual attitudes.

✓ **Partial panel** — How to control the aircraft without the attitude indicator and heading indicator.

✓ **Timed turns** — How to make timed turns on partial panel using the magnetic compass.

✓ **Logging instrument time** — What requirements you must meet in order to log instrument flight time.

SECTION

C

INSTRUMENT NAVIGATION

The purpose of this presentation is to provide a sound basis for your advanced navigation training. In your VFR flying, you have already developed a familiarity with many of the subjects in this section. Now you will be exposed to the finer points of radio navigation. You will learn to use various navigation indicators to intercept and track courses to or from navaids. You also must learn to visualize your position without the benefit of normal outside references. As an instrument pilot, you must develop the ability to orient to your position and quickly select headings that will intercept and maintain your desired course. This section lays important groundwork you will need when you start flying instrument approaches. Let's begin with a review of the VOR navigation system.

The Federal airway system is based primarily upon the very high frequency omnidirectional range (VOR). This system consists of several hundred VOR stations that transmit course guidance signals in the VHF range between 108.00 and 117.95 MHz. Because of this, VORs are relatively free from precipitation static and interference caused by storms or other weather phenomena. When your airborne equipment is properly calibrated, it provides a course accuracy of plus or minus one degree.

VOR FACILITIES

Each VOR is classified according to its standard service volume (SSV), which defines the reception limits in terms of altitude and distance. This is important information, particularly when you are planning direct flights between navaids over random or unpublished routes. Published routes are shown on instrument enroute charts as Federal airways, and VOR reception is guaranteed above charted minimum altitudes.

The first VOR classification is the **high altitude VOR** (HVOR). HVOR stations are effective to various ranges, depending on the altitude of use. The **low altitude VOR** (LVOR) is an essential part of the IFR low altitude enroute structure, while **terminal VORs** (TVORs) are used primarily as approach aids for airports and are not normally used as part of the enroute structure. [Figure 1-21]

VOR ACCURACY CHECKS

For IFR flight, VOR equipment must be checked for accuracy every 30 days and found to be within tolerances.

For flight in the IFR environment, FARs require that the accuracy of your airborne VOR equipment be within specified tolerances. VOR equipment must either be maintained, checked, and inspected under an approved procedure, or it must have been operationally checked for accuracy

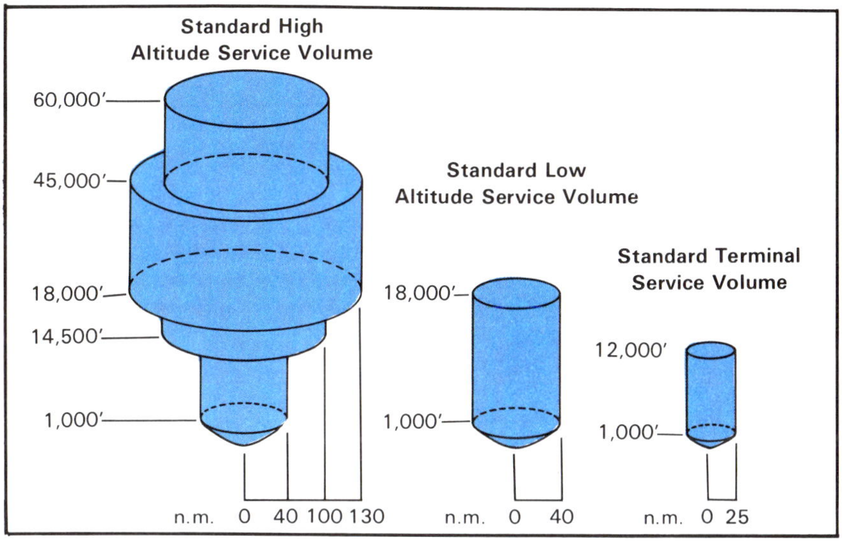

Figure 1-21. The standard service volumes for the three classes of VOR facilities vary considerably. Note that below 18,000 feet, the maximum reception distance of an HVOR is 100 n.m. This is why two of these facilities forming an unpublished direct airway in the U.S. should be no more than 200 n.m. apart. All altitudes shown here are AGL heights based on the elevation of the station site.

within the 30 days preceding the flight. This applies whether you are using a basic VOR, HSI, or RMI. These checks are normally made by the pilot in command.

If you are completing a VOR check, regulations require that you make an entry in the aircraft log or other reliable record. Your entry must show the date, location, bearing error, and your signature. If you are preparing for an IFR flight, you should check the aircraft records to ensure that the VOR equipment meets FAR requirements.

Specific information regarding VOR operational checks must be noted on a permanent record.

There are five ways to make VOR operational checks. The first is to use an FAA VOR test facility (VOT) signal. To use the VOT, tune in the VOT frequency. With the CDI centered on a basic VOR indicator or an HSI, you should read 0° with the TO-FROM indicator showing FROM, or 180° with a TO indication. You can determine the exact error by turning the course selector until the CDI is centered and then noting the degrees difference between the course selected and 180° or 0°. The airborne use of certain VOTs may be permitted; however, their use is strictly limited to those areas/altitudes specifically authorized in the *Airport/Facility Directory* excerpt for the specific VOT. The maximum error permitted for a VOT check is ± 4° (air or ground).

For a VOT test, the OBS should read either 360° with a FROM indication or 180° with a TO indication. The permissible bearing error is ± 4°.

At certain airports, you can taxi the aircraft to a point on the airport surface designated as a VOR **ground checkpoint**. This system uses a VOR radial from a station that is located on or near the airport. If neither a test signal nor a designated checkpoint on the surface is available, you may use a designated **airborne checkpoint**. These checkpoints are on specific radials over specific geographic locations in the immediate vicinity of the airport.

If no designated checkpoints are available, you may use an **airway checkpoint** to test your VOR receiver while in flight. To use this type of checkpoint, first select a VOR radial that lies along the centerline of an established Victor airway. Then, while flying along this airway, select a prominent ground point on the radial, preferably one more than 20 miles from the VOR facility. Maneuver the aircraft directly over the point at a reasonably low altitude and note the VOR bearing indicated by the receiver. This check must be performed in VFR conditions, and it normally requires the use of a sectional chart.

If dual VOR systems, independent of each other except for the antenna, are installed in your aircraft, you may check the equipment by comparing one system against the other. This is called a **dual VOR check**, and it may be performed on the ground or while airborne. Simply tune both VORs to the same VOR facility and note the indicated bearings to that station. [Figure 1-22]

When you are conducting a VOR sensitivity check, a 10° to 12° course selector change should cause full-scale movement of the CDI.

In addition to the accuracy checks, you also can verify the course sensitivity of a VOR indicator. First, rotate the course selector so the CDI moves from the center to the last dot on either side. Then, note the number of degrees the course changed. This figure should be between 10° and 12°.

SIGNAL STRENGTH

When a VOR is undergoing maintenance, the identification feature is removed or a test code is broadcast; you should disregard any navigation indications.

Sugar Factory Stack 215° 7½ miles

During cross-country navigation, you usually tune to the next VOR facility when you reach the halfway point between VORs. A station may be identified by a three-letter Morse code signal or by a combination of code

Maximum Error	Type of Check
Ground: ± 4°	1. VOT or Approved Radio Repair Station Test Signal
	2. Designated VOR System Checkpoint on Airport Surface *
Flight: ± 6°	3. Designated Airborne VOR Checkpoint *
	4. Made-up Check
Ground and Flight: 4° Differential	5. Dual VORs, Both Tuned to the Same VOR

Figure 1-22. The methods for performing the checks are listed in their order of preference, along with the maximum permissible bearing error for each. Locations for VOT, VOR receiver ground, and VOR receiver airborne checkpoints are found in the *Airport/Facility Directory* (A/FD). The dual VOR check, although listed last, is an acceptable substitute for the others.

and a repetitive voice transmission that gives the name of the VOR. You should also use positive TO-FROM indications to verify that the VOR is within reliable reception range. Even though positive audio identification has been verified, the absence of a positive TO-FROM flag indicates weak signal strength. This condition may also be accompanied by a wavering CDI, which appears to be "hunting" the course signal. Whereas the audio verifies frequency selection, a positive TO-FROM flag verifies reliable navigation signal reception. When a station is shut down for maintenance, it may radiate a T-E-S-T code (— • ••• —), or the identifier may be removed. However, the CDI may be responsive to VOR signals. You should not rely on these facilities for navigation.

INTERPRETING VOR INDICATORS

There are various types of indicators for VOR navigation, including the basic VOR indicator, the horizontal situation indicator (HSI), and the radio magnetic indicator (RMI). Although these instruments are somewhat different from a functional standpoint, they all provide similar information. For example, all provide a means of orienting to a desired course, tracking that course, and showing your direction of travel to or from the station. HSI and RMI are discussed later in this section.

BASIC VOR INDICATOR

Many general aviation aircraft are equipped with basic, or conventional, VOR indicators. They give you position information relative to your aircraft and the VOR. This information, if interpreted correctly, includes the angular difference between the aircraft heading and the course to or from the station. To understand the information the indicator provides, it is important to review its components briefly. [Figure 1-23]

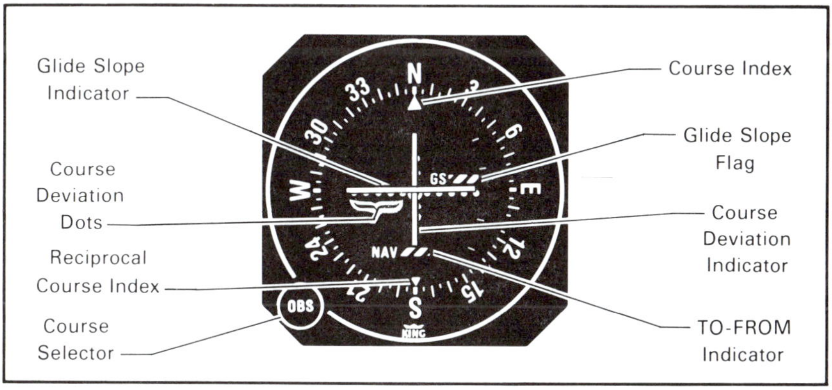

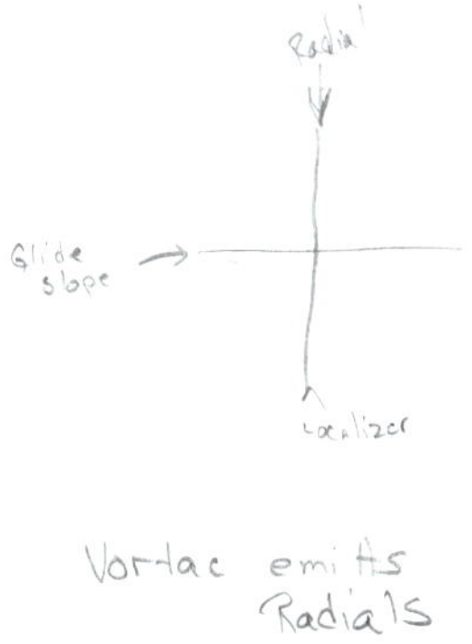

Figure 1-23. The course selector, also called the omnibearing selector (OBS), allows you to choose a particular radial by setting it under the course index. The course deviation needle (CDI) shows whether you are on or off course. The dots indicate a course deviation of 2° per dot. The TO-FROM, or ambiguity, indicator tells you whether the course selected is to or from the VOR station. This instrument also is designed to receive localizer and glide slope information.

VOR ORIENTATION

To determine your position relative to a VOR, do a VOR orientation by rotating the OBS until the CDI centers with a FROM indication. With the needle centered, read the radial from the station under the index. To determine the course to a station, use the same procedure, only center the needle with a TO indication. [Figure 1-24]

Full-scale deflection of the CDI represents a minimum of 10° deviation from the selected course centerline.

When using VOR, you will find it helpful to visualize your position, the location of the station, and where you want to go. Keep in mind that the farther you are from the VOR, the slower the needle moves; the needle does not begin to move until you are within 10° of the course selected. The CDI can indicate up to 10° of error on either side of the selected radial. Since each dot represents 2° off course, a deflection of four dots means your aircraft is 8° from the course you have selected. You also can use the course deviation dots to help you anticipate the interception of a desired radial. [Figure 1-25]

A constant off-course CDI deflection while tracking away from a VOR actually means you are flying away from the desired radial.

When you are tracking away from a station, it is important to realize that any off-course indication is multiplied as you get further from the station. For example, a one-half scale deflection that remains constant for a period of time actually means you are flying farther and farther away from the course you have selected.

BRACKETING

As you know from your VFR cross-country experience, VOR tracking is accomplished by using correct course selection and by maintaining a magnetic heading which results in a centered CDI. Under IFR, you must

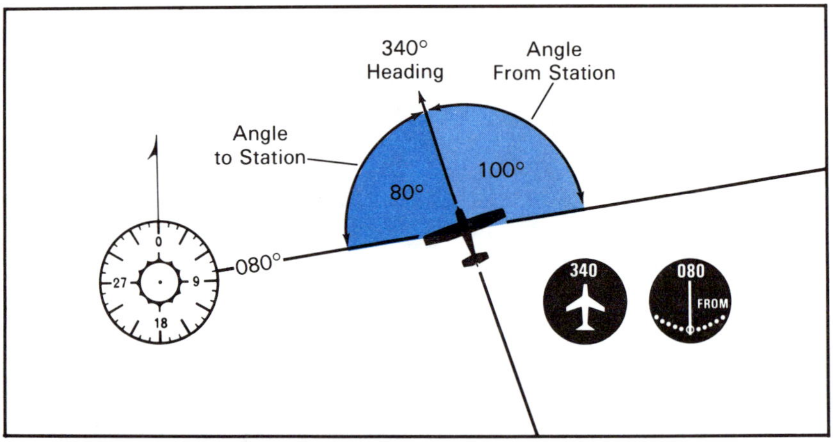

Figure 1-24. If you are flying a heading of 340° and want to fly from the VOR, center the CDI with a FROM indication. In this example, you are on the 080° radial from the station. To fly outbound on the 080° radial, you can see that a right turn of 100° is required. You also can see that a left turn of 80° is needed to fly inbound to the station.

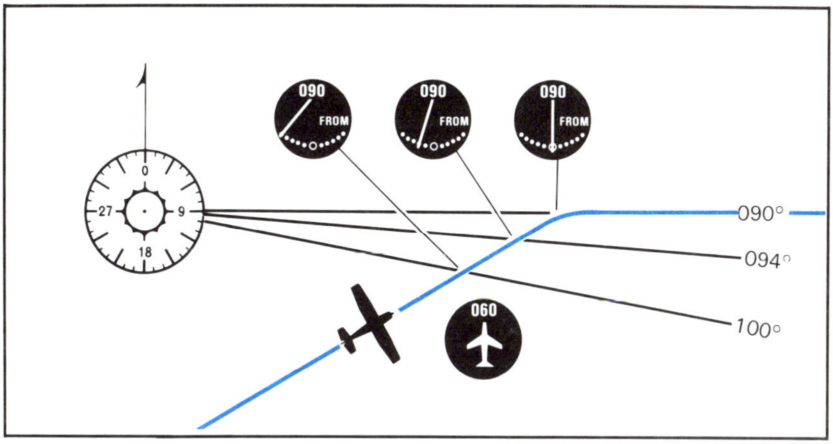

Figure 1-25. Assume you are southeast of the station on a heading of 060° and you want to track outbound on the 090° radial. The CDI shows a full left deflection until you are within 10° of the desired course. As you cross the 100° radial, the CDI begins to move in. As you cross 098°, it shows a four-dot deflection; over 094°, it shows a two-dot deflection, and so on. You can use this information to make smooth course interceptions.

navigate more precisely; during basic navigation exercises, you will soon learn to bracket your desired course quickly and select a heading which will compensate for wind. [Figure 1-26]

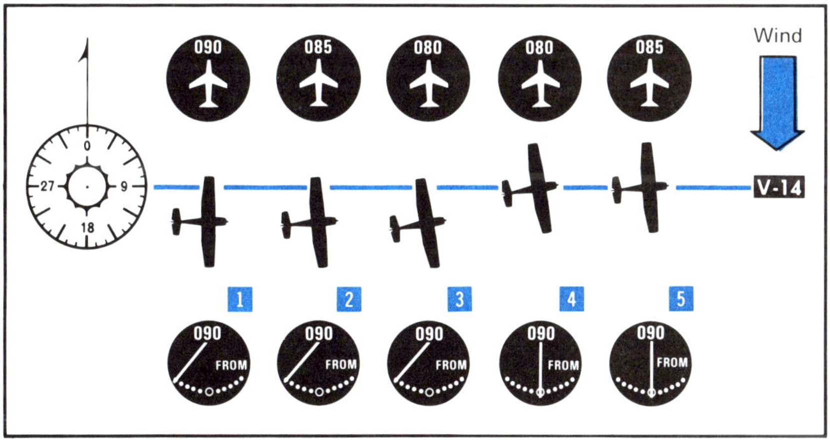

Figure 1-26. Assume you are flying eastbound on V-14 with a moderate north wind at cruising altitude. At position 1, your aircraft is drifting right of course while you maintain a heading of 090°. At position 2, you apply a 5° left heading correction (085°) to stop drift, resulting in a flight path which nearly parallels the selected course. By doubling the correction angle in position 3 to 10° (080°), your aircraft begins to return to course. The centered CDI at position 4 indicates course interception. Finally, at position 5, you return the heading to 085°, and your aircraft maintains course. Note that 085° was the heading required to fly roughly parallel to the original course in position 2.

When flying cross-country in variable wind conditions, you may have to repeat the process several times to maintain course. Therefore, the earlier you see a need for a course correction and apply it, the more precisely you will maintain course.

INTERCEPTS

During instrument flight, you will need to make a variety of intercepts to establish your aircraft on specific courses defined by navaids. Intercepts require you to visualize your present position in relation to where you want to go. Then, you select a magnetic heading which will intercept the course or radial at a specific angle. The angle of interception usually depends on a given situation. Your groundspeed and proximity to the navigation aid are important factors. Other factors include whether you will track to or from the facility. [Figure 1-27]

A wide range of intercept angles may be suitable, depending on the situation. In many cases, you may not need to make an inbound intercept because an outbound intercept is more appropriate. [Figure 1-28]

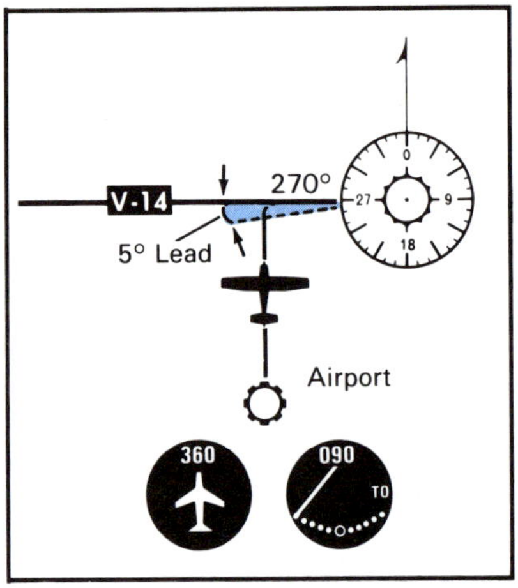

Figure 1-27. Assume you want to fly inbound on the 270° radial after departing the airport. In this situation, a heading of 360° could be used, resulting in a 90° inbound intercept. Since the turn inbound will take 30 seconds at standard rate, you should lead the turn by about 5°. You can do this by setting the course selector to 085° and beginning your turn when the CDI centers. With the course selector reset to 090°, the actual roll-out heading depends on CDI deflection and rate of movement as your heading approaches 090°.

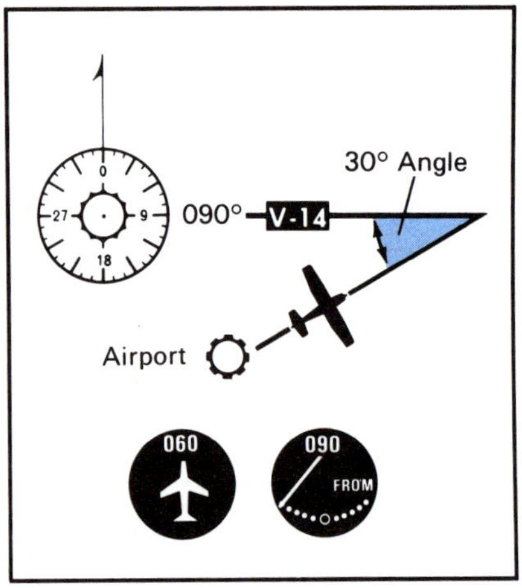

Figure 1-28. Assume you want to fly outbound on the 090° radial after departing the airport. In this situation, you could use a heading of 060° for a 30° outbound intercept. As you approach the radial, you can judge when to start the turn based on CDI deflection and rate of movement. A standard-rate turn of 30° requires only 10 seconds, so the amount of lead is small. Although a different intercept angle may be used, 30° is common for enroute operations. The greater the angle, the more lead required for the turn onto course. Other factors affecting lead include groundspeed and navaid proximity.

Another example of an outbound intercept which you will use is a course or airway change at a navaid. In these situations, you approach the navaid from one direction and depart in another. [Figure 1-29]

During intercepts, you also may use the course deviation dots to estimate the lateral distance to or from a selected radial when you are relatively close to a VOR. To make this computation, remember that each dot equals 200 feet for each nautical mile from the VOR station. [Figure 1-30]

TIME AND DISTANCE

Minutes to Station = Time in Seconds ÷ Number of Degrees.

As part of VOR tracking exercises, you may perform time and distance checks, which are exercises that tell you the time to the station in minutes and the distance in nautical miles. These exercises will help you in VOR orientation and position visualization. Essentially, you time your passage across a predetermined radial span while on a perpendicular flight path. As you cross the first radial and again as you cross the second radial, you note the time when the CDI centers. Then, divide the time in seconds by the number of degrees between the two radials used. This is the number of minutes it will take you to reach the station if you turn inbound. Keep in mind that the formula is accurate only in a

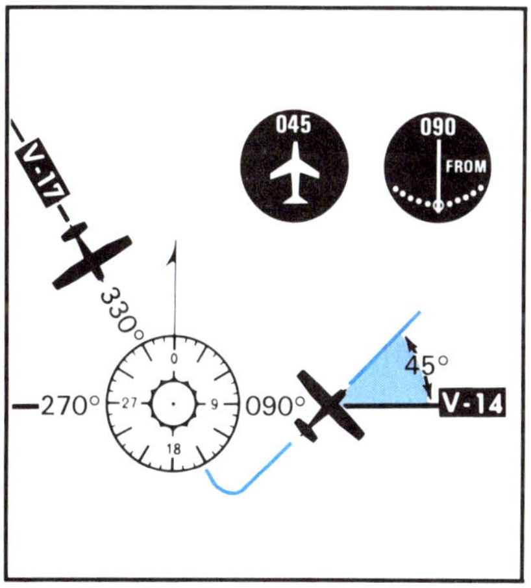

Figure 1-29. Assume you are approaching the VOR on V-17 but will depart on V-14. For this 60° course change, a 45° outbound intercept angle is appropriate. After station passage, turn left to 045°, reset the course selector to 090°, and prepare to turn on course when the CDI begins to center. The lead used in this situation may be determined by observing the rate of CDI movement. The faster the movement, the greater the lead required.

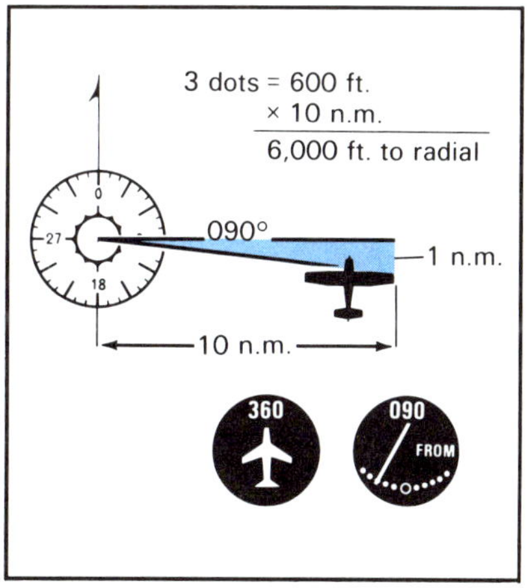

Figure 1-30. Assume you are 10 miles from the VOR and the CDI is deflected 3 dots. A quick math computation (3 dots x 200 feet x 10 miles = 6,000) shows that your aircraft is 6,000 feet, or approximately one mile, from the selected radial.

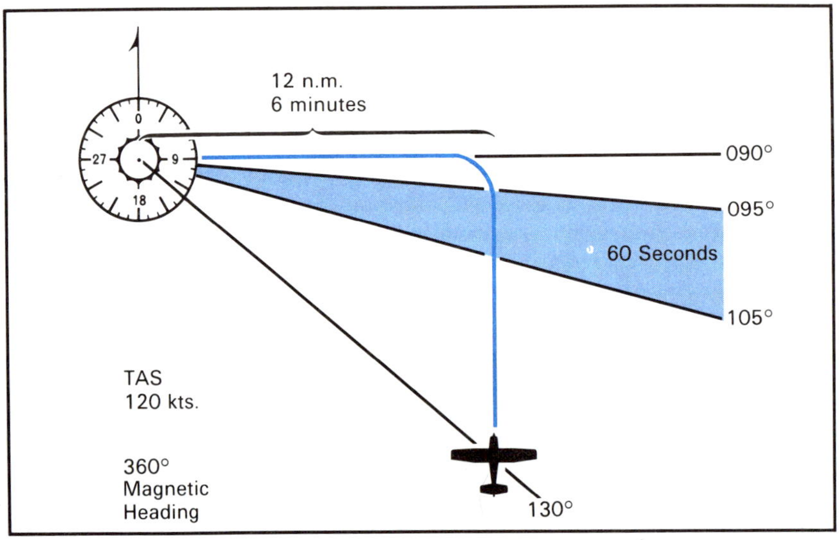

Figure 1-31. If your aircraft is on the 130° radial of a VOR and your heading is 360°, you are flying perpendicular to the 090° radial. As you continue flying on the heading of 360°, note the time crossing the 105° and 095° radials. If 60 seconds elapse during the 10° of radial span, the time to the VOR will be 6 minutes (60 ÷ 10 = 6). At 120 knots (two miles per minute), your distance from the station is approximately 12 n.m.

no-wind condition. The formula also can be used with ADF tracking exercises. [Figure 1-31]

STATION PASSAGE

VOR station passage is indicated by the first complete reversal of the TO-FROM indicator.

Although the principles of VOR navigation remain the same whether you are operating under VFR or IFR, there are some special considerations for instrument flight. For example, recognition of station passage is very important when you are flying in IFR conditions. There are some definite cues that tell you when passage is imminent and then when it is complete. The main signal occurs when you enter the area around a VOR where, for a brief period of time, there are no signals received from the station and there is a momentary OFF flag. The extent of this area increases with altitude, so the time it takes to fly through it varies. Positive station passage is shown by the first complete reversal of the TO-FROM indicator. In addition to station passage, you will frequently need to determine your arrival at a position fix along your route. VOR fixes or intersections are shown on aeronautical charts. They often are identified by the intersection of two VOR radials. [Figure 1-32]

VOR LIMITATIONS

As you know from your VFR experience, the primary limiting factor of VOR navigation is line-of-sight reception. Obstacles, such as buildings, terrain features, and earth curvature, block reception at lower altitudes. In mountainous regions, there may be momentary CDI oscillations which

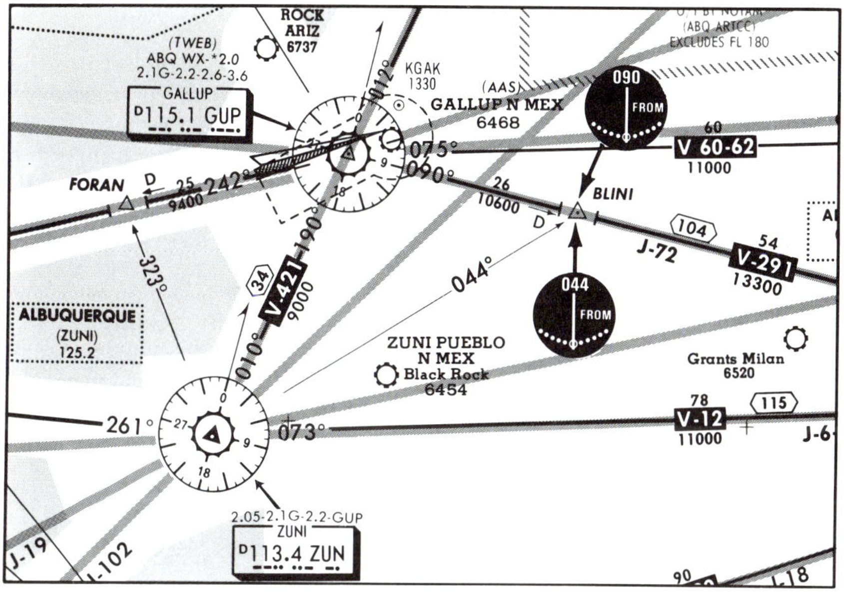

Figure 1-32. If you are flying east of Gallup on V-291, you could use the 044° radial from Zuni VORTAC and the 090° radial from Gallup VORTAC to identify Blini Intersection. Notice that both VOR indicators show centered CDIs with FROM indications. You also can identify this intersection using only one VOR receiver.

should not be misinterpreted as a station passage indication. Normally, the VOR facility listing in the *Airport/Facility Directory* will designate such areas as unusable.

DISTANCE MEASURING EQUIPMENT

Distance measuring equipment is a valuable navigation aid for any flight, but it is particularly useful for IFR operations. With DME, you are provided with direct readouts of groundspeed and distance from selected navaids. DME operates on frequencies between 962 MHz and 1,213 MHz in the UHF spectrum.

Depending on the airborne equipment, navaids which provide distance information as well as azimuth information include VOR/DME, VORTACs, TACANs, and ILS/DME facilities. VOR/DME and VORTAC facilities provide azimuth and distance/groundspeed information from collocated components under a frequency pairing plan. Because of paired frequencies, DME receivers usually display the VHF frequency of the paired VOR for channeling rather than the actual UHF frequency of the DME.

TACANs provide both azimuth and DME information on UHF frequencies and are used primarily by military aircraft. DME has been installed at many ILS locations and uses frequency pairing like that of VOR/DME or VORTAC facilities.

OPERATION

DME operates by transmitting paired pulses of radio signals from the aircraft to the ground station. The interrogation from the airborne equipment causes the ground station to respond with radio transmissions on a different frequency but at the same pulse rate. The time required for the round trip of this signal exchange is measured in the airborne DME equipment and translated into distance information in nautical miles and/or the groundspeed of the aircraft in knots.

DME indicators display slant range distance in nautical miles.

Although UHF radio waves are limited to line-of-sight reception, DME may be accurately received at distances up to 199 nautical miles. Accuracy is within one-half mile or three percent of the distance, whichever is greater. Since the airborne interrogator is at altitude, DME distance is the slant range distance between the aircraft and the ground station rather than the actual horizontal distance measured on the earth's surface. Slant range error is negligible at maximum range and altitudes, but it may be significant at close range. The greatest error is when you are at high altitudes close to the VORTAC. [Figure 1-33]

IDENTIFICATION FEATURE

VOR azimuth and DME identifiers are transmitted on a time-share basis.

VOR/DME and VORTAC facilities are identified by synchronized identifications which are transmitted on a time-share basis. The VOR azimuth position of the facility broadcasts a coded tone modulated at 1,020 Hz or by a combination of code and voice. As stated previously, the only positive method of identifying the facility is by its Morse code identification or by the voice identification. During periods of maintenance, the

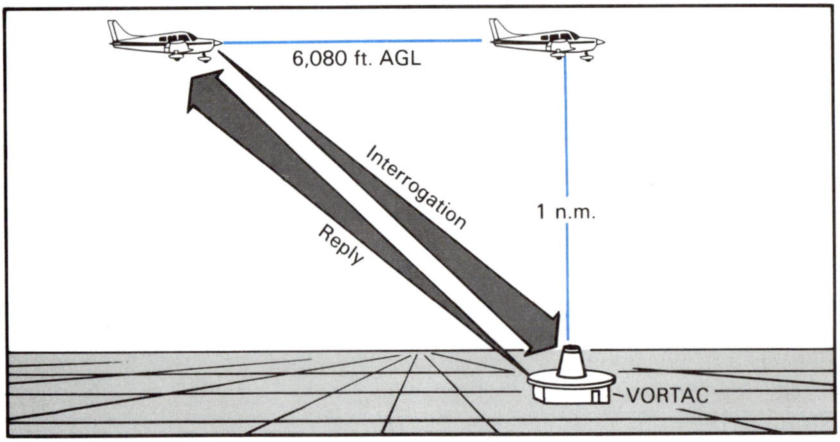

Figure 1-33. If you cross a DME facility at 6,080 feet AGL, one nautical mile, you will observe the distance decrease to one mile, rather than zero, as you pass the station. Slant-range error is negligible if an aircraft is one mile from the DME facility for each 1,000 feet of altitude above the facility.

coded facility identification is removed, or it may radiate a T-E-S-T (— • ••• —) code.

The TACAN or DME is identified by a coded tone modulated at 1,350 Hz. The DME or TACAN coded identification is transmitted one time for each three or four times that the VOR coded identification is transmitted. When either the VOR or the DME is inoperative, it is important to recognize which identifier is retained for the operative facility. A single, coded identification with a repetition interval of approximately 30 seconds indicates the DME is operative. If the DME is inoperative, the coded identification will be removed even though the distance indications may appear normal.

When only the DME portion of a navaid is operating, the identifier is transmitted approximately every 30 seconds.

HORIZONTAL SITUATION INDICATOR

When compared to the basic VOR indicator, the horizontal situation indicator (HSI) presents more navigation information in a somewhat pictorial form, making interpretation of the information easier for you. The main advantage of an HSI is that it combines the information available from both the VOR and heading indicators into one display. [Figure 1-34]

VOR ORIENTATION WITH HSI

Position-finding procedures using the HSI are essentially the same as those using a basic VOR indicator. As was the procedure with VOR,

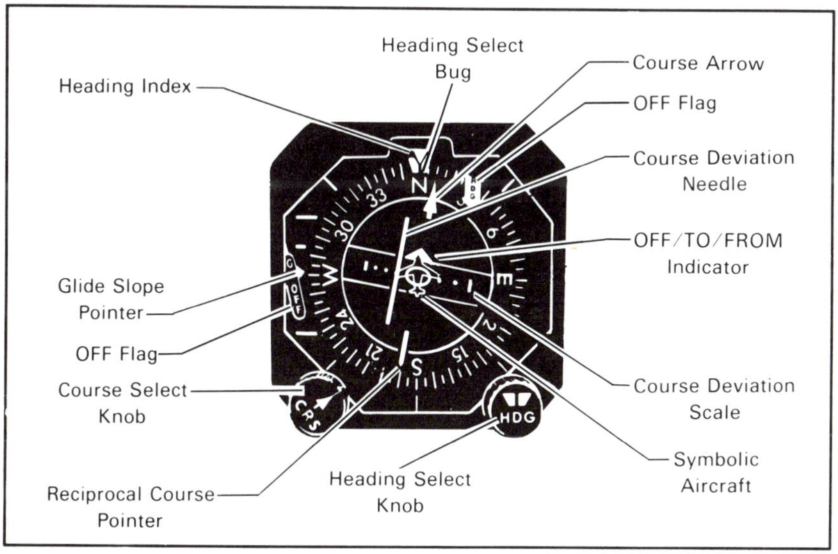

Figure 1-34. The course arrow and the reciprocal course pointer move when you turn the course select knob. The symbolic aircraft in the center represents your aircraft's position with respect to the desired course. Three OFF flags may be used to alert you to unreliable signals or equipment malfunctions. Glide slope information is also presented in a vertical display on the left side.

turn the course select knob to get a centered course needle with the desired TO-FROM indication. Read the selected course over the course arrow pointer and the reciprocal course beneath the reciprocal pointer.

To determine the angle to or from a VOR station, center the deviation needle and note the number of degrees between the aircraft heading and the course arrow. Since the HSI incorporates VOR and heading information together, the direction of turn is pictorially displayed. If the course arrow is to the right of the heading, you should turn to the right, and vice versa. As with basic VOR, you can visualize your aircraft position with respect to the radial or the VOR by using the HSI indications. The course needle shows the general direction of the selected radial; the TO-FROM indicator shows the general direction of the VOR. [Figure 1-35]

For every nautical mile from the station, one dot of deflection of the HSI course needle represents 200 feet that you are off course.

The lateral deviation scale also is used in the same way as the course deviation dots on the VOR indicator. The HSI displays the number of degrees from the selected radial (if less than 10°), and can be used to estimate the distance in nautical miles from the desired radial. Each dot represents a 2° deviation, which represents a lateral distance of 200 feet per nautical mile from the selected radial. For example, 1 dot at 10 nautical miles represents 2,000 feet in lateral distance.

The heading indicators of some HSIs use a **remote indicating compass**. A slaving control button is used to select either the "slaved" or "free gyro" mode. When the magnetic compass and the HSI heading (remote indicating compass) do not agree, you can adjust the HSI heading by selecting "free gyro" and pressing the appropriate drive button (clockwise or

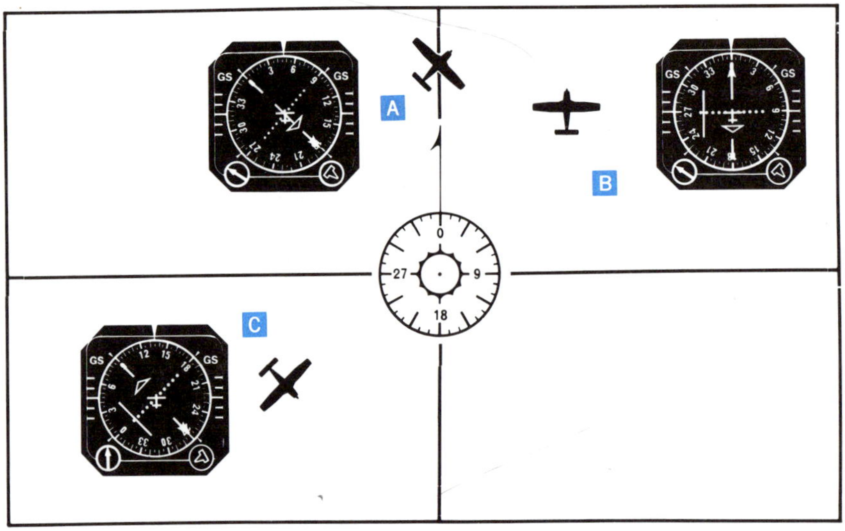

Figure 1-35. Aircraft A is on the selected radial and the VOR lies to the south. Aircraft B is east of the selected course, and the VOR is to the south of the aircraft. Aircraft C is south of the selected radial, and the VOR is located to the north.

counterclockwise). For example, if your magnetic compass reads 100° and the HSI heading is 110°, you would switch to "free gyro" and push the "clockwise" button until the heading indicator rotates clockwise to 100°.

ADF FACILITIES

Automatic direction finding equipment normally is used as a secondary navigation system in the conterminous United States. However, in parts of Alaska, as well as in Canada and many other countries, you may find heavier reliance on ADF. Familiarity with this system provides you with useful alternatives in the event of VHF navigation equipment failure. In addition to instrument approach operations, you may use ADF to monitor your position while enroute. It also may be used in areas where VHF signals are blocked by terrain or the earth's curvature. Within the United States, the navigational facilities which you may use primarily with ADF are NDBs, ILS compass locators, and commercial broadcast stations.

NONDIRECTIONAL RADIO BEACONS

Nondirectional beacons (NDBs) transmit in the low/medium frequency (L/MF) range between 190 and 535 kHz, using a continuous signal at 1,020 Hertz modulation. These signals travel both as ground waves that penetrate obstacles and as sky waves that are refracted by the ionosphere. As a result, L/MF waves are not limited to line-of-sight like VOR, so they can be received at lower altitudes.

NDBs are classified by their power output and reception range. NDBs are further classified according to voice capability and weather broadcast services in the *Airport/Facility Directory*. The power output and IFR reception ranges of the various classes at all altitudes are as follows:

1. The MH radio beacon has a power output of less than 50 watts and a usable range of 25 miles.
2. The H classification has a power output of 50 through 1,999 watts, with a usable range of up to 50 miles.
3. The HH radio beacon has a power output of 2,000 watts or more and a usable range of 75 miles.
4. The compass locator is collocated with the outer marker (LOM) or middle marker (LMM) of an instrument landing system. It normally has a power output of less than 25 watts and a usable range of 15 miles.

An NDB has a three-letter Morse code identification, except when it is used as an ILS compass locator. In this case, a continuous two-letter identification code is normally used. The identification feature of an NDB is particularly important, because ADF receivers do not have a "flag" to warn you that erroneous bearing information is being received. Therefore, you should continuously monitor the NDB identification during critical phases of operation.

COMMERCIAL BROADCAST STATIONS

You may also receive commercial broadcast stations of the AM (amplitude modulation) class with ADF equipment. Although these facilities are not approved for normal IFR operation or air traffic control, you may use them for VFR operations. Selected commercial broadcast stations and frequencies are shown on aeronautical charts. One of the disadvantages of using commercial broadcast stations is that station identification may be provided as infrequently as once each hour.

INTERPRETING ADF INDICATORS

Bearing indicators are of two types — movable and fixed card. Both have a bearing pointer and azimuth ring. When using the movable card indicator, you may rotate the azimuth ring to reflect the heading of the aircraft. With the fixed card indicator, the nose of the aircraft is always oriented to 0°, and the tail of the aircraft is always 180°. The fixed card bearing pointer measures, in a clockwise direction, the number of degrees between the nose of the aircraft and the station (relative bearing). To determine the magnetic bearing (course) to the station using a fixed card ADF, you must use the ADF formula: magnetic heading + relative bearing = magnetic bearing to the station. [Figure 1-36]

RB Relative To The nose of the Plane

BRACKETING

The following discussion refers to a fixed card indicator. ADF tracking in a crosswind is somewhat more involved than VOR, since ADF does not automatically provide wind correction. Consequently, it is more difficult to interpret ADF during bracketing maneuvers. [Figure 1-37]

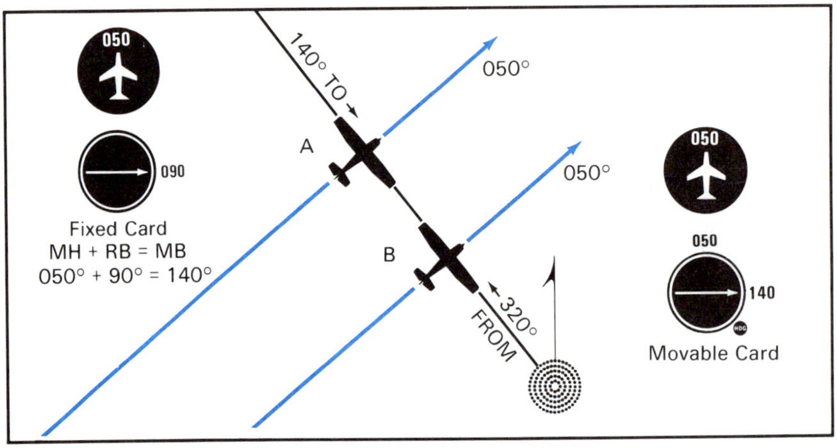

Figure 1-36. If the magnetic heading (MH) of aircraft A is 050° and the relative bearing (RB) is 090°, the magnetic bearing (MB) to the station is 140°. However, your location is on the reciprocal of the 140° bearing to the station, which is the 320° bearing from the station. Aircraft B has a movable card indicator, so you can set 050° (aircraft heading) below the index and the needle will point to the 140°, the magnetic bearing to the station.

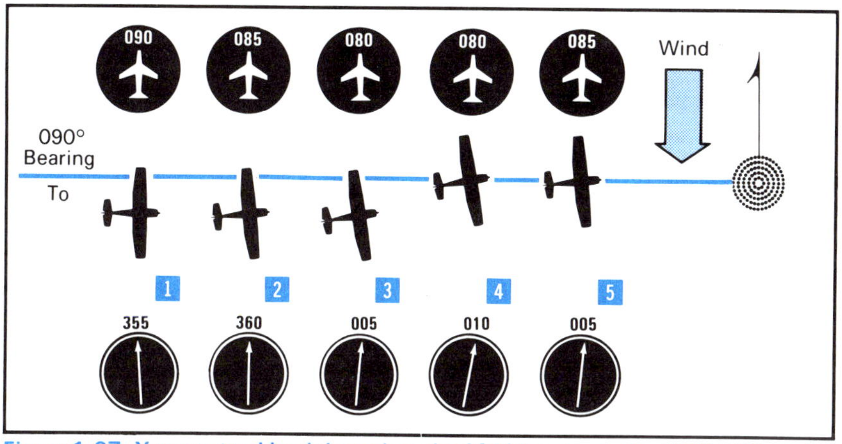

Figure 1-37. You are tracking inbound on the 090° bearing to the NDB with a wind from the north. At position 1, your aircraft is drifting right of course while on a heading of 090°. At position 2, you apply a 5° left heading correction to stop drift, resulting in a flight path which nearly parallels the desired course. Your heading and relative bearing will remain constant. By doubling the correction angle in position 3 to 10°, your aircraft begins to return to course. At position 4, your magnetic heading (080°) plus your relative bearing (010°) equals the desired magnetic bearing to the station (090°). You have intercepted the original course and can return to 085° (the heading that paralleled the course in position 2). At position 5, you are maintaining course with a 5° left correction angle and a 005° relative bearing (085° + 005° = 090°).

During tracking exercises, the relative bearing should equal the wind correction angle if you are maintaining the course. Regardless of the number of degrees you are correcting left or right, you are on course when the wind correction angle equals the number of degrees a station bears to the left or right of the aircraft's nose. When you are crossing an NDB, the ADF indicator may show side-to-side needle deflections or a steady movement toward a wingtip position. You should maintain a constant aircraft heading and avoid "chasing" the needle. This will result in a straight ground track, and you may resume normal navigation procedures after you have passed the station. When the needle stabilizes at or near the 180° position, you have passed the station. If you do not pass directly over the facility, station passage occurs when the needle is steady and points to either wingtip position.

ADF station passage is indicated when the needle shows either wingtip position or settles at or near the 180° position.

ADF INTERCEPTS

ADF intercepts are similar to tracking, except you use larger heading changes. One method of intercepting a bearing to an NDB is to first determine your position relative to the station by turning to a heading that parallels the bearing to be intercepted. Note whether the station is to the left or right of the aircraft. Determine the number of degrees of needle deflection from the 0° position and double this amount for the interception angle. Then, turn the aircraft that same number of degrees toward the desired magnetic bearing. Like VOR, there are several ways to intercept an NDB bearing. [Figure 1-38]

During inbound intercepts, maintain the intercept heading until the ADF needle approaches the same number of degrees from the nose as the intercept angle. Then, using an appropriate amount of lead, turn inbound and continue with normal tracking procedures. During outbound intercepts, your intercept angle is reflected by the needle position with respect to the tail of the aircraft. As in VOR operations, you may use various intercept angles. [Figure 1-39]

TIME-TO-STATION ESTIMATES

Minutes to the station equals time in seconds divided by degrees of wingtip bearing change. When not on a heading perpendicular to a bearing or radial, time to station equals the time to double the relative bearing.

You can estimate time and distance to an NDB by using the same formula as shown for VOR when the relative bearing is approximately 90° or 270°. Other methods involve variations of the isosceles triangle principle, and do not require you to be on a perpendicular heading to a bearing or radial. For example, if you maintain a constant heading and the relative bearing to the station doubles from 15° to 30° in six minutes, you are six minutes from the station, assuming a uniform groundspeed. This method also works if you turn so that the ADF needle is on 45° (or 315°). Maintain heading until the needle is at 90° (or 270°). The time flown equals time to the station. Apply groundspeed to find the distance to the NDB.

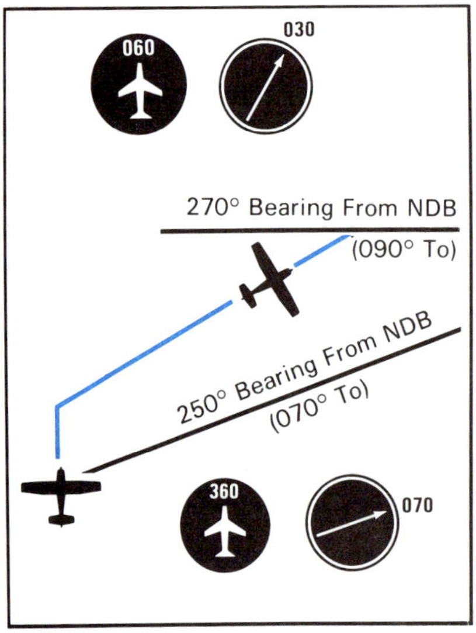

Figure 1-38. Assume you're on the 250° bearing from an NDB and you want to track inbound on the 270° bearing from the station. A heading of 060° will produce a 30° intercept angle; therefore, you will intercept the 270° bearing when the relative bearing is 030°. As the relative bearing approaches that value, turn to track inbound to the NDB.

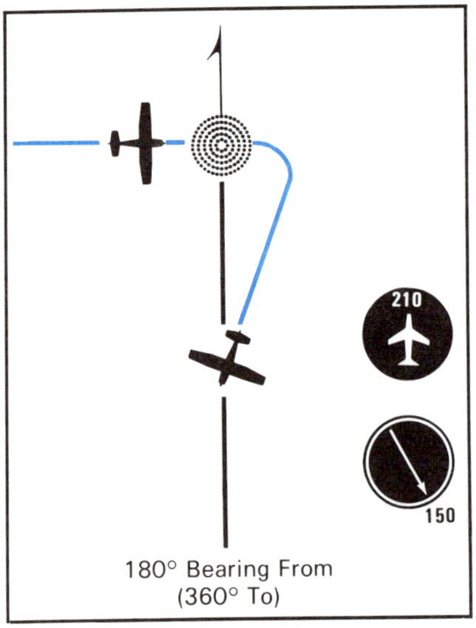

Figure 1-39. In this example, you plan to depart an NDB on the 180° bearing from the station with a 30° intercept heading of 210°. Start the turn on course when the bearing from the tail approaches a relative bearing of 150° or a 30° angle to the tail. After turning to 180°, a continuous 180° relative bearing indicates you are maintaining course.

ADF LIMITATIONS

You should be aware that reliable ADF navigation depends on the accuracy of the heading indicator. If the heading indicator has precessed 10°, the tracking procedures and intercepts will be in error by 10°. Although this will not be so significant during inbound tracking, during outbound tracking over large distances the error is multiplied. You can be miles off course and yet be completely unaware of your predicament unless you confirm agreement between the magnetic compass and the heading indicator.

L/MF waves are subject to some disturbances which result in erroneous ADF indications. These disturbances are caused by twilight, terrain, precipitation, and thunderstorms.

Twilight effect is caused by L/MF wave refraction from the ionosphere and is most pronounced just before and just after sunrise or sunset. The height of the ionosphere changes during periods of sunset and sunrise. Wave refraction occurring during these periods is caused by the elliptical-shaped ionosphere. Wave refraction introduces errors which are greatest at maximum range and result in ADF needle fluctuations. [Figure 1-40]

Mountainous terrain has the ability to reflect L/MF waves, resulting in false courses or indefinite indications. The best procedure in mountainous regions is to select high power stations, provided these stations are not separated from the aircraft by high terrain features. Shorelines also disturb L/MF wave patterns, because of the variations in the reflective characteristics of land and water. [Figure 1-41]

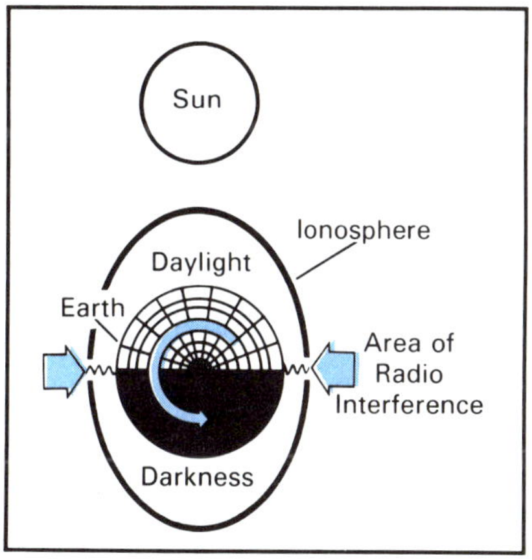

Figure 1-40. You can generally minimize twilight effect by flying at higher altitudes. You can also minimize twilight effect if you are able to select NDBs with frequencies lower than 350 kHz.

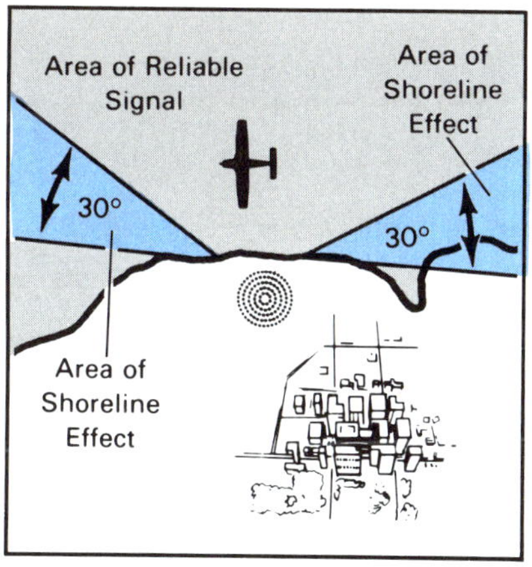

Figure 1-41. Shoreline effect is most pronounced when the radio waves cross the shoreline at less than a 30° angle. Avoid using facilities in this category until the aircraft is in a more perpendicular position to the station and shoreline.

Saint Elmo's Fire is caused by static discharges from the airplane.

ADF frequencies are subject to static when an airplane is flying in heavy precipitation. Precipitation static caused by rain, snow, or cloud droplets may be enough to cause the ADF needle to wander aimlessly. Reception may be so hampered that it is impossible to identify a station. Precipitation static is the result of an accumulation of static charges on various parts of the airplane such as the wing, tail, and propeller. As these charges increase in intensity, they are discharged into the surrounding air. Much like lightning, static discharge looks like lights "dancing" on the exterior of the aircraft. This effect is known as **Saint Elmo's Fire**. Static dischargers, installed at the trailing edges of the control surfaces, promote the discharge of static electricity. Flying at lower airspeed will also lessen precipitation static.

Thunderstorms contain tremendous charges of static electricity. When electrical charges in one portion of the storm differ greatly from another portion, lightning is generated. Lightning radiates radio waves in the L/MF band and causes the ADF indicator to point momentarily toward the area of electrical discharge. You should disregard the ADF indicator during lightning flashes.

RADIO MAGNETIC INDICATOR

RMI provides continuous position orientation from two navaids at the same time.

A single navigation display which features two bearing pointers superimposed over a slaved compass card is the **radio magnetic indicator** (RMI). Because it allows you to monitor your position constantly, the RMI does not have deviation dots or a TO-FROM indicator. [Figure 1-42]

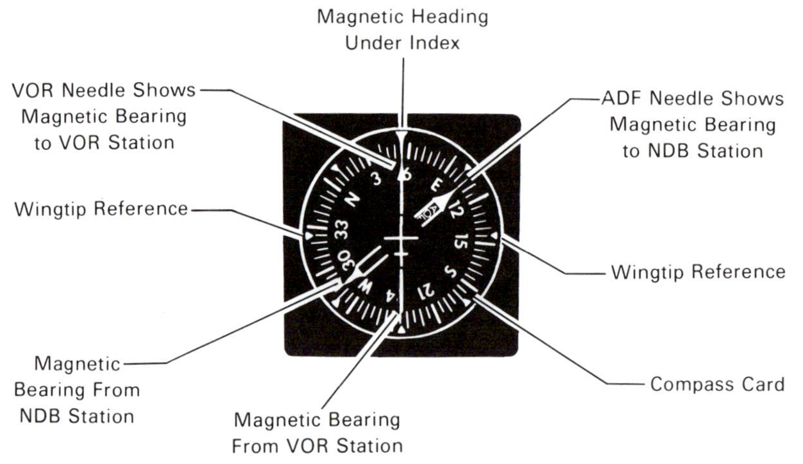

Figure 1-42. One bearing pointer indicates the magnetic bearing in relation to a VOR station, and the other shows the magnetic bearing to an NDB station. Therefore, the RMI provides you with magnetic heading and simultaneous VOR and ADF navigation information. In addition, you do not need to compute the magnetic bearing to or from the respective navaids, since it is displayed directly under the needles.

Each bearing pointer is governed by the frequency set in its associated receiver. However, some of the more advanced RMIs are designed to access either two different VOR or two NDB facilities. Always refer to the pilot's operating handbook (POH) or manufacturer's description for the equipment you are using.

When you are determining station passage with RMI, the needle will start to oscillate, then swing from the top toward the reciprocal bearing at the bottom of the indicator. You also can use RMI to identify position fixes. [Figure 1-43]

Since the VOR bearing pointer of an RMI is part of the VOR equipment, it also must meet the required tolerances. When you have tuned the associated VOR receiver to a VOT frequency, an accurate VOR needle will indicate 180°, with a maximum difference of four degrees.

DME ARCS

Many instrument approach procedures incorporate DME arcs for transition from the enroute phase of flight to the final approach course. DME arcs require DME equipment in conjunction with VOR, HSI, or RMI navigation indicators. However, the FAA recommends the use of RMI for DME arcs unless you are highly proficient in the use of other airborne equipment and in performing the specific approach procedure. While

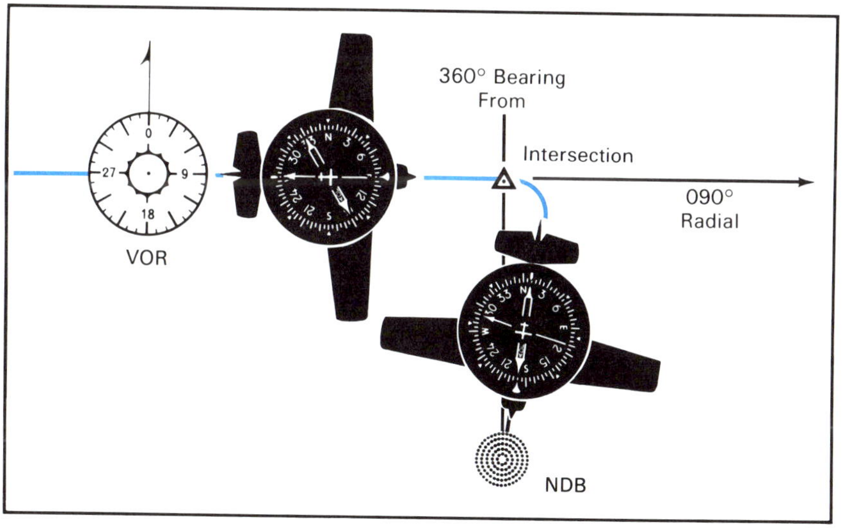

Figure 1-43. Assume you are tracking outbound from the VOR on the 090° radial, with the number one bearing indicator (single needle) tuned to the VOR and the number two bearing indicator (double needle) tuned to the NDB. You want to proceed direct to the NDB after passing the intersection. Since the slaved compass card reflects the magnetic heading, turn right and proceed direct to the NDB when the number two bearing indicator reflects a magnetic bearing of 180°.

intercepting and flying an arc, it is important for you to maintain a continuous mental picture of your position on the arc and relative to the VOR station.

INTERCEPTING THE ARC

You may intercept a DME arc while flying to or from a VORTAC or VOR/DME facility. To do this, you should plan an initial turn of 90°. Since a 90° turn at standard rate (three degrees per second) requires 30 seconds, you will find it necessary to lead the initial turn to the arc to achieve accurate arc interception. If you correctly determine the leadpoint, your chances of overshooting or undershooting the arc diminish. For groundspeeds below 150 knots, a leadpoint of one-half nautical mile usually is sufficient. Before reaching the leadpoint, you should determine the correct direction of turn and the magnetic heading for the roll-out. [Figure 1-44]

MAINTAINING THE ARC

You will find that larger arcs are easier to fly because of their gradual rate of curvature. High groundspeeds in conjunction with arcs of small radii are more difficult because of the higher rate of curvature. Maintaining the arc is simplified if you plan to navigate slightly inside the curve. Thus, the arc is always turning toward your aircraft, and you may intercept it by holding a straight course. If your aircraft is outside the curve, the arc is "turning away" from you and a greater correction is required.

With a high enough groundspeed, it is theoretically possible to fly an exact circle around a facility in a no-wind condition by maintaining the bearing pointer on the wingtip reference. In actual practice, however, you normally fly a series of short legs. With the bearing pointer on the appropriate left or right wingtip reference and the aircraft at the desired DME range, maintain the heading and allow the bearing pointer to move 5° to 10° behind the wingtip (item 2). This will cause the DME distance to increase slightly.

Next, turn toward the facility to place the bearing pointer 5° to 10° ahead of the wingtip reference (item 3). Then, maintain the heading until the pointer is again behind the wingtip. Repeat this procedure, as necessary, to maintain the approximate arc.

OFF-COURSE CORRECTIONS

As a guide in making heading corrections, you may change the relative bearing 10° to 20° for each one-half mile deviation from the desired arc. For example, if you are one-half mile outside the arc and the bearing pointer is on the wingtip reference (item 4), you should turn 20° toward the facility to return to the arc. If the off-course position is one-half mile inside the arc and the bearing pointer is on the wingtip reference (item 5), turn 10° away from the facility.

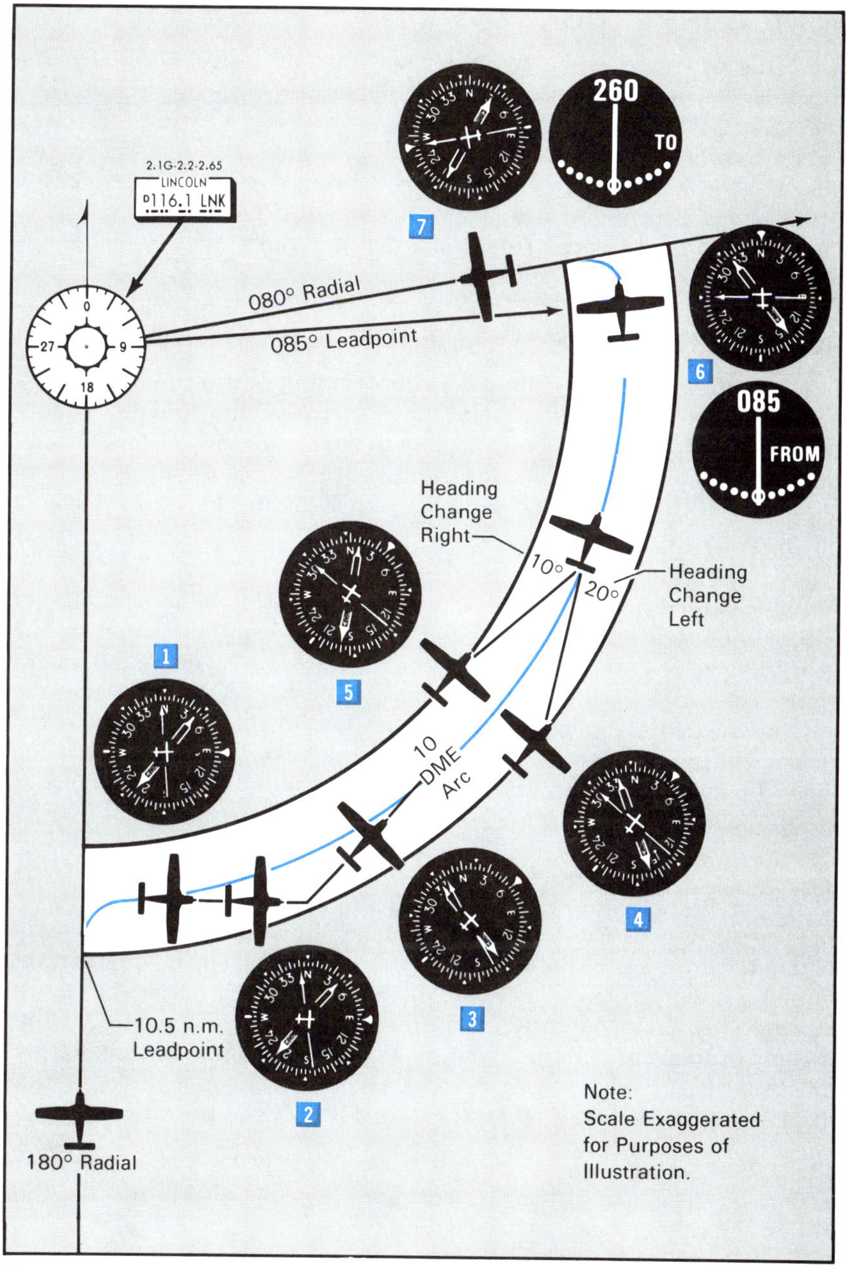

Figure 1-44. With RMI equipment, you can monitor the VOR bearing pointer during the turn and roll out with the pointer on or near the appropriate left or right wingtip reference point (item 1). Regardless of the equipment used for the arc, if it appears the turn will place your aircraft outside the arc, you should roll out early. If it is apparent your aircraft will be positioned inside the arc, you should continue the turn past the roll-out point originally planned. The numbered callouts in the remaining discussion refer back to this illustration.

WIND CORRECTION

If the wind is blowing you away from the navaid, keep the bearing pointer ahead of the indicator's wingtip reference. Keep it behind the wingtip reference if the wind is pushing you toward the facility.

Since the wind correction angle is changing constantly throughout the arc, wind orientation is important. You must use a reference other than the wingtip when operating in a crosswind. If a crosswind is causing your aircraft to drift away from the facility, maintain the bearing pointer ahead of the wingtip. If a crosswind is drifting you toward the facility, establish the bearing pointer behind the wingtip. The selected reference point should be displaced from the wingtip by an amount equal to the required wind correction angle.

INTERCEPTING A RADIAL FROM A DME ARC

Intercepting a radial from an arc requires a different method for determining your leadpoint since your DME won't provide you with distance to the desired radial. The actual lead will vary with the arc's radius and the aircraft's groundspeed. A lead of five degrees or less is sufficient at speeds up to 150 knots when you are flying DME arcs like those depicted on most approach charts.

With an RMI, you should monitor the rate of bearing movement closely while flying the arc. Determine the approximate lead radial, and set the bearing selector of the VOR navigation indicator to the outbound course of the lead radial. When the CDI shows your aircraft is crossing the appropriate radial (item 6), initiate the turn to intercept the inbound course. During the turn, reset the OBS to the inbound course. Following the roll-out, the CDI should be centered and the VOR bearing indicator should point directly ahead of the aircraft (item 7).

AREA NAVIGATION

Area navigation, or RNAV, permits flight operations on any desired course within the coverage of station-referenced navigation signals or within the limits of a self-contained system capability. Random RNAV routes are direct routes, based on area navigation capability, between waypoints defined in terms of latitude/longitude coordinates, degree/distance fixes, or offsets from published or established routes/airways at a specified distance and direction. The major types of RNAV equipment include:

VORTAC referenced or course line computer (CLC) systems have been widely used for many years. To function, the CLC RNAV unit must be within the service range of a VORTAC.

OMEGA/VLF, although two separate systems, can be considered as one unit. It is a long-range navigation system based on very low frequency radio signals transmitted from a network of 17 stations worldwide.

An inertial navigation system (INS) is self-contained and requires no information from external references. INS provides aircraft position and navigation information in response to signals resulting from inertial effects on components within the system.

MLS area navigation (MLS/RNAV) is a system which provides area navigation with reference to an MLS ground facility.

LORAN-C is a long-range radio navigation system that uses ground waves transmitted at low frequency to provide position information at ranges of up to 600 to 1,200 nautical miles at both enroute and approach altitudes. The usable signal coverage areas are determined by the signal-to-noise ratio, the envelope-to-cycle difference, and the geometric relationship between the positions of the user and the transmitting equipment.

The three basic types of LORAN receivers in use include one for VFR use only, while the other two are appropriate for specific kinds of IFR operations. One of these is limited to enroute and terminal area IFR operations but cannot be used for instrument approaches. The other has the capability for all LORAN navigation including enroute, terminal, and approach operations.

In order to determine which operations are permitted for a particular installation, you need to know where to look. In some cases, placards provide this information or you may need to read section nine of the pilot's operating handbook (POH). In other cases, an approved airplane flight manual supplement may be provided or the information may be available on an FAA Form 337, Major Repair and Alteration.

Global positioning system, or GPS, is a space-based radio positioning, navigation, and time-transfer system. The system provides highly accurate position and velocity information, and precise time, on a continuous global basis, to an unlimited number of properly equipped users. The system is unaffected by weather, and provides a worldwide common grid reference system. Additional descriptive information on GPS, as well as limitations and operational capabilities, is included in Chapter 4.

CHECKLIST

After studying this section, you should have a basic understanding of:

- ✓ **VOR accuracy checks** — What the various types of VOR equipment checks are, including their tolerances and priority of use.
- ✓ **VOR navigation** — How to use VOR navigation equipment for VOR intercepts, tracking, and time/distance checks.
- ✓ **Distance measuring equipment** — How it operates and what causes slant-range error.

✓ **Horizontal situation indicator** — How the HSI simplifies VOR navigation and helps in position orientation.

✓ **ADF navigation** — How ADF equipment is used to intercept bearings, establish wind correction, and track to or from an NDB.

✓ **Radio magnetic indicator** — How the RMI presents navigation information from two navaids to continuously reflect aircraft position.

✓ **DME arcs** — How to intercept and maintain an arc using RMI, and how to compensate for wind.

✓ **Area navigation** — What the basic types of equipment are and how they are used.

CHAPTER 2

IFR FLIGHT ENVIRONMENT

INTRODUCTION

As pilot-in-command of an aircraft operating under IFR, you must have a thorough knowledge of the IFR flight environment and how to operate safely within it. This chapter contains a general description of the air traffic control system, as well as the types of clearances you may be issued during IFR operations. It also includes a review of airspace, airports, and flight information, with emphasis on instrument flight considerations.

SECTION A

AIR TRAFFIC CONTROL SYSTEM

To operate proficiently within the IFR environment, it is important that you understand the **air traffic control** (ATC) system. In a broad sense, the ATC system may be thought of as a network of radar and nonradar facilities that provides nationwide traffic separation during all phases of IFR flight. These facilities include the air route traffic control centers, which are the main ATC facilities for enroute operations. In terminal areas, facilities include clearance delivery, tower, departure control, and approach control.

AIR ROUTE TRAFFIC CONTROL CENTER

The facilities which provide air traffic control service to aircraft operating on IFR flight plans in controlled airspace are the **air route traffic control centers** (ARTCCs). They are also the central authority for issuing IFR clearances, and they provide nationwide monitoring of each IFR flight, primarily during the enroute phase. Each center, due to its size, is divided into **sectors**. Each sector is manned by one or more controllers, who maintain lateral and/or vertical separation of aircraft within its airspace boundaries. In addition, they coordinate traffic arriving and departing their assigned areas. Frequently, sectors are further stratified by altitude. For example, there may be low altitude sectors that extend from the floor of controlled airspace to altitudes of 18,000 to 24,000 feet MSL. Above these levels, one or more high altitude sectors may be established. [Figure 2-1]

ARTCC TRAFFIC SEPARATION

ATC's primary responsibilities are the separation of known IFR aircraft and the issuance of safety alerts.

Although the ATC system provides numerous services, the highest priorities are the safe separation of known IFR traffic and the issuance of safety alerts. Separation of other aircraft is conducted on a workload-permitting basis. There are many methods of traffic separation, but the elimination of traffic conflicts actually begins when you file your IFR flight plan.

PROCESSING THE IFR FLIGHT PLAN

IFR flight plans should be filed at least 30 minutes before departure.

Prior to departure from within, or prior to entering controlled airspace, you must submit a complete flight plan and receive an air traffic clearance, if weather conditions are below VFR minimums. Instrument flight plans may be submitted to the nearest FSS or air traffic control tower either in person or by telephone (or by radio if no other means are available). You should file your IFR flight plan at least 30 minutes prior to estimated time of departure to preclude possible delay in receiving a departure clearance from ATC.

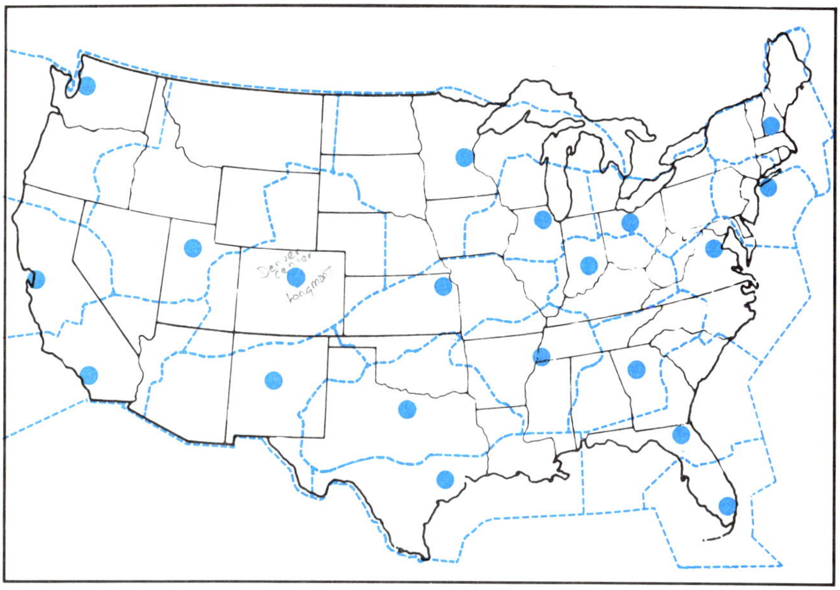

Figure 2-1. The dashed lines portray the various ARTCC boundaries within the United States, while the circles indicate the physical location of each center. Appropriate radar and communication sites are connected to the centers by microwave links and telephone lines.

In addition, to minimize your delay in entering a Class B, Class C, Class D, and Class E surface area at destination when IFR weather conditions exist or are forecast at that airport, an IFR flight plan should be filed before departure. Otherwise, a 30 minute delay is not unusual in receiving an ATC clearance because of time spent in processing flight plan data. Traffic saturation frequently prevents control personnel from accepting flight plans by radio. In such cases, the AIM recommends that you contact the nearest FSS for the purpose of filing the flight plan.

When you are departing an airport with an operating control tower, you normally request your clearance through ground control or, if appropriate, clearance delivery. If you are departing an airport served only by a flight service station (FSS), the FSS will request your clearance and forward it to you before your departure. In the event you are departing an airport without an ATC or FSS facility on the field, you may request your clearance by contacting the local FSS by telephone prior to departure. If necessary, you can obtain your clearance after departing VFR, if weather conditions permit. In this situation, you could contact the ARTCC sector where your flight originated, or you could request your clearance from the local FSS.

IFR flight plans are usually deleted from the ARTCC computer if they are not activated within one hour of the proposed departure time.

To prevent computer saturation, most centers will delete an IFR flight plan a minimum of one hour after the proposed departure time. To ensure your flight plan remains active, advise ATC of your revised departure time if you will be delayed one hour or more.

ENROUTE TRAFFIC SEPARATION

During the enroute phase of the IFR flight, certain situations may require the controller to amend your clearance so adequate separation is maintained between aircraft. Among these are deviations due to weather, unplanned pilot requests, flow control restrictions, and aircraft emergencies. If required, the controller will employ various techniques to ensure adequate separation standards. Some of the most common of these include route changes, radar vectoring, altitude crossing restrictions for navaids and intersections, altitude changes, and speed adjustments.

An ATC request for a speed reduction means you should maintain the new indicated airspeed within 10 knots.

Speed adjustments are most commonly used during the arrival phase of a flight. For example, ATC may advise you to *"reduce speed to 100"* as you near your destination. This means ATC has requested you to decrease your indicated airspeed to 100 knots and to maintain that speed within 10 knots.

PILOT RESPONSIBILITIES

Although ATC has strict requirements with regard to the separation of IFR aircraft, keep in mind there are certain pilot responsibilities as well. For example, you must know the requirements for IFR flight, and you must know when an IFR clearance is required. Regulations state that you may not act as pilot in command of a flight conducted under IFR unless you hold an instrument rating and meet the recent experience requirements for instrument flight as specified under FAR Part 61. In addition, your aircraft must meet the equipment and inspection requirements of FAR Part 91. You must file a flight plan and obtain an IFR clearance before conducting an IFR flight in controlled airspace and prior to any flight within positive controlled airspace.

ADDITIONAL ARTCC SERVICES

In addition to separation of all IFR traffic, ARTCCs provide additional safety-related services. These services include separating IFR aircraft from other traffic known to the controller, weather information, safety alerts, and emergency assistance.

SEPARATION FROM OTHER TRAFFIC

Regardless of whether operating VFR or IFR, it is your responsibility to see and avoid other aircraft whenever weather conditions permit.

ATC's first priority is the separation of all IFR aircraft from one another. However, if workload permits, the controller may advise you of other aircraft which affect your flight. It is important for you to realize the controller is not obligated to provide traffic advisory service. In addition, some aircraft in your area may not appear on the controller's radar display. For this reason, FARs require every pilot to see and avoid other aircraft whenever possible, even when they are operated under positive radar control, as in Class B airspace. When you are operating under IFR in VFR conditions, you must continually search for all other aircraft, regardless of the radar service being provided.

WEATHER INFORMATION

Although providing weather information is of lower priority than the separation of IFR traffic, center controllers will make every effort to update you on current conditions on your route of flight. ARTCCs are phasing in computer-generated digital radar displays. This system provides the controller with two distinct levels of weather intensity by assigning radar display symbols for specific precipitation densities. To the extent possible, controllers will issue pertinent information on significant weather areas and assist you in avoiding such areas on request. Keep in mind that frequency congestion and limitations of ATC radar may affect the controller's ability to provide this type of service.

ATC radar limitations and frequency congestion may limit a controller's capability to provide in-flight weather avoidance assistance.

To ensure that timely weather information is available, each center has a meteorologist who monitors the weather within the airspace of that center. One of the tools the center meteorologist uses for providing "real-time" weather is the **radar remote weather display system** (RRWDS). This system uses color to show the various levels of weather intensity. When appropriate, weather information is distributed to each controller and to other facilities. If the center meteorologist finds it necessary, a **center weather advisory** (CWA) will be issued. This report is used to alert pilots to existing or forecast adverse weather conditions that may affect terminal or enroute operations. These reports may later be issued as part of an in-flight advisory. A CWA is passed to the sector controllers who broadcast the information to all affected aircraft. The center weather advisory is an important subject and will be discussed in detail in Chapter 6.

SAFETY ALERTS

A safety alert will be issued by a center controller when it becomes apparent that your flight is in unsafe proximity to terrain, obstructions, or other aircraft. A **terrain** or **obstruction alert** is issued when your Mode C altitude readout indicates your flight is below the published minimum safe altitude for that area. In general, you will be requested to check your altitude immediately. The controller will then provide the minimum altitude required in your area of flight. [Figure 2-2]

A safety alert is issued when, in the controller's judgment, an aircraft is in unsafe proximity to terrain, an obstruction, or another aircraft.

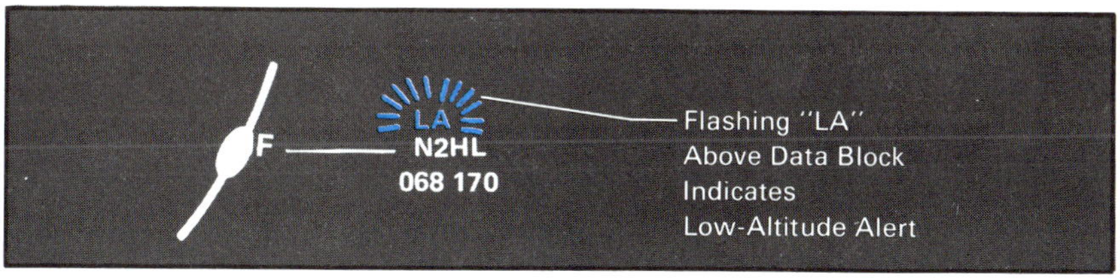

Figure 2-2. Based on the Mode C information provided by this aircraft's transponder, the computer has determined the aircraft is below the minimum safe altitude for this area. This is called a minimum safe altitude warning (MSAW). The flashing "LA" alerts the controller to this potential danger; the controller, in turn, alerts the pilot.

The second type of safety alert is called an **aircraft conflict alert**. This service is provided when the controller determines that the minimum separation between an aircraft being controlled and another aircraft could be compromised. If a conflict alert is issued to you, the controller will advise you of the position of the other aircraft and a possible alternate course of action. Keep in mind that for either of these alert services to be available, your aircraft must be under radar control and your Mode C transponder must be fully operational. Both types of safety alerts may also be issued by terminal radar facilities.

EMERGENCY ASSISTANCE

One advantage of IFR flight is the continual radio contact with ATC. In addition, throughout most of the United States, your flight will be in continuous radar contact. If a problem arises, ATC is immediately available to render a wide variety of services, from simply clearing conflicting traffic to providing radar vectors and, if required, a radar approach to the nearest suitable airport. If a serious situation requiring an immediate landing develops, ATC will alert Search and Rescue (SAR) agencies in the area. [Figure 2-3]

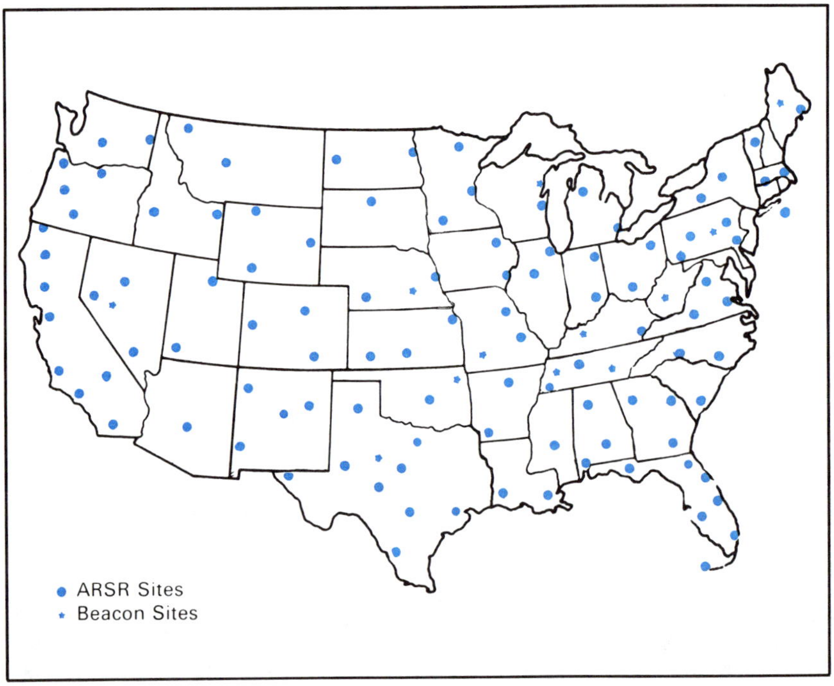

Figure 2-3. This map shows the air route surveillance radar (ARSR) sites used by ARTCC to monitor enroute operations. Except for areas where radar signals are blocked by terrain, you can see that radar coverage extends throughout the conterminous United States. Some of these facilities can detect only transponder-equipped aircraft and are referred to as beacon-only sites.

TERMINAL FACILITIES

Within the air traffic control system, each terminal facility is closely linked with the associated ARTCC to integrate the flow of IFR departures and arrivals. Terminal services include ATIS, clearance delivery, control tower, approach and departure control, and FSS.

ATIS

At busy airports, airport advisory information is provided by automatic terminal information service (ATIS). This continuous, recorded broadcast of noncontrol information helps to improve controller effectiveness and to reduce frequency congestion. At larger airports, there may be one ATIS frequency for departing aircraft and another one for arriving aircraft. ATIS is updated whenever any official weather is received, regardless of content change. It is also updated whenever airport conditions change. When a new ATIS is broadcast, it is changed to the next letter of the phonetic alphabet, such as *"Information Bravo"* or *"Information Charlie"* and so on.

Regardless of content change, ATIS broadcasts are updated upon receipt of any official weather.

CLEARANCE DELIVERY

In order to relieve congestion on the ground control frequency at busier airports, a discrete clearance delivery frequency may be provided. This facility allows you to receive an IFR clearance prior to contacting ground control for taxi. Additionally, this service may be used by VFR pilots to receive an ATC clearance when departing an airport within Class B or Class C airspace. They also may use it to receive a departure control frequency and transponder code when departing an airport with a radar departure control.

CONTROL TOWER

Towers are responsible for the control of all traffic which is landing, taking off, and operating within their Class D airspace area. Aircraft that are departing IFR are integrated into the departure sequence by the tower. Prior to takeoff, the tower controller coordinates with departure control to assure adequate aircraft spacing. Under VFR and IFR, pilots are required to state their position when ready for takeoff from a runway intersection. After takeoff, you are required to remain on the tower frequency until you are instructed to contact departure control. Pilots of VFR aircraft are required to remain on the tower's frequency while in the Class D airspace unless directed otherwise.

During a takeoff in IFR conditions, contact departure control only after you are advised to do so by the tower controller.

When arriving IFR at a controlled airport, you will be sequenced by approach control for spacing and then advised to contact the tower for landing clearance. The tower controller will issue your landing clearance which may include wind direction, wind velocity, current visibility and, if appropriate, special instructions. This information may be omitted at an airport served by ATIS. If you are arriving VFR, you should contact the tower approximately 15 miles from the airport.

APPROACH AND DEPARTURE CONTROL

Approach and departure control coordinate very closely with the ARTCC to integrate arrival traffic from the enroute stage to the terminal area, and transition departure traffic to the enroute phase. In addition to coordinating IFR traffic, radar-equipped terminal areas also provide optional radar service to VFR aircraft. These services are designated as either basic or TRSA (terminal radar service area) service. In addition, radar service, called Class B or Class C service, is provided to all aircraft operating within Class B or C airspace areas.

RADAR SERVICE FOR VFR AIRCRAFT

Basic radar service provides safety alerts, traffic advisories and limited vectoring, depending on the controller's workload. When you are approaching an airport for landing, you should contact approach control for this service. At some locations, sequencing for VFR traffic is also available. The radar controller provides sequencing between participating VFR aircraft and all IFR traffic to adjust the flow into the traffic pattern. For this service, contact approach control when you are approximately 25 miles from the airport, and state you "have numbers" or have the current ATIS. At airports within a terminal radar service area (TRSA), air traffic controllers provide **TRSA service** which includes radar vectoring, sequencing, and separation for all IFR and participating VFR aircraft.

Basic and TRSA radar services are offered on a voluntary basis, but pilot participation is strongly encouraged.

Use of basic and TRSA radar service is not mandatory, but pilot participation is strongly urged. Participation does not relieve you of your responsibilities regarding terrain/obstruction clearance, vortex exposure, and to see and avoid other aircraft when operating under VFR. In addition, an ATC instruction to follow a preceding aircraft does not authorize you to comply with any ATC clearance or instruction issued to the preceding aircraft. A clearance issued by the radar controller which will cause you to violate a rule, such as entering a cloud under VFR, must be declined. In this situation, you should advise the controller of your inability to comply with the issued clearance and request a revised clearance or instruction.

CLASS C AIRSPACE AREAS

You must establish two-way radio communications before you enter Class C airspace and maintain it at all times when operating within it. Participation is mandatory.

Class C airspace areas are established by regulation at locations where traffic conditions warrant. You are not permitted to operate within Class C airspace unless you have established two-way radio communications with the ATC facility having jurisdiction over the area. In addition, you must maintain radio contact while operating within its limits, and you can expect ATC to provide sequencing of all arriving aircraft to the primary airport. Also, ATC will provide standard separation between all IFR aircraft, traffic advisories and conflict resolution between IFR and VFR aircraft, and traffic advisories between VFR aircraft. This includes safety alerts, if appropriate. The requirement to participate in air traffic services is specified in FAR Part 91.

CLASS B AIRSPACE AREAS

A specific clearance to enter Class B airspace must be obtained prior to entry.

Class B airspace areas are established to accommodate arrivals and departures at the nation's busiest terminal areas. You may not operate within Class B airspace unless you obtain a specific ATC clearance to enter the area. Thereafter, all operations of the aircraft must be in accordance with an ATC clearance. To operate within Class B airspace, both you and the aircraft must meet certain minimum requirements specified in the FARs.

TRAFFIC ADVISORIES

No matter what type of ATC radar facility you work with, the controller will follow certain conventions when calling traffic to your attention. Normally, you will be told the position (azimuth) of the traffic relative to your aircraft, its distance in nautical miles, its direction of movement, the type of aircraft, and its altitude, if known. When calling out traffic, controllers describe the position of the traffic in terms of the 12-hour clock. For example, "*traffic at 3 o'clock*" indicates the aircraft lies off your right wing. Keep in mind that the issuance of traffic information is based on the observation of your ground track and the position of the traffic. [Figure 2-4]

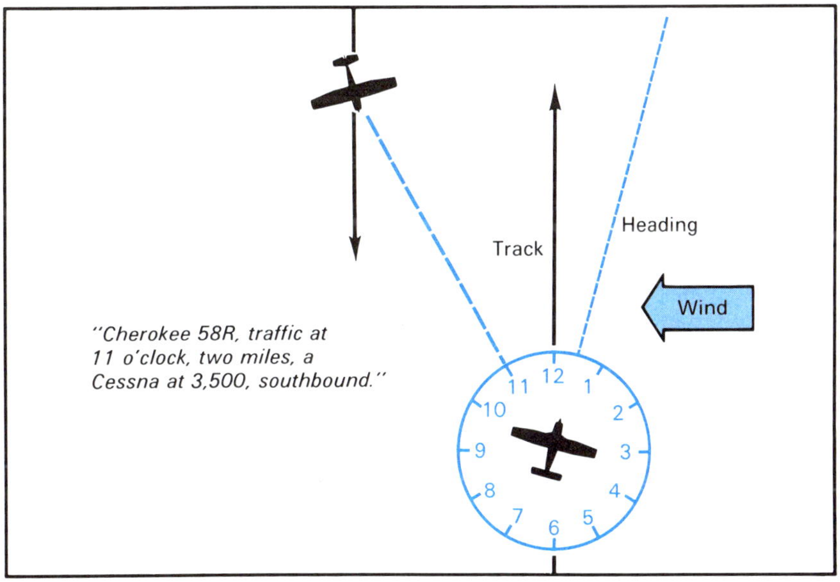

Figure 2-4. A radarscope cannot account for the amount of wind correction you may be using to maintain your track over the ground. Since the controller is unable to determine your actual heading, you must adjust the traffic callout for any wind correction angle you may be using. In this example, the controller advises you about traffic at your 11 o'clock position, but your wind correction angle places it closer to your 10 o'clock position.

FLIGHT SERVICE STATIONS

LAA is provided by flight service at FSS airports not served by an operating control tower.

Flight service stations (FSSs) at selected locations also provide a number of essential functions for both IFR and VFR aircraft. In addition to conducting weather briefings and handling flight plans, they also provide local airport advisory (LAA). This service is provided by a nonautomated FSS that is located on an airport which does not have a control tower or where the tower operates on a part-time basis. At these locations, the FSS provides official weather information, and also relays clearances from ATC. When inbound under VFR, you should report when you are approximately 10 miles from the airport and provide your altitude and aircraft type. Also, state your location relative to the airport, whether landing or overflying, and request a local airport advisory. Departing aircraft should state the aircraft type, full identification number, type of flight (VFR or IFR), and the planned destination or direction of flight.

While LAA is useful, when available, you should be aware that airport advisory areas and the associated services are being reduced as automated flight service stations replace existing nonautomated facilities. Currently, if you fly into an airport with an automated FSS, you will not be able to receive local airport advisory on the frequency, 123.6 MHz.

CHECKLIST

After studying this section, you should have a basic understanding of:

✓ **Air route traffic control centers** — What services they offer and what their responsibilities are.

✓ **Separation criteria** — Under what circumstances ATC provides separation between aircraft and what your own responsibilities are for collision avoidance.

✓ **Terminal facilities** — How approach and departure control facilities work with centers to achieve an orderly flow of IFR traffic.

✓ **Radar service** — What the various levels of radar service are and what unique operating requirements apply to Class B and C airspace areas.

SECTION B

ATC CLEARANCES

An ATC clearance constitutes an authorization for you to proceed under a specified set of conditions within controlled airspace. It is the means by which ATC exercises its responsibility to provide separation between aircraft. It is not, however, an authorization for you to deviate from any regulation or minimum altitude, nor to conduct unsafe operations.

PILOT RESPONSIBILITIES

When ATC issues a clearance, regulations specify that you are not to deviate from it, except in an emergency, unless an amended clearance is received or unless complying with that clearance will cause you to violate a rule. As pilot in command, you must determine if you can safely comply with each clearance issued to you. Points to consider include whether compliance will cause you to violate a rule (such as being vectored into a cloud when you are operating VFR) or to exceed the performance capabilities of yourself or your aircraft. If, in your opinion, a clearance is unsafe or not appropriate, it is your responsibility to take whatever action is necessary to avoid a hazardous situation or a rule violation and promptly request an amended clearance. Your failure to request an amendment promptly may compromise the safety of your aircraft. If you find it necessary to deviate from a clearance due to an emergency, you must notify ATC as soon as possible. When ATC has given you priority, as pilot in command, you may be requested to submit a written report within 48 hours to the chief of that ATC facility.

You may not deviate from a clearance unless you experience an emergency or the clearance will cause you to violate an FAR.

When meteorological conditions permit, regardless of the type of flight plan and whether or not you are under the control of an ATC facility, you are responsible to see and avoid other aircraft, terrain, or obstacles. This rule has been established solely in the interest of safety. ATC radar is not capable of detecting every aircraft which may pose a hazard to your flight. Also, ATC's primary responsibility is the separation of IFR traffic. Traffic advisories are provided for pilots operating under VFR conditions on a workload-permitting basis only.

In VFR conditions, you are required to see and avoid all aircraft, regardless of the type of flight plan you have filed.

WHERE A CLEARANCE IS REQUIRED

You are required to have an ATC clearance to fly within Class A and Class B airspace, regardless of weather conditions. In addition, you must receive a clearance prior to operating within Class B, C, D, or E airspace when the weather is below VFR minimums, and during any flight in controlled airspace under **instrument meteorological conditions**, or **IMC**.

IFR CLEARANCES

A flight plan and a clearance is required prior to IFR flight within controlled airspace. The appropriate navigation equipment also is required.

Prior to flying in controlled airspace during IMC, you are required to file an IFR flight plan and to obtain a clearance. Keep in mind that you may not file an IFR flight plan unless you hold an instrument rating for the category of aircraft you are flying and you are instrument current as specified in FAR Part 61. In addition, the aircraft to be used must be approved for IFR flight and must have the navigation equipment appropriate to the navigation aids to be used.

You may cancel IFR anytime you are operating in VFR conditions outside of Class A airspace.

You may cancel the IFR flight plan anytime you are operating under VFR conditions below 18,000 feet MSL. Remember that once you cancel IFR, the flight must be conducted strictly in VFR conditions from that point on; should you again encounter IFR weather, you must remain in VFR conditions while you file a new flight plan and obtain an IFR clearance.

ELEMENTS OF AN IFR CLEARANCE

An IFR clearance is made up of one or more instructions. Knowing the order in which ATC will issue these instructions makes it easier to understand a clearance. The following items will be contained in an initial IFR clearance, as appropriate, in the order shown:

- Aircraft identification
- Clearance limit
- Departure procedure or SID
- Route of flight
- Altitudes, in order to be flown
- Holding instructions
- Any special information
- Frequency and transponder code information

Standard instrument departures (SIDs) and standard terminal arrival routes (STARs) will be explained in Chapter 3. They are essentially charted procedures that help simplify the issuance of a clearance. ATC assumes you have all applicable SIDs and STARs and will issue them, as appropriate, without request. If you do not possess SIDs and STARs, or do not wish to use them, you should include the phrase "No SID/-STAR" in the remarks sections of your IFR flight plan.

ABBREVIATED IFR DEPARTURE CLEARANCE

In order to decrease radio congestion and controller workload, ATC will issue an abbreviated IFR departure clearance whenever possible. This type of clearance uses the phrase "*cleared as filed*" to indicate you have been cleared to fly the route as contained in your IFR flight plan. This technique is particularly useful when numerous navigation fixes and Victor airways are contained in the flight plan, and ATC can accommodate the routing as you filed it. However, if ATC finds it necessary to change your requested routing, a full route clearance will be issued and the abbreviated clearance procedure will not be used. An example of an abbreviated clearance is, "*. . . cleared to the Boston Logan Airport as filed, maintain 9,000.*"

In the event you have filed a SID procedure in your flight plan or a SID is in use at the departure airport, it will be included in the abbreviated clearance, since it is not considered to be a route component. For example, your clearance may be, "*. . . cleared to the Los Angeles International Airport, Denver seven departure, then as filed, maintain 10,000, expect flight level 350 within 10 minutes.*" An abbreviated clearance will always contain the name of the destination airport or a clearance limit; any applicable SID name, number, and transition; and your assigned enroute altitude.

An abbreviated clearance contains the name of your destination airport or clearance limit, SID information, and the assigned enroute altitude.

Although SID procedures are included in the abbreviated clearance, standard arrival routes are considered a portion of the routing and, therefore, are not stated in the body of the clearance. If, for example, you filed a flight plan which included a STAR and ATC did not amend it in your clearance, you should plan to fly the entire route, including the STAR, when you are cleared as filed.

CRUISE CLEARANCE

A cruise clearance may be issued by ATC in situations where the route segment is relatively short and traffic congestion is not a consideration. In this type of clearance, the controller will use the word "*cruise*" instead of the word "*maintain*" when issuing an altitude assignment; for example, "*. . . cleared to Goodland Airport, cruise 8,000.*" The significance of a **cruise clearance** is that you may operate at any altitude, from the minimum IFR altitude up to and including the altitude specified in the clearance. You may climb, level off, descend, and cruise at an intermediate altitude at any time. Each change in altitude does not require a report to ATC. However, once you begin a descent and report leaving an altitude, you may not climb back to that altitude without obtaining an ATC clearance. Another important aspect of a cruise clearance is that it also authorizes you to proceed to and execute an approach at the destination airport. In other words, you do not need to request, and ATC will not issue, a separate approach clearance at the destination airport when you have been issued a cruise clearance.

A cruise clearance is an authorization to execute an instrument approach at the destination airport.

When you are operating in uncontrolled airspace on a cruise clearance, ATC does not exercise control; you are responsible for determining the minimum IFR altitude. In addition, your descent and landing at an airport in uncontrolled airspace is governed by the applicable FARs for VFR flight.

APPROACH CLEARANCES

You should be aware of a few peculiarities with the issuance of an instrument approach clearance. First, if only one approach procedure exists or if you are authorized by ATC to execute the approach procedure of your choice, you will be issued a clearance, such as "*. . . cleared for approach.*" If more than one approach procedure is available at the destination airport or if ATC restricts you to a specific approach, the controller will specify, "*. . . cleared for ILS runway 35 right approach.*"

If you are established on a route or approach segment that has a published minimum altitude, your approach clearance generally will not specify an altitude to maintain. However, if you are being radar vectored to the final approach course, your approach clearance should always include an altitude to maintain, such as "*. . . maintain 2,000 until established on the localizer. Cleared for ILS runway 36 approach.*"

When the landing will be made on a runway that is not aligned with the approach being flown, the controller may issue a **circling approach** clearance. In this case, the controller will specify, "*. . . cleared for VOR runway 17 approach, circle to land runway 23.*"

A contact approach may not be initiated by ATC.

A **contact approach** clearance must be requested by you and authorized by the controller. It may not be initiated by ATC. In order for ATC to approve your request for a contact approach, the airport must have a standard or special instrument approach procedure, the reported ground visibility must be at least one statute mile, and you must be able to remain clear of clouds with at least one mile flight visibility throughout the approach. A contact approach procedure may be used instead of the published procedure to expedite your arrival, as long as the previously mentioned criteria can be met.

A visual approach may be initiated by either ATC or the pilot.

The **visual approach** clearance criteria are slightly different from those of the contact approach, in that the controller can initiate the clearance. When it is operationally beneficial, ATC may authorize you to conduct a visual approach to the airport in lieu of the published approach procedure. Before issuing a visual approach clearance, the controller will verify that you have the airport in sight, or a preceding aircraft which you are to follow to the airport. In the event you have the airport in sight but do not see the aircraft you are to follow, ATC may issue the visual approach clearance but will also be responsible for aircraft separation. Once you report the aircraft in sight, you assume the responsibilities for your own separation and wake turbulence avoidance.

Keep in mind that the visual approach clearance is issued to expedite the flow of traffic to the airport. It is authorized when the weather is reported or expected to be at least a 1,000-foot ceiling and 3 miles visibility. You must remain clear of the clouds at all times while conducting a visual approach. At a controlled airport, you may be cleared to fly a visual approach to one runway while others are conducting VFR or IFR approaches to another parallel, intersecting, or converging runway. Also, when radar service is provided, it is automatically terminated when the controller advises you to change to the tower or advisory frequency.

VFR ON TOP

When flying VFR on top, you are required to comply with the rules of both VFR and IFR flight.

In some situations, it may be to your advantage to request **VFR on top** during an IFR flight. This type of clearance does not cancel your IFR flight plan. Instead, it allows you more flexibility with regard to altitude assignments. Basically, a VFR-on-top clearance allows you to fly in VFR conditions and at appropriate VFR cruising altitudes of your choice. You may only request this clearance if you are in VFR conditions below 18,000 feet. ATC may not initiate a VFR-on-top clearance. Once ATC approves your request, you must maintain VFR flight conditions at all times. Altitude selection must comply with the VFR cruising altitude rules, which are based on the magnetic course of the aircraft. You may not, however, select an altitude which is less than the minimum enroute altitude prescribed for the route segment.

An ATC authorization to maintain VFR on top does not literally restrict you to "on top" operations. You may operate VFR on an IFR flight plan when you are above, below, or between layers, or in the clear. This type of clearance simply allows you to change altitude in VFR conditions after advising ATC of the intended altitude changes. Keep in mind, however, that all the rules applicable to instrument flight must still be followed. If at any time VFR conditions cannot be maintained, you must inform ATC and receive a new clearance before you enter IFR conditions.

TO VFR ON TOP

A variation of the VFR-on-top clearance is a request to climb to VFR on top. You would request this type of clearance when departing an airport where you wanted to climb through an area of restricted visibility, such as a haze or fog layer, and then proceed enroute under VFR. If you are departing a controlled airport, it may not be necessary to file a complete IFR flight plan with a request to VFR on top. You can generally make your request directly with ground control or clearance delivery. However, if an ATC facility is not available, your request should be made with a filed flight plan.

When ATC issues a clearance to climb to VFR on top, you should expect a clearance limit and an ATC request to report reaching VFR on top. Also, if you elect to file an IFR flight plan, the term "*VFR on top*" will be used instead of an altitude assignment when you receive your clearance. If necessary, ATC may restrict your climb to maintain traffic separation. A typical clearance would be, "*. . . cleared to the Waterloo Airport as filed, climb to and report reaching VFR on top. If not on top at 5,000, maintain 5,000 and advise. Maintain VFR on top.*"

VFR RESTRICTIONS TO AN IFR CLEARANCE

VFR restrictions to an IFR flight can only be initiated by the pilot.

During the issuance of an ATC clearance, the controller may direct you to "*maintain VFR conditions.*" However, this restriction is not issued routinely. It is issued only when you request it. For example, a VFR restriction is issued when you request a VFR climb or descent. This is the case when you are departing or arriving in VFR conditions and you wish to avoid a complicated, time-consuming departure or arrival procedure. In the case of a VFR departure, for example, you may request a VFR climb, which would allow you to avoid the departure procedure and climb on course. However, you should fully understand that while operating on an IFR flight plan with a VFR restriction, you must remain in VFR conditions and maintain your own traffic separation during the VFR portion of your clearance.

COMPOSITE FLIGHT PLAN

A composite flight plan is a request to operate IFR on one portion of a flight and VFR for another portion. When you file a composite flight plan, the route should include all normal IFR route segments, as well as the clearance limit fix where you anticipate the IFR portion of the flight will terminate. You may file a composite flight plan any time a portion of the flight will be flown under VFR. For example, if the first portion of your flight is VFR, you should activate your VFR flight plan with the nearest FSS after departure. As you near the point where you planned to activate your IFR flight plan, you must contact the nearest FSS, close your VFR flight plan, and request your IFR clearance. Keep in mind that you must remain in VFR conditions until you receive your IFR clearance. In the event your flight dictates IFR for the first portion and VFR for the latter portion, you will normally be cleared IFR to the point where the change is proposed. As you near this point and are operating in VFR conditions, you should cancel your IFR flight plan and contact the nearest FSS to activate your VFR flight plan. If you want to continue past your clearance limit on an IFR flight plan, you must contact ATC five minutes before you reach the limit to request a further clearance. Should you reach your clearance limit without receiving a clearance to continue, you are expected to enter a holding pattern and wait for the clearance.

TOWER ENROUTE CONTROL CLEARANCE

Tower enroute control (TEC) is an alternative IFR procedure which permits you to fly short, low altitude routes between terminal areas. TEC routes are published for certain portions of the United States in the *Airport/Facility Directory* and in Jeppesen's *Airway Manual* service. Essentially, a flight is transferred from departure control at one airport to successive approach control facilities. In most cases, TEC routes are intended for flights below 10,000 feet and are limited to less than two hours. The service is primarily designed for nonturbojet aircraft. Normal flight plan filing procedures are used. If you want to fly a TEC route, include the acronym TEC in the remarks section of the flight plan.

HOLD FOR RELEASE

Occasionally, you may be advised to "*hold for release.*" This type of restriction is generally a result of traffic saturation, weather, or ATC departure management procedures. When ATC issues a hold for release, you may not depart until you receive a release time or you are given additional instructions. Generally, the additional instructions will include the expected release time and the length of the departure delay. Once the local conditions and traffic permit, you will be released for departure.

CLEARANCE VOID TIME

uncontrolled Airport (GXY)

If you are operating at an airport not served by an operating control tower, ATC may find it necessary to issue a **clearance void time** in conjunction with your IFR departure clearance. The wording, "*clearance void if not off by . . . ,*" indicates that ATC expects you to be airborne by a certain time. A common situation for the issuance of a void time is when inbound traffic is expected to arrive at approximately the same time as your departure. In this case, the required traffic separation cannot be achieved without restricting your departure. In the event you do not depart by the void time, you must advise ATC of your intentions as soon as possible, but no later than 30 minutes after the void time.

You may not depart an airport after the clearance void time when one has been issued to you.

CLEARANCE READBACK

Although there is no requirement that an ATC clearance be read back, you are expected to read back those parts of any clearance which contain altitude assignments, radar vectors, or any instructions requiring verification. Additionally, controllers may request that you read back a clearance when the complexity of the clearance or any other factors indicate a need.

You should read back that portion of a clearance containing altitude assignments, radar vectors, or any instruction requiring clarification.

Read Back

As the pilot in command, you should read back the clearance if you feel the need for confirmation. Even though it is not specifically stated, it is generally expected that you will read back the initial enroute clearance you receive from clearance delivery, ground control, or a flight service station.

When you receive your IFR clearance, the phraseology may be slightly different, depending upon the facility issuing the clearance. The term "*ATC clears*" is used only when a facility other than air traffic control is used to transmit an IFR clearance. For example, when you receive a clearance from a flight service station, the clearance will be preceded by "*ATC clears.*"

CLEARANCE SHORTHAND

To operate efficiently in the IFR environment, you must be able to copy and thoroughly understand clearances. Copying IFR clearances becomes easy with practice. Although numerous changes have been made since the acceptance of the first shorthand used by early instrument pilots, many of the original symbols have been retained. The shorthand symbols in this chapter are considered by the FAA and experienced instrument pilots to be the best. However, the most important consideration is not what symbol you use, but that you can still interpret it after a period of time. [Figure 2-5]

Proficiency in copying ATC clearances is the result of knowing clearance terminology and practicing it. The following samples of ATC clearances are presented as they might be issued. Each is followed by its appropriate clearance shorthand. If you feel you need additional practice copying clearances, Jeppesen offers an audio cassette program called *ATC Clears*. In this program, Side One provides an explanation of the various types of clearances you can expect during IFR flight. Side Two provides additional practice copying clearances of various complexities and at different rates of speed.

"*Commander 480S cleared to the Abilene Municipal Airport as filed. Maintain 12,000. After departure turn right heading 340 for radar vectors. Squawk 2021.*"

80S C ABI A AF M120 AFT DP ↱ HDG 340 RV SQ 2021

"*Cessna 1351F cleared to the Ardmore Municipal Airport, radar vectors Blue Ridge, Victor 15. Maintain 5,000. Departure control frequency 124.5. Squawk 0412.*"

51F C ADM A RV BUJ V15 M50 DP 124.5 SQ 0412

"*ATC clears Aztec 103MC to the Addison Airport, Victor 18 Dallas-Ft. Worth, direct. Maintain VFR on top. If not VFR on top at 5,000, maintain 5,000 and advise.*"

C 3MC ADDISON A V18 DFW DR M <u>VFR</u> (or M50 & ADV)

"*Cessna 1351F, descend and maintain 8,000. Report reaching 8,000.*"

51F ↓ M80 RR 80

"*Piper 43532, radar contact 15 miles southeast of the Love VORTAC. Turn left heading 345. Intercept the Love 182 radial. Cleared for the VOR Alpha approach to the Addison Airport, contact Addison Tower on 121.1 at the VOR inbound.*"

532 RADAR CT 15 SE LUE Ⓣ ↰ HDG 345 ⩤ LUE 182R C ⊙ A

AP ADS A CT ADS Z 121.1 ⊙ IB

SHORTHAND SYMBOLS

Words and Phrases	Shorthand
ABOVE	ABV
ABOVE ("*Above Six Thousand*")	$\underline{60}$
ADVISE	ADV
AFTER (Passing)	< or AFT
AIRPORT	A
(ALTERNATE INSTRUCTIONS)	( )
ALTITUDE 6,000 — 17,000	60—170
AND	& or &
APPROACH	AP
FINAL	F
INSTRUMENT LANDING SYSTEM	ILS
LOCALIZER BACK COURSE	LBC
LOCALIZER ONLY	LCO
NONDIRECTIONAL BEACON	NDB
PRECISION (Approach) RADAR	PAR
SURVEILLANCE RADAR	ASR
VOR	VOR or ⊙
APPROACH CONTROL	APC
AS FILED	AF
AS PUBLISHED	APUB
AT	@
(ATC) ADVISES	CA
(ATC) CLEARS OR CLEARED	C
(ATC) REQUESTS	CR
BEARING	BR
BEFORE (Reaching, Passing)	>
BELOW	BLO
BELOW ("*Below Six Thousand*")	$\overline{60}$
CENTER	CTR
CLEARED AS FILED	CAF
CLIMB (TO)	↑
CONTACT	CT
COURSE	CRS
CROSS (Crossing)	X
CRUISE	
DEPART (Departure)	DP
DESCEND (TO)	↓
DIRECT	DR
DME FIX (15 DME Mile Fix)	(15)
EACH	ea
EASTBOUND	EB
EXPECT	EX
EXPECT FURTHER CLEARANCE (Time or Location)	EFC
FLIGHT LEVEL	FL

Words and Phrases	Shorthand
FLIGHT PLANNED ROUTE	FPR
FOR FURTHER CLEARANCE	FFC
HEADING	HDG
HOLD (Direction) ("*Hold West*")	H-W
HOLDING PATTERN	⬭
IF NOT POSSIBLE	or
ILS LOCALIZER	L
INBOUND	IB
INTERCEPT	△✓
INTERSECTION	△ or XN
MAINTAIN (or Magnetic)	M
MIDDLE MARKER	MM
LOW FREQUENCY BEACON LOCATED AT MIDDLE MARKER	LMM
OUTBOUND	OB
OUTER MARKER	OM
LOW FREQUENCY BEACON LOCATED AT OUTER MARKER	LOM
OVER (Ident, Over the Line)	$\underline{\text{OKC}}$
PROCEDURE TURN	PT
RADAR VECTOR	RV
RADIAL (092 Radial)	092R
REPORT	R
REPORT LEAVING	RL
REPORT ON COURSE	RC
REPORT OVER	RO
REPORT PASSING	RP
REPORT REACHING	RR
REPORT STARTING PROCEDURE TURN	RSPT
RUNWAY (Number)	RY26
SHUTTLE CLIMB	S↑
SQUAWK	SQ
STANDBY	STBY
TAKEOFF	T-O
TOWER	Z
TRACK	TR
TURN LEFT, OR TURN LEFT AFTER DEPARTURE	↰ or LT
TURN RIGHT, OR TURN RIGHT AFTER DEPARTURE	↱ or RT
UNTIL	til or U
UNTIL ADVISED (By)	UA
UNTIL FURTHER ADVISED	UFA
VFR ON TOP	$\underline{\text{VFR}}$
VICTOR (Airway Number)	V 294
VOR	⊙
VORTAC	Ⓣ

Figure 2-5. The purpose of these shorthand symbols is to allow you to copy IFR clearances as fast as they are read. The more clearances you copy, the easier the task becomes.

CHECKLIST

After studying this section, you should have a basic understanding of:

- ✓ **Clearances** — What a clearance is and when an IFR clearance is required. What clearances you must request and what clearances are issued by ATC automatically.
- ✓ **Cancelling IFR** — The conditions under which you may cancel an IFR flight plan.
- ✓ **Elements of a clearance** — The items which make up an IFR clearance and the order in which they are issued.
- ✓ **Abbreviated clearance** — When an abbreviated clearance is issued and the items which are always contained in an abbreviated clearance.
- ✓ **Cruise and TEC clearances** — The differences between each type of clearance and the meaning of each to an IFR flight.
- ✓ **Contact and visual approaches** — What the differences are between the contact and visual approaches. What your responsibilities are and the operational requirements for each.
- ✓ **VFR on top** — The significance of a request for VFR on top during IFR flight and the meaning of a climb to VFR on top from a departure airport.
- ✓ **Composite flight plan** — When a composite flight plan may be filed and how to transition from one portion of the flight plan to the other.
- ✓ **Clearance void time** — What a clearance void time is and its significance to an IFR departure.
- ✓ **Clearance readback** — When to read back a clearance and what portions should be read back.

SECTION C

AIRPORTS, AIRSPACE, AND FLIGHT INFORMATION

Much of the information in this section is a review of subject areas you have studied in previous training programs. If you need additional information, you may want to refer back to your initial training materials.

THE AIRPORT ENVIRONMENT

An important aspect of IFR flight operations is a sound knowledge of the airport environment, which includes runway markings, airport signs, runway lighting, approach lighting systems, and associated airport lighting. The types of markings, signs, and lighting systems installed may vary from airport to airport, depending on size, traffic volume, and the types of operations and approaches authorized.

RUNWAY MARKINGS

Markings vary between runways used solely for VFR operations and those used for IFR operations. A basic VFR runway usually is marked only with the runway numbers and a centerline, but threshold and fixed distance markings may be included. IFR runways have additional markings and are categorized as either precision or nonprecision.

Precision instrument runways are served by nonvisual precision approach aids, such as the instrument landing system (ILS). The ILS uses an electronic glide slope to provide glide path information during the approach. The associated runways are marked so you can receive important visual cues during periods of extremely low visibility. [Figure 2-6]

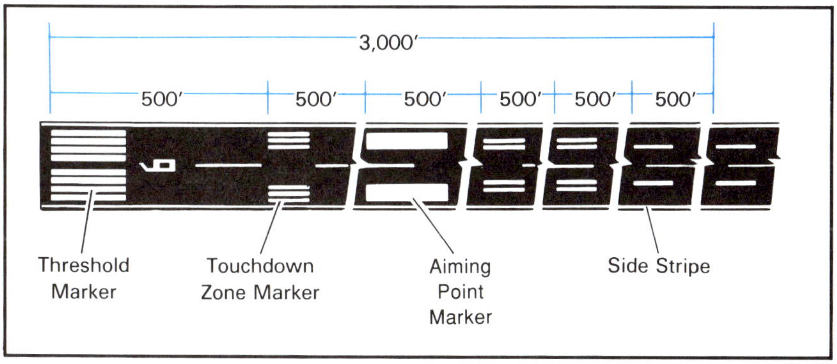

Figure 2-6. A precision instrument runway incorporates a threshold marker, touchdown zone marker, aiming point marker, side stripes, and additional markings. The aiming point marker is located approximately 1,000 feet from the threshold and provides a prominent reference for pilots of landing aircraft.

A **nonprecision instrument runway** is used with an instrument approach that does not have an electronic glide slope for approach glide path information. This type of runway has the basic runway markings, plus the threshold marker. If the runway is at least 4,000 feet long and used by jet aircraft, an aiming point marker is included.

SPECIAL PURPOSE AREAS

You may encounter a variety of special markings on runways and taxiways which place restrictions on their use. The following paragraphs provide a brief review of some typical special purpose areas. [Figure 2-7]

A **displaced threshold** is marked by a solid white line extending across the runway perpendicular to the centerline. It marks the point at which all normal takeoff and landing operations are permitted. The operations permitted prior to this point depend on the type of restriction imposed. **Taxi, takeoff, and rollout areas** are marked by white arrows leading to a displaced threshold. When landing, you must touch down beyond the displaced threshold. The paved area prior to the displaced threshold is not designed for landing. A **taxi-only area** is a designated portion of a runway to be used only for taxi operations. It is marked by a yellow taxi line leading to a displaced threshold. When departing such a runway, you may use the entire length. Also, you must plan your landing beyond the displaced threshold marking. When departing from or arriving on the opposite end, you may not consider this area as usable for takeoff or landing, except as an overrun during an aborted takeoff.

Blast pad/stopway areas are marked by yellow chevrons and may not be used for taxiing, takeoffs, or landings. The blast pad area allows propeller or jet blast to dissipate without creating a hazard to others. In the event

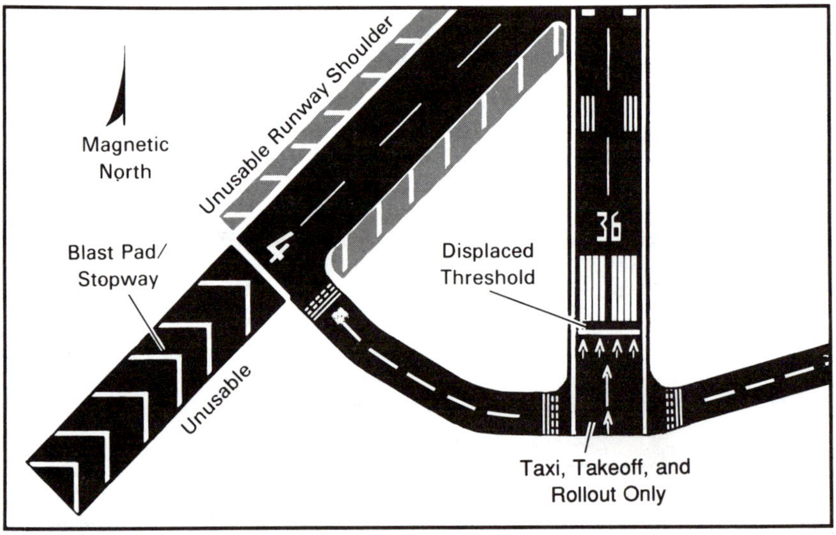

Figure 2-7. This illustration shows some samples of different types of runway restrictions. You may wish to refer to it as each element is discussed.

you must abort a takeoff, the stopway provides additional paved surface for you to decelerate and stop. A **closed runway** is marked with a large white "X" at each end. Although the closed runway may not be used, other runways and taxiways which cross it are not affected unless specifically marked. A closed taxiway may be marked by "Xs," or it may simply be blocked off. **Taxiway hold lines** are used to keep aircraft clear of runways and, at controlled airports, serve as the point that separates the responsibilities of ground control from those of the tower. Taxiway hold lines are usually placed between 125 and 250 feet from the runway centerline.

AIRPORT SIGNS

There are six types of airport signs which are designed to assist you while operating on the movement areas of an airport. The ball flags in the following paragraphs are keyed to the illustrated examples. [Figure 2-8]

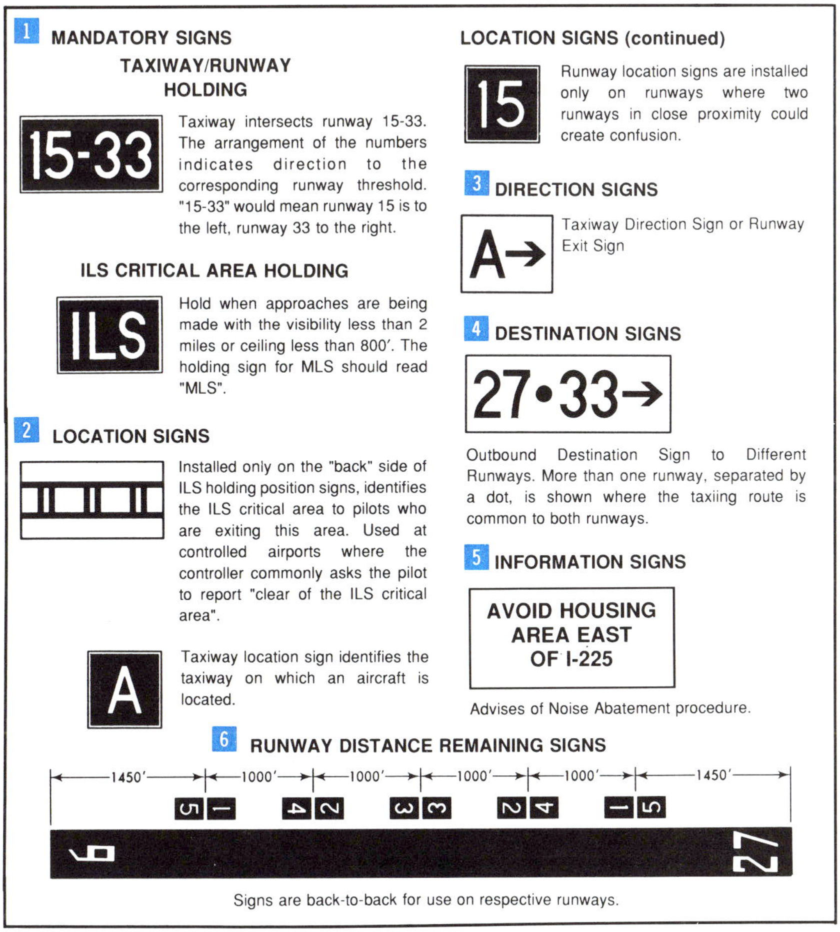

Figure 2-8. This illustration shows selected examples of the six types of airport signs as depicted in the AIM and AC 150/5340-18C.

1 **Mandatory Instruction Signs** denote an entrance to a runway, a critical area, or an area prohibited to aircraft. These signs are red with white letters or numbers.

2 **Location Signs** identify either the taxiway or runway where your aircraft is located. They also identify the boundary of the runway safety area/obstacle free zone (OFZ) or ILS critical area for aircraft exiting the runway. These signs are black with yellow inscriptions and a yellow border and do not have arrows.

3 **Direction Signs** indicate directions of taxiways leading out of an intersection. They have black inscriptions on a yellow background and always contain arrows which show the approximate direction of turn.

4 **Destination Signs** indicate the general direction to a remote location such as for inbound or outbound traffic. They have black inscriptions on a yellow background and always contain an arrow.

5 **Information Signs** advise you of such things as areas that cannot be seen from the control tower, applicable radio frequencies, and noise abatement procedures. These signs use yellow backgrounds with black inscriptions.

6 **Runway Distance Remaining Signs** are used to provide distance remaining information to pilots during take-off and landing operations. The signs are located along the sides of the runway, and the inscription consists of a white numeral on a black background. The signs indicate the distance remaining in thousands of feet.

LIGHTING SYSTEMS

Airport lighting systems range from the simple lighting needed for VFR night landings to sophisticated systems which guide you to the runway in IFR conditions. You should familiarize yourself with each type of lighting and its significance to VFR, as well as IFR, operations.

APPROACH LIGHT SYSTEMS

The approach light system (ALS) helps you transition from instrument to visual references during the approach to landing. It makes the runway environment more apparent in low visibility conditions and helps you maintain correct alignment with the runway. Approach light systems use a configuration of lights starting at the landing threshold and extending into the approach area. Normally, they extend outward to a distance of 2,400 to 3,000 feet from precision instrument runways and 1,400 to 1,500 feet from nonprecision instrument runways. [Figure 2-9]

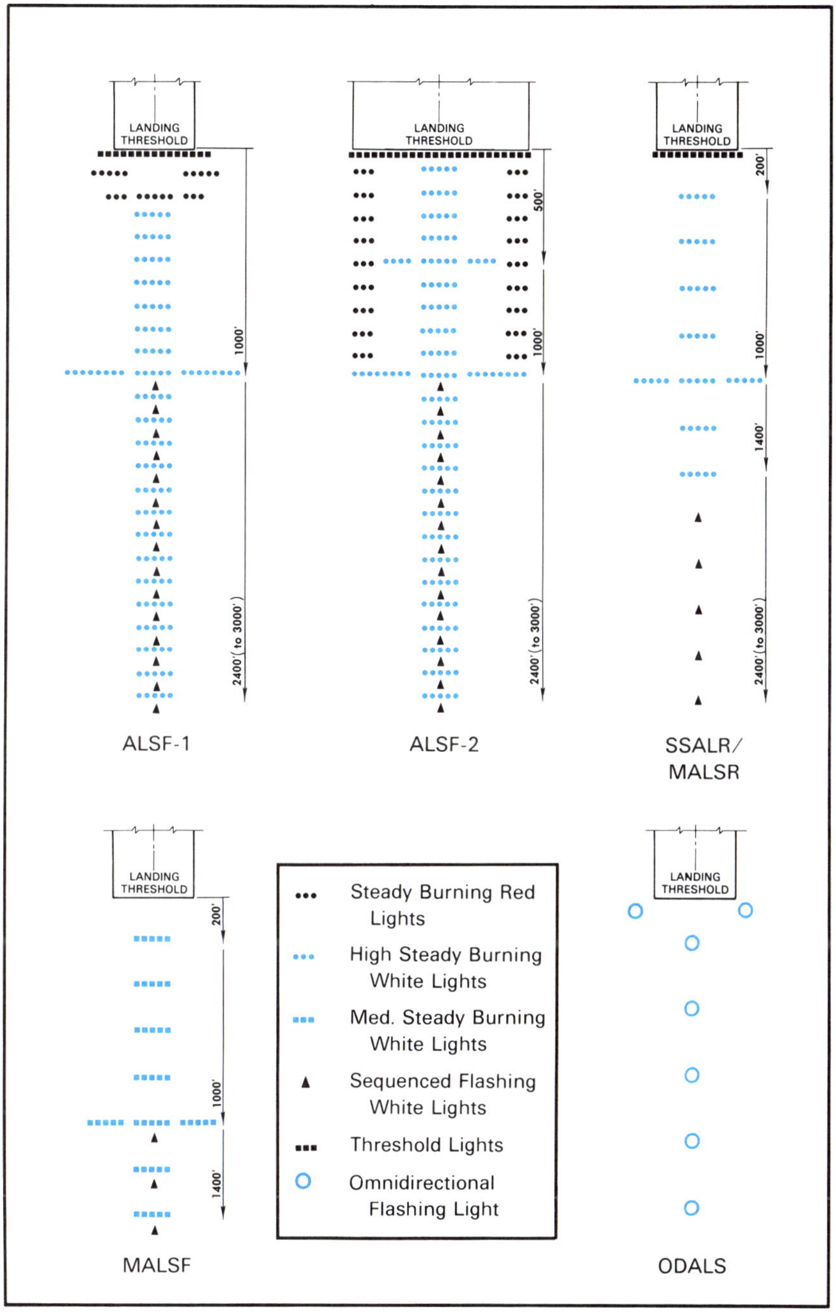

Figure 2-9. This illustration shows several approach light systems in various configurations. ALSF-1 and ALSF-2 configurations are typical for runways that have instrument landing systems, although SSALR/MALSR also may be used. The MALSF configuration and ODALS are used in conjunction with nonprecision instrument runways.

Some approach light systems include **sequenced flashing lights** (SFL) or **runway alignment indicator lights** (RAIL). SFL and RAIL consist of a series of brilliant blue-white bursts of flashing light. From your viewpoint, these systems give the impression of a ball of light traveling at high speed toward the approach end of the runway. SFL and RAIL usually are incorporated into other approach light systems, although RAIL can also be used as an independent approach light system.

REIL refers to a pair of synchronized flashing lights which provide rapid identification of the approach end of the runway during low visibility conditions.

At some locations, a high intensity white strobe light is placed on each side of the runway to mark the threshold. These lights, called **runway end identifier lights** (REIL), can be used in conjunction with the green threshold lights. Their purpose is to provide you with a means of rapidly identifying the approach end of the runway during reduced visibility. They are normally aimed 10° up and 15° away from the runway centerline.

VISUAL GLIDE SLOPE INDICATORS

If you remain on the proper glide path of a VASI, you are assured safe obstruction clearance in the approach area. Two-bar VASIs, which normally have an approach angle of three degrees, show red over white when you are on the glide path.

Once you have the runway environment in sight, visual glide slope indicators help you establish and maintain a safe descent path to the runway. Their purpose is to provide a clear visual means to determine if you are too high, too low, or on the correct glide path. These indicators are extremely useful during low visibility, or at night, when it may be difficult to judge the descent angle accurately due to a lack of runway contrast. Several different visual glide slope indicator systems are used, but one of the most common is the two-bar **visual approach slope indicator** (VASI). The two-bar system provides one visual glide path, normally set to three degrees. Staying on the VASI glide path assures you of safe obstruction clearance within ± 10 degrees of the extended runway centerline and out to four n.m. from the threshold. When using VASI, you should attempt to fly your approach so the far bars indicate red and the near bars show white. These are the proper light indications for maintaining the glide slope. [Figure 2-10]

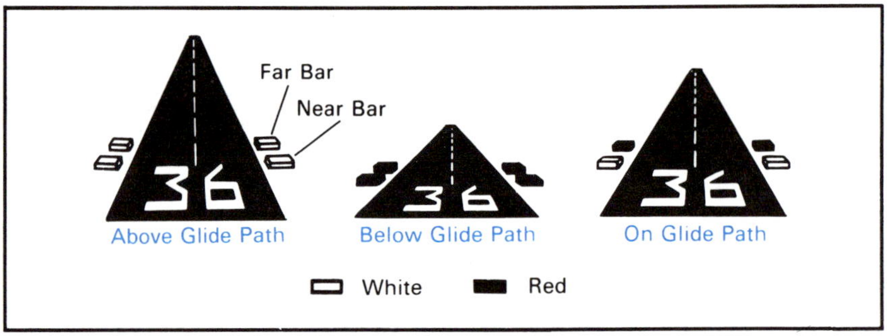

Figure 2-10. The two-bar VASI system consists of two light boxes on each side of the runway. An abbreviated version, called AVASI, may use only a single set of boxes, usually installed on the left side. If you are approaching a runway and all of the VASI lights appear to be red, as shown in the center, you should level off momentarily to intercept the proper approach path.

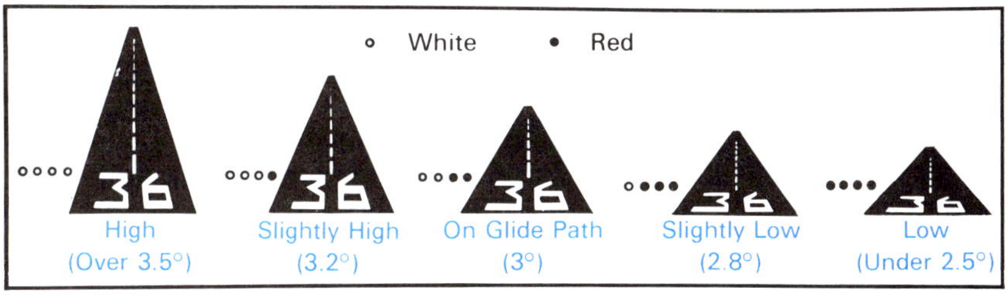

Figure 2-11. Like VASI, an all-white indication shows that you are too high, while all-red indicates you are low. The four-light PAPI system also gives you intermediate information to show if you are slightly above or below glide slope. With a two-light system, two white lights indicate you are high, one white and one red light means you are on the glide path, and two red lights show that you are low.

Some airports are equipped with three-bar VASI systems consisting of three sets of light sources forming near, middle, and far bars. These systems provide two visual glide paths to the same runway. The first uses the near and middle bars. This glide path is the same as that provided by a standard two-bar VASI installation. The second uses the middle and far bars. This upper glide path must be used by pilots of high-cockpit aircraft and is about one-quarter of a degree steeper than the first. The far bars are located approximately 700 feet beyond the middle bars. When on the upper glide path, which should be used only by high-cockpit aircraft, the pilot will see red, white, and white.

The middle and far bars of a three-bar VASI can be used like a two-bar VASI to descend on the upper glide path.

The **precision approach path indicator** (PAPI) uses lights similar to VASI, but the lights are installed in a single row of two- or four-light units. PAPI is normally located on the left side of the runway. [Figure 2-11]

TRI-COLOR VASI

Another system, which consists of a single light unit projecting a three-color visual glide path into the final approach area of the runway, is referred to as **tri-color VASI**. Depending on visibility conditions, this type of approach slope indicator has a useful range of approximately one-half to one mile during the day and up to five miles at night. [Figure 2-12]

Tri-color VASIs normally consist of one light projector with three colors — amber, green, and red.

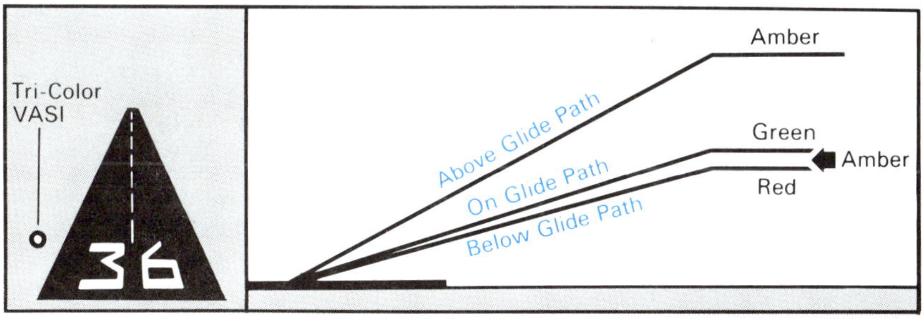

Figure 2-12. The tri-color VASI uses amber, green, and red lights to show your position with respect to the glide path. As you descend below the proper glide path, you may see dark amber during the transition from green to red, so don't be deceived into thinking that you are too high.

RUNWAY LIGHTING

Runway lights outline the landing area by clearly defining its boundaries. Some of these systems have bidirectional features which help you judge your position from the ends of the runway. A thorough understanding of runway lighting is important, particularly during low-visibility IFR operations.

Runway edge lights are used to outline the runway during periods of darkness or restricted visibility. They are classified according to their brightness — **high intensity runway lights** (HIRL), **medium intensity runway lights** (MIRL), and **low intensity runway lights** (LIRL). The HIRL and MIRL systems have variable intensity controls which may be adjusted from the control tower or by the pilot using the CTAF or UNICOM frequency. The LIRL system normally has only one intensity setting. Runway edge lights are white, except on instrument runways where amber replaces white on the last 2,000 feet or half the runway length, whichever is less. The amber lights along the final 2,000 feet or one half of the instrument runway indicate a caution zone.

Green lights arranged on the left and right sides of the runway centerline identify the runway threshold.

Bidirectional **threshold lights** mark the ends of each runway. As you approach for landing, the lights are green, indicating the beginning of the runway. As you take off, the lights are red, marking the departure end of the runway. These lights are inside from the runway edge lights and perpendicular to the runway centerline.

Lights also help you identify a displaced threshold during low visibility conditions or at night. **Displaced threshold lights**, which appear green during an approach to a landing, are located on each side of the runway. They are outboard from the runway edges and perpendicular to the runway centerline. No landings are permitted short of the green displaced threshold lights. Additionally, the absence of runway edge lights in this area indicates that no operations are authorized short of the displaced threshold. If the displaced runway area is usable for specific operations (taxi, takeoff, and rollout), runway edge lights are installed to outline this area short of the displaced threshold.

You may use the area short of the displaced threshold lights for taxi, takeoff, or rollout purposes when the runway edge lights appear in one of the following combinations.

Red runway edge lights signify a displaced threshold, where taxi, takeoff, and rollout operations are permitted.

1. Red runway edge lights located with visible displaced threshold lights, as viewed from the displaced threshold end of the runway
2. White runway edge lights located with no visible threshold lights, as viewed from the runway end opposite the displaced threshold

Touchdown zone lighting helps you identify the touchdown zone when visibility is reduced. It consists of a series of white lights flush-mounted in the runway. They begin approximately 100 feet from the landing threshold and extend 3,000 feet down the runway or to the midpoint of the runway, whichever is less. These lights are visible only from the approach end of the runway.

Runway centerline lights are flush-mounted in the runway to help you maintain the centerline during takeoff and landing. They are spaced at intervals of 50 feet, beginning 75 feet from the landing threshold and extending to within 75 feet of the opposite end of the runway. As you approach the runway, the centerline lights first appear white. They change to alternating red and white lights when you have 3,000 feet of remaining runway, then they show all red for the last 1,000 feet of runway. These lights are bidirectional, so you will see the correct lighting from either direction.

When 3,000 feet of runway remain, runway centerline lights change to alternating red and white lights. All-red lights indicate that 1,000 feet of runway remain.

Taxiway turnoff lights are similar to runway centerline lights. They generally are flush-mounted green lights spaced at 50-foot intervals. They define the curved path of an aircraft from a point near the runway centerline to the center of the intersecting taxiway. When installed, **taxiway centerline lights** are green and **taxiway edge lights** are blue.

Pilot-controlled lighting (PCL) is designed primarily to conserve energy and may be found at some airports which do not have a full-time tower or an FSS. Typically, you control the lights by keying the aircraft microphone a specified number of times in a given number of seconds. For example, you may key the microphone seven times in five seconds to turn on the lights to maximum intensity. To reduce the lighting level, key the microphone the number of times specified. However, you should be aware that using the lower intensity on some installations may turn the runway end identifier lights completely off. The lights normally turn off automatically 15 minutes after they were last activated. You can find information on PCL and the airports which have it in the *Airport/Facility Directory*.

AIRPORT BEACON AND OBSTRUCTION LIGHTS

Some of the other types of lights that are located at or near airports include the airport beacon and obstruction lighting. The beacon is designed to help you locate the airport at night and during conditions of reduced visibility. Operation of the beacon during daylight hours at an airport within controlled airspace may indicate that the ground visibility is less than three miles and/or the ceiling is less than 1,000 feet. However, since they are often turned on by photoelectric cells or time clocks, you must not rely on the airport beacon to indicate that the weather is below minimums. Remember, an ATC clearance is required if you wish to take off or land when the weather is below VFR minimums.

If the airport is located within controlled airspace, operation of the airport beacon during daylight hours may indicate that the airport is below VFR landing minimums. An ATC clearance is required for takeoffs and landings if the weather conditions are less than VFR.

Obstruction lights are installed on prominent structures such as towers, buildings and, occasionally, even powerlines. Bright red and high intensity white lights are typically used and may flash on and off to warn you of obstructions.

AIRSPACE

Within the United States, airspace is classified as either controlled or uncontrolled. Special use airspace and other airspace areas are additional classifications which may include both controlled and uncontrolled segments. As pilot in command, you must know which flight restrictions or aircraft equipment requirements are applicable in these different airspace classifications.

CONTROLLED AIRSPACE

Flight within controlled airspace under IFR requires an ATC clearance, and you must turn your transponder on, including Mode C if installed.

Controlled airspace means an airspace of defined dimensions within which air traffic control service is provided to IFR flights and to VFR flights in accordance with the airspace classification. Controlled airspace is a generic term which covers Class A, Class B, Class C, Class D, and Class E airspace. As a routine measure, when you are operating under IFR, your flight must conform with ATC clearances from takeoff to touchdown, and your transponder must be on, including Mode C if installed. During this time, ATC provides separation between your aircraft and all other IFR flights. If workload permits, ATC also provides traffic advisories for VFR operations. It is very important to remember that controllers are not required to separate your aircraft from VFR flights and cannot provide separation from aircraft which do not appear on their radar display. Therefore, it is your responsibility to see and avoid other aircraft when you are operating under IFR in VFR weather conditions. [Figure 2-13]

CLASS A AIRSPACE

Over most of the U.S., the Class A airspace extends from 18,000 feet MSL to FL 600.

Within the conterminous United States, Class A airspace extends from 18,000 feet MSL up to and including FL 600. Since VFR flight is not permitted in this area, your instrument training may provide your first opportunity to fly in Class A airspace. Instrument high altitude enroute charts must be used for flights in Class A airspace.

VFR IN CONTROLLED AIRSPACE		
Altitude	**Flight Visibility**	**Distance From Clouds**
Class A	Not Applicable	Not Applicable
Class B	3 statute miles	Clear of Clouds
Class C and Class D	3 statute miles	500 feet below 1,000 feet above 2,000 feet horizontal
Class E: Less than 10,000 feet MSL	3 statue miles	500 feet below 1,000 feet above 2,000 feet horizontal
At or above 10,000 feet MSL	5 statute miles	1,000 feet below 1,000 feet above 1 statute mile horizontal

Figure 2-13. Basic VFR weather minimums which apply in Class A, B, C, D, and E controlled airspace are listed in FAR 91.155. Notice VFR flight is not authorized in Class A airspace.

Because aircraft in Class A airspace operate at such high speeds, it would be impractical for pilots to reset their altimeters every 100 n.m. So, within Class A airspace, you are required to use a standard setting of 29.92 in. Hg. This means that all pilots are maintaining their assigned altitudes using the same altimeter reference. In addition, altitudes are prefaced by the letters "FL," meaning **flight level**, with the last two zeros omitted. For example, 35,000 feet is referenced as FL 350.

To fly in Class A airspace, you must adhere to the following guidelines:

To fly in Class A airspace, you must be instrument rated, current, and on an IFR flight plan. The aircraft must be IFR equipped and, in most cases, DME is required at or above 24,000 feet MSL.

1. If acting as pilot in command, you must be rated and current for instrument flight.
2. You must operate under an IFR flight plan and in accordance with an ATC clearance at specified flight levels.
3. Your aircraft must be equipped with instruments and equipment required for IFR operations, including an encoding altimeter and transponder. You are also required to have a radio providing direct pilot/controller communications on the frequency specified by ATC for the area concerned. In addition, you must have navigation equipment appropriate to the ground facilities to be used.
4. When VOR equipment is required for navigation, your aircraft must also be equipped with distance measuring equipment (DME) if the flight is conducted at or above 24,000 feet MSL. If the DME fails in flight, you must immediately notify ATC. Then, you may continue to operate at or above 24,000 feet MSL and proceed to the next airport of intended landing where repairs can be made.

To fly in controlled airspace within the contiguous United States, your aircraft must meet certain equipment requirements. For example, if your aircraft has a functioning transponder which has been properly inspected, you must operate it at all times during flight within all controlled and uncontrolled airspace. In addition, if your aircraft is equipped with a properly inspected encoding altimeter, you must operate it as well. Mode C is required in all airspace (controlled or uncontrolled) within the 48 contiguous states and the District of Columbia at and above 10,000 feet MSL, excluding the airspace at and below 2,500 feet AGL. For flights under IFR, you must operate Mode C at all times unless ATC directs otherwise. Now, let's look at some specific classes of airspace which are described in the following paragraphs. [Figure 2-14]

An operable transponder with altitude reporting capability is required in all controlled and uncontrolled airspace within the 48 states at and above 10,000 feet MSL, excluding the airspace at or below 2,500 feet AGL.

CLASS B AIRSPACE

At some of the country's busiest airports, Class B airspace has been established to separate all arriving and departing traffic. Each Class B airspace is unique since it is designed to facilitate traffic separation at a particular terminal. Pilot participation is mandatory, and an ATC clearance must be received before you enter a Class B area. Some of the Class B airspace areas have VFR corridors to allow pilots of VFR aircraft to pass through them without contacting ATC.

You must have ATC authorization to operate in Class B airspace, unless you are in a VFR corridor, where certain exemptions may apply.

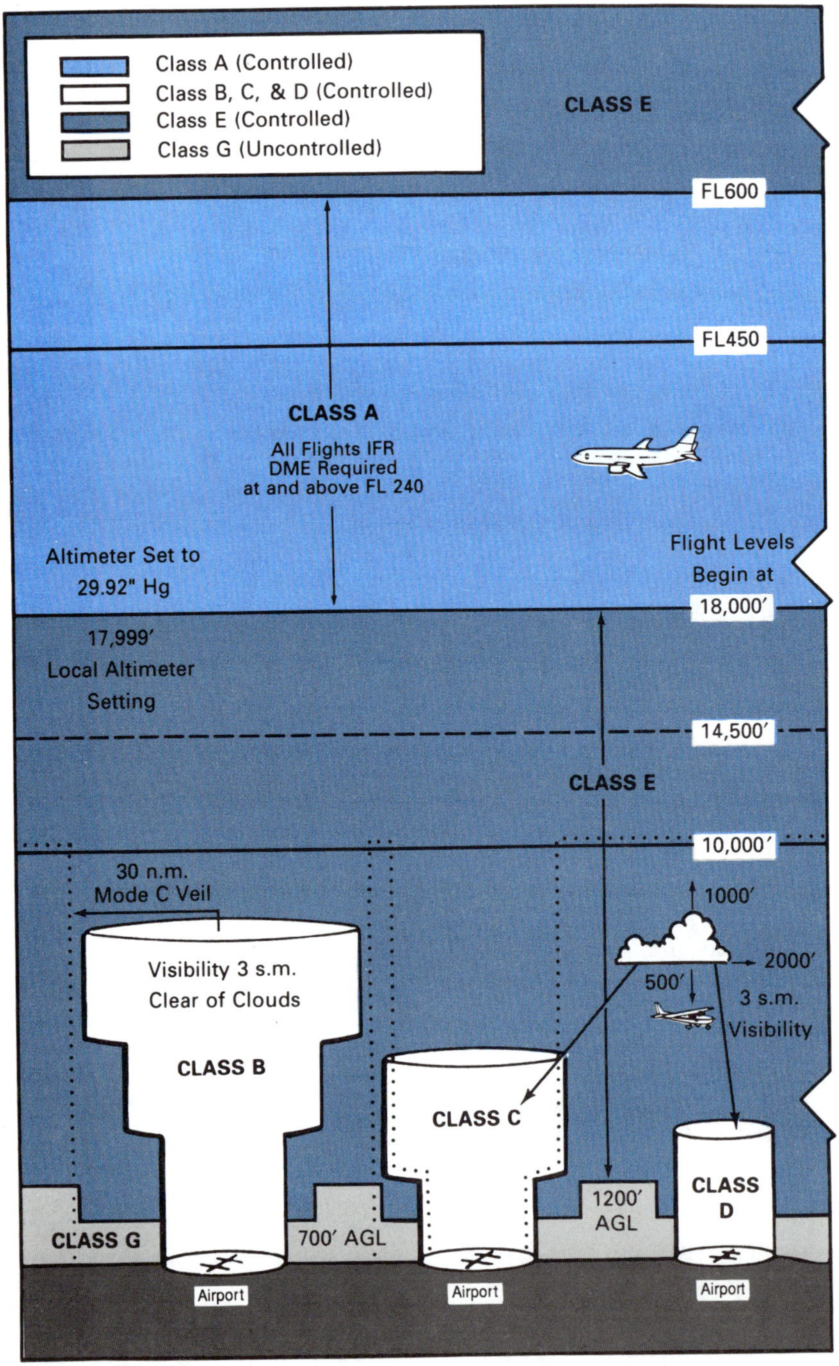

Figure 2-14. To enhance safety for all aircraft, the airspace that covers the United States is divided into controlled and uncontrolled airspace. Operational requirements depend on the class of airspace you are flying in, as well as your altitude.

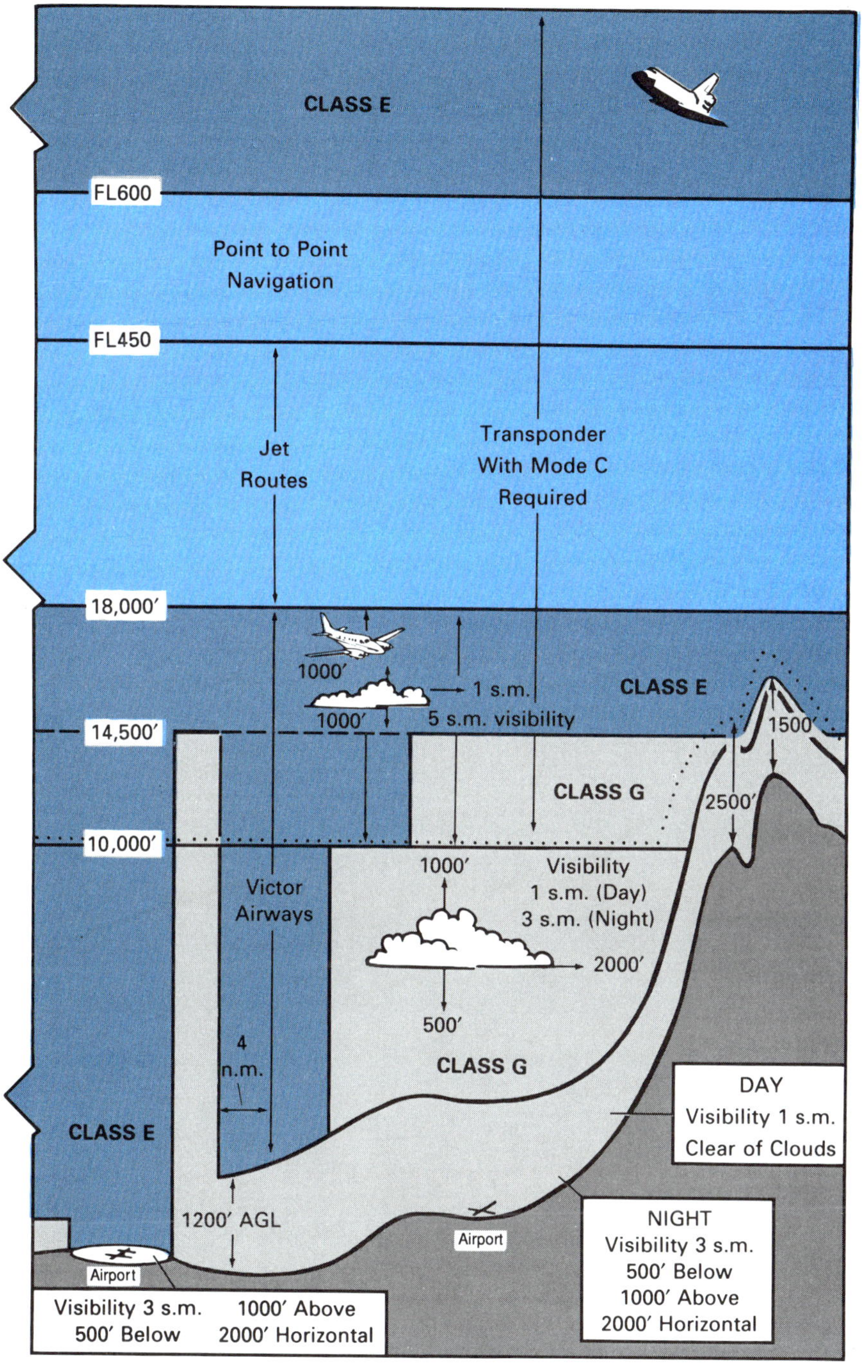

Along with these requirements, you also must be aware of the minimum day/night flight visibilities and cloud clearance requirements that apply at various altitudes in controlled and uncontrolled airspace.

Aircraft operating in all airspace of the 48 contiguous States and the District of Columbia at and above 10,000 feet MSL must be equipped with an operable transponder with Mode C except when operating at and below 2,500 feet AGL. See FAR Part 91 for other Mode C transponder requirements. The dotted line in the airspace diagram shows where these requirements prevail.

Among other requirements for flight within Class B airspace, your aircraft must be equipped with either a Mode S or a 4096-code transponder that has Mode C altitude reporting equipment.

To operate under VFR in Class B airspace, your aircraft must have a two-way radio, and either a Mode S or a 4096-code transponder with Mode C automatic altitude reporting. For IFR operations, you must also have a VOR receiver. In addition, to take off or land at certain large airports listed in Appendix D of FAR 91, you must hold at least a private pilot certificate. With certain exceptions, a transponder with altitude reporting capability is required anytime you are operating within 30 nautical miles of the primary airport from the surface upward to 10,000 feet MSL.

CLASS C AIRSPACE

Two-way radio communications equipment is required for flight operations within Class C airspace.

Class C airspace areas are designated at certain airports where ATC is equipped to provide radar service for all aircraft. Normally, Class C airspace designations consist of two circular areas which extend outward from the primary airport and are referred to as the inner circle and outer circle. In addition, an outer area extends 10 n.m. beyond the outer circle. Although it is not required, pilot participation for flights within the outer area is strongly encouraged. Before operating within the inner and outer circles, you must establish two-way communications with the ATC facility having jurisdiction over the area and maintain radio contact at all times. If you depart a satellite airport located within Class C airspace, you must establish two-way communications with ATC as soon as practicable. [Figure 2-15]

All aircraft operating in Class C airspace, and in all airspace above the ceiling and within the lateral boundaries extending upward to 10,000 feet

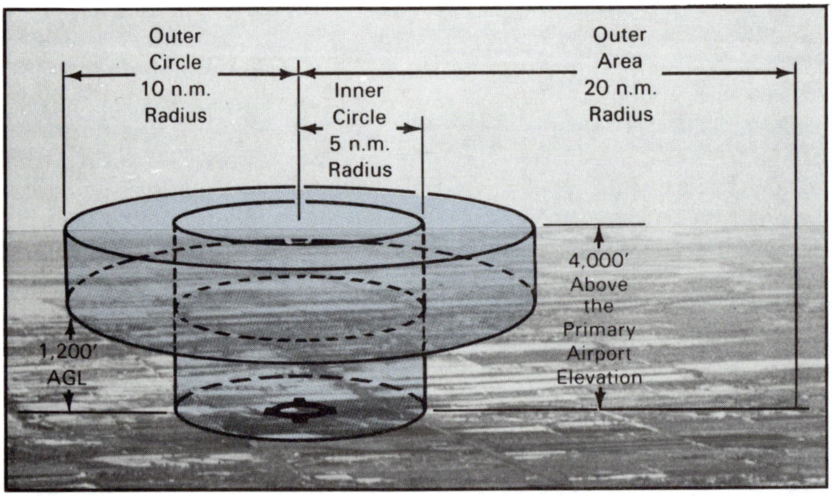

Figure 2-15. The inner circle typically begins at the surface and has a 5 n.m. radius from the primary airport, while the outer circle normally starts at 1,200 feet AGL and has a 10 n.m. radius. Both have upper limits of 4,000 feet above the primary airport. The outer area usually has a radius of 20 n.m. and extends from the lower limits of radar/radio coverage up to the ceiling of approach control's delegated airspace, excluding the Class C airspace itself.

MSL, must be equipped with an operable transponder with Mode C. Aircraft operating in the airspace beneath Class C airspace will not be required to have a transponder with Mode C. Most of the airports where Class C airspace areas have been designated were once **terminal radar service areas**, or TRSAs. Within a TRSA, radar service provides separation between all IFR aircraft and participating VFR aircraft; but participation is not mandatory if you are operating under VFR.

CLASS D AIRSPACE

Class D airspace areas are designated at airports with operating control towers which are not associated with Class B or C airspace. Before you enter Class D airspace you must establish and maintain two-way radio communications with the control tower. When departing the primary airport within Class D airspace, you also must establish and maintain communications with the tower. It is important to note that airspace at an airport with a part-time control tower is classified as a Class D airspace only when the control tower is in operation.

At part-time tower locations, Class D airspace normally becomes Class E airspace when the tower is closed. However, if weather observations and reporting are not available, the airspace becomes Class G.

At some locations, there may be a satellite airport within the same Class D airspace designated for the primary airport. If the satellite airport also has a control tower, similar radio communications requirements with that tower prevail for arrivals and departures. If the satellite airport is a nontower field, arriving aircraft must establish contact with the primary airport's control tower. Departures from a nontower satellite must establish communication with the ATC facility (tower) having jurisdiction over the Class D airspace as soon as practicable after departing. To the maximum extent practical and consistent with safety, satellite airports have been excluded from Class D airspace. For example, a satellite airport without an operating control tower might have airspace carved out of the Class D airspace, or it could be placed under a shelf of the Class D airspace.

DIMENSIONS OF CLASS D AIRSPACE

Normally, the ceiling of Class D airspace areas (designated for airports) are specified at 2,500 feet above the surface of the airport converted to mean sea level, and rounded to the nearest 100-foot increment. If conditions of a particular airspace area warrant, the ceiling may be raised or lowered as appropriate. For example, if local conditions warrant, the ceiling can be designated at more than 2,500 feet above the surface. Conversely, some airports with a limited volume of nonturbine-powered aircraft may have a lower vertical limit. The ceiling of Class D airspace is shown in hundreds of feet MSL on sectional charts. Lateral dimensions of Class D airspace (also depicted as blue dashed lines on sectional charts) will be based on the instrument procedures for which the controlled airspace is established. Therefore, the lateral dimensions may be irregular as opposed to circular in many cases. Arrival extensions will either be charted as part of the basic surface area with blue dashed lines or as a separate surface area indicated with a dashed magenta line. ATC communications are not required within the magenta lines (Class E airspace).

CLASS E AIRSPACE

One type of Class E airspace extends upward from 14,500 feet MSL up to 18,000 feet MSL.

Several types of airspace may be designated as Class E. One Class E designation consists of the airspace over the 48 contiguous states, District of Columbia, and that area of Alaska east of 160° west longitude. This particular Class E designation begins at 14,500 feet MSL, but does not include the airspace within 1,500 feet of the surface, or restricted and prohibited areas. It extends up to, but not including, 18,000 feet MSL.

Another Class E designation includes domestic airspace areas which extend upward from 1,200 feet or more above the surface when designated in conjunction with an airway or route. Class E designations for airways or routes coincide with the lateral dimensions of the Federal airway and extend upward from 1,200 feet or higher unless otherwise specified. [Figure 2-16]

Still another Class E designation includes domestic airspace areas which extend upward from 700 feet or more above the surface. These areas are designated at non-tower airports which have an approved instrument approach procedure. At part-time tower locations when the tower is closed, the surface area is designated as Class E airspace if a weather observer is available at the airport. When no weather reporting is available, the Class E airspace begins at 700 feet AGL, and the surface area up to 700 feet reverts to Class G. These airspace designations are being incorporated into the *Airport/Facility Directory*. The lateral dimensions of Class E airspace designated for airports are depicted on sectional charts with a segmented magenta line. The magenta line denotes controlled airspace

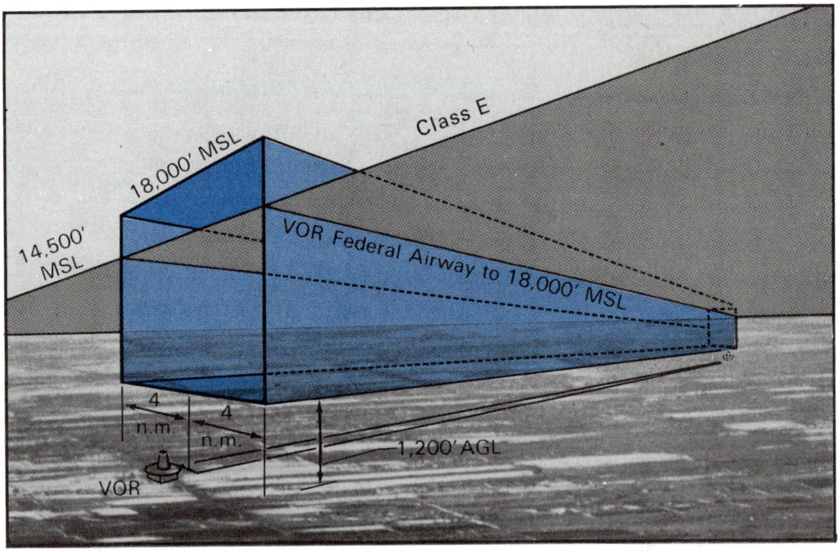

Figure 2-16. Federal airways are designated as Class E airspace and usually are eight nautical miles wide, begin at 1,200 feet AGL, and extend up to, but not including, 18,000 feet MSL.

extending upward from the surface to the overlying or adjacent controlled airspace; therefore, the vertical limit is not depicted. Communication with air traffic control is not required within airspace encompassed by the magenta lines (Class E airspace).

SPECIAL VFR

You may conduct normal VFR operations (traffic pattern entry, takeoffs, or landings) below 10,000 feet MSL within the airspace contained by the upward extension of the lateral boundaries of the controlled airspace designated to the surface for an airport. You may obtain an ATC clearance for special VFR during the day if the visibility is at least one mile and you can remain clear of clouds. Special VFR is not permitted between sunset and sunrise unless you have a current instrument rating and the aircraft is equipped for instrument flight. In addition, special VFR clearances are not issued to fixed-wing aircraft (day or night) at the nation's busier airports which are listed in Section 3 of Appendix D of FAR 91. The table on the next page provides a summation of each class of airspace. [Figure 2-17]

Special VFR clearances require you to maintain a minimum in-flight visibility of one mile and remain clear of clouds.

AIRCRAFT SPEED LIMITS

As you move into faster aircraft, you must also be concerned with aircraft speed limitations. For example, unless otherwise authorized by air traffic control or exceptions to the regulations, you may not operate an aircraft below 10,000 feet MSL at a speed greater than 250 knots indicated airspeed (KIAS). Further, unless otherwise authorized or required by ATC, you may not operate an aircraft at or below 2,500 feet above the surface within four nautical miles of the primary airport of a Class C or Class D airspace area at a speed greater than 200 KIAS. The 200 KIAS limit also applies to the airspace underlying a Class B airspace area or in a VFR corridor designated through such airspace.

Regulations specify the maximum permissible indicated airspeeds and the airspace where the speeds apply.

SPECIAL USE AIRSPACE

Large segments of airspace, both controlled and uncontrolled, have been designated as special use airspace. Activities conducted within these areas are considered hazardous to civil aircraft, so civil operations may be limited or prohibited. The various types of airspace may be designated as prohibited, restricted, warning, alert, military operations areas, and controlled firing areas. Information about hours of operation and effective altitudes may be listed directly on aeronautical charts or indexed by area number on a chart panel. Let's look briefly at each of these.

Prohibited areas contain airspace of defined dimensions identified by an area on the surface of the earth within which the flight of aircraft is prohibited. Such areas are established for security or other reasons associated with the national welfare. These areas are published in the *Federal Register* and are depicted on aeronautical charts.

You may not fly in a prohibited area.

Restricted areas include airspace where flight operations are subject to certain limitations. They denote the existence of unusual, often invisible, hazards to aircraft, such as artillery firing, aerial gunnery, or flight of

Restricted areas indicate the existence of unusual and often invisible hazards to aircraft.

AIRSPACE CLASSIFICATIONS						
Airspace features	**Class A airspace**	**Class B airspace**	**Class C airspace**	**Class D airspace**	**Class E airspace**	**Class G airspace**
Operations Permitted	IFR	IFR and VFR	IFR and VFR	IFR and VFR	IFR and VFR	IFR and VFR
Entry Prerequisites	ATC clearance	ATC clearance	ATC clearance for IFR, Radio contact for all.	ATC clearance for IFR, Radio contact for all.	ATC clearance for IFR, Radio contact for all IFR	None
Minimum Pilot Qualifications	Instrument rating	Private or student certificate.	Student certificate.	Student certificate.	Student certificate.	Student certificate.
Two-way radio communications	Yes	Yes	Yes	Yes	Yes for IFR operations.	No
VFR Minimum Visibility	Not applicable	3 statute miles	3 statute miles	3 statute miles	* 3 statute miles	** 1 statute mile
VFR Minimum Distance from Clouds.	Not applicable	Clear of Clouds	500 feet below, 1,000 feet above, and 2,000 feet horizontal.	500 feet below, 1,000 feet above, and 2,000 feet horizontal.	* 500 feet below, 1,000 feet above, and 2,000 feet horizontal.	** 500 feet below, 1,000 feet above, and 2,000 feet horizontal.
Aircraft separation	All	All	IFR, SVFR, and runway operations.	IFR, SVFR, and runway operations.	IFR, SVFR	None
Conflict Resolution	Not applicable	Not applicable	Between IFR and VFR operations.	No	No	No
Traffic Advisories	Not applicable	Not applicable	Yes	Workload permitting	Workload permitting	Workload permitting
Safety Advisories	Yes	Yes	Yes	Yes	Yes	Yes

* Different visibility minima and distance from cloud requirements exist for operations above 10,000 feet MSL.

** Different visibility minima and distance from cloud requirements exist for night operations, operations above 10,000 feet MSL, and operations below 1,200 feet AGL.

NOTE: Two-way radio communications are required for all operations where temporary or non-federal control towers are established and operating within Class E and G airspace. Refer to FARs 91.126 and 91.127.

Figure 2-17. This chart outlines several features of each class of airspace. You can use this as a review of the various airspace classifications.

guided missiles. Penetration of restricted areas without authorization from the controlling agency can be extremely hazardous. If ATC issues you an IFR clearance which will take you through restricted airspace, such a clearance constitutes authorization to penetrate the airspace. In this case, you need take no further action other than to comply with the clearance, as issued, and maintain normal vigilance.

A **warning area** is airspace of defined dimensions, extending from three nautical miles outward from the coast of the United States, that contains activity that may be hazardous to nonparticipating aircraft. The purpose of such warning areas is to warn nonparticipating pilots of the potential danger. A warning area may be located over domestic or international waters or both.

Warning areas may contain hazards to nonparticipating aircraft.

Alert areas are shown on sectional charts to inform you of areas that may contain a high volume of pilot training or an unusual type of aerial activity. Flight within alert areas is not restricted, but you are urged to exercise extreme caution. Pilots of participating aircraft, as well as pilots transiting the area, are equally responsible for collision avoidance.

All pilots flying in alert areas are equally responsible for collision avoidance.

Military operations areas, or MOAs, are established to separate certain military training activities from civilian flight operations. When you are flying IFR, you may be cleared through an active MOA if ATC can provide separation. Otherwise, ATC will reroute or restrict your flight operations. If you are flying VFR, you should exercise extreme caution within an active MOA. Information regarding route activity is available from any FSS within 100 n.m. of the area. Established MOAs are shown on sectional, VFR terminal area, and low altitude enroute charts. Before entering an active MOA under VFR, you should contact the controlling agency for traffic advisories.

VFR flight within an active MOA calls for extreme caution.

OTHER AIRSPACE AREAS

Other airspace areas include airport advisory areas, military training routes, and areas where temporary restrictions or limitations/prohibitions apply. Parachute jump areas are an example.

An **airport advisory area** encompasses the airspace within 10 s.m. of an airport where a nonautomated FSS is located and there is no operating control tower. At these locations, the FSS provides **local airport advisory (LAA)** service. When inbound with an operable radio, you should monitor the appropriate CTAF beginning approximately 10 miles from the airport. Recommended radio procedures are outlined in the AIM.

An airport advisory area is established within 10 s.m. of an airport with an FSS where no control tower is operating.

The centerlines of **military training routes** (MTRs) are depicted on sectional and NOS low altitude enroute charts. They involve both IFR and VFR high-speed operations. Generally, MTRs are established below 10,000 feet MSL for operations at speeds in excess of 250 knots. The IFR routes (IRs) may be operated in either IFR or VFR conditions, while VFR routes (VRs) are operated only under VFR conditions. You should contact an FSS within 100 n.m. of a particular MTR for current information on route usage.

Regardless of weather conditions, military traffic on MTRs may be operating at speeds in excess of 250 knots.

Temporary flight restrictions are imposed by the FAA to protect persons and property on the surface or in the air. For example, the FAA will normally issue a NOTAM to provide a safe environment for rescue/relief operations and to prevent unsafe congestion above an incident or event which may generate high public interest. These situations include events such as forest fires, toxic spills, nuclear incidents, volcanic eruptions, and aircraft hijackings. The restricted airspace is usually limited to 2,000 feet above the surface within a two nautical mile radius. Incidents near controlled airports are handled through existing procedures and normally do not require issuance of a NOTAM. However, NOTAMs are issued to restrict flight in the vicinity of space flight operations and in the proximity of the President, Vice President, and other public figures. These **flight limitations/prohibitions** are put into place to reduce the hazards associated with aircraft operating in the vicinity of large assemblies of people. When such NOTAMs are issued, they are considered to be regulatory.

The *Airport/Facility Directory* contains a list of **parachute jumping areas,** with their times of operation and MSL altitudes. Frequently used sites are depicted on sectional charts.

Aircraft entering U.S. domestic airspace from points outside must provide identification prior to entry. **Air defense identification zones** (ADIZs) have been established to facilitate this early identification. You must file a flight plan to penetrate or operate within a coastal or domestic ADIZ. Your flight plan must be filed with an appropriate facility, such as an FSS. If flying VFR, you file a defense VFR (DVFR) flight plan. It contains information similar to local flight plans, but helps to identify your aircraft as you enter the country. Unless otherwise authorized by ATC, a transponder with Mode C (or Mode S) capability is required, and the transponder must be turned on and operable. You are also required to have a two-way radio and periodically give ATC reports of your location while inbound toward the ADIZ. Failure to follow these steps may result in your aircraft being intercepted by U.S. security. The Alaskan ADIZ is similar to a contiguous U.S. ADIZ, except that it lies along the coastal waters of Alaska. In addition, the Alaskan ADIZ has different operating rules. If you are thinking of flying into the Alaskan ADIZ, you should refer to the *Aeronautical Information Manual* or the *International Flight Information Manual* for detailed procedural information.

During a defense emergency or during air defense emergency conditions, special security instructions may be issued in accordance with the Security Control of Air Traffic and Air Navigation Aids (SCATANA) Plan. Under the provisions of this plan, the military will direct the necessary actions to land, ground, divert, or disperse aircraft and take over control of navaids in the defense of the United States. If SCATANA goes into effect, ATC facilities will broadcast instructions over available frequencies.

CLASS G AIRSPACE (UNCONTROLLED)

Class G airspace is that area which has not been designated as Class A, B, C, D, or E airspace and is essentially uncontrolled by ATC. For example, the airspace below a Class E airspace area or below a Victor airway is normally uncontrolled. Most Class G airspace terminates at the base of Class E airspace at 700 or 1,200 feet AGL, or at 14,500 feet MSL. An exception to this rule occurs when 14,500 feet MSL is lower than 1,500 feet AGL. In this situation, Class G airspace continues up to 1,500 feet above the surface. The amount of uncontrolled airspace has steadily declined because of the expanding need to coordinate the movement of aircraft.

Although ATC does not have responsibility for or authority over aircraft in Class G airspace, most of the regulations affecting pilots and aircraft still apply. For example, although a flight plan is not required for IFR operations in Class G airspace, both pilot and aircraft must still be fully qualified for IFR flight. Day weather minimums for VFR flight in uncontrolled airspace are also reduced. [Figure 2-18]

FLIGHT INFORMATION

Regulations require you to familiarize yourself with all available information concerning each flight. The following review of flight information publications is designed to help you fulfill this requirement. The publications in this section include the *Airport/Facility Directory*, *Aeronautical Information Manual*, *Notices to Airmen*, the *International Flight Information Manual*, *Advisory Circulars*, and the *Jeppesen Pilot Resource Services*.

AIRPORT/FACILITY DIRECTORY

The *Airport/Facility Directory* (A/FD) is a series of regional books which includes a tabulation of all data on record with the FAA for public-use civil airports, associated terminal control facilities, air route traffic control

VFR CLASS G AIRSPACE (UNCONTROLLED)		
Altitude	**Flight Visibility**	**Distance From Clouds**
1,200 ft. or less above the surface (regardless of MSL altitude)	Day: 1 s.m.	Clear of clouds
	Night: 3 s.m.	500 ft. below 1,000 ft. above 2,000 ft. horizontal
More than 1,200 ft. above the surface, but less than 10,000 ft. MSL	Day: 1 s.m. Night: 3 s.m.	500 ft. below 1,000 ft. above 2,000 ft. horizontal
More than 1,200 ft. above the surface and at or above 10,000 ft. MSL	Day and night: 5 s.m.	1,000 ft. below 1,000 ft. above 1 s.m. horizontal

Figure 2-18. Weather minimums for VFR flight in Class G airspace are shown in this illustration. The VFR minimums at or above 10,000 feet MSL (and more than 1,200 feet AGL) are the same for Class G and Class E.

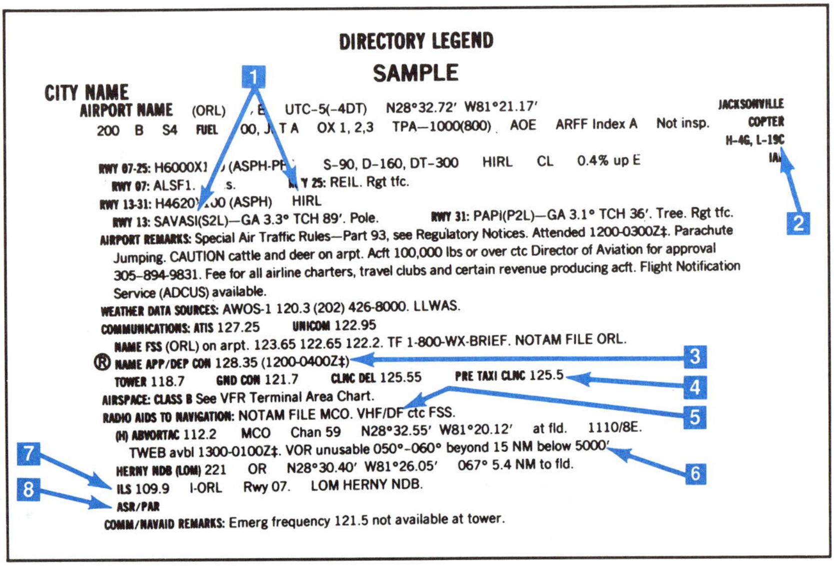

Figure 2-19. Runway 13-31 has high intensity runway lights, and runway 13 has a simplified abbreviated VASI with two light boxes on the left side of the runway (item 1). This airport is depicted on the Jacksonville Sectional, a helicopter chart, NOS high altitude enroute chart H-4, panel G, and low altitude enroute chart L-19, panel C (item 2). The hours for approach control may affect the IFR arrival procedure (item 3). Your IFR clearance for departure should be obtained on the pretaxi clearance frequency of 125.5 MHz (item 4). The FSS has direction-finding equipment (item 5). Below 5,000 feet, the VOR is restricted between the 050° and 060° radials beyond 15 n.m. (item 6). The airport is equipped with an instrument landing system (item 7), and ASR or PAR approach minimums are published (item 8).

centers, and radio aids to navigation. Each book also contains additional data such as special notices, operational procedures, and preferred IFR routes relevant to the coverage area. A comprehensive legend sample is printed in the first few pages of each regional book. It provides you with a breakdown of all the information in the A/FD booklet. [Figure 2-19]

The normal traffic pattern altitudes for turbine and reciprocating engine aircraft are 1,500 and 1,000 feet AGL, respectively.

The *Airport/Facility Directory* listing provides several other important items, depending on the particular airport. For example, you may find information on runways with right-hand traffic patterns. Remember, a left-hand traffic pattern is considered to be standard at both controlled and uncontrolled airports. At controlled airports, however, the tower advises what pattern to expect, while at uncontrolled airports you must determine this information for yourself. You can do this by checking the A/FD or by observing traffic pattern indicators located on the ground at the field. If you fly over the airport to determine the traffic direction, you should do so at least 500 feet above pattern altitude. The A/FD may also list the traffic pattern altitude (TPA) for some airports. Although traffic pattern altitudes vary, you should generally fly at 1,000 feet above the airport elevation if you are operating a small reciprocating-engine aircraft. Large or turbine-powered aircraft should be flown at a pattern altitude of 1,500 feet. If it becomes necessary to adjust your spacing to follow another aircraft in the pattern at a tower-controlled airport, you are permitted to make shallow S-turns without ATC approval.

OTHER FLIGHT PLANNING INFORMATION

Although airport and facility information make up the bulk of the directory, there are several other sections that contain essential information. Many of these pertain to IFR flight operations. Each directory has a special notices section which lists "hard to find," but significant, information. For example, civil use of military fields, newly certified airports, and continuous power facilities are explained, and special flight procedures are identified. Another section provides telephone numbers for the FSS and NWS outlets, as well as numbers for PATWAS and TWEB. You will also find a listing of VOR receiver checkpoints and VOT facilities for each region. If you have not used the A/FD recently, you should get a copy and thoroughly familiarize yourself with it. [Figure 2-20]

You can find the location of VOR receiver checkpoints and the VOT facilities for a particular airport in the A/FD.

AERONAUTICAL INFORMATION MANUAL

The *Aeronautical Information Manual* (AIM) contains fundamental information required for both VFR and IFR flight operations within the national airspace system. It is revised several times each year and is an excellent source of operational information which you should study and review periodically. For example, the AIM describes the capabilities, components, and procedures required for each type of air navigation aid and includes a discussion of radar services, capabilities, and limitations. You will also find a comprehensive description of current airport lighting and runway markings.

The coverage of operating requirements and ATC procedures described includes a discussion of the various forms of controlled, uncontrolled, and

AIR ROUTE TRAFFIC CONTROL CENTERS

Ⓡ **ATLANTA CENTER** H-3-4, L-14-18-19-20-21-22-27, A-1 (KZTL)

Albermarle -133.15
Anniston -134.95
Athens -127.5 124.45
Atlanta (A) -135.0
Augusta -128.1

PREFERRED IFR ROUTES

LOW ALTITUDE

Terminals	Route	Effective Times (UTC)
ATLANTA METRO AREA		
Chicago Midway	(60-170 incl) V97 NELLO V51W HCH V51 CGT	1200-0300
Chicago O'Hare	(60-170 incl) V97 NELLO V51W HCH V51 CGT V7 BEBEE	1200-0300
Cincinnati	(80-170 incl) V97 HYK V57 FLM	1200-0300

Figure 2-20. Areas of the A/FD you may not have used before include the ARTCC sector frequencies, shown at the top, and the low altitude preferred IFR routes, depicted at the bottom.

special use airspace, as well as the weather requirements for VFR flight operations within each type. In addition, a thorough coverage of the ATC system describes the facilities and services available for both VFR and IFR flight. These include air route traffic control centers, control towers, flight service stations, and aeronautical advisory stations. Another important related area is the Pilot/Controller Glossary which is intended to promote a common understanding of the terms used in the ATC system. International terms that differ from the FAA definitions are listed after their U.S. equivalents. The safety of flight section covers altimetry, wake turbulence, and potential flight hazards, as well as safety reporting programs, medical facts for pilots, and aeronautical charts.

NOTICES TO AIRMEN

Notices to airmen or NOTAMs contain time-critical, aeronautical information that could affect your decision to make a flight. The information is either temporary in nature or unknown in time for publication on aeronautical charts and/or in other documents. Airport or primary runway closures, changes in the status of navigational aids, radar service availability, and other data essential to enroute, terminal, or landing operations are examples of information that may be included in NOTAMs. Dissemination is via the National Notice to Airmen System, a telecommunications network. NOTAMs are divided into three categories, NOTAM-D (distant), NOTAM-L (local), and Flight Data Center (FDC) NOTAMs.

NOTAM-D information is disseminated for all navigational facilities that are part of the National Airspace System, all public use airports, seaplane bases, and heliports listed in the *Airport/Facility Directory*. The complete file of all NOTAM-D information is maintained in a central computer data base in Atlanta, Georgia. This category of information is distributed automatically via the Service A telecommunications system. Air traffic facilities, primarily FSSs, with Service A capability have access to the entire data base of NOTAMs. These NOTAMs remain available via Service A for the duration of their validity or until published. Once published, the NOTAM data is deleted from the system.

The latest status of airport conditions can be determined from the *Airport/Facility Directory*, as well as NOTAM-D and NOTAM-L information.

NOTAM-L information includes local items like taxiway closures, construction activities near runways, snow conditions, and changes in the status of airport lighting, such as VASI, that do not affect instrument approach criteria. NOTAM-L information is only distributed locally and is not appended to hourly weather reports. A separate file of local NOTAMs is maintained at each FSS for facilities in their area. You must specifically request NOTAM-L information directly from other FSS areas that have responsibility for the airport concerned. To find the current status of airport conditions, you should consult the A/FD, NOTAM-Ds, and NOTAM-Ls.

FDC NOTAMs are used to disseminate information that is regulatory in nature. Examples are amendments to aeronautical charts, changes to instrument approach procedures, and temporary flight restrictions. FDC NOTAMs are kept on file at the FSS until published or cancelled. FSSs are responsible for maintaining unpublished FDC NOTAMs concerning conditions within 400 nautical miles of their facilities. FDC NOTAM information that affects conditions more than 400 miles from the FSS, or that is already published, is provided to a pilot only upon request. FDC NOTAMs, including Center Area NOTAMs (CANs), are published in the *Notices to Airmen* publication. CANs are in FDC format. All FDC NOTAMs are available from appropriate air traffic control facilities with telecommunications access.

FDC NOTAMs are issued to advise pilots of changes in flight data which affect instrument approach procedures, aeronautical charts, and flight restrictions prior to normal publication.

An important part of your preflight planning should be a review of the *Notices to Airmen* publication. Data is included in this publication to reduce congestion on telecommunications circuits. Included are current NOTAM-Ds and FDC NOTAMs available at time of publication. *Notices to Airmen* is issued every 28 days and it covers notices that are expected to remain in effect for at least seven days after the effective date of the publication. Once published in the *Notices to Airmen*, this information is taken off the distribution circuits and will not be provided in a pilot weather briefing unless you specifically request it. Data of a permanent nature is sometimes printed in the *Notices to Airmen* as an interim step prior to publication in the appropriate aeronautical chart or *Airport/ Facility Directory*.

You may obtain FDC NOTAMs from an FSS when they are issued too late for inclusion in the *Notices to Airmen* publication.

INTERNATIONAL FLIGHT INFORMATION MANUAL

The *International Flight Information Manual* contains the requirements and instructions for flying outside of the United States. It is intended to be used as a preflight planning guide by nonscheduled operators. Airport of entry, passport/visa, and customs procedures for each country are detailed. In addition, routes, such as the established North American Routes (NARs), minimum navigation equipment, long-range navigation information, and other planning data are listed.

ADVISORY CIRCULARS

To provide current aviation information on a recurring basis, the Department of Transportation publishes and distributes **Advisory Circulars**. These circulars provide information and procedures which are necessary for good operating practice, but which are not binding on the public unless they are incorporated into a regulation.

For ease of reference, Advisory Circulars use a coded numbering system that corresponds to the subject areas of the FARs. An *Advisory Circular Checklist*, which is issued periodically as AC 00-2, contains the subjects covered and the availability of each circular. It also provides you with pricing information. While many advisory circulars are free of charge, you must purchase others.

JEPPESEN PILOT RESOURCE SERVICES

The *Jeppesen Pilot Resource Services* includes data for both VFR and IFR flight operations. It contains the Jeppesen Aeronautical Information Manual (AIM), Pilot FARs, and the Jeppesen Airport Directory, all printed in loose-leaf format for ease of revision. Also included is Jeppesen's GPS/LORAN Coordinate Directory, an important reference for users of area navigation equipment. Because of the nature of the material and the frequency of revision, FAR Part 135.81 authorizes its use as a substitute for the AIM and A/FD for air taxi and commercial operations.

CHECKLIST

After studying this section, you should have a basic understanding of:

✓ **Runway markings** — What the markings on a precision instrument runway mean and how the markings for nonprecision instrument runways differ.

✓ **Special purpose areas** — How airport surface areas are marked and what operations are permitted.

✓ **Lighting systems** — What the different types of approach, VASI, and runway lighting are, and how to activate those which are pilot controlled.

✓ **Controlled airspace** — What the IFR requirements are to fly within Class A airspace, and the significance of Class B, C, D, and E airspace areas.

✓ **Special use airspace** — How the operating considerations vary in each of these types of airspace and what the possible hazards are.

✓ ***Airport/Facility Directory*** — Where to locate the different types of information available in each regional volume.

✓ ***Aeronautical Information Manual*** — What basic flight information and ATC procedures it contains.

✓ ***Notices to Airmen*** — How NOTAMs are used to update changes to aeronautical charts and other flight information.

✓ **Advisory Circulars** — What they are, how they are indexed, and how you can obtain them.

✓ **Jeppesen Pilot Resource Services** — The type of information included and how it provides you with current aeronautical and airport data.

CHAPTER 3

CHARTS FOR INSTRUMENT FLIGHT

INTRODUCTION

Instrument flight may be broken down into three broad phases. The departure phase takes you from the airport to the enroute structure. The enroute phase is used to travel from one location to another, and the arrival phase permits you to transition from the enroute structure to your destination. The arrival phase may begin with a published arrival procedure and usually ends with an instrument approach.

Instrument charts for each of these phases are published by Jeppesen Sanderson, Inc., a private firm, and the United States Department of Commerce, which produces the National Ocean Service (NOS) charts. Jeppesen charts are used by most civil operators, including air carriers. The FAA Instrument Rating Airmen Knowledge Test is primarily based on NOS charts but also presents several examples of Jeppesen charts. Therefore, it is important to be familiar with both chart formats. In this chapter, we present a side-by-side comparison and analysis of both charts. The depicted examples are representative of current formats for both Jeppesen and NOS charts. However, they may not be the most current editions and should not be used for navigation.

SECTION A

INSTRUMENT APPROACH CHARTS

The standard **instrument approach procedure**, or IAP, allows you to descend safely from the enroute altitude to a relatively low altitude near a runway at your destination. In the case of an instrument landing system (ILS) approach, you typically arrive at a point located approximately one-half mile from the runway threshold at an altitude of 200 feet above the elevation of the touchdown zone. At this point, if certain visual requirements are met, you may continue to a landing using visual references. If the visual requirements cannot be met at this point, or if the required visual cues are lost after passing it, you must discontinue the approach and follow the missed approach procedure.

APPROACH CHART

Although there are many different types of approaches in use, most incorporate common procedures and chart symbology. Therefore, your ability to read one approach chart generally means you will be able to read others. This section is designed to help you become proficient at reading charts. An ILS approach chart has been selected, since this type of chart includes most of the features found on other approach charts.

APPROACH SEGMENTS

Before looking at approach chart symbology, it is helpful to have a basic understanding of an approach procedure. An instrument approach may be divided into as many as four segments: initial, intermediate, final, and missed approach. These segments often begin and end at designated fixes. However, they also can begin and end at the completion of a particular maneuver, such as a course reversal. [Figure 3-1]

Feeder routes, although technically not considered approach segments, are an integral part of many instrument approach procedures. They provide a link between the enroute and approach structures. Although an approach procedure may have several feeder routes, you generally use the one closest to your enroute arrival point. When a feeder route is shown, the chart provides the course or bearing to be flown, the distance, and the minimum altitude.

INITIAL APPROACH SEGMENT

The purpose of the initial approach segment is to provide a method for aligning your aircraft with the approach course. This is accomplished by using an arc procedure, a course reversal such as a procedure turn or

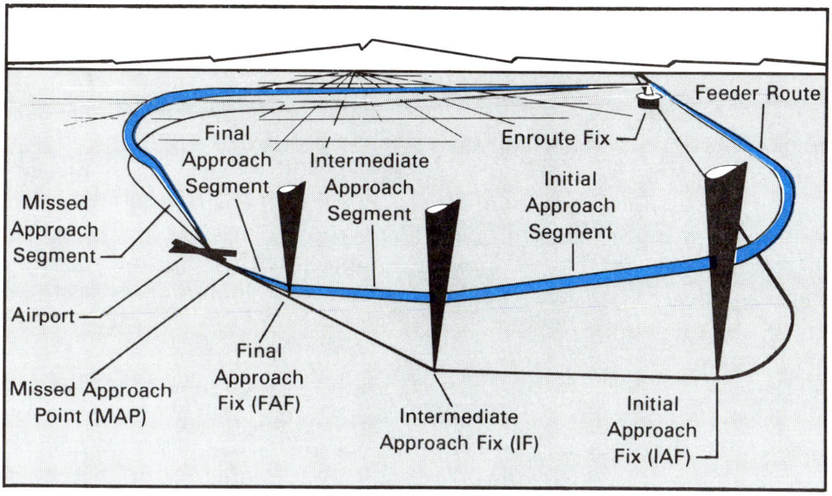

Figure 3-1. A feeder, or terminal, route may be used to take you from the enroute structure to an initial approach fix (IAF). Next, you follow an initial approach segment to an intermediate fix (IF). From here, you follow the intermediate segment to the final approach fix (FAF). The final approach segment ends at the runway, airport, or missed approach point (MAP). Upon reaching the MAP, if you are unable to continue the approach to a landing, you follow the missed approach segment back to the enroute structure.

holding pattern, or by following a route which intersects the final approach course. These procedures are discussed later.

An initial approach segment begins at the **initial approach fix** (IAF) and usually ends where it joins the intermediate approach segment. A given procedure may have several initial approach segments. Where more than one exists, each will join a common intermediate segment, although not necessarily at the same location.

The letters "IAF" indicate the location of an initial approach fix.

Occasionally, a chart may depict an IAF, even though there is no initial approach segment for the procedure. This usually occurs where the intermediate segment begins at a point located within the enroute structure. In this situation, the IAF signals the beginning of the intermediate segment. Course, distance, and minimum altitudes are also provided for initial approach segments.

INTERMEDIATE APPROACH SEGMENT

The intermediate segment is designed primarily to position your aircraft for the final descent to the airport. During it, you typically reduce your airspeed to at or near the approach airspeed, complete the landing checklist (except for extending the landing gear and making the final flap selection), and make a final review of the approach procedure and applicable minimums. Like the feeder route and initial approach segment, the chart depiction of the intermediate segment provides you with course, distance, and minimum altitude information. [Figure 3-2]

The intermediate segment is normally aligned within 30° of the final approach course, begins at the **intermediate fix** (IF), or point, and ends at the beginning of the final approach segment. In some cases, an intermediate fix is not shown on an approach chart. In this situation, the intermediate segment begins at a point where you are proceeding inbound to the final approach fix, are properly aligned with the final approach course, and are located within the prescribed distance from the final approach fix.

During a vector, you must remain continually aware of your position.

ATC may provide you with radar vectors to any point along the initial or intermediate approach segments. During a radar vector, it is your responsibility to continually be aware of your position in case of communications failure. It is also your responsibility to see and avoid other aircraft when operating in VFR conditions.

Two common situations can arise when you are radar vectored to an approach. The first occurs when ATC assigns you an altitude that is below the minimum altitude charted for your area. It is important for you to realize that the controller's display is divided into sectors. Within each sector, a **minimum vectoring altitude** (MVA) is established, which may be lower than the minimum charted altitude.

The second situation occurs when ATC vectors you through the final approach course without giving you an approach clearance. In this situation, you must continue to maintain the last assigned heading and not

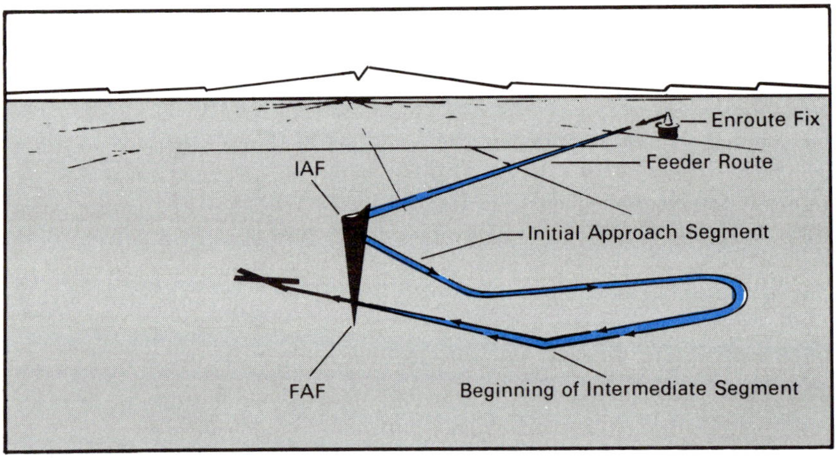

Figure 3-2. An instrument approach that incorporates a procedure turn is the most common example of an approach which may not have a charted intermediate approach fix. The intermediate segment, in this case, begins when you intercept the inbound course after completing the procedure turn.

intercept the final approach course until specifically advised to do so by ATC. Such a procedure is commonly used by ATC to provide separation between you and another aircraft.

Although each of these is commonly encountered, it also can signal a potentially dangerous situation. As the pilot in command, you should immediately question any clearance that will compromise the safety of the flight or that is different from what you expected to receive under the prevailing conditions.

FINAL APPROACH SEGMENT

The purpose of the final approach segment is to allow you to navigate safely to a point at which, if the required visual references are available, you can continue the approach to a landing. If you cannot see the required cues at the missed approach point, you must execute the missed approach procedure. The final approach segment for a precision approach begins where the glide slope is intercepted at the minimum glide slope intercept altitude specified by the approach chart. If ATC authorizes a lower intercept altitude, the final approach segment begins upon glide slope interception at that altitude. For a nonprecision approach, the final approach segment begins either at a designated **final approach fix** (FAF) or at the point where you are aligned with the final approach course. When an FAF is not designated, this point is typically where the procedure turn intersects the final approach course inbound. The final approach segment ends either at the designated missed approach point or when you land.

Although the charted final approach segment provides you with course and distance information, many factors influence the minimum altitude to which you can descend. These include the type of aircraft being flown, the aircraft's equipment and approach speed, the operational status of navaids, the airport lighting, the type of approach being flown, and local terrain features.

MISSED APPROACH SEGMENT

The purpose of the missed approach segment is to allow you to safely navigate from the MAP to a point where you can attempt another approach or continue to another airport. Every instrument approach will have a missed approach segment along with appropriate heading, course, and altitude information.

The missed approach segment begins at the **missed approach point** (MAP) and ends at a designated point, such as an initial approach or enroute fix. The actual location of the missed approach point depends upon the type of approach you are flying. For example, during a precision approach the MAP occurs when you reach a designated altitude called the decision height (DH). For nonprecision approaches, the missed

The MAP for a precision approach is the decision height.

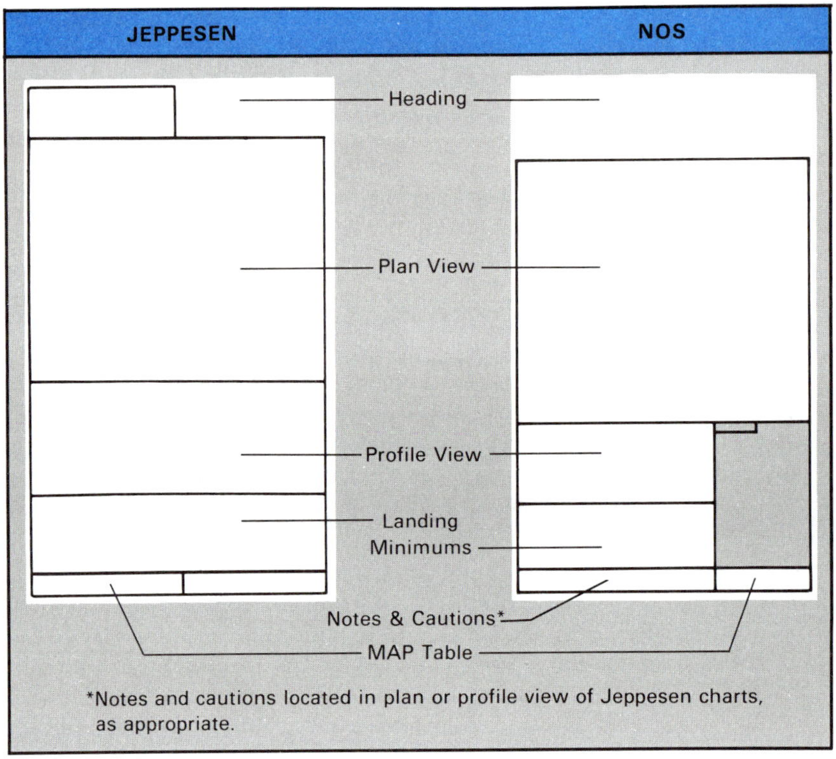

Figure 3-3. Here is the general layout for both Jeppesen and NOS approach charts. Knowing where information is located is just as important as interpreting the chart symbology. The information contained in each one of these sections will be covered in the following pages.

approach point occurs either at a fix defined by a navaid or after a specified period of time has elapsed since you crossed the final approach fix.

CHART LAYOUT

Both Jeppesen and NOS use charts to portray the instrument approaches which are available at a given airport. Generally, both systems present the same information; however, the symbology and chart layout vary. [Figure 3-3]

HEADING SECTION

On Jeppesen charts, the heading section identifies the airport, the primary approach facility, communications frequencies, and if available, minimum safe altitude information. On NOS charts, some of this information is located in the heading section, with the rest located on the plan view or airport sketch. [Figure 3-4]

Jeppesen lists the city and state where the airport is located (item 1), followed by the airport name (item 2) on the next line. These charts are

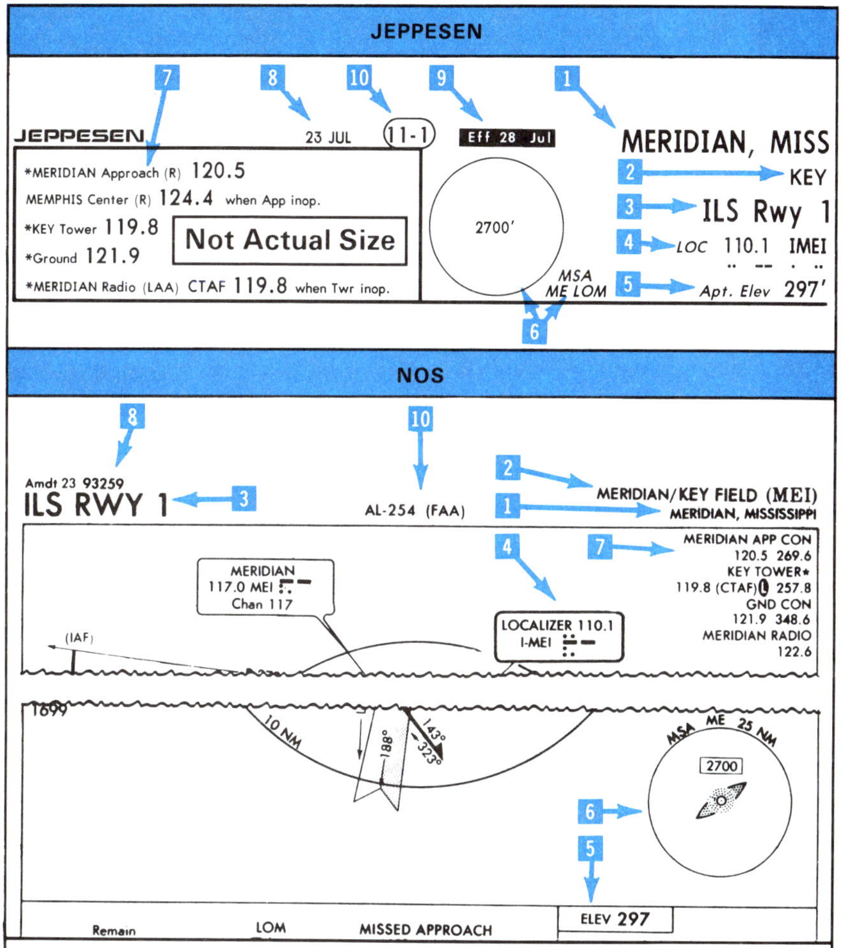

1. Geographic Name
2. Airport Name
3. Procedure Name
4. Primary Facility, Frequency, and Morse Code
5. Airport Elevation
6. MSA Sector and Altitude Information
7. Communications Data (* Indicates part-time facility)
8. Chart Date
9. Effective Date
10. Chart Index Number

Figure 3-4. These are excerpts of the Jeppesen and NOS approach charts for the ILS Runway 1 approach to Key Field at Meridian, Mississippi. Some of these elements are addressed more fully in the accompanying text.

filed in a loose-leaf format by state, then by city within each state. NOS usually lists the name of the airport and its identifier on the first line, followed by the city and state on the next line. The city and airport name are also repeated at the bottom of the chart. NOS charts are published in regional volumes, with each airport filed alphabetically by the name of the associated city.

The procedure name indicates the type of approach system used.

The procedure name (item 3) indicates the type of approach system used and the runway served if it is a straight-in landing procedure. The minimum equipment required to fly the approach is indicated by the procedure name and/or notes on the approach chart. This approach uses an instrument landing system (ILS) and serves runway 1 at Meridian. If your aircraft is capable of flying an ILS approach, you may use this procedure.

A procedure name such as "ILS DME Rwy 1" requires you to have a DME receiver in addition to ILS equipment to fly the approach. When a letter appears in the procedure name, such as "VOR-A," it means the procedure does not meet the criteria for a straight-in landing. In this case, a turning maneuver, called a circling approach, may be required to complete the landing.

The MSA provides 1,000 feet of obstruction clearance within a specified distance, usually 25 n.m., from the facility.

The **minimum safe altitude**, or MSA (item 6), provides 1,000 feet of obstruction clearance within 25 n.m. of the indicated facility (unless some other distance is specified). The MSA may be divided into several sectors which are defined by radials or bearings from the navaid. At Meridian, however, there is only one sector, and the MSA is 2,700 feet. Each MSA is applicable only to the approach chart on which it is displayed. It may not be used for any other approach. The minimum safe altitude circle is contained in the heading section of Jeppesen charts and on the plan view of NOS charts.

The MSA provides only obstruction clearance and does not guarantee navigation or communication signal coverage.

There are several important features of the MSA. First, it provides only obstruction clearance within the sector. Neither navigation nor communication coverage is guaranteed. Second, the MSA is designed only for use in an emergency or during VFR flight, such as during a VFR approach at night. And third, an MSA is not listed for every approach. Its omission may be due to the lack of an easily identifiable facility upon which to orient the MSA circle.

Communications frequencies are listed on approach charts.

Both Jeppesen and NOS present communications data (item 7) in the sequence in which you normally use them. For example, your first contact is usually with Meridian Approach Control, followed by Key Tower, then Ground Control. Meridian Approach Control and Key Tower are part-time facilities, as indicated by the asterisk. On the NOS chart, an asterisk is not used with ground control, since that facility is associated with the tower.

Jeppesen provides an alternate communications contact for part-time facilities. For example, when Meridian Approach Control is shut down, you should contact Memphis Center. When the tower is closed, contact Meridian FSS on the common traffic advisory frequency of 119.8 MHz. Another feature of Jeppesen charts is the capital letter "R" in parentheses, which indicates the facility is equipped with radar. In this case, both Meridian Approach and Memphis Center have radar.

NOS charts list both VHF and, where available, UHF communications frequencies. UHF is used by the military and is, therefore, not included on Jeppesen charts.

Jeppesen charts use an effective date (item 9) if a chart is issued before it can be used. In this situation, you should continue to use the previous chart until the effective date. If there's no effective date, you may use the chart upon receipt.

Both Jeppesen and NOS use chart index numbers (item 10) to identify the chart. The number on Jeppesen charts helps you file a chart and identify certain features. For example, the first digit represents an airport number. Normally, this is the number "1." However, where more than one airport shares the same city and state name, the first airport is given the number "1," the second the number "2," and so on. The second digit identifies the type of chart, as shown on the following list.

- 0 — Area, SID, STAR, Class B, etc.
- 1 — ILS, LOC, MLS, LDA, SDF
- 2 — GPS (sole use)
- 3 — VOR
- 4 — TACAN
- 5 — Reserved
- 6 — NDB
- 7 — DF
- 8 — PAR, ASR
- 9 — RNAV, Vicinity Chart, Visual Arrival or Visual Departure Chart, LORAN

The third digit is used to index charts with the same first and second index numbers. For example, the first ILS approach at the second airport is given an index number of "21-1." The second ILS at the same airport receives the number "21-2." The first VOR approach at this airport receives the number "23-1."

PLAN VIEW

The plan view displays an overhead presentation of the entire approach procedure. Since approach charts are intended for use during instrument weather conditions, they show only limited terrain and obstruction information. However, when the procedure is flown as depicted, obstruction clearance is provided throughout the approach. [Figure 3-5]

Navaid facility boxes (items 1, 7, and 14) identify various navaids which appear on the approach chart. On Jeppesen charts, a rectangular box is used to show a VOR, NDB, or LOM facility, while an oval shape is used to depict an ILS, LOC, LDA, or SDF facility. Jeppesen uses shading around the primary facility upon which the approach is based. A heavy outline is used on NOS for the same purpose. In this case, the primary facility is an ILS localizer aligned on a magnetic course of 008°. Its frequency is 110.1 MHz. The Morse code identification also is provided. The letter "I" precedes the identifier to indicate an ILS or LDA localizer. If the frequency is underlined on an NOS chart, it indicates the navaid is without voice capabilities. The Jeppesen chart also identifies the VOR, VORTAC, or VOR/DME class. The "H" listed for Meridian indicates it is a high altitude class of VORTAC.

Adherence to published minimum altitudes on approach charts provides clearance from terrain and obstructions.

Some, but not all, reference points, such as natural or man-made objects, are depicted on the chart. These reference points with their elevations cannot be relied on for terrain or obstruction avoidance since there may be higher uncharted terrain or obstructions within the same vicinity. Generally, reference points less than 400 feet above the airport elevation are not depicted. The highest reference point (item 3) is indicated by an arrow on Jeppesen charts and by large type and a bold dot on NOS charts. The highest point on this NOS chart (item 3) is the 1699-foot obstruction shown near the lower left corner. A variety of symbols are used to signify structures and terrain. Minimum altitudes on the approach procedures provide clearance of terrain and obstructions along the depicted flight tracks.

A heavy line arrow indicates a flyable route.

Feeder routes (item 4) are indicated with a heavy line arrow on both Jeppesen and NOS charts. Each flyable route lists the radial (or bearing), the distance, and the minimum altitude you must fly to reach an IAF. If approaching Key Field from the east, you follow the Kewanee 242° radial to the Savoy LOM. In this case, the LOM is the IAF. The distance from Kewanee to Savoy is 17.5 n.m., and the minimum altitude for this feeder route is 2,000 feet MSL. Another feeder route is from the Meridian VORTAC. In this case, the distance from Meridian to Savoy is 8 n.m. along the 163° radial, and the minimum altitude is 2,000 feet MSL. Marker beacons are shown by lens-shaped symbols. The middle marker beacon (item 5) is indicated by the letters "MM" adjacent to the symbol. If an NDB is collocated with a middle marker, it is a middle compass locator, or LMM. When an NDB is collocated at the outer marker (OM) beacon, the combination is an outer compass locator, or LOM (item 6).

A procedure turn is a standard method of reversing your course. When the procedure turn is depicted, as it is in this example (item 9), it means you may reverse course any way you desire as long as the turn is made on the same side of the approach course as the symbol and within the specified airspace. The procedure turn shown here suggests an outbound

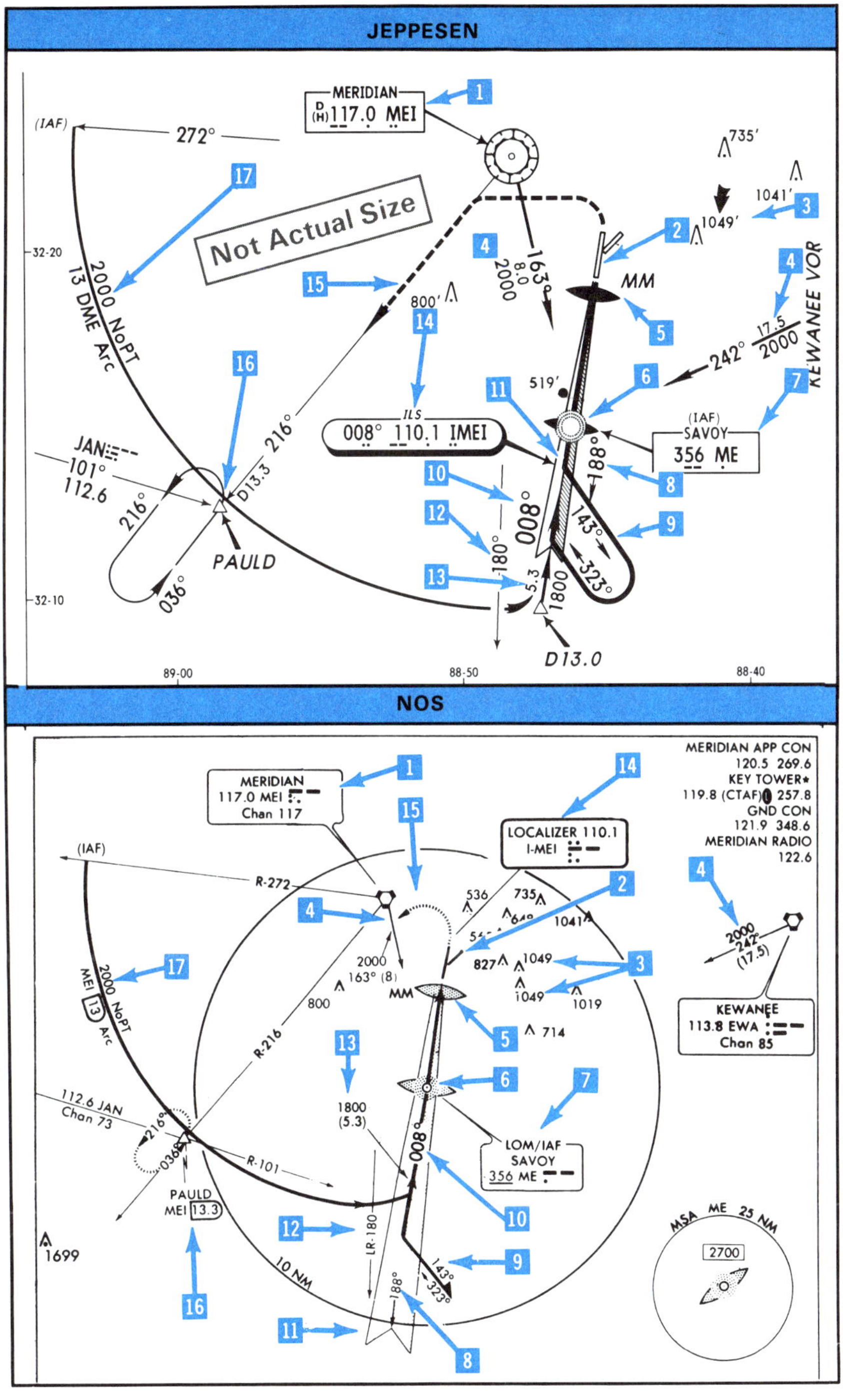

1 Navaid ID
2 Runway Layout
3 Highest Portrayed Reference Point
4 Feeder Routes
5 MM
6 OM or LOM
7 Facility Box
8 Outbound Course
9 Procedure Turn
10 Inbound Course
11 Localizer
12 Lead Radial
13 Transition
14 Primary Approach Facility
15 Missed Approach Path
16 Missed Approach Holding Fix
17 Arc Transition

Figure 3-5. The callouts identify many of the features on the plan view. Some of these elements are discussed further in the accompanying text.

Absence of a PT barb indicates a procedure turn is not authorized.

heading of 142°. After the appropriate time, turn right to 322° to intercept the localizer inbound. If a holding or teardrop pattern is shown instead of a procedure turn, it indicates the only approved method of course reversal. If a procedure turn barb is not shown, a procedure turn is not authorized.

A thin line is used to indicate nonflyable radials or bearings.

A lead radial (item 12) is a common example of a nonflyable radial. It is used to indicate the beginning of the intermediate segment when you are flying the arc transition. Notice that a thin line is used to depict the nonflyable lead radial, and it does not have a minimum altitude. These features help you distinguish it from a flyable route. The transition from the point where the 13-DME arc intercepts the localizer to the LOM is 5.3 miles (item 13). This segment has a minimum altitude of 1,800 feet MSL.

The missed approach path (item 15) and the associated holding pattern (item 16) also are shown. In this case, the holding pattern is shown at Pauld Intersection along with its inbound and outbound courses. The intersection may be identified two ways. If your airplane is DME-equipped, you can identify it as being on the Meridian 216° radial at a distance of 13.3 DME. You can also identify it as being at the intersection of the Meridian 216° radial and the 101° radial of the Jackson (JAN) VORTAC. Although this facility is located off the plan view, all the information required to identify and use it is included. The actual missed approach procedure is included in the profile view.

PROFILE VIEW

The profile view shows the approach from the side and displays flight path, facilities, and minimum altitudes in feet MSL. On Jeppesen charts, an additional height is shown in parentheses and it can be either a height above the touchdown zone or height above the airport elevation. The **height above touchdown** (HAT) is measured from the touchdown zone elevation of the runway served by the approach. The **height above airport** (HAA) is measured above the official airport elevation. On the profile view of Jeppesen charts, HAT values are shown when a touchdown zone elevation (TDZE) is given. When a TDZE is not given, the numbers represent HAA. [Figure 3-6]

The procedure turn must be completed within the prescribed distance from the facility.

For this approach, the procedure turn must be completed within 10 n.m. of the LOM (item 1). The minimum altitude is 2,000 feet MSL (item 2). This is 1,711 feet above the touchdown zone elevation (HAT). After completing the turn, and when established on the inbound course, you may descend to the minimum glide slope intercept altitude of 1,800 feet MSL (item 3).

At this point, it is important to emphasize that all the altitudes given on Jeppesen approach charts represent minimum altitudes, unless noted as "max," "maximum," "mandatory," or "recommended." On NOS charts, a minimum altitude is shown with a line below the number, while a line above the number indicates a maximum altitude. A line above and below

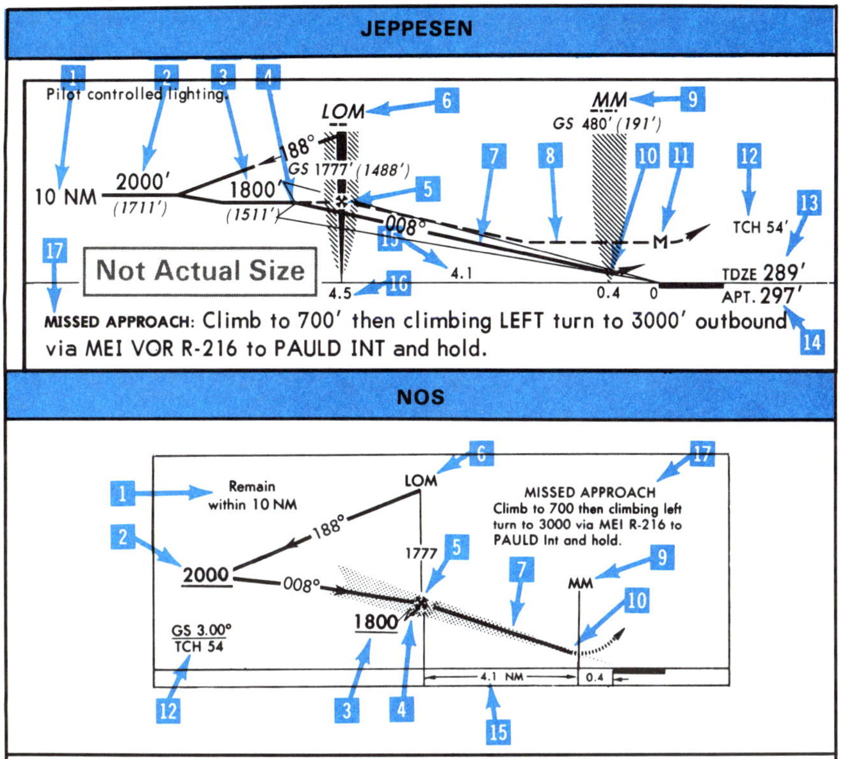

1 Procedure Turn Distance Limit
2 Procedure Turn Altitude
3 Minimum Glide Slope Intercept Altitude
4 Glide Slope Intercept Point
5 Nonprecision Final Approach Fix
6 Outer Marker
7 Precision Approach Path
8 Nonprecision Approach Path
9 Middle Marker
10 Precision Missed Approach Point
11 Nonprecision Missed Approach Point
12 Threshold Crossing Height
13 Touchdown Zone Elevation
14 Airport Elevation
15 Distance Between Fixes
16 Distance From Threshold
17 Missed Approach Procedure

Figure 3-6. The callouts identify many features on the profile view. Some of these elements are discussed further in the accompanying text.

the altitude indicates a mandatory altitude. If the number does not have a line above or below, it is a recommended altitude.

The glide slope intercept point (item 4) shows where you will intercept the glide slope when operating at the minimum glide slope intercept altitude. This also represents the final approach fix for a precision approach. The nonprecision final approach fix is shown by a cross (item 5). An aircraft executing a nonprecision approach must maintain at least the minimum altitude of 1,800 feet MSL until reaching this point.

The precision approach FAF is located at the minimum glide slope intercept altitude. It is identified in the profile view by the lightening bolt symbol on NOS charts and the glide slope/path intercept point on Jeppesen charts.

The outer marker (item 6) on this approach is identified on both charts by the letters LOM, since it is collocated with a compass locator. Both Jeppesen and NOS provide the altitude at which you cross this fix when established on the glide slope centerline. In this case, it is 1,777 feet MSL. Jeppesen also provides an HAT elevation and the outer marker Morse code.

The precision approach path (item 7) is shown on both Jeppesen and NOS charts, along with the magnetic course of the localizer. Since the approach can be flown using the localizer only (no glide slope), Jeppesen also depicts the nonprecision approach path (item 8). Notice how this path levels off for a period of time. The significance of this will become more apparent when approach minimums are discussed.

Jeppesen also provides the glide slope crossing altitude for the middle marker (item 9). In this case, it is 480 feet MSL, or 191 feet above the runway touchdown zone. The Morse code identification for the middle marker is also shown.

The TCH is the altitude at which you cross the runway threshold when established on the glide slope centerline.

The missed approach point for the precision approach (item 10) is shown on both charts. Jeppesen also shows the approximate nonprecision missed approach point (item 11). If you are able to continue the approach to a landing, your **threshold crossing height** (TCH) is 54 feet above the runway (item 12). This assumes you continue to remain on the glide slope centerline. The glide slope angle of 3.00° is shown on the NOS chart. This information is listed in the timing block of the landing minimums section at the bottom of a Jeppesen chart.

The TDZE is the highest elevation for the first 3,000 feet of the landing runway.

Jeppesen provides the touchdown zone and airport elevation on the profile view, while NOS charts them on the airport sketch. The **touchdown zone elevation**, or TDZE (item 13), is the highest centerline altitude for the first 3,000 feet of the landing runway. It is used to compute all HAT figures. The airport elevation (item 14) is the official elevation of the airport. It is the highest point of an airport's usable runways. This value is used to compute all HAA elevations.

The distance between fixes (item 15) is shown on both Jeppesen and NOS charts. In this example, the distance from the outer marker to the middle marker is 4.1 n.m. Jeppesen also depicts, where appropriate, a distance to runway threshold (item 16). This value allows you to quickly determine the outer marker to be 4.5 n.m. from the runway threshold. A textual description of the missed approach procedure (item 17) is contained just below the profile view on Jeppesen charts and within the profile view on NOS charts.

Stepdown Fix and VDP

Many approaches incorporate one or more **stepdown fixes**. They are commonly used along approach segments to allow you to descend to a lower altitude as you overfly various obstacles. However, only one stepdown fix is normally permitted between the final approach fix and the missed approach point. If you cannot identify a stepdown fix, the altitude given just prior to the fix becomes your minimum altitude for the approach. [Figure 3-7]

When you cannot identify a stepdown fix, you must use the minimum altitude given just prior to the fix.

LANDING MINIMUMS

The landing minimum section is an extremely important part of an approach chart. Minimums have been established for each approach at a given airport and can vary from runway to runway. Factors which affect these minimums include the type of approach equipment installed, runway lighting, and obstructions in the approach or missed approach paths. Landing minimums are also affected by the equipment on board your

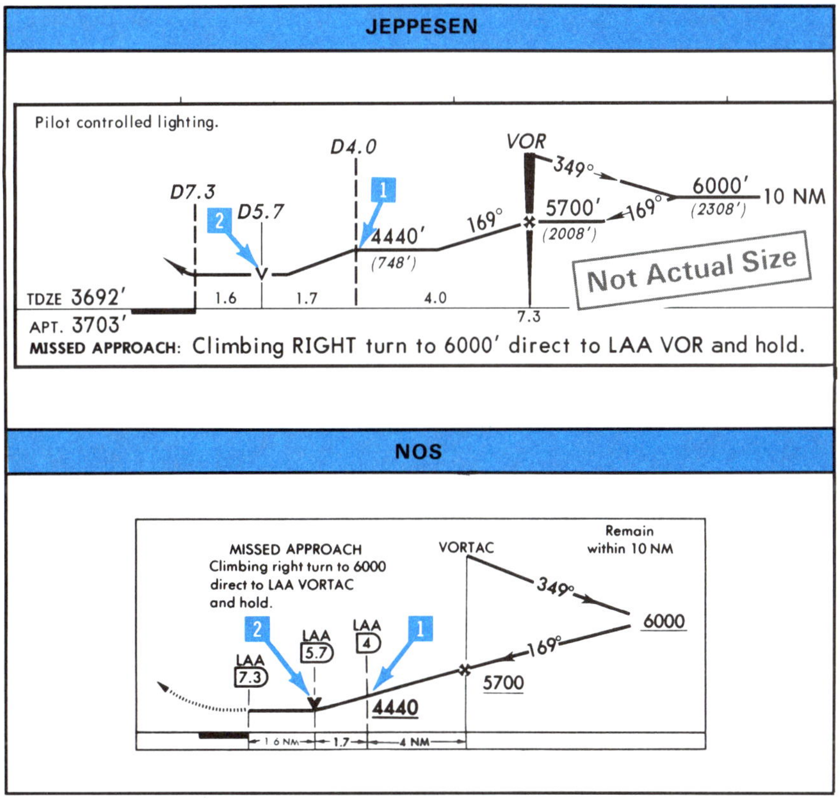

Figure 3-7. Prior to this stepdown fix (item 1), you must maintain a minimum altitude of 4,440 feet MSL. When you reach the 4.0 DME fix, you may descend to the appropriate minimum altitude for the approach. A visual descent point, or VDP (item 2), represents the point from which you can make a normal descent to a landing, assuming you have the runway in sight and you are starting from the minimum descent altitude. A descent from the minimum altitude should not be started prior to reaching the VDP.

aircraft and your approach speed. Approach minimums contain both minimum visibility and minimum altitude requirements. [Figure 3-8]

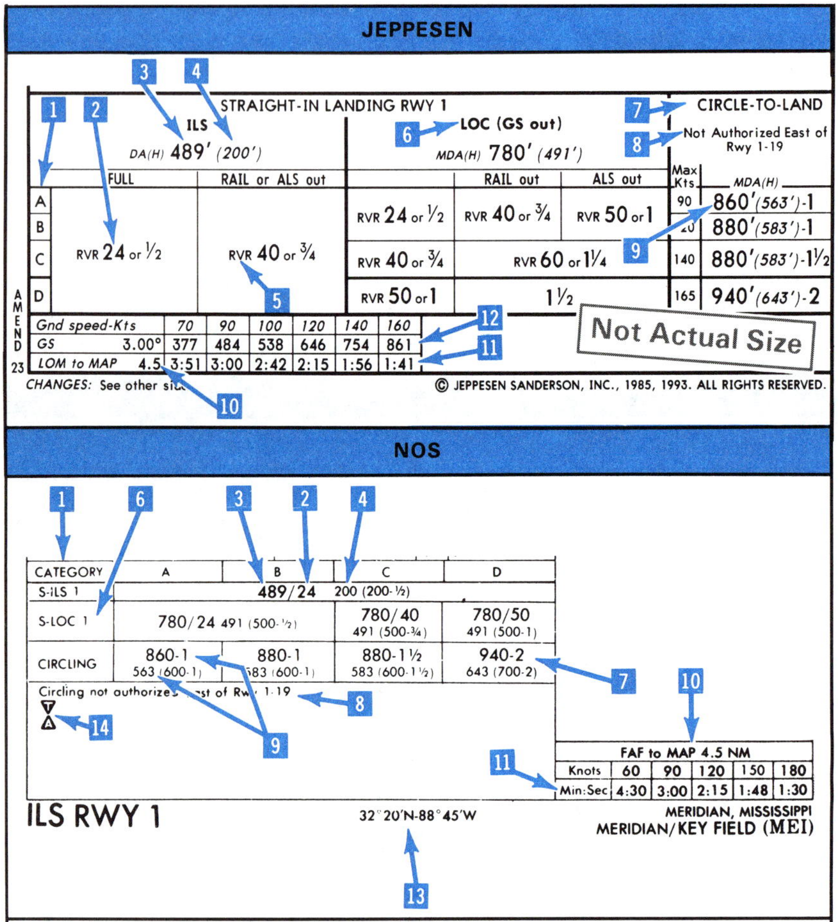

1. Aircraft Approach Categories
2. Visibility Requirement
3. Decision Altitude (Height) or DA(H)
4. Height Above Touchdown (HAT)
5. Inoperative Components Minimums
6. Localizer Approach Minimums
7. Circling Approach Minimums
8. Circling Restrictions
9. Circling Minimum Descent Altitude (Height) (MDA(H)), Height Above Airport (HAA) and Visibility
10. Location of MAP
11. Time from FAF to MAP
12. Glide Slope Angle and Rate of Descent
13. Latitude and Longitude
14. Nonstandard Takeoff and Alternate Minimums

Figure 3-8. This is an excerpt of the minimums section for both Jeppesen and NOS charts. Some of the general features and callouts are discussed further in the accompanying text.

Aircraft Approach Categories

Each aircraft is placed into an approach category (item 1) based on its computed approach speed. This speed equals 130% of the aircraft's power-off stall speed in the landing configuration at the maximum certificated landing weight ($1.3V_{S0}$). For example, if the stalling speed of your aircraft is 65 knots, its computed approach speed is 85 knots. The following table defines the range of the various aircraft approach categories based on approach speed.

Aircraft approach categories are based on $1.3V_{S0}$.

Approach Category	Approach Speed
A	Up to 90 knots
B	91 to 120 knots
C	121 to 140 knots
D	141 to 165 knots
E	Above 165 knots

Although Category E is shown, it usually is not included on approach charts, since it applies only to certain military aircraft. In this case, Category D applies to all civil aircraft with a computed approach speed of 141 knots or more.

Once you have determined your aircraft's approach category, you can use the landing minimums section of the approach chart to determine your visibility and descent minimums for that approach. During a circling approach, you often have to maneuver at a higher speed than normal, or ATC may request a higher speed to expedite traffic flow. When an increase in speed exceeds your normal approach category limit, you should use the minimums that apply to the higher category. As an aid to help you determine the correct category for a circling approach, Jeppesen charts include the maximum speeds for each aircraft approach category in the circle-to-land minimums.

If you fly a Category B aircraft at an airspeed that falls into Category C, you should use Category C minimums.

Visibility Requirements

To continue your approach past the missed approach point, the visibility must be equal to or better than that listed on the approach chart. In addition, FAR Part 91 specifies that you must be able to clearly identify at least one of the following visual references for the intended runway.

Both a minimum visibility and specific references must be visible for you to continue past the MAP.

1. Approach light system
2. Threshold
3. Threshold markings
4. Threshold lights
5. Runway end identifier lights

6. Visual approach slope indicator
7. Touchdown zone or touchdown zone markings
8. Touchdown zone lights
9. Runway or runway markings
10. Runway lights

Visibility is listed on approach charts in either statute miles or hundreds of feet (item 2). When it is expressed in miles or fractions of miles, it is usually a prevailing visibility that is reported by an accredited observer (tower or weather personnel). When the visibility is expressed in hundreds of feet, it is determined through the use of runway visual range (RVR) equipment. On the chart, an RVR listing of 24 means the minimum visibility for the approach is 2,400 feet. The visibility minimum for the full ILS runway 1 approach at Key is RVR 24 (2,400 feet) or, if RVR is not reported, one-half statute mile. Jeppesen provides both the RVR and the equivalent prevailing visibility in the approach minimums section. NOS provides a separate table in the front of each volume to convert RVR to miles. The ceiling and visibility figures in parentheses on NOS charts are used for planning purposes by the military.

Minimum Descent Requirements

The descent limits shown in the minimums section indicate how low you are permitted to descend in an attempt to identify the required visual cues. This can either be a decision height or a minimum descent altitude depending on the type of approach. For precision approaches, the height where you must make the decision to continue the approach or execute a missed approach on NOS charts is referred to as the MSL **decision height**, or DH (item 3). Jeppesen refers to this MSL altitude as the **decision altitude (height)**, or DA(H). The figure in parentheses on Jeppesen charts (item 4) is the height above touchdown (HAT). On NOS charts this same figure is shown after the visibility. The numbers in parentheses on NOS charts are used by the military. For the ILS runway 1 approach at Key, the lowest MSL altitude you can descend to is 489 feet. This is 200 feet above the touchdown zone elevation for the runway (HAT).

When the ILS glide slope is inoperative, the procedure becomes a nonprecision , localizer approach (item 6). Nonprecision approaches use a **minimum descent altitude (height)** MDA(H). Upon crossing the final approach fix or stepdown fix, as appropriate, you descend to the MDA. You can remain at this altitude until you reach the missed approach point. However, you may not descend below the MDA unless the visual requirements can be met. You also must continuously be in a position from which a descent to a landing on the intended runway can be made at a normal descent rate using normal maneuvers. In this case, the missed approach point is determined by timing from the final approach fix.

For timed approaches, both Jeppesen and NOS identify the MAP in the conversion table located at the bottom of the chart (item 10). The table provides various elapsed times to the MAP based on the aircraft's groundspeed. If your groundspeed on the localizer approach to runway 1 is 90 knots, the elapsed time from the LOM to the MAP is three minutes. The absence of timing information means the MAP cannot be determined by timing and a timed approach is not authorized.

The conversion table on Jeppesen charts also provides the glide slope angle and a recommended rate of descent to maintain the glide slope during an ILS approach (item 12). For example, if your groundspeed is 90 knots, your rate of descent should be 484 feet per minute to follow the glide slope. This information is also provided by NOS in a separate table contained in each approach chart volume.

One feature of many instrument approaches is the ability to fly an approach to one runway, then land on another. This procedure is called a **circling approach** (item 7), and separate circle-to-land minimums are published for this procedure (item 9). You must comply with any restrictions applicable to the procedure. For example, at Meridian, it is not permissible to circle to the east of the runway 1-19 centerline (item 8). Since a circling approach does not terminate at a specific runway, the number in parentheses following each MDA is an HAA elevation.

Another procedure listed on some charts is a variation of the circling maneuver. In this procedure, you will be cleared for an approach to one runway with a clearance to land on a parallel runway. This is called a **sidestep maneuver**. The minimums for this procedure are usually higher than the minimums for the straight-in runway, but lower than the circling minimums. When executing a sidestep maneuver, you are expected to begin the sidestep as soon as possible after sighting the runway environment.

When flying a sidestep maneuver, you are expected to begin the sidestep maneuver as soon as possible after sighting the runway.

Inoperative Components

Landing minimums usually increase when a required component or visual aid becomes inoperative. Jeppesen provides inoperative component minimums directly on each chart (item 5). For example, if the approach light system (ALS) is inoperative during the ILS runway 1 approach, the same decision altitude (height) of 489 feet applies, but the visibility minimums increase to RVR 40 or three-quarters of a mile. If more than one component is inoperative, each minimum is raised to the highest minimum required by any single component that is inoperative. ILS glide slope inoperative minimums are published on instrument approach charts as localizer minimums.

NOS provides a separate table in each bound volume. You must consult this table to determine the appropriate corrections for inoperative components or visual aids. [Figure 3-9]

Regulations permit you to make substitutions for certain components when the component is inoperative or is not utilized during an approach. For example, on an ILS approach, a compass locator or precision radar may be substituted for the outer marker. This allows you to fly the approach without increasing the landing minimums. When DME, VOR, or NDB fixes are authorized in the approach procedure, or where surveillance radar is available, their use also may be substituted for the outer marker. Effective 15 October 1992, the FAA removed the penalty for inoperative middle markers from all U.S. ILS approach procedures. The MM out column on Jeppesen charts has been removed as charts are revised for other reasons. In the meantime, the MM out penalty can be ignored.

OTHER APPROACH PROCEDURES

The most common approaches you will fly are the ILS, localizer, VOR, and NDB. However, you must also be familiar with some of the other approaches, such as LDA, SDF, RNAV, and MLS. LDA and SDF are similar to an ILS localizer approach, except the approach course is not aligned with the runway and the course width may not be as accurate as a localizer. If your aircraft is equipped to fly an ILS, you also can fly an LDA or SDF approach.

A **localizer-type directional aid** (LDA) can be thought of as a localizer approach system that is not aligned with the runway centerline. Some LDA approaches have an electronic glide slope, although this is not a requirement for the system. Straight-in landing minimums may be available if the final approach course is aligned to within 30° of the runway centerline. The **simplified directional facility** (SDF) approach system offers less accuracy than the LDA. The typical ILS or LDA localizer is between 3° and 6° wide. SDF localizers are either 6° or 12° wide. In addition, SDF systems do not incorporate an electronic glide slope. Both RNAV and MLS approaches require your aircraft to have additional equipment. An **area navigation** (RNAV) approach is very similar to a VOR approach; however, instead of the approach being oriented around fixed navaids, the RNAV computer allows you to use electronically designated waypoints for the approach procedure. The **global positioning system** (GPS), which is covered in Chapter 4, may also be used for approaches if required procedures are followed.

The MLS three-letter identifier is preceded by the letter "M."

The **microwave landing system** (MLS) approach was designed as a precision approach similar to ILS. The Morse code for the letter "M" precedes the three-letter Morse code identifier for an MLS to distinguish it from an ILS which uses the letter "I." MLS provides an approach azimuth, elevation, and range (DME). A back azimuth may be included for guidance during the missed approach and departure. The microwave landing system is less susceptible to interference and is,

INOP COMPONENTS

91262

INOPERATIVE COMPONENTS OR VISUAL AIDS TABLE

Landing minimums published on instrument approach procedure charts are based upon full operation of all components and visual aids associated with the particular instrument approach chart being used. Higher minimums are required with inoperative components or visual aids as indicated below. If more than one component is inoperative, each minimum is raised to the highest minimum required by any single component that is inoperative. ILS glide slope inoperative minimums are published on instrument approach charts as localizer minimums. This table may be amended by notes on the approach chart. Such notes apply only to the particular approach category(ies) as stated. See legend page for description of components indicated below.

(1) ILS, MLS, and PAR

Inoperative Component or Aid	Approach Category	Increase Visibility
ALSF 1 & 2, MALSR, & SSALR	ABCD	¼ mile

(2) ILS with visibility minimum of 1,800 RVR.

1 ALSF 1 & 2, MALSR, & SSALR	ABCD	To 4000 RVR
2 TDZL RCLS	ABCD	To 2400 RVR
RVR	ABCD	To ½ mile

(3) VOR, VOR/DME, VORTAC, VOR (TAC), VOR/DME (TAC), LOC, LOC/DME, LDA, LDA/DME, SDF, SDF/DME, RNAV, and ASR

Inoperative Visual Aid	Approach Category	Increase Visibility
ALSF 1 & 2, MALSR, & SSALR	ABCD	½ mile
SSALS, MALS, & ODALS	ABC	¼ mile

(4) NDB

ALSF 1 & 2, MALSR & SSALR	C	½ mile
	ABD	¼ mile
MALS, SSALS, ODALS	ABC	¼ mile

CORRECTIONS, COMMENTS AND/OR PROCUREMENT

FOR CHARTING ERRORS:
Contact National Ocean Service
NOAA, N/CG31
6010 Executive Blvd.
Rockville, MD. 20852
Telephone Toll-Free 800-626-3677

FOR CHANGES, ADDITIONS, OR RECOMMENDATIONS ON PROCEDURAL ASPECTS:
Contact Federal Aviation Administration, ATM 613
800 Independence Avenue, S.W.
Washington, D.C. 20591
Telephone (202) 267-9288

TO PURCHASE CHARTS CONTACT:
National Ocean Service
NOAA, N/CG33
Distribution Branch
Riverdale, MD. 20737
Telephone (301) 436-6993

Requests for the creation or revisions to Airport Diagrams should be in accordance with FAA Order 7910.4B.

INOP COMPONENTS

Figure 3-9. Using NOS approach charts, you consult this table to adjust the approach minimums when any component of the system is inoperative. For example, if both approach lights (item 1) and the touchdown zone lights (item 2) are inoperative, the visibility must be increased to 4,000 RVR.

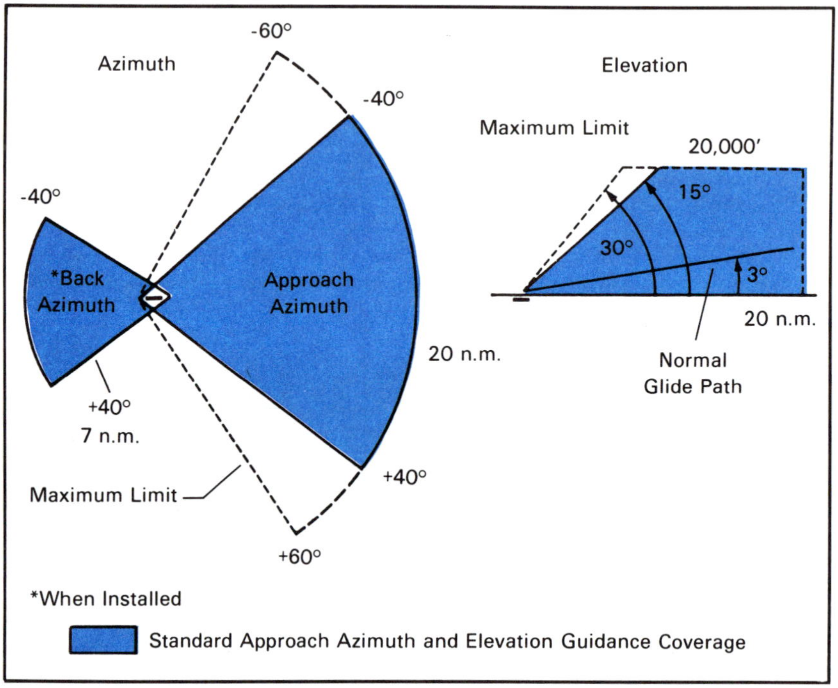

Figure 3-10. The standard and maximum azimuth and elevation coverage volumes for MLS are shown in this illustration. Within this coverage area, MLS can accept multiple final approach paths, including curved paths, and multiple glide slope angles.

therefore, well suited to areas in which the installation of ILS is difficult or impossible. [Figure 3-10]

AIRPORT CHART

Both Jeppesen and NOS include a diagram of each airport for which they publish an instrument approach procedure. NOS places an airport sketch in the lower right-hand corner of each approach chart. They also provide a full-page airport diagram for selected airports to assist the movement of ground traffic where complex runway and taxiway configurations exist. Jeppesen uses an entire page, called the airport chart, for each airport. This chart is usually located on the reverse side of the first approach chart for a given airport. At larger airports, Jeppesen provides even more detail than the standard airport chart. [Figure 3-11]

HEADING SECTION

The heading section of the Jeppesen airport chart identifies the airport, its location, elevation, magnetic variation, and outbound communication frequencies. You will notice many similarities between this section and the heading section of a Jeppesen approach chart. [Figure 3-12]

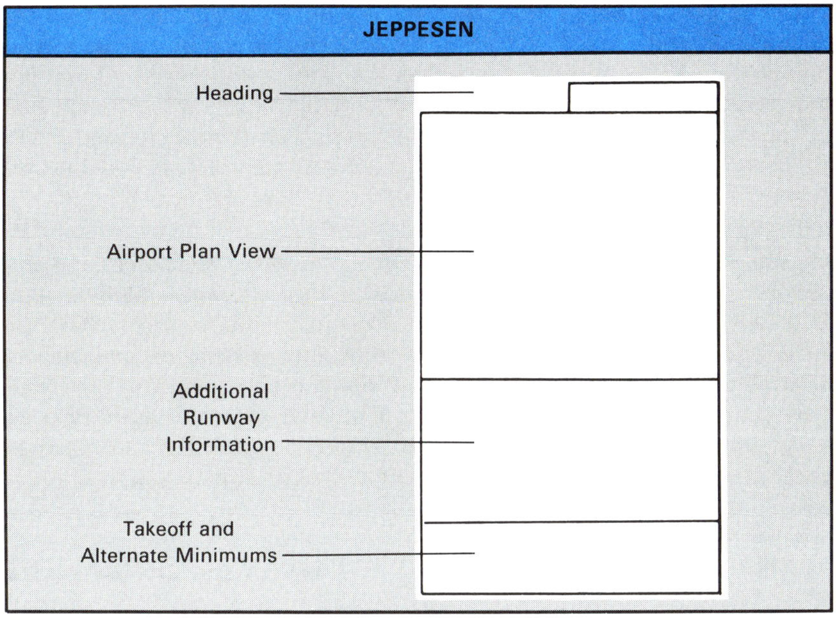

Figure 3-11. Like Jeppesen's approach chart, the airport chart is divided into several sections. Much of the information contained on the Jeppesen airport chart is located on the NOS approach chart or the airport sketch.

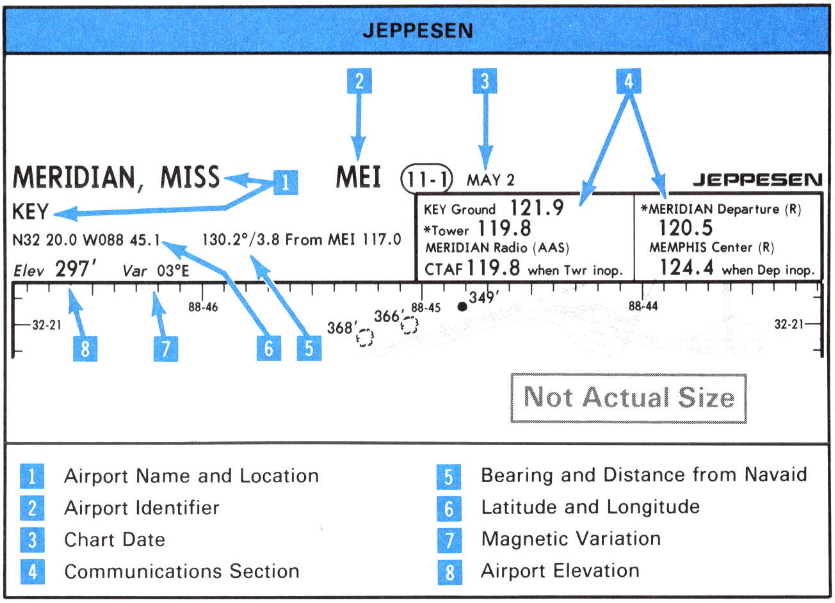

Figure 3-12. Here is an example of the Jeppesen heading section for Key Field at Meridian, Mississippi. Except as noted in the following text, most of the information contained on this chart is listed on an NOS approach chart.

The communications section (item 4) lists the frequencies in the order they are normally used when you depart the airport. In this case, you first contact Key Ground Control, then Key Tower, followed by Meridian Departure Control. Alternate communication frequencies are provided for operations when the tower and approach control are closed. NOS charts do not list these alternate frequencies, nor do they list the frequency for departure control.

Meridian's geographic location is given in two formats. The bearing and distance from a VORTAC or VOR/DME within 40 n.m. is shown (item 5). These can be used by DME- or RNAV-equipped aircraft to locate the airport. The latitude and longitude coordinates of the official airport location or the airport reference point also are provided (item 6). NOS provides these coordinates at the bottom of each approach chart. Bearing and distance information is usually provided on the NOS airport sketch. Local magnetic variation (item 7) for the airport is included on the Jeppesen airport chart and only on the full-page NOS diagram.

PLAN VIEW AND ADDITIONAL RUNWAY INFORMATION

The plan view portrays an overhead view of the airport. Its purpose is to provide you with information about the airport, such as its runways and lighting systems. Jeppesen lists some of this information below the airport diagram. This allows a large amount of information to be presented in an uncluttered fashion. [Figure 3-13]

Jeppesen provides latitude/longitude tick marks (item 1) for aircraft that have advanced navigation equipment. This allows you to update your position more accurately on the ramp before flight and improves equipment accuracy during flight. NOS includes these tick marks on the large airport diagrams.

Both Jeppesen and NOS provide a representative selection of reference points appropriate to the airport (item 2). The elevation of the depicted reference points is above mean sea level.

Jeppesen's airport charts usually lists the elevation of each end of the runway (item 3). This allows you to estimate the average runway slope or gradient. Runway gradient information appears on the NOS airport diagram/sketch when it exceeds 0.3%.

The location and elevation of the airport's beacon (item 4) may be shown, along with the location and elevation of the airport's control tower (item 5). Runway numbers (item 6) are located at the end of each runway. Jeppesen also lists the actual magnetic direction of the runway. This information allows you to make a final cross-check of the magnetic compass and heading indicator prior to flight.

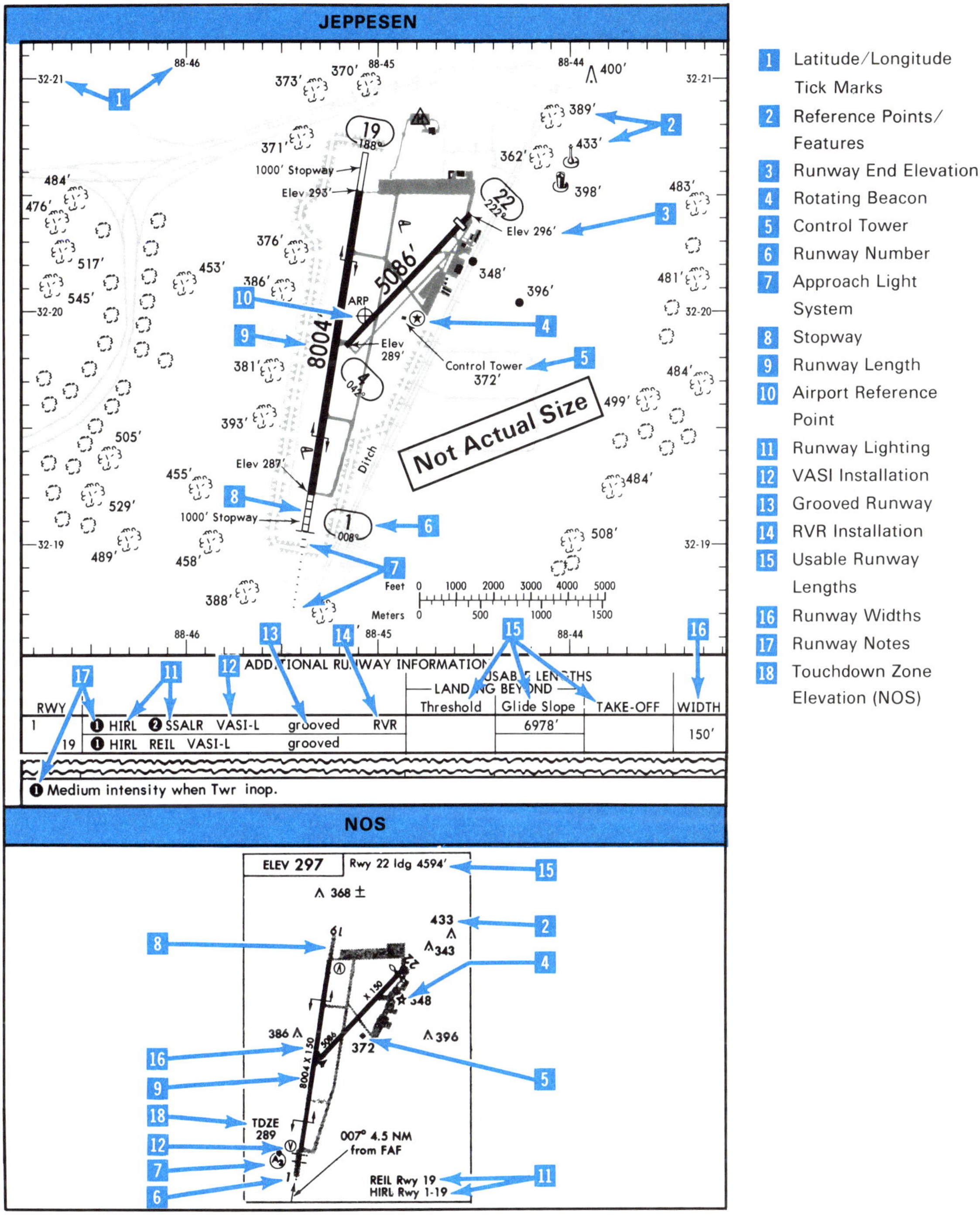

Figure 3-13. The callouts identify many of the features found on the plan view of the Jeppesen and NOS airport diagram/sketch.

Approach lighting systems (item 7) are shown on both charts by an approach lighting system diagram. NOS also uses a coded symbol placed adjacent to the runway to indicate the type of lighting system. Jeppesen specifies the system type in the additional runway information section. An illustration of each approach lighting system is included in the legend pages for Jeppesen charts and in each NOS volume.

Both Jeppesen and NOS show stopways (item 8) associated with a runway. Jeppesen includes the length of the stopway. The runway length (item 9) also is shown on both charts. The **airport reference point**, or ARP, (item 10) is where the official latitude and longitude coordinates are derived. The location of this point is shown on Jeppesen charts.

Runway and approach lighting systems (item 11), as well as VASI installations (item 12), are shown on both charts. Jeppesen includes this information in the additional runway information section. This section contains other data, such as grooved runways (item 13), RVR equipment (item 14), usable runway lengths (item 15), and runway widths (item 16). Some of this information also is shown on NOS charts or contained in the *Airport/Facility Directory*. The last item, notes applicable to runways (item 17), is an additional feature of Jeppesen charts. Here, you will find any notes of importance about the runways. For example, when Key Tower is closed, medium intensity runway lights are the only runway lights available on runways 1 and 19.

TAKEOFF AND ALTERNATE MINIMUMS

The FAA has established rules which provide takeoff minimums for large aircraft and all commercially operated aircraft. As a Part 91 operator, you are not required to comply with published IFR takeoff minimums. However, you should realize that these minimums have been established for the safe operation of commercial aircraft. These aircraft are piloted by highly trained and skilled individuals, and are usually operated by two pilots. Although legal, it is unwise to initiate a takeoff in weather conditions that would keep a commercial flight grounded.

The FAA has also established rules for selecting an alternate airport for an IFR flight. If the forecast weather at your estimated time of arrival, plus or minus one hour, indicates a ceiling of less than 2,000 feet or a visibility of less than 3 miles, you must list an alternate airport on your IFR flight plan. To qualify as an alternate, the airport you select and the forecast weather must meet certain conditions. Although the rules for takeoff and alternate minimums will be explained in greater detail in Chapter 5, you should be familiar with the way related information is presented on IFR charts. [Figure 3-14]

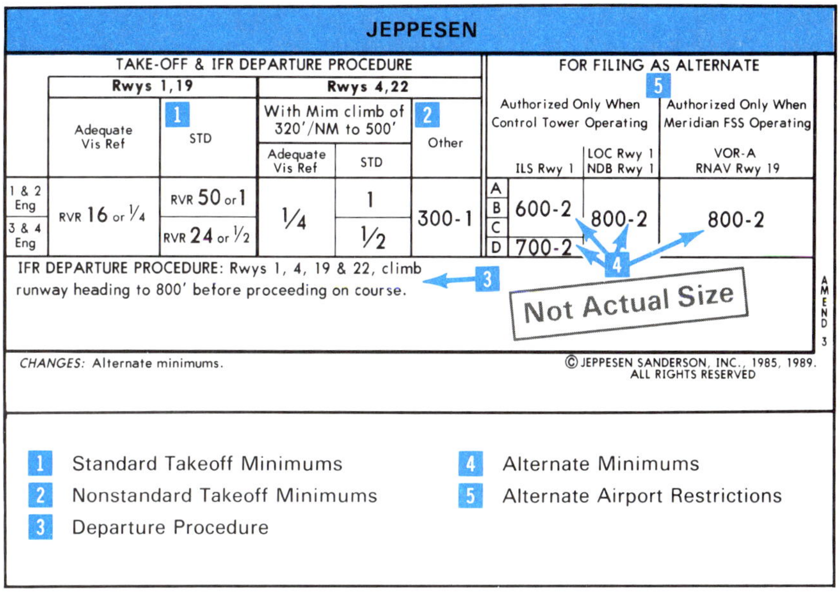

Figure 3-14. This section of the Jeppesen airport chart can be thought of as three separate sections. The takeoff minimums, departure procedures, and alternate minimums section. Some of the elements listed here are discussed further in the accompanying text.

Jeppesen lists the takeoff minimums for each airport, whether they are standard or nonstandard minimums. NOS uses a ▼ symbol on an approach chart to indicate when nonstandard takeoff minimums apply and/or IFR departure procedures are published. These minimums are listed in separate sections of the NOS chart booklet. Standard takeoff minimums are based on the number of engines on the aircraft. If your aircraft is equipped with one or two engines, the standard takeoff minimum is a visibility of one statute mile. For aircraft with three or more engines, the requirement is one-half statute mile.

Standard takeoff minimums for aircraft with one or two engines is a visibility of one statute mile.

Standard takeoff minimums (item 1) are applicable to runways 1 and 19 at this airport. Since runway 1 is equipped with RVR equipment, RVR minimums also are listed. The minimums listed under adequate visual reference (Vis Ref) are applicable to commercial operators who have been approved for reduced visibility takeoffs.

Nonstandard takeoff minimums (item 2) are applicable to aircraft departing runways 4 and 22. Remember, for NOS charts, you must consult a separate section of the booklet to determine these minimums. Notice that standard minimums still apply if you can climb to 500 feet MSL at a rate of at least 320 feet per nautical mile. If you cannot, the more restrictive minimums of a 300-foot ceiling and one statute mile visibility must be used.

IFR departure procedures are designed to keep you clear of obstructions in the departure path.

IFR departure procedures (item 3) are published only when they are necessary to keep aircraft clear of obstructions. In this case, you must climb on the runway heading to 800 feet before proceeding on course, unless cleared otherwise. This procedure applies to takeoffs from all runways at this airport. NOS uses the ▼ symbol to alert you to check the nonstandard takeoff minimums section of the approach booklet for this information.

Standard alternate minimums for a precision approach are 600 and 2. For a nonprecision approach, the minimums are 800 and 2.

Jeppesen lists the alternate minimums for each airport, whether standard or nonstandard minimums apply (item 4). NOS uses an ▲ symbol on the approach chart to indicate nonstandard minimums. You must consult the alternate airport minimums listed in the NOS booklet. Jeppesen prints the applicable notes (item 5) for alternate minimums on the airport chart. For NOS, these notes are included in the alternate minimums section.

The standard alternate minimums for a precision approach are a ceiling of 600 feet and a visibility of two statute miles. For nonprecision approaches, they are an 800-foot ceiling and two statute miles visibility. For Key Field, standard alternate minimums apply for the nonprecision approach and for Category A and B aircraft for the precision approach. Nonstandard minimums apply for Category C and D aircraft. In addition, you can use this airport as an alternate on your flight plan only when the control tower is operating.

CHECKLIST

After studying this section, you should have a basic understanding of:

- ✓ **Approach segments** — What the various approach segments are, where each begins and ends, and the importance of each segment to the approach.

- ✓ **Approach chart** — The information contained on approach charts and where it is located.

- ✓ **Stepdown fix and VDP** — What these two special points are and how they are used during an approach.

- ✓ **Landing minimums** — How to determine the landing minimums for an approach and the importance of aircraft categories, visibility, altitudes, and inoperative components.

- ✓ **LDA and SDF** — The difference between ILS, LDA, and SDF approach facilities.

✓ **Airport chart —** The information contained on airport charts and where it is located.

✓ **Takeoff and alternate minimums —** What the standard takeoff and alternate minimums are and where nonstandard minimums can be found.

SECTION B

ENROUTE AND AREA CHARTS

This section presents an overview of the information and symbology contained on enroute and area charts. Enroute charts are primarily used for IFR cross-country flights and contain essential information that helps you keep track of your position, as well as maintain obstacle clearance and navigation signal reception. Area charts portray more detail than enroute charts and are primarily used during the transition to or from the enroute structure. This section compares and analyzes the symbology of both Jeppesen and NOS charts. In some cases, the symbols appear in the margin with an explanation in the accompanying text.

ENROUTE CHARTS

Airways below 18,000 feet are called Victor airways. Those at and above 18,000 feet are Jet routes.

In the United States, 18,000 feet MSL is the altitude that separates the Federal airway system from Class A airspace. Therefore, it is a convenient place to establish the division between the low and high altitude airway structures. Airways located below 18,000 feet MSL are depicted on **low altitude enroute charts** and are referred to as **Victor airways.** Those at and above 18,000 feet MSL up to FL 450 are shown on **high altitude enroute charts** and are called **Jet routes**. This section concentrates on low altitude enroute charts, since your initial IFR experience probably will be within this airspace. These charts display vast amounts of information about the airways, routes of flight, radio facilities used for enroute navigation, identification of instrument fixes, airport information, communications available, altitudes that guarantee proper reception of navigation signals, and terrain clearance. If you have an enroute chart for your local area, you may wish to locate each symbol as it is presented. High altitude enroute chart symbology is not presented, since these charts are similar to low altitude charts.

FRONT PANEL

Both Jeppesen and NOS portray the area covered by each enroute chart. Generally, both systems present the same information; however, the symbology and coverage areas vary. [Figure 3-15]

The front panel contains valuable information concerning the use of the chart. The effective date at the top tells you when the old chart becomes obsolete for navigation and the information on the new chart becomes effective. A bold outline is used to indicate the chart you are holding, as

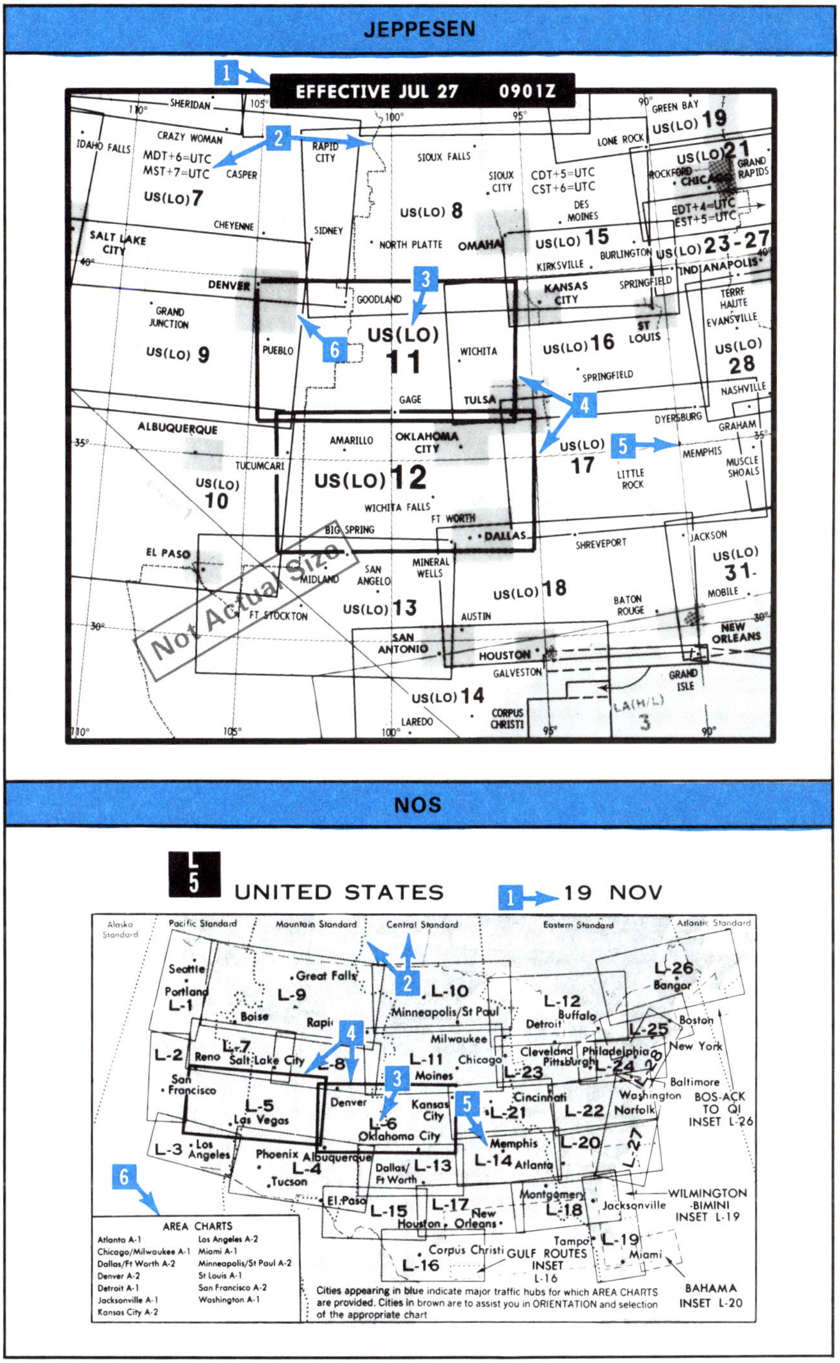

1. Effective Date
2. Time Zone and Conversion Factor
3. Enroute Chart Name
4. Chart Coverage
5. City Reference
6. Area Charts

Figure 3-15. These are excerpts from the front panels of both Jeppesen and NOS enroute charts. Some elements shown here are addressed more fully in the accompanying text.

well as its coverage area. The locations of major cities are included to help you find the appropriate chart quickly.

Jeppesen uses blue shading to show where area charts are available. NOS provides a list of the area charts in a box in the lower part of the diagram. Time zone boundaries are included on both charts. NOS uses a series of dots, while Jeppesen uses a series of "T's." Jeppesen includes the formula to convert both standard and daylight time to Coordinated Universal Time on the front panel, while NOS shows this information on the chart itself.

Both NOS and Jeppesen provide legends to help you interpret the symbols on enroute charts. Jeppesen uses a comprehensive legend included in a separate introduction section. The NOS legend is located on each chart. You should maintain a working knowledge of chart symbology and review the appropriate legend periodically for updates and improvements. Now let's look at some of the individual symbols. [Figure 3-16]

JEPPESEN NOS

NAVIGATION AIDS

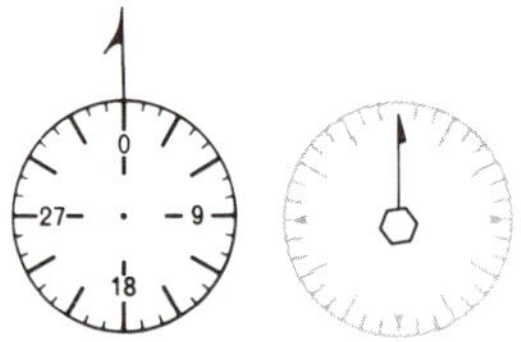

A VOR is represented as a compass rose oriented to magnetic north. The center of the compass rose shows the location of the VOR station. Jeppesen charts use the symbol itself to indicate a VOR facility, while NOS uses a hexagon symbol.

Jeppesen uses a serrated circle to portray both TACAN and DME facilities. NOS uses a symbol resembling a triangle with the points cut off to show a TACAN location.

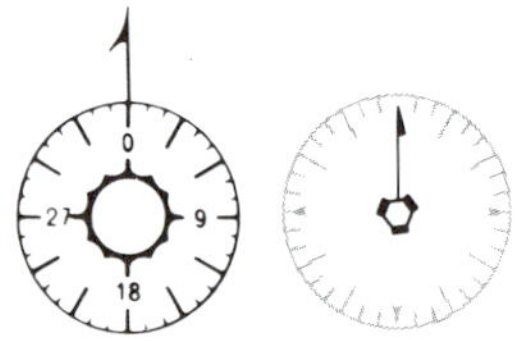

When a VOR and a TACAN are collocated, the facility is called a VORTAC. A collocated VOR and DME is called a VOR/DME. Since a VORTAC and a VOR/DME are functionally identical for civil aircraft, Jeppesen uses a single symbol to indicate both types of facilities. NOS uses the symbol shown in the margin for a VORTAC, but changes the center of the compass rose to for a VOR/DME.

The nondirectional radio beacon (NDB) is a low frequency navaid used for instrument approaches, as well as for enroute navigation. On both charts, a nondirectional radio beacon is shown as a circle of dots. The arrow extending out of the top is aligned with magnetic north, which makes it possible to measure the magnetic bearing with a plotter. This symbol is also similar to the ones used to depict compass locators. On Jeppesen charts, they are shown only when the facility provides an enroute function or TWEB information.

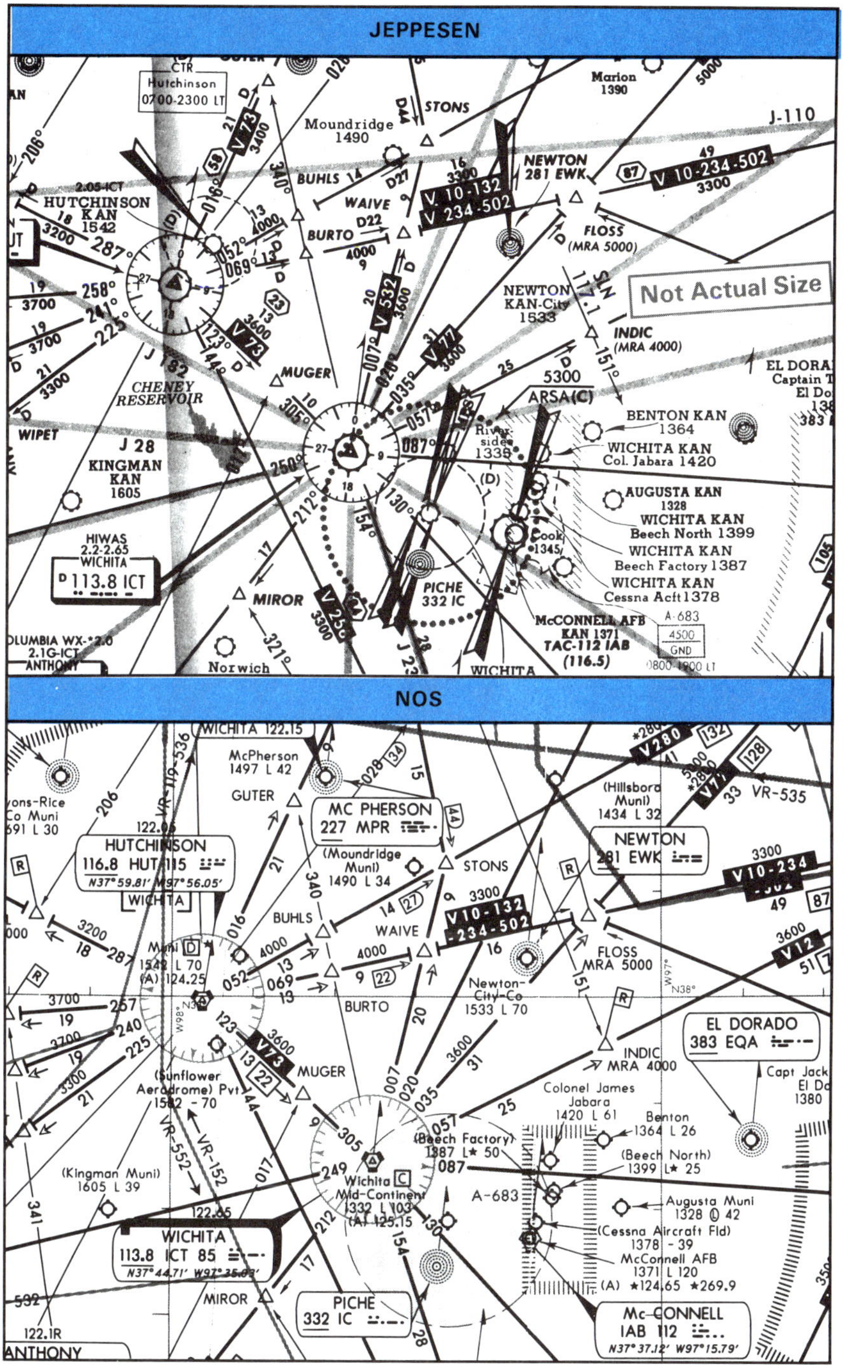

Figure 3-16. Here are sample excerpts from Jeppesen and NOS enroute charts for the same area. Many of the symbols discussed in the following text appear on these excerpts.

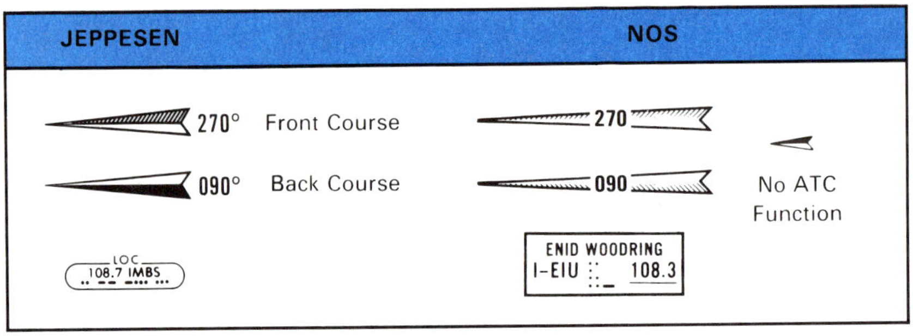

Figure 3-17. Besides showing front course symbols, both charts depict back course localizers, where appropriate. Where the localizer course serves an enroute (ATC) function, the facility identifier, frequency, and inbound course are shown.

Jeppesen and NOS show the availability of ILS, MLS, LDA, and SDF approaches with a localizer symbol. NOS uses two types of symbols to indicate the approach courses. The large symbol is used to show locations where the localizer course provides an ATC function, such as identifying a fix or intersection. [Figure 3-17]

JEPPESEN NOS

Selected commercial broadcast stations are also included on Jeppesen enroute charts. These stations may be used as VFR navigation aids by aircraft equipped with ADF receivers. Since they do not transmit a continuous identification, they are considered unreliable for IFR navigation.

The information in a VOR or VORTAC facility box provides you with the facility's name, frequency, and identifier (both three-letter and Morse code). The letter "D" preceding the frequency in the Jeppesen excerpt indicates DME information is available. The channel number in the NOS box indicates the availability of DME to civil aircraft and is used by military aircraft. The facility box for an NDB includes the name, frequency, and identification code. Jeppesen often groups NDB information with an adjacent airport if it has the same name.

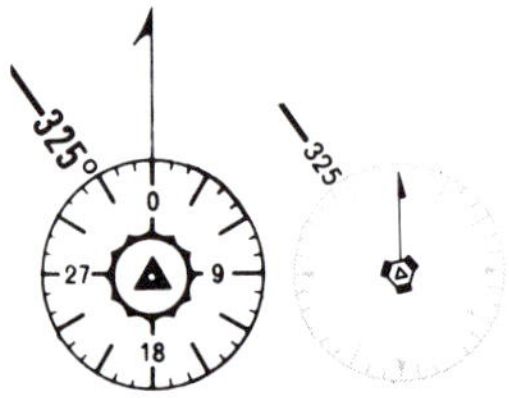

VICTOR AIRWAYS

The designated routes between VOR or VORTAC stations are marked as Victor airways. These airways are formed by radials extending out from the facilities, such as the 325° radial from this VORTAC.

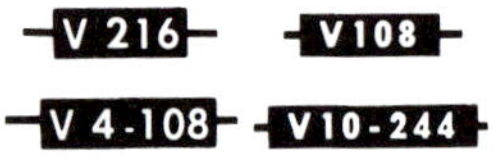

Victor airways are identified by a number that indicates their general direction. Even-numbered airways usually are oriented in an east-west direction, while odd-numbered airways are generally oriented north-south. When more than one airway shares a common route segment, all identification numbers are shown.

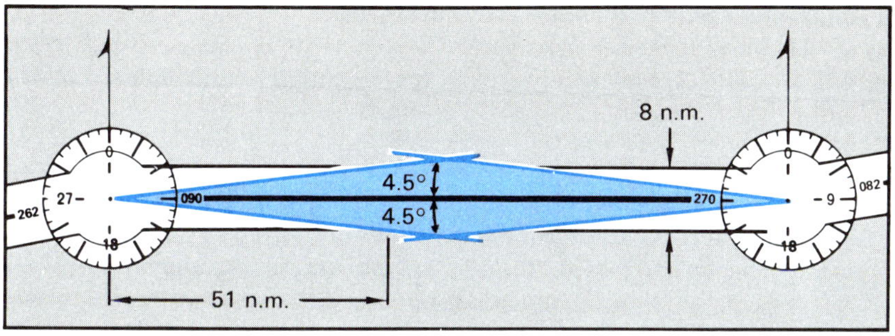

Figure 3-18. When the airway extends beyond 51 n.m. from the nearest navaid, it includes the airspace between lines diverging at angles of 4.5° from the center of each navaid. These lines extend until they intersect the diverging lines from the other navigational aid.

The width of an airway is normally four nautical miles on each side of its centerline. This provides a total airway width of eight nautical miles. When the total length of an airway exceeds 102 n.m., additional airspace is allocated. [Figure 3-18]

The standard airway width is eight nautical miles.

All distances on IFR enroute charts are expressed in nautical miles. On Jeppesen charts, a number enclosed within an outlined box indicates total mileage between navaids. NOS uses the outlined box to show total mileage between navaids and/or compulsory reporting points. A number that is not enclosed in an outlined box indicates the mileage between intersections, navigation aids, or mileage break points.

JEPPESEN NOS

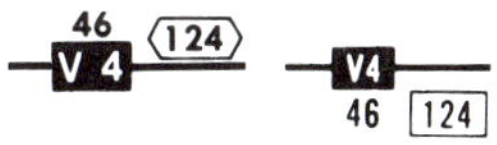

A **mileage break point** is shown on a chart by a small "x" on the airway. Generally, this symbol indicates a point on the airway where the course changes direction and where an intersection is not designated.

Intersections are checkpoints along an airway that provide a means for you and ATC to check the progress of a flight. They are often located at points where the airway turns or where you need a positive means of establishing your position. Intersections are given five-letter names. The actual location of an intersection may be based on two VOR radials, DME, or other navaids, such as a localizer or a bearing to an NDB. Arrows are placed next to the intersection with the stems pointing toward the navaids that form the intersection. In the excerpt shown, the intersection is formed by the airway and the 355° radial of the DDC VOR.

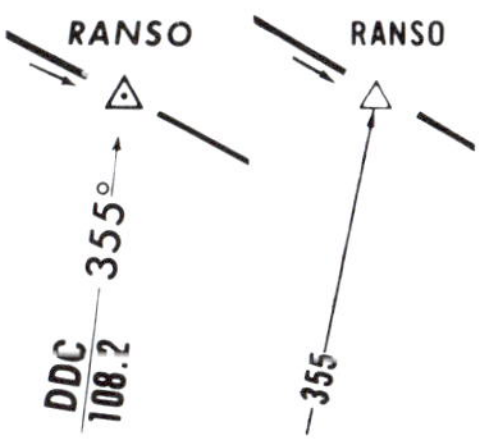

On Jeppesen charts, an intersection that can be defined by DME is depicted by an arrow with the letter "D" above it, while NOS uses an open arrow. If it is the first intersection from the navaid, the mileage is found along the airway as a standard mileage number.

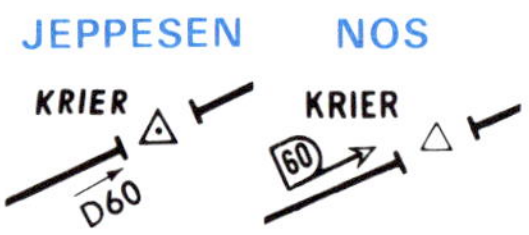

When not otherwise obvious, DME mileage to an intersection follows the letter "D" on Jeppesen charts. On NOS charts, it is enclosed within the letter "D." This number represents the cumulative distance from the navaid to the fix.

Intersections are designated as either compulsory or noncompulsory reporting points. A **compulsory reporting point** is identified by a solid triangle. In a nonradar environment, you are required to make a position report when you pass over this point. **Noncompulsory reporting points** are identified by open triangles, and position reports are not required, unless requested by ATC.

An MEA guarantees obstruction clearance and navigation signal coverage.

The **minimum enroute altitude** (MEA) is usually the lowest published altitude between radio fixes that guarantees adequate obstruction clearance (2,000 feet in mountainous areas and 1,000 feet elsewhere) and navigation signal reception. It is normally the lowest altitude you use during a cross-country flight. Usually you can expect adequate communications at the MEA, although it is not guaranteed. The MEA for the route in this excerpt is 5,500 feet MSL.

A **minimum obstruction clearance altitude** (MOCA) is shown for some route segments. On Jeppesen charts, it is identified by the letter "T" following the altitude. On NOS charts, an asterisk precedes the altitude. In the previous excerpt, the MOCA for the route segment is 4,000 feet MSL. The major difference between an MEA and a MOCA is that the MOCA ensures a reliable navigation signal only within 22 n.m. of the facility; conversely, the MEA guarantees reliable navigation signals throughout the entire segment.

A MOCA does not guarantee you will receive a reliable navigation signal if you are more than 22 n.m. from the facility.

When you are more than 22 n.m. from the station and are at the MOCA, adequate terrain clearance is assured, but you may not be able to receive the facility and navigate along the airway. Under certain conditions, ATC may issue the MOCA as an assigned altitude, but only when you are within 22 n.m. of the facility.

You must determine the minimum safe altitude for off-route operations. Normally, this requires 1,000 feet of obstacle clearance within 4 n.m. of the course to be flown.

When you are not following an airway such as during a direct segment of an IFR flight, you are responsible for determining the minimum altitude in accordance with FAR Part 91. Basically, you must remain at least 1,000 feet above the highest obstacle within a horizontal distance of four nautical miles from the course to be flown. In designated mountainous areas, the minimum altitude is increased to 2,000 feet and the distance from the course remains the same. You should also consider the range limitations of the navigation facilities and communications requirements when you establish your minimum altitude.

Occasionally, it is necessary to establish a **maximum authorized altitude** (MAA) for a route segment. MAA is the highest altitude you can fly based on the transmitting distance of VOR or VORTAC stations which use the same frequency. The MAA is established when the distance between two stations using the same frequency is such that both signals can be received at the same time, giving you an unreliable navigation signal. An MAA is shown with the letters "MAA," followed by the altitude.

Minimum reception altitude (MRA) is the lowest altitude that ensures adequate reception of the navigation signals forming an intersection or other fix. Operating below an MRA does not mean you will be unable to maintain the airway centerline, only that you may not be able to identify the intersection or fix.

An MRA is designated where a minimum altitude is needed to receive an off-airway navaid forming an intersection.

The MRA is important when a position report is required by ATC or when you need to establish your position. It is also important to remember that an MRA ensures reception of navigation signals, not terrain clearance. On both Jeppesen and NOS charts, the letters "MRA" precede the minimum altitude. NOS also alerts you to an MRA by enclosing the letter "R" in a flag. In the excerpt, the MRA to identify Ranso Intersection is 10,000 feet MSL.

JEPPESEN NOS

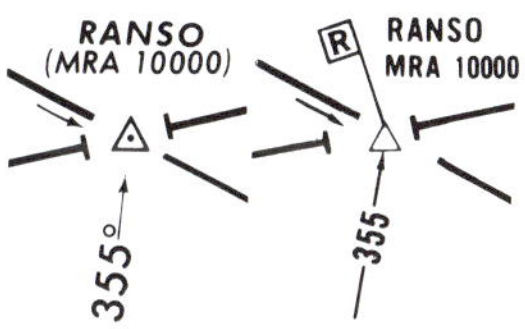

A bar symbol crossing an airway at an intersection indicates a change in MEA. On Jeppesen charts, this symbol can also be used to indicate a change in MAA or to show a change in the MOCA when an MEA is not published for the route. NOS charts use the bar symbol to indicate a change in MEA, MAA, or MOCA. When you see this symbol, be sure to compare MEAs and MOCAs along the entire route and look for an MAA to determine the basis for the change.

When an MEA changes to a higher altitude, obstruction clearance normally requires you to begin your climb upon reaching the fix where the change occurs. However, in some cases, you must begin the climb before you reach the fix. In this case, a **minimum crossing altitude** (MCA) is usually established at the fix. In certain circumstances, the MCA is used to guarantee the reception of navigation signals, not for obstruction clearance. [Figure 3-19]

Unless an MCA is published, you begin your climb to a higher MEA at the fix where the MEA change occurs.

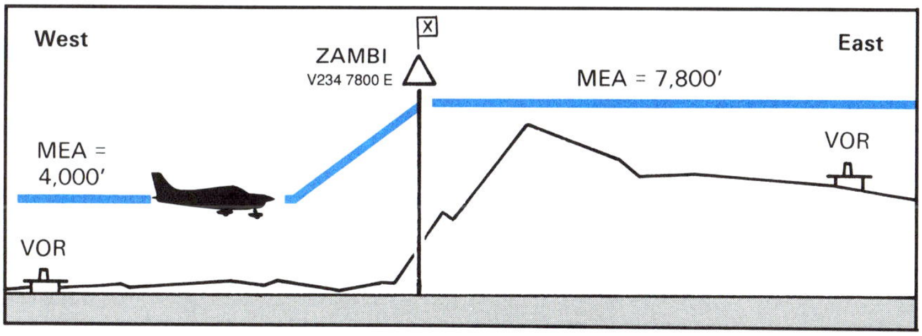

Figure 3-19. On an eastbound flight, you normally would not begin your climb to the higher MEA until you reach Zambi Intersection. Because of terrain, however, it is apparent that you must begin your climb to the higher MEA before you reach the intersection. The MCA of 7,800 feet applies to eastbound flights on Victor 234. Notice that a westbound flight is not affected by the MCA.

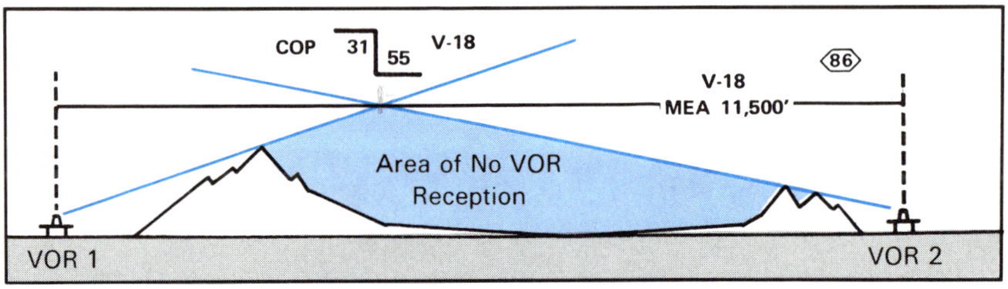

Figure 3-20. During a flight between VOR 1 and VOR 2, the navigation signals cannot be received to the midpoint of the route, so a changeover point is depicted. COPs indicate the point where a frequency change is necessary and indicate the distance in nautical miles to each navigation aid.

Jeppesen usually shows an MCA next to the intersection, along with the airway number, altitude, and direction. To reduce clutter, a number enclosed in a circle is sometimes used to indicate an MCA. When this occurs, a box located near the fix has the additional information. NOS uses an "X" enclosed in a flag to alert you to an MCA restriction. Do not confuse this with the "X" used to show a mileage break point.

You normally change frequencies midway between navaids, unless a COP is designated.

When you are following an airway, you normally change frequencies midway between navigation aids. However, there are times when this is not practical. When a change must be made at other than the midpoint, a **changeover point** (COP) is established. [Figure 3-20]

JEPPESEN NOS

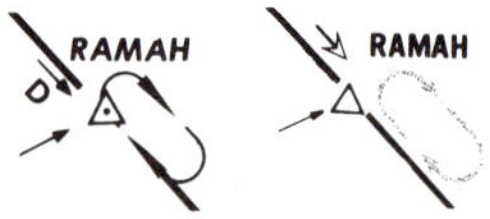

If your clearance limit is to an intersection with a designated holding pattern and you have not received an authorization to continue, ATC expects you to enter the pattern shown on the chart. Holding patterns are covered in greater depth in Chapter 5.

The MEA along Jet routes is 18,000 feet MSL, unless otherwise specified.

To help you make the transition between low altitude and high altitude airways, Jeppesen includes the high altitude enroute structure on low altitude charts. These airways are printed in green and include the appropriate Jet route number. On high altitude charts, Jet routes frequently do not list MEAs. This is because the MEA for all Jet routes is established at 18,000 feet (FL 180), unless otherwise specified.

COMMUNICATIONS

On Jeppesen charts, FSS communications frequencies are found above navaid facility boxes. Since these frequencies begin with "12," Jeppesen displays only the last two or three digits. NOS charts show the complete frequency. NOS also uses a heavy lined box to indicate the standard FSS frequency of 122.2 MHz, while Jeppesen includes this frequency with the others. AT the Gallup VORTAC, you can communicate with either the Gallup FSS or the Albuquerque FSS. This is indicated on Jeppesen charts by either the name or the three-letter identifier following the frequencies. On NOS charts, the FSS name appears below the facility box, and where

JEPPESEN NOS

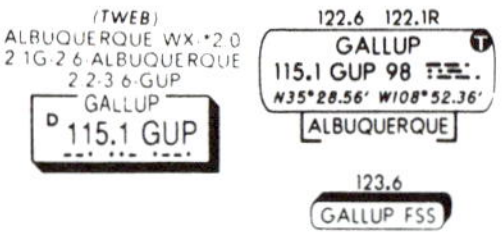

the FSS is not associated with the NAVAID, it is shown separately. In this excerpt, Albuquerque FSS is available on 122.1 and 122.6 MHz, but 122.1 MHz is a guard, or receive-only, frequency. The emergency frequency, 121.5, is not shown on either Jeppesen or NOS, since it is available at most FSSs.

An additional feature of Jeppesen charts is the identifier and frequency of the FSS offering enroute flight advisory service (EFAS) in the area. In this example, you would contact Albuquerque radio on 122.0. The asterisk preceding the frequency indicates this service is available on a part-time basis. If HIWAS or TWEB is available at a particular facility, it is indicated above the box on Jeppesen charts. NOS places a small square in the upper left-hand corner of the facility box to indicate HIWAS is available, and a circled "T" in the upper right-hand corner signifies a TWEB at this NAVAID.

JEPPESEN NOS

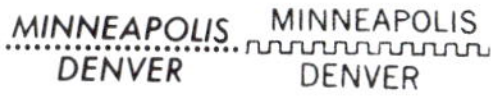

The boundary of an air route traffic control center (ARTCC) is designated by a line, along with the name of the controlling center. **Remote communications outlets** (RCOs) have been set up by each center to provide adequate communications coverage throughout the area served by the center. An RCO is shown with a box that lists the name of the center, such as Denver, followed by the location of the remote site (Goodland), and the frequencies used (132.5 MHz). NOS charts include the UHF frequencies. RCOs are also used by flight service stations to extend their area of coverage. On NOS charts, an RCO not associated with a navaid is shown with a thin-lined box. The name of the FSS and the appropriate frequency are listed inside the box. On Jeppesen charts, an RCO at an airport will be listed above the airport data. In this example, the RCO at Fairview is controlled by McAlester (MLC) FSS and the frequency is 122.5 MHz.

AIRPORTS

Jeppesen lists airports with approaches in capital letters; NOS prints the airport in blue.

Airports are divided into two basic categories — those with a published instrument approach procedure and those without such a procedure. If an airport has a published instrument approach, Jeppesen prints the airport, city, and state name in capital letters; those without an approach are printed in upper and lower case. NOS uses color to distinguish between the two types. Airports with an instrument approach are printed in blue and green; airports without an instrument approach are printed in brown.

Jeppesen and NOS provide basic information about each airport. For example, Huron Regional has an instrument approach procedure, and its elevation is 1,288 feet. Where local airport advisory service is available, Jeppesen includes the letters "LAA." The FSS frequencies are depicted nearby, usually above a NAVAID or FSS frequency box. NOS indicates the availability of airport lighting with the letter "L." An "L" enclosed in a circle means the lighting is pilot controlled. NOS also lists the length of the longest runway in hundreds of feet which, in this case, is 7,200 feet.

JEPPESEN NOS

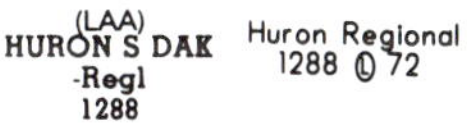

NOS lists VOTs on the communications panel.

Additional information about an airport with an instrument approach is found on the chart panel of both Jeppesen and NOS charts. This information generally includes the name of the approach and departure facility and its frequencies. NOS also includes the frequencies of VOR test facilities (VOTs).

AIRSPACE

Within the conterminous United States, all airspace at and above 14,500 feet MSL (excluding the airspace within 1,500 feet AGL) is controlled airspace. Below this altitude, the airspace may be either controlled or uncontrolled. Both Jeppesen and NOS use color to indicate the different types of airspace. Areas in white show controlled airspace, which includes Class B, C, D, and E airspace. Uncontrolled airspace (Class G) is colored blue on Jeppesen charts and brown on NOS charts.

JEPPESEN NOS

Class B airspace is outlined on Jeppesen charts by a light blue waffle screen pattern, and the upper limit is shown. NOS uses solid blue lines and light blue shading to identify these areas. In addition, Mode C areas are shown on NOS charts by blue and white striped shading.

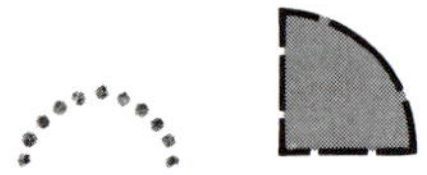

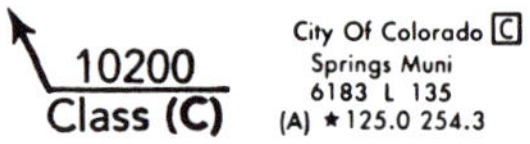

NOS also depicts Class C airspace with light blue shading, but the lateral limits are shown with dashed lines. A box with a "C" or "D," where applicable, following the airport name designates the class of airspace surrounding each airport within the area. Jeppesen depicts the boundaries of Class C areas with blue dots, and the ceiling is shown in feet MSL. Both Jeppesen and NOS provide a separate set of area charts that describe Class B and some Class C areas in greater detail.

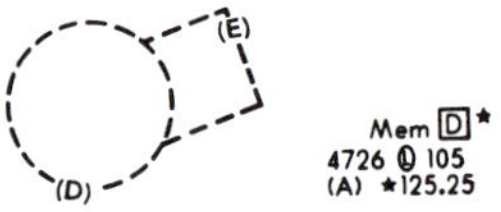

Jeppesen shows the boundaries for Class D and Class E airspace surrounding airports. These are thin green dashed lines which include the letter class of the airspace. While these boundaries are not depicted on NOS charts, Class D airspace is indicated by a "D" in a box following the airport name. Open areas in white designate Class E airspace, unless otherwise indicated.

At airports where special VFR operations are prohibited, NOS adds the note "NO SVFR" above the airport name on both the enroute and area charts. Jeppesen depicts the boundaries of airspace where special VFR is prohibited with a line of small, dark green squares. These symbols are shown only on Jeppesen's area charts.

Special use airspace is designated and identified on enroute charts as either prohibited, restricted, warning, alert, or military operations areas. Military training routes (MTRs) are also depicted on NOS low altitude enroute charts. On Jeppesen charts, all special use airspace is outlined by a green border. These areas on NOS charts are defined with a blue border, except for MOAs, which are outlined in brown. Information concerning each area is listed in or near the airspace, or it can be found on a chart panel.

AREA CHARTS

JEPPESEN NOS

On both Jeppesen and NOS enroute charts, a heavy blue dashed line signifies that an area chart is published for that terminal area. The information within the outlined area is normally limited to that necessary for through flights. If your departure or destination airport is within the boundary, you should refer to the appropriate area chart for complete information.

Most of the symbology on area charts is identical to that found on enroute charts. Because of the scale used, more detail is possible, and readability is improved. Jeppesen includes several additional features that are not found on NOS charts. For example, major airports are shown with their runway layout. On charts without area minimum altitude envelopes, man-made reference points that are 1,000 feet or more AGL are portrayed, along with their MSL elevation.

Area charts that have reference point elevations in excess of 4,000 feet above the main airport may contain contour information. Terrain is portrayed by **area minimum altitude** (AMA) envelopes. When charted, AMA envelopes portray the general layout of terrain rather than detailed contour lines. Terrain rise between two charted envelope lines is indicated by gradient tints. The area between each envelope line includes an AMA figure which represents the reference point clearance altitude for the envelope area. [Figure 3-21]

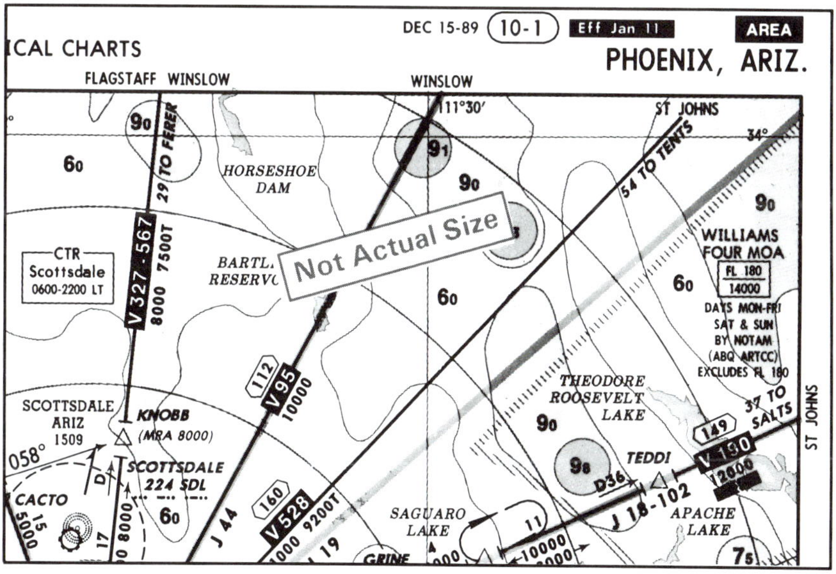

Figure 3-21. AMA values clear all reference points by 1,000 feet in areas where the highest reference points are 5,000 feet MSL or lower. AMA values clear all reference points by 2,000 feet in areas where the highest reference points are 5,001 feet or higher. Where appropriate, DME arcs and radials are included for relating your position to an AMA envelope.

CHECKLIST

After studying this section, you should have a basic understanding of:

✓ **Front panel —** The information which can be found on the enroute chart front panel.

✓ **Navigation aids —** The basic symbology used to portray the various navigation facilities and the contents of their information boxes.

✓ **Airways —** The width of a standard airway and the meaning of the acronyms MEA, MOCA, MAA, MRA, MCA, and COP.

✓ **Intersections —** How an intersection is identified and which are compulsory and noncompulsory reporting points.

✓ **Communications —** How to determine which frequency to use to communicate with a flight service station or a center.

✓ **Airports —** Which airports have instrument approaches and where additional information about the airport is located.

✓ **Airspace —** Which airspace is controlled and which is uncontrolled.

✓ **Area charts —** How they are indicated on enroute charts and what information is portrayed.

SECTION

C

DEPARTURE AND ARRIVAL CHARTS

Departure and arrival charts are published to help simplify complex clearances, reduce frequency congestion, and control the flow of traffic around an airport. In some cases, they help reduce fuel consumption and often include noise abatement procedures. The **standard instrument departure** (SID) is used after takeoff to provide a transition between the airport and the enroute structure. The **standard terminal arrival route** (STAR) performs the opposite function. It provides a standard method for departing the enroute structure and navigating to your destination. STARs usually terminate with an instrument or visual approach procedure. This section analyzes both SID and STAR charts to help you become better acquainted with the symbols and information they contain. The practical application of flying these procedures will be covered in Chapter 5.

One of the reasons for establishing SIDs and STARs is to simplify clearance delivery procedures.

Both Jeppesen and the National Ocean Service (NOS) publish SIDs, STARs, and Profile Descents for airports with procedures authorized by the Federal Aviation Administration. Jeppesen includes the applicable departure and arrival charts in the basic approach chart subscription and these charts are filed with the airport's approach charts. NOS groups SIDs with the associated approach charts in a Terminal Procedures Publication. STARs and Profile Descents are included at the front of each NOS booklet.

When you accept a SID or STAR in a clearance, or you file for one in your flight plan, you are required to have a graphic or, at least, a textual description of the procedure in your possession. If you don't have the appropriate charts or don't want to use them, you should include the notation "NO SID — NO STAR" in the remarks section of your IFR flight plan. This advises ATC not to issue a clearance which contains one of these procedures.

To avoid being issued SIDs and STARs, enter the phrase "NO SID — NO STAR" in the remarks section of your flight plan.

To illustrate how these procedures can be used to simplify a complex clearance and reduce frequency congestion, consider the following departure clearance issued to a pilot who was about to depart from Terps, California. "*Cessna 32G, cleared to the Oakville Airport via the Terps 170° radial Terps, Victor 195 Thor, Thor 285° radial and Barnsdall 103° radial to Barnsdall, then as filed. Maintain 8,000 feet. After departure, maintain runway heading until reaching 3,000 feet, then turn right to 120° and intercept the Terps 170° radial. Cross Terps at or above 5,000 feet.*"

Now consider how this same clearance is issued when a SID exists for this departure. *"Cessna 32G, cleared to Oakville Airport, Thor One Departure, Barnsdall Transition, then as filed. Maintain 8,000 feet."* This brief transmission conveys the same information as the longer example.

STANDARD INSTRUMENT DEPARTURE

It is your responsibility to ensure your aircraft can comply with the requirements of a particular SID.

Any clearance you are issued that contains a SID takes precedence over any IFR departure procedures published for the airport. Since you are obligated to comply with the provisions listed for the SID, you must ensure your aircraft is capable of achieving the performance requirements. For example, if a SID ends along a Jet route, most light aircraft will have difficulty complying with it, since Jet routes have MEAs of at least 18,000 feet. Some SIDs require you to cross a fix at or above a specified altitude. If the climb performance of your aircraft does not allow you to reach the altitude by the fix, you cannot use the SID. Your responsibility as pilot in command is to review each SID, make sure your aircraft can comply with the procedures, and refuse any SID that is beyond the limits of your aircraft.

SIDs are divided into two general classifications: pilot nav and vector SIDs. **Pilot nav SIDs** are designed to allow you to navigate along a specified route with minimal communications with ATC. These SIDs usually contain an initial set of instructions that apply to all aircraft, followed by one or more transition routes that require you to navigate to the appropriate fix within the enroute structure. Many pilot nav SIDs include a radar vector segment that helps you join the SID. **Vector SIDs** are established where ATC provides radar navigation guidance. The chart usually contains an initial set of instructions, such as a heading to fly and an altitude to maintain. When ATC establishes radar contact, they will provide vectors to help you reach one of several fixes portrayed on the chart.

PILOT NAV SID

The symbols used on both Jeppesen and NOS SIDs are very similar to those used on their other charts. Both chart publishers include a textual description of the initial takeoff and transition procedures, and a plan view of the routing. In some cases, a textual description may not be provided for a simple transition. [Figure 3-22]

An additional feature on NOS SID charts is the listing of applicable enroute charts below the facility box for each navaid. This information helps you locate the facility closest to your planned route of flight. NOS

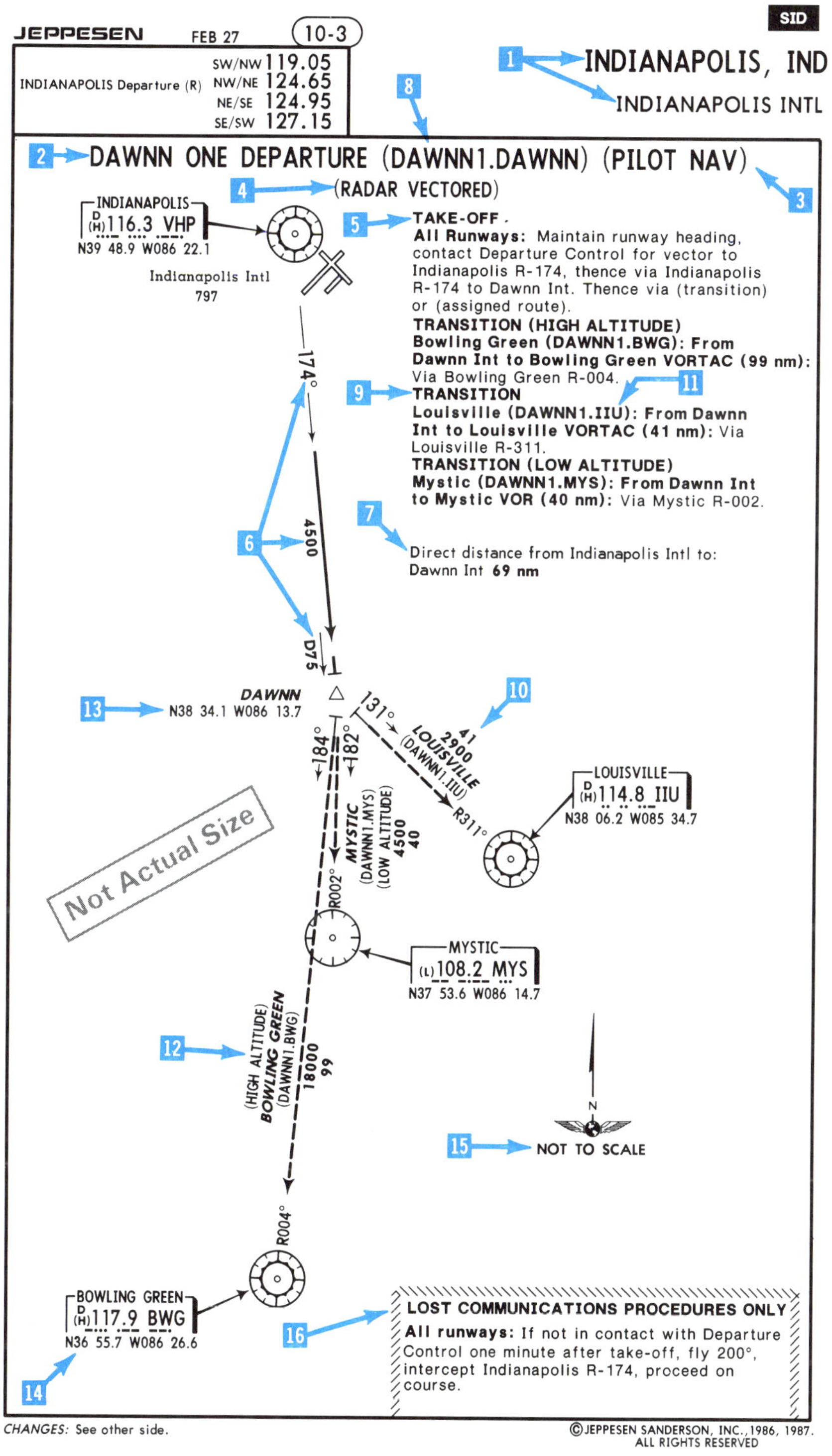

Figure 3-22. This is a typical example of a Jeppesen pilot nav SID. As with most Jeppesen charts, appropriate communications information is shown at the top of the chart. The numbered callouts which appear on the Dawnn One Departure are discussed in the accompanying text.

SIDs, like Jeppesen SIDs, often are not drawn to scale. For this reason, it is important that you study the milage symbols carefully. [Figure 3-23]

Maintain runway heading means the actual magnetic value of the centerline. If Runway 4 centerline is 044°, maintain 044°.

Both Jeppesen and NOS list the airport served by the procedure at the top of the chart (item 1). In this case, it is Indianapolis International in Indiana. The name and the type of SID also are listed at the top of the chart. This is the Dawnn One Departure (item 2) which is a pilot nav SID (item 3). The chart also indicates radar vectors are used during some portions of the departure (item 4). When cleared to fly or maintain runway heading, pilots are expected to maintain the heading that corresponds with the extended runway centerline and drift correction should not be applied. The initial takeoff procedure (item 5) applies to all runways. The textual instructions are used immediately after takeoff to navigate to the Dawnn Intersection. The graphic, or plan view, of the departure route portrays the radial, altitude, and DME distance from Indianapolis to Dawnn (item 6). On both Jeppesen and NOS charts, this route is shown by a bold line. Thin lines are used to indicate the appropriate radial for navigating or identifying fixes. The actual mileage between a given runway and the intersection varies with aircraft performance, pilot technique and, in this case, the length of the radar vector. Therefore, Jeppesen includes the direct distance of 69 n.m. from the airport to Dawnn Intersection (item 7).

When you file for a SID, be sure to use the appropriate ATC code in your flight plan.

If you intend to use only the basic portion of this SID that ends at the Dawnn Intersection, list the ATC code DAWNN1.DAWNN (item 8) as the first part of your routing on your IFR flight plan. This lets ATC know that you intend to use the SID to get from the airport to Dawnn Intersection. The remainder of the route from Dawnn is listed following the SID code. Your ATC clearance might sound like this: "*Cessna 32G, cleared to Athens Airport, Dawnn One Departure, then as filed. Maintain 9,000 feet.*"

If your planned route allows you to follow one of the transitions, it is usually to your benefit to file for the transition. For example, suppose your route of flight takes you over the Louisville VORTAC. In this case, you can include the Louisville Transition in your flight plan. The textual description (item 9) includes the radial to use from Dawnn to Louisville. The plan view (item 10) also lists this information, as well as the mileage, MEA, and inbound course for the route. The distance from Dawnn to the Louisville VORTAC is 41 n.m., the MEA is 2,900 feet MSL, and the course follows the 311° radial from Louisville (131° inbound). Notice that transition routes are shown with dashed lines on Jeppesen charts and with light, solid lines on NOS charts.

Using the ATC code for a transition in your flight plan informs ATC you intend to fly both the SID and the appropriate transition.

When you file for this transition, the first part of your routing should list the ATC code DAWNN1.IIU (item 11). This tells ATC you plan to fly the Dawnn One Departure (DAWNN1) and the Louisville transition (.IIU). Listing the code exactly as it appears on the SID helps ATC enter it into the computer and reduces the time required to process your flight plan. The remainder of your route, starting at Louisville, is entered following

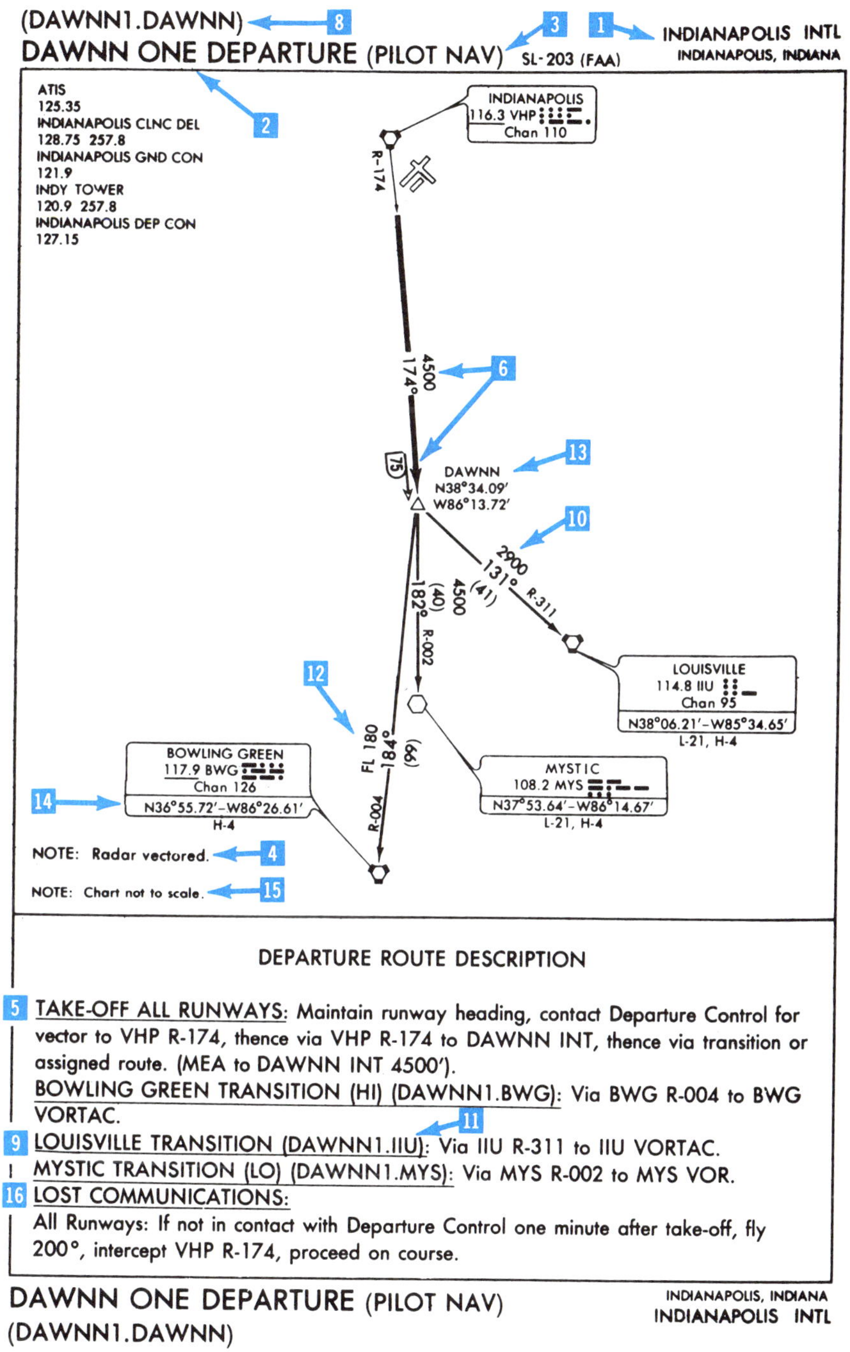

Figure 3-23. Here is the NOS version of the Dawnn One Departure. Notice that the applicable NOS enroute charts are listed below the navaid information boxes. The numbered callouts on this chart are discussed in the accompanying text. They correspond to those shown on the Jeppesen chart.

this code. Your clearance from ATC might sound like this: "*Cessna 34G, cleared to Athens Airport, Dawnn One Departure, Louisville Transition, then as filed. Maintain 9,000.*"

The Bowling Green Transition (item 12) illustrates why you should examine a SID carefully to make sure you can comply with its requirements. This transition has an MEA of 18,000 feet (FL 180). Unless you can comply with this altitude requirement, you must not file for this transition, nor accept it if it is issued by ATC.

Both Jeppesen and NOS provide the latitude and longitude coordinates of intersections (item 13), mileage break points, and navigation facilities (item 14) used in the procedure. Jeppesen also shows the VOR facility class. For example, the Bowling Green VORTAC is a high altitude class facility, as indicated by the letter "H" in the facility box. Because of the large area covered, most SIDs are not usually drawn to scale. When this is the case, the notation "not to scale" is printed on the chart (item 15).

When special lost communications procedures are necessary for a SID, they are included on the chart.

Since it is possible to experience two-way radio communications failure during a SID, you must be familiar with the lost communications procedures specified in FAR 91.185. The details for complying with this regulation are covered in more detail in Chapter 7. When the lost communications procedures are different from those specified in the FARs, they are clearly stated on the chart (item 16).

VECTOR SID

The vector SID chart is similar to the pilot nav SID, except for the absence of departure routes and transitions. This is because ATC provides radar vectors that start just after takeoff and continue until you reach your filed or assigned route or one of the fixes shown on the chart. The textual description provides initial procedures to follow after takeoff. IFR departure procedures, including pilot nav and vector SIDs, are based on a minimum climb gradient of 200 feet per nautical mile. This requirement is established to ensure you can clear obstacles in the departure path. When obstacle clearance cannot be guaranteed with this gradient, the procedure may specify a minimum ceiling and visibility to allow you to see and avoid the obstacle, a climb gradient greater than 200 feet per mile, detailed flight maneuvers, or a combination of these procedures. [Figure 3-24]

Minimum climb gradients are given in feet per nautical mile and must be converted to feet per minute for use during departure.

Since climb gradients are given in feet per nautical mile, they must be converted to feet per minute to be usable during departure. Jeppesen provides a conversion table on each SID chart that has a climb gradient restriction. According to the table shown, a gradient of 240 feet per nautical mile at a groundspeed of 100 knots requires you to maintain a climb

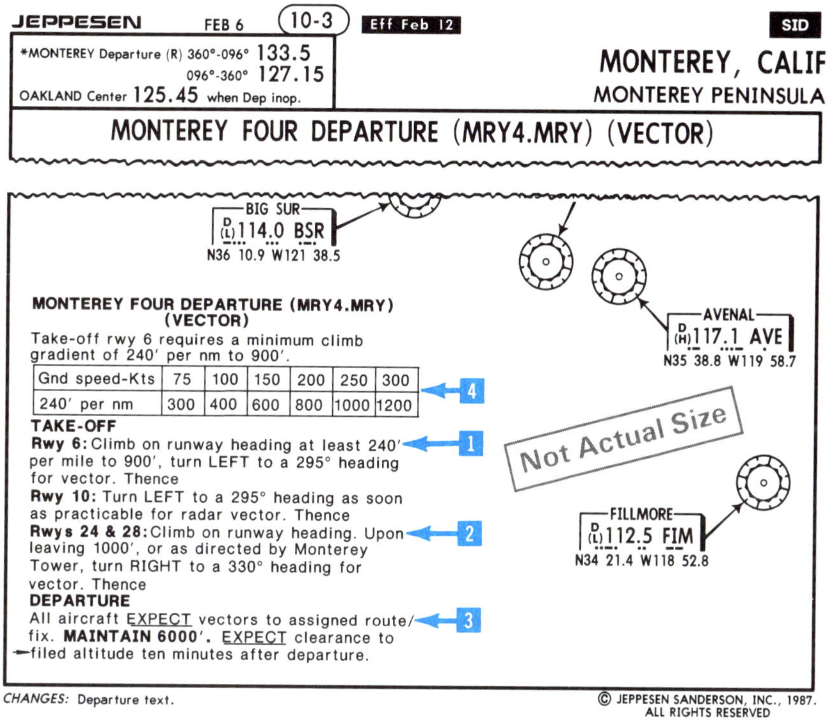

JEPPESEN FEB 6 (10-3) Eff Feb 12 SID

*MONTEREY Departure (R) 360°-096° 133.5
096°-360° 127.15
OAKLAND Center 125.45 when Dep inop.

MONTEREY, CALIF
MONTEREY PENINSULA

MONTEREY FOUR DEPARTURE (MRY4.MRY) (VECTOR)

BIG SUR
D (L) 114.0 BSR
N36 10.9 W121 38.5

AVENAL
D (H) 117.1 AVE
N35 38.8 W119 58.7

Not Actual Size

FILLMORE
D (L) 112.5 FIM
N34 21.4 W118 52.8

MONTEREY FOUR DEPARTURE (MRY4.MRY) (VECTOR)

Take-off rwy 6 requires a minimum climb gradient of 240′ per nm to 900′.

Gnd speed-Kts	75	100	150	200	250	300
240′ per nm	300	400	600	800	1000	1200

4

TAKE-OFF

Rwy 6: Climb on runway heading at least 240′ per mile to 900′, turn LEFT to a 295° heading for vector. Thence 1

Rwy 10: Turn LEFT to a 295° heading as soon as practicable for radar vector. Thence

Rwys 24 & 28: Climb on runway heading. Upon leaving 1000′, or as directed by Monterey Tower, turn RIGHT to a 330° heading for vector. Thence 2

DEPARTURE

All aircraft EXPECT vectors to assigned route/fix. **MAINTAIN 6000′.** EXPECT clearance to filed altitude ten minutes after departure. 3

CHANGES: Departure text.

Figure 3-24. This Jeppesen vector SID shows that if you are taking off from runway 6 (item 1), you must maintain the runway heading and a minimum climb of 240 feet per nautical mile to an altitude of 900 feet MSL then turn to a heading of 295°. The takeoff procedure for runways 24 and 28 (item 2) is to maintain the runway heading until you reach 1,000 feet or the altitude specified by Monterey Tower. At this point, turn right to 330°. The departure instructions (item 3) tell you to expect vectors to your assigned route or fix. They also require you to maintain an initial altitude of 6,000 feet and to expect clearance to your filed altitude 10 minutes after departure. The table (item 4) helps you convert feet per nautical mile to feet per minute, depending on your groundspeed.

of 400 feet per minute. NOS includes a conversion table that is located in the front portion of the SID booklet.

STANDARD TERMINAL ARRIVAL ROUTE

The standard terminal arrival route (STAR) uses symbols similar to those found on a SID chart. Since the procedure may begin at more than one enroute fix, several transitions may be listed which join into one common arrival route. The arrival route may terminate at an approach fix or with radar vectors to the final approach course. On the accompanying Comic Two arrival charts, three transitions are portrayed from the Pulaski, Roanoke, and Lynchburg VORTACs. These routes converge at the Danville VOR. From here, they follow a common transition route to Comic Intersection. [Figure 3-25 and Figure 3-26]

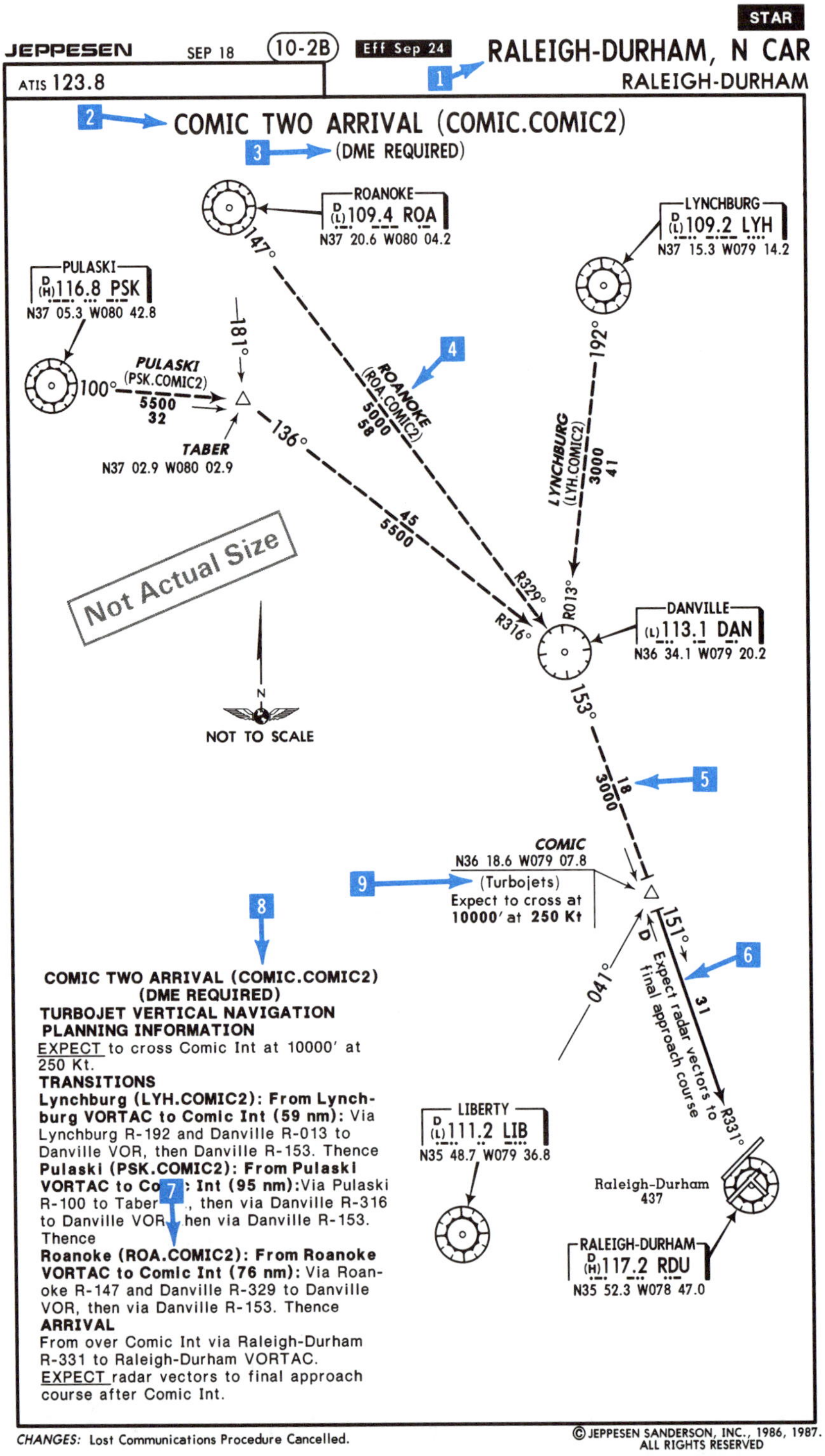

Figure 3-25. This Jeppesen STAR is for the Comic Two Arrival at Raleigh-Durham, North Carolina. The callouts in this illustration are discussed further in the accompanying text.

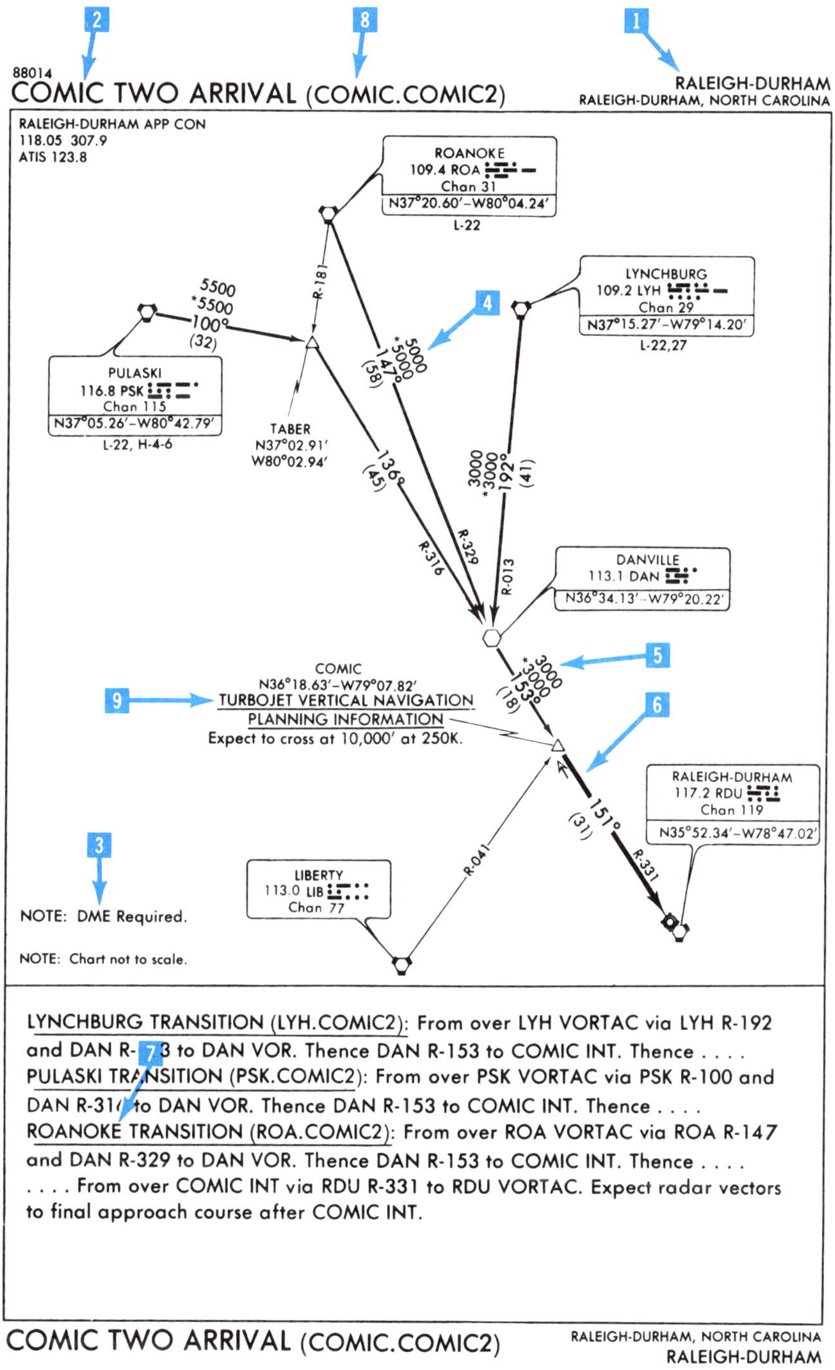

Figure 3-26. Here is the NOS version of the Comic Two Arrival. The numbered callouts on this chart match those shown on the Jeppesen chart.

Like SIDs, STARs include the name of the airport served (item 1) and the name of the procedure (item 2) near the top of the chart. This STAR requires the use of DME to fly the procedure (item 3). If you do not have DME equipment in your aircraft, you cannot file for or accept this STAR. The Roanoke Transition (item 4) begins at the Roanoke VORTAC and follows the 147° radial and the Danville 329° radial to the Danville VOR. The MEA for this route segment is 5,000 feet MSL, and the distance is 58 n.m. The NOS chart lists both an MEA and a MOCA for each route segment. Since they are the same altitude, you are assured of both navigation signal reception and obstacle clearance for the entire route when you fly at or above the MEA.

From the Danville VOR, the transition follows the 153° radial and ends at the Comic Intersection. The MEA is 3,000 feet MSL, and the distance is 18 n.m. (item 5). Comic can be identified by either a DME distance from the Raleigh-Durham VORTAC or the 041° radial from the Liberty VORTAC. Jeppesen portrays transition routes with a dashed line, while NOS uses a solid line.

The arrival route (item 6) from over Comic Intersection lies along the Raleigh-Durham 331° radial, and it is 31 n.m. long. The inbound course for this route is 151°. Since an MEA is not shown for this route, you should maintain 3,000 feet until you are assigned a lower altitude by ATC. Both Jeppesen and NOS use a heavy solid line to depict an arrival route. The Jeppesen plan view also notes that you should expect radar vectors to the final approach course during this segment.

The textual explanation for this STAR includes a description of each transition and the arrival route. The appropriate ATC codes are also provided. These codes are used on your flight plan to tell ATC which procedure you want to fly. If the enroute portion of your flight ends at the Roanoke VORTAC, you should add the Roanoke Transition of the Comic Two Arrival (item 7) to the end of the route description in your IFR flight plan. This transition is entered on your flight plan as ROA.COMIC2. You should use transitions whenever possible. However, if you are able to navigate directly to the Comic Intersection and your route does not follow a designated transition, you can file COMIC.COMIC2 (item 8) on your flight plan.

Some STARs list information that helps pilots of turbojet aircraft plan their descents (item 9). Vertical navigation planning information normally does not preclude you from using a STAR. In this case, the note applies only to turbojet aircraft, although some charts specify a crossing altitude that may limit your ability to comply with the instructions. Always make sure the procedure is within the capabilities of your aircraft.

A variation of a STAR is a **profile descent.** It is designed to reduce the amount of low altitude flying time for high-performance aircraft, such as turbojets and turboprops. The procedure allows an uninterrupted descent from cruising altitude to the interception of the glide slope, or to a minimum altitude specified for the initial or intermediate segment of an instrument approach. It may also terminate with radar vectors to an approach. When crossing altitudes and speed restrictions are shown, ATC expects you to descend to the crossing altitude first, then reduce speed. You usually do not file for a profile descent in your flight plan.

Issuance of a profile descent or a STAR does not constitute a clearance to fly an approach. The last "*maintain altitude*" specified in the procedure or by ATC constitutes the lowest altitude to which you can descend until you receive an approach clearance or a lower altitude assignment.

The issuance of a profile descent or a STAR does not constitute an approach clearance.

CHECKLIST

After studying this section, you should have a basic understanding of:

✓ **Standard instrument departures** — How they are used to simplify clearances and reduce frequency congestion.

✓ **Pilot nav and vector SIDs** — What their differences are and how each type is flown.

✓ **Standard terminal arrival routes** — Their purpose and how they should be used.

✓ **Flight plan coding** — How to use the ATC code in your flight plan to specify a SID or STAR.

CHAPTER 4

INSTRUMENT APPROACHES

INTRODUCTION

The instrument approach procedures discussed in this chapter include ILS, VOR, and NDB. In Section A, you will become familiar with the various ground components of the instrument landing system and how they are used. You also will learn how to conduct an approach chart review and extract the information you need to fly the procedure. Other important features involve the use of supplemental DME fixes and allowable substitutes for inoperative components. Methods for making pitch and power adjustments to maintain the localizer and glide slope centerline also are included. Section B provides you with a similar comprehensive analysis of VOR and VOR/DME approaches. Examples of both on-airport and off-airport facilities are included, and straight-in landings and circling approaches are discussed in detail. In Section C, you learn how to use ADF equipment to execute NDB approaches. NDB procedures often provide the only approaches to remote airport locations. In addition, they may function as backup procedures in terminal areas. The final section will familiarize you with the global positioning system. You will learn the components of the system and gain a basic understanding of how to use GPS for navigation. The approach overlay program and stand alone GPS approaches are explained.

SECTION A

ILS APPROACHES

ILS minimums normally include a DH of 200 feet and a visibility of one-half statute mile.

Decision HeighT

The instrument landing system (ILS) has several advantages over other approach procedures because it provides highly accurate course, glide slope, and distance (DME) guidance to a given runway. Therefore, ILS approaches allow safer descents to lower minimums than most other procedures. ILS approaches normally provide for descents to 200 feet above the runway in visibilities as low as one-half statute mile.

ILS COMPONENTS AND VISUAL AIDS

The basic ground components of an ILS are the localizer, glide slope, outer marker, and middle marker. A typical ILS has several associated visual aids, such as runway lighting systems and RVR equipment. The placement of the ground components results in the precise course guidance that the ILS provides. [Figure 4-1]

Runway visual Range

LOCALIZER

A primary piece of equipment for any ILS system is the **localizer.** The localizer transmitter is placed at the far end of the runway, opposite the approach end, and emits a navigational array that provides course guidance to the runway centerline. When these signals are used, the approach is said to be a front course approach. This transmitter can also provide a signal in the direction opposite the front course for back course approach guidance. Do not navigate using back course signals unless a back

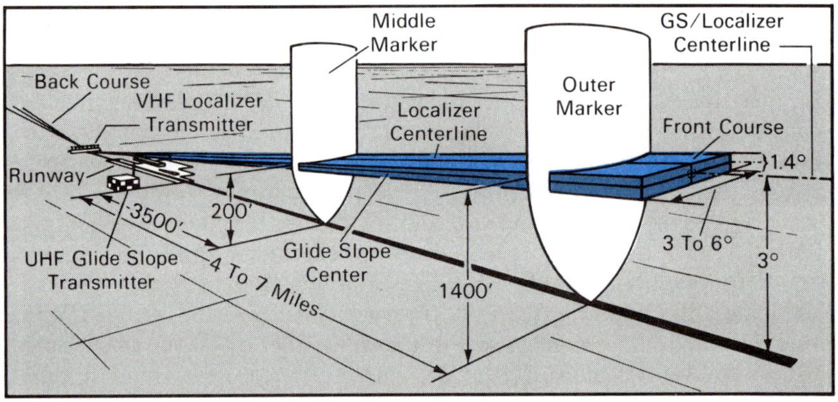

Figure 4-1. A typical ILS installation follows this general arrangement. Specific installations vary somewhat in terms of glide slope elevation, localizer width, or marker utilization. You may want to refer back to this illustration during the following discussion on individual components.

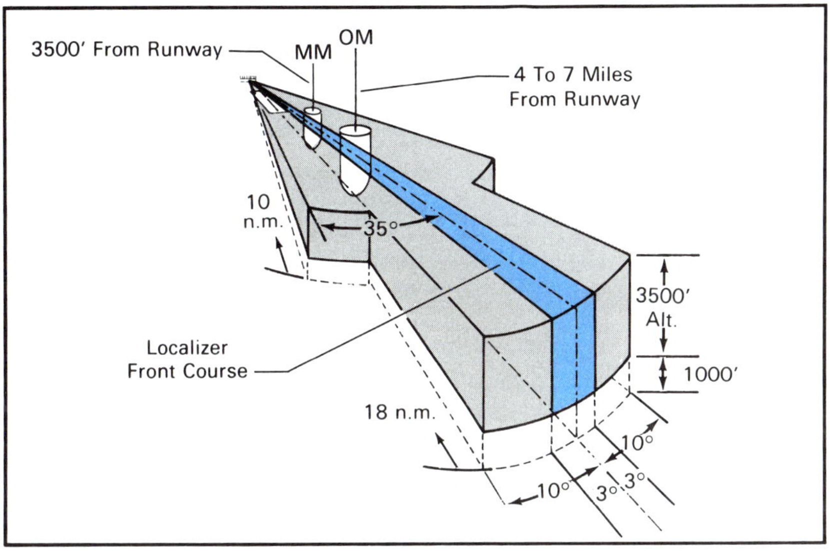

Figure 4-2. Within a specific altitude block, course information is accurate to at least 18 n.m. from the antenna. At 18 n.m., the minimum reception altitude is 1,000 feet above the highest terrain along the approach course. The maximum reception altitude throughout the approach course is 4,500 feet above the elevation of the antenna. The operational service volume of a localizer extends through an angular area 35° on either side of the course, out to a range of 10 n.m. From 10 to 18 n.m., this area is reduced to 10° on either side of the course.

Outer marker is associated the NDB
middle
200 Above AGL ½ mile from the Threshold

course approach is established and ATC has authorized you to use the back course for guidance. The angular width of localizers varies between three and six degrees in order to provide a signal width of approximately 700 feet at the runway threshold. [Figure 4-2]

With 50 kHz spacing, 40 ILS channels using odd-tenth frequencies are available (108.10, 108.15, 108.30, 108.35 MHz through 111.95 MHz). Today, 200-channel navigation receivers incorporate 40-channel ILS capability. Localizer audio identification consists of a three-letter Morse code identifier preceded by the letter "I." Since the Morse code for the letter "I" is two dots, localizers have a distinctive identification feature. For example, Pueblo Memorial Airport in Colorado has two separate ILS approaches. One localizer identifier is "IPUB" and the other is "ITFR."

Another distinctive feature of the ILS is that the CDI has greater sensitivity. For instance, full-scale deflection of the CDI represents a 20° radial span, or 10° to each side of course, with VOR signals. When using localizer signals, the total span of the CDI is 5°, or 2.5° each side of the course. The advantage to greater receiver sensitivity is that a localizer course can be tracked with a theoretical accuracy four times that of a VOR radial. Because of this, however, CDI movement is more rapid during localizer tracking. You must make smaller course corrections than with VOR, and you must make these corrections more promptly.

When receiving localizer signals, the CDI is sensitive to 2.5° either side of the course.

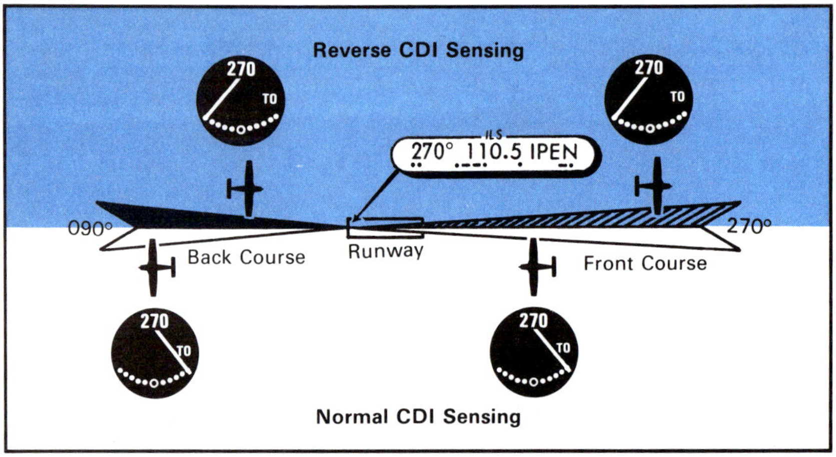

Figure 4-3. When using a basic VOR indicator, normal sensing occurs inbound on the front course and outbound on the back course. Reverse sensing occurs inbound on the back course and outbound on the front course. With HSI, you can avoid reverse sensing by setting the published front course under the course index. This applies regardless of your direction of travel, whether inbound or outbound on either the front or back course.

The signal from the localizer transmitter represents only one magnetic course to the runway. Therefore, the course selected on the OBS of a basic VOR indicator does not affect course tracking. Regardless of what course you select, the CDI senses off-course position only with respect to the localizer course. However, you may find it helpful to set the published course of the ILS on the navigation indicator as a reminder of the inbound course during tracking and heading corrections. If you are using an HSI for the approach, you normally must set it to the specific ILS front course for guidance. When you are tracking the front course of an ILS toward the runway using a basic VOR indicator, CDI sensing is normal; that is, right heading corrections are applied to right deflections of the CDI. Reverse CDI sensing occurs whenever the aircraft travels on the reciprocal heading of the localizer course. [Figure 4-3]

GLIDE SLOPE

The horizontal needle of a VOR indicator or HSI provides the vertical guidance you need to maintain the **glide slope**. The glide slope transmitter is usually placed 750 to 1,250 feet down from the approach end of the runway and is offset 250 to 650 feet from the runway centerline. Normally, glide slope signals are directed only to the front course approach. Glide slope transmitters operate on a UHF frequency which is paired to the associated VHF localizer frequency. When you select the localizer frequency, you automatically select the associated glide slope frequency. Since the transmitter is offset and somewhat elevated from the runway centerline, glide slope guidance may not be provided all the way to touchdown.

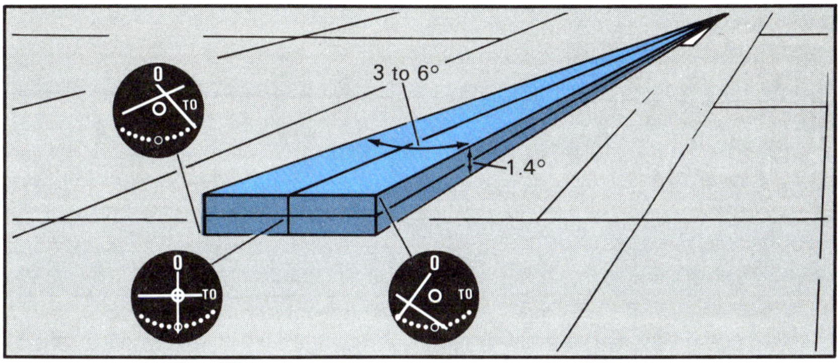

Figure 4-4. A position only slightly off the glide slope or localizer centerline will produce large needle deflections on the navigation indicator. When navigating on an ILS, you must respond immediately to needle movement with heading, pitch and/or power changes.

The center of the glide slope is normally adjusted to three degrees above the horizontal. At this angle, the glide slope, which is 1.4° thick, intercepts the middle marker at about 200 feet and the outer marker at about 1,400 feet above the runway elevation. False signals and reverse sensing can occur at high angles above the normal three-degree glide slope projection. You should disregard the glide slope indications until you are approaching the glide slope intercept altitude (GSIA) shown on the approach chart. [Figure 4-4]

The glide slope centerline normally intersects the middle marker at an altitude of approximately 200 feet.

ILS MARKER BEACONS

In addition to the glide slope and localizer, **ILS marker beacons** provide you with range information with respect to the runway during the approach. All ILS marker beacons project an elliptical array upward from the antenna site. At about 1,000 feet above the antenna, the array is 2,400 feet thick and 4,200 feet wide. At 1,000 feet AGL, with a groundspeed of 120 knots, you will receive the signal for about 12 seconds.

Usually, there are two marker beacons associated with an ILS: the **outer marker** (OM) and **middle marker** (MM). The placement of the OM varies from four to seven miles from the runway, depending on the installation. It usually is placed inside the point where an aircraft flying the ILS intercepts the glide slope. The OM is aurally identified by a continuous series of Morse code dashes at the rate of two per second. The MM is usually located 3,500 feet from the landing threshold, with its signal array intercepting the glide slope at approximately 200 feet above the touchdown zone. The aural identification feature of the MM is 190 alternating Morse code dots and dashes per minute. At some locations where Category II ILS operations have been certified, an **inner marker** (IM) indicates the decision height on the CAT II glide slope. The CAT II DH is normally between the middle marker and the landing threshold. The IM is aurally identified with Morse code dots at the rate of six per second.

Marker Receiver	Color	Label	Code
Outer	Blue	O	— — —
Middle	Amber	M	• — • —
Inner	White	A	• • • •

Figure 4-5. A blue light designates passage of the OM, an amber light indicates the MM, and a white light identifies the IM. When the aircraft passes through the signal array of the respective marker beacons, these lights flash as the audio identification sounds.

Occasionally, marker beacons may be used on the localizer back course as the final approach fix. Back course markers are identified with two Morse code dots at a rate of 144 to 190 per minute.

In most general aviation aircraft, the **marker beacon receiver** is incorporated into the audio control console for the avionics. Marker beacon receivers have three separate lights to correspond to the three types of marker beacons. The audio identification of each marker beacon can be heard over the speaker during station passage. Usually, a control is provided for high or low receiver sensitivity. [Figure 4-5]

COMPASS LOCATORS

Many ILS systems use an NDB as a compass locator, which normally is collocated with the outer marker. The combined facility is called a **compass locator at the outer marker** (LOM). Compass locators usually have a power output of less than 25 watts, resulting in a reception range of at least 15 miles. The frequency range for compass locators is 190 to 535 kHz. At a few locations, compass locators are installed at the middle marker (LMM). Compass locators transmit a two-letter Morse code identifier taken from the last three letters of the ILS identifier group. For example, the LOM transmits the first two letters of the localizer ident; the LMM, when used, transmits the second two. At other locations, high powered NDBs of up to 400 watts, may be utilized. These facilities usually carry TWEB broadcasts. They may be collocated with the outer marker or placed farther out on the approach.

The first two letters of the localizer identification group designate the LOM, and the last two letters are used for the LMM.

ILS WITH DME

On many ILS procedures, a DME transmitter is placed at or near the localizer or glide slope transmitter. ILS with DME provides direct runway distance information. When installed on an ILS approach, you may use DME in lieu of the outer marker or to establish other published fixes on a localizer front or back course. In some cases, you also may use DME information from a separate facility, such as a VORTAC. An ILS approach which uses DME will be discussed later in this section.

ILS VISUAL AIDS

Approach lighting systems are visual aids which normally are associated with the ILS. In addition, whenever the minimum landing visibility for an ILS approach is specified as 1,800 feet runway visual range (RVR), other visual aids are included. They are high intensity runway lights, touchdown zone lights, centerline lights and markings, and RVR. These visual aids were described in Chapters 2 and 3.

The lowest landing minimums on an approach are authorized when all components and visual aids are operating. Inoperative components may require higher landing minimums, but they are not cumulative when more than one component is inoperative. In this case, you apply only the greatest increase in altitude and/or visibility required by the failure of a single component.

If more than one component of an ILS is unusable, use the highest minimum required by any single component that is unusable.

According to FAR 91.175, certain substitutions for equipment outages are authorized. A compass locator or precision radar may be substituted for the outer marker or middle marker. DME, VOR, or NDB fixes authorized in the standard instrument approach procedure or surveillance radar may be substituted for the outer marker. However, when the glide slope becomes inoperative, localizer-only minimums must be used. When a substitution for the outer marker cannot be made, or when the localizer is out, an ILS approach is not authorized.

A compass locator, precision radar, surveillance radar, or published DME, VOR, or NDB fixes may be substituted for the outer marker.

When any basic ILS ground component (except the middle marker or the localizer) or required visual aid is inoperative, unusable, or not utilized, the standard straight-in landing minimums prescribed by the approach are raised. As discussed earlier, Jeppesen charts show the increase in minimums directly on the approach chart. NOS users consult a separate tabulation entitled "Inoperative Components or Visual Aids Table." If a component is inoperative, you must then compute the higher minimums based on the applicable equipment failure.

If an ILS visual aid is inoperative, the visibility requirements are raised on approaches where RVR 1,800 is authorized.

Remember, your aircraft must be equipped with the proper navigation equipment to fly the approach. For example, if the approach title is "LOC/DME BC," you need a LOC receiver and DME to fly the procedure. Additionally, if the airspace you are flying in has special equipment requirements, your aircraft must also meet those requirements.

Before beginning an approach procedure, your aircraft must have the equipment specified in the procedure, in addition to any other equipment required by regulations.

ILS CATEGORIES

The basic ILS approach is termed Category I, because it requires only that you be instrument rated and current and that your aircraft be equipped appropriately. Normally, the minimum visibility required to fly the approach is one-half statute mile or RVR 2,400 feet. It may be reduced to RVR 1,800 feet when centerline and touchdown zone lights are provided. Normally, the decision height is 200 feet above the touchdown zone elevation. Category II and Category III ILS minimums require special certification for operators, pilots, aircraft and air/ground equipment. In return for this stringent compliance, CAT II ILS approach procedures provide for a DH of not less than 100 feet above touchdown and an RVR of not less than 1,200 feet. CAT III ILS approaches are subdivided into three groups, all of which allow descents to touchdown, since decision heights are not specified. Because of the complexity and high cost of the equipment required, this category pertains mainly to air carrier and military operations. CAT IIIa approaches require an RVR of 700 feet, CAT IIIb require an RVR of 150 feet, and CAT IIIc approaches are authorized without an RVR minimum.

FLYING THE ILS APPROACH

ILS approach training introduces vertical guidance by radio navigation. When your aircraft is on an ILS approach course, the CDI of the VOR indicator is centered, regardless of your altitude. On the other hand, the glide slope indicator senses the vertical movement of your aircraft in relation to an inclined plane projected from the glide slope transmitter. [Figure 4-6]

Prior to ILS glide slope interception, your primary concern is to stabilize airspeed and altitude and arrive at a magnetic heading which will maintain the aircraft on the localizer centerline. Flying at a constant airspeed is not only desirable, but essential for smooth, accurate descents to DH. At glide slope interception, you will need to initiate a descent. In fixed-gear aircraft, you will usually need to reduce power, although the

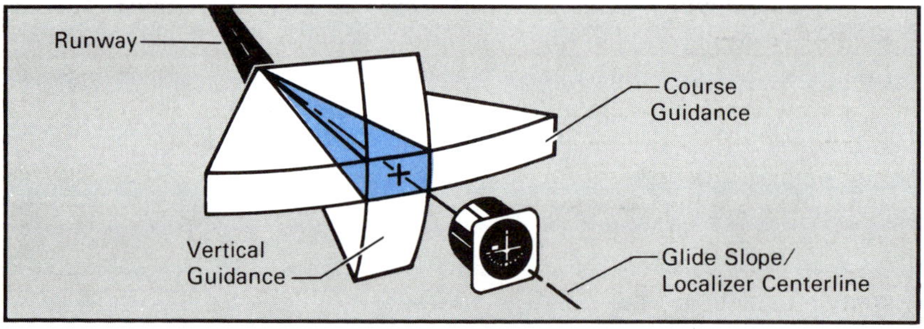

Figure 4-6. When flying an ILS, you actually track the line formed by the intersection of the glide slope and localizer courses. While tracking the localizer, your CDI senses horizontal movement of the aircraft away from the course. Glide slope indications provide a precise descent path to the runway.

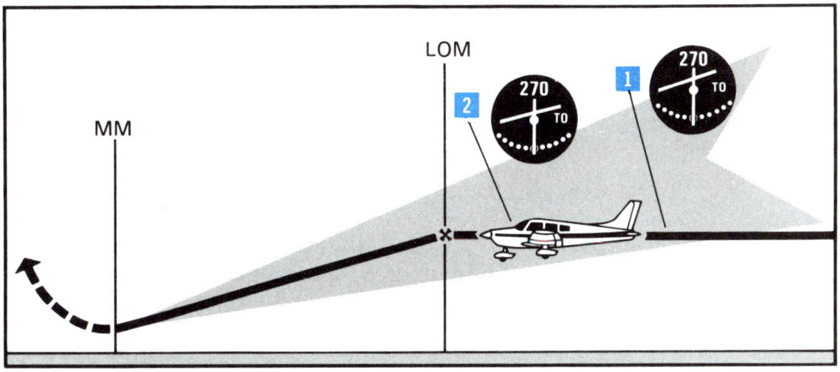

Figure 4-7. Since your aircraft is usually below the glide slope during the intermediate approach segment, the glide slope indicator will display a full-up needle deflection (item 1). You should observe the initial downward movement of the indicator (item 2) and lead the descent to intercept the glide slope centerline accordingly.

amount of reduction depends on whether you lower the flaps. In retractable gear aircraft, you may not need to make pitch and power changes when you lower the landing gear as you approach the glide slope. However, when such adjustments are required, you may need to retrim the aircraft. [Figure 4-7]

If the glide slope and localizer are centered but your airspeed is too fast, your initial adjustment is to reduce power.

Intercepting the glide slope at the proper speed will make the descent more stable. However, if your airspeed is too high after glide slope interception, a further power reduction is required. As the airspeed decreases, you will need to make a pitch adjustment to keep the aircraft from going below the glide slope. After the descent rate stabilizes, you should use power as necessary to maintain a constant approach speed. You can use small pitch changes to maintain the glide slope. However, if the glide slope indicator approaches full scale deflection, you should respond immediately with pitch and power adjustments to reintercept the glide slope. As your aircraft approaches the MDA or DH, you should be prepared to lead the leveloff by smoothly increasing back pressure on the yoke and increasing power to stop the descent without causing a reduction in airspeed.

During actual IFR weather conditions, it is usually apparent when you can continue the approach visually. However, prior to DH or MDA, you should continue your instrument cross-check with only brief glances outside until you are sure that positive visual contact with the runway environment is established. It is not unusual on an ILS approach to establish visual contact at 500 to 600 feet AGL and then lose outside visual references as the descent continues. This may be due to a very low fog layer that allows visual contact from above but causes loss of visual cues on a horizontal plane within the layer. For this reason, you should avoid descents below the glide slope, even though you have visual contact with the runway before you reach DH.

NONRADAR ILS PROCEDURES

As with the other types of approaches, each ILS approach is designed for the individual airport. A representative procedure is the ILS runway 26 approach for Pueblo Memorial Airport. Although radar transitions to the final approach course are available at Pueblo, this discussion is based on a nonradar arrival. [Figure 4-8]

The transition from the enroute phase of flight to the approach phase can be made using only radio navigation. Several transitions using DME arcs, VOR, ADF, or a combination of these are available for this procedure. When you begin the transition from enroute to the approach phase of flight, your workload increases rapidly. You must acknowledge and comply with controller instructions, adjust the aircraft speed and configuration, and set up the navigation equipment for the approach. You also need to make a final review of the approach chart, concentrating on the transition routes you will use, as well as the applicable DH and missed approach procedure. [Figure 4-9]

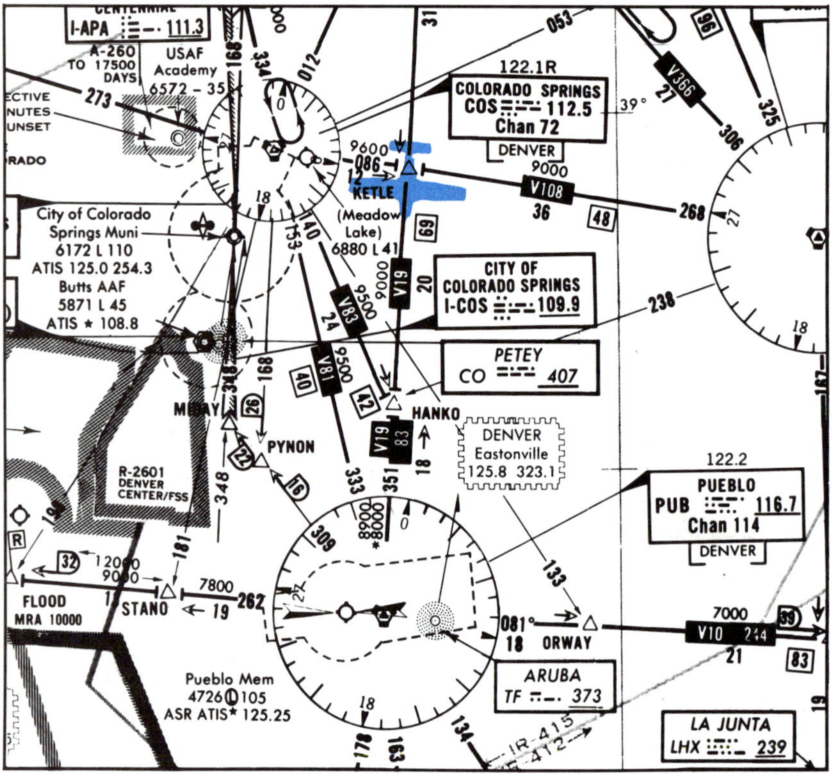

Figure 4-8. Assume you are enroute to Pueblo, Colorado, via V-19 in IFR conditions at the MEA altitude of 9,000 feet. Near Hanko Intersection, the controller advises, *"Arrow 9433P, expect handoff to Pueblo Approach Control at Hanko Intersection, ILS runway 26R approaches in progress."*

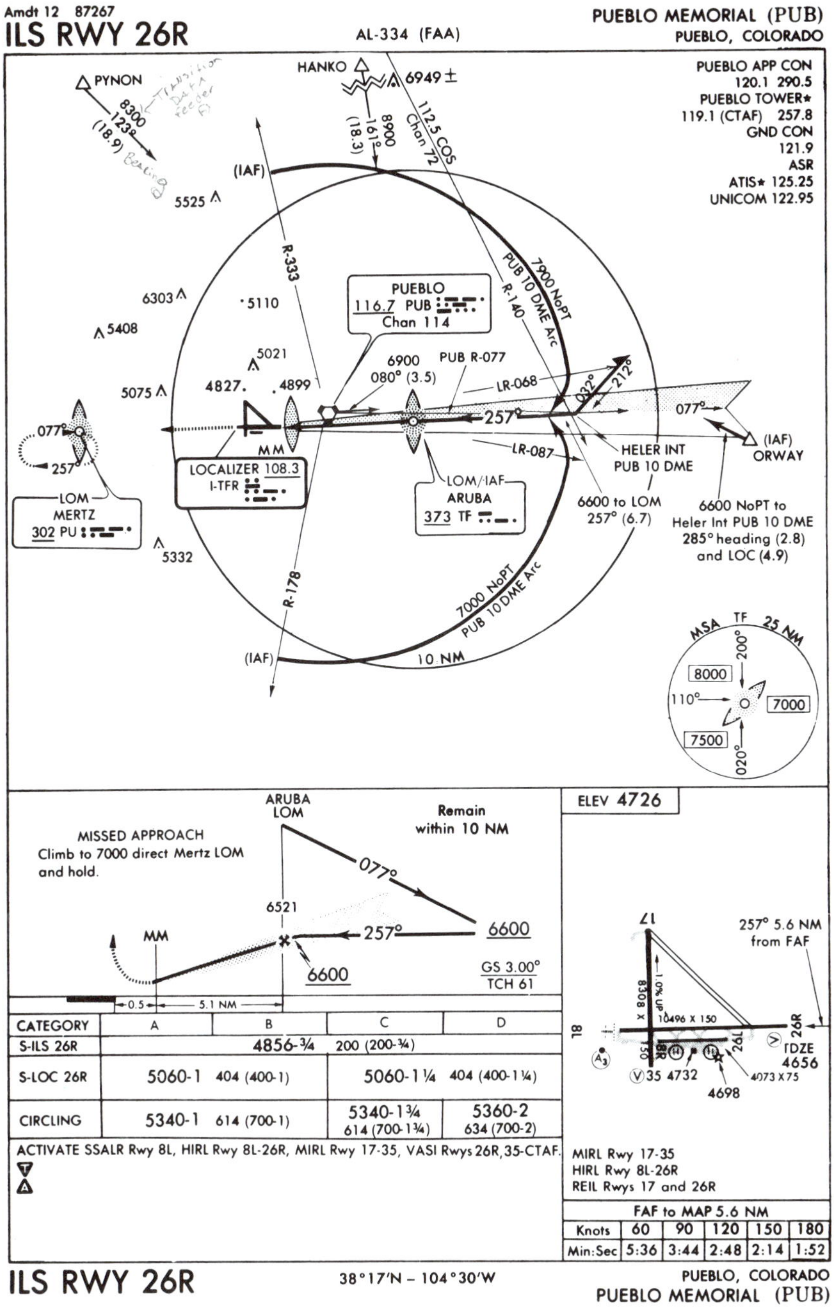

On NOS charts, an underlined navaid frequency indicates that there is no voice capability on that frequency.

Figure 4-9. Notice the variety of feeder routes and other transitions. Several of these do not require or permit a procedure turn. Refer back to this illustration during the accompanying discussion.

APPROACH CHART REVIEW

The ILS runway 26R approach chart shows a 10-mile DME arc beginning at the 333° radial from the Pueblo VORTAC. This point is the designated initial approach fix (IAF), which marks the beginning of the initial approach segment. In a nonradar environment, and assuming no other clearance or instructions from ATC, you are required to fly the complete instrument approach procedure as published. In this case, you cannot substitute interception of the DME arc for beginning at a published IAF. If a feeder route to an IAF is part of the published procedure, it is considered a mandatory part of the approach.

On the north DME arc, the minimum altitude is 7,900 feet MSL. Remember, the notation "NoPT" means that a procedure turn is neither required nor expected by the controller. In fact, you must not make a procedure turn without first obtaining specific ATC approval when you are flying a route which has the "NoPT" notation.

Your descent to the decision height can be started when established on the localizer course and after intercepting the glide slope.

After flying the 10 DME arc, you must make a right turn to intercept the localizer after you have crossed the Pueblo 068° lead radial. Lead radials are advisory points for turning to the final approach course. When you are established on the localizer course inbound, you may descend to 6,600 feet MSL. Begin your descent to the DH after glide slope interception, which will occur before reaching the Aruba LOM at an altitude of 6,600 feet MSL. The visibility minimum is three-fourths of a mile, and the altitude you can descend to with all ILS components operative is 4,856 feet MSL, or 200 feet above the touchdown zone elevation. Unless the controller issues alternate instructions, the missed approach procedure requires a climb to 7,000 feet direct to the Mertz LOM, frequency 302 kHz. A standard holding pattern west of the LOM is prescribed. Holding patterns and entry procedures will be discussed in Chapter 5.

The NOS airport sketch includes such items as runway lighting and the TDZE for the runway associated with the approach procedure.

The airport sketch shows that the airport elevation is 4,726 feet MSL. High intensity runway lights (HIRL), runway end identifier lights (REIL), and VASI are available for runway 26R, which is 10,496 feet long and 150 feet wide. The touchdown zone elevation (TDZE) is 4,656 feet MSL. The remarks section adjacent to the airport diagram indicates you can activate HIRL and VASI on the CTAF of 119.1 MHz.

Listening to the Pueblo ATIS on 125.25 reveals the weather is 300 overcast, three-fourths of a mile in fog and light rain, wind 280° at 8, and the altimeter setting is 29.97. Note that the visibility is at the minimum required for the approach.

SETTING UP THE APPROACH

You should reset your altimeter and tune in the frequencies for the approach. Set the ADF to 373 kHz for the LOM and the VOR/LOC receiver to 108.30 MHz for the localizer. Set the number two communications radio to the tower frequency of 119.1 MHz; leave the other VOR receiver and the DME tuned to the Pueblo VORTAC. Verify all navaid frequencies, including the DME, by listening to the identification feature, then turn on the marker beacon receiver and test it for proper operation. Once the navigation equipment has been properly set, you should consider completing the prelanding checklist to avoid any additional workload during the DME arc procedure. After the handoff to Pueblo Approach Control, you receive the following clearance. *"Arrow 9433P, cleared for the ILS 26R approach via the 10 DME arc, report . . . . ."*

TRANSITION VIA DME ARC

In a radar environment, a clearance directly to the 10 DME arc via V-19 could be expected. In this case, start the initial turn to intercept the 10 DME arc at your leadpoint, which should be 10.5 DME since your groundspeed is less than 150 knots. As you approach the leadpoint, begin a 90° turn to intercept the arc. The best equipment for maintaining the arc is an RMI, since it provides continuous position orientation. If you have only a basic VOR indicator or an HSI, maintaining the arc requires a much higher level of proficiency. After the initial turn, you will need to set the course selector to a heading of 360° and initiate a slow turn to 090°. As the aircraft crosses the 360° radial, set the course selector ahead another 10° to 010°, and turn the aircraft to the 90° intercept heading of 100°. Repeat this procedure as the aircraft progresses around the 10 DME arc.

While navigating the DME arc, you may need to adjust your headings to compensate for the wind from the west. If you undershoot or overshoot the initial turn for intercepting the arc, you will need to make additional heading changes to return to the arc.

Localizer interception is indicated by movement of the CDI to the center of the indicator.

After crossing the 068° lead radial from PUB, you should turn to intercept the localizer course and establish an appropriate approach speed. In this case, the approach speed will be 90 knots. You will know you have intercepted the course when the CDI moves to the center position. In addition, the ADF equipment will indicate a 0° relative bearing as the aircraft rolls out on a heading of 257°. After localizer interception, you should apply wind correction to keep the needle centered and report to the controller, *"Pueblo Approach, Arrow 9433P, localizer inbound."* An example of a typical controller response is, *"Arrow 33P, roger, winds 280° at 12, report Aruba to the tower on 119.1."*

On a well-executed ILS in calm wind, you should not need heading corrections greater than 2° after you have passed the outer marker.

While inbound on the localizer, establish required drift corrections before you reach the OM. Make small drift corrections and reduce them proportionately as the course narrows. By the time you reach the outer marker, your drift correction should be established accurately enough to permit you to complete the approach with heading corrections no greater than two degrees. Of course, wind shear can easily prevent this degree of accuracy.

If your groundspeed decreases, the rate of descent required to stay on the glide slope must also decrease, and vice versa.

If you maintain the glide slope for this approach, the DH will be reached at approximately the middle marker. Keep in mind that your rate of descent will change if your groundspeed changes. Although DH may be reached at or near the MM, the charted MAP for an ILS approach is the point where the glide slope intercepts the decision height. This point may not necessarily be at the middle marker.

The DH is the MAP on a precision approach if you have not established the required visual references.

The conditions under which you can descend below DH or MDA on an approach are specified in FAR 91.175. First, your aircraft must continuously be in a position from which a descent to a landing on the intended runway can be made at a normal rate of descent using normal maneuvers. Second, the flight visibility must not be less than that prescribed for the procedure you are executing. Third, at least one of the listed visual references for the intended runway must be distinctly visible and identifiable to you. If these conditions are not met at DH, a missed approach is mandatory. This is why the MAP on a precision approach is the DH. A missed approach also is required if you cannot maintain these requirements all the way to touchdown. During a missed approach, you must comply with the published procedure unless an alternate one has been specified by ATC.

ADF TRANSITION

When an NDB is used on an ILS approach, you may also use ADF navigation to make transitions to the approach. For example, if DME equipment were not available for the previous approach, the following procedure would be appropriate. [Figure 4-10]

FLYING THE APPROACH

When using ADF, remember to identify the NDB and check the bearing indicator for proper operation.

Before you reach Hanko, tune your ADF receiver to 373 kHz for the Aruba LOM, identify the station, and check the bearing indicator for proper operation. When you report at Hanko Intersection, Pueblo Approach will issue an approach clearance such as, "*Arrow 9433P, cleared for ILS runway 26R approach, report Aruba outbound.*" Begin an immediate left turn until the ADF bearing indicator points directly ahead of the nose of the aircraft; then start a descent to 8,900 feet MSL.

You should report leaving your last assigned altitude and make a check of the ADF bearing indicator for any apparent wind drift while stabilizing the descent. It may be necessary to apply wind correction while you transition to the LOM.

APPROACH COURSE OUTBOUND

Since your aircraft will be approaching the localizer course outbound, reverse CDI sensing will be apparent. Watch for movement of the localizer needle and the ADF bearing indicator. As the localizer needle moves

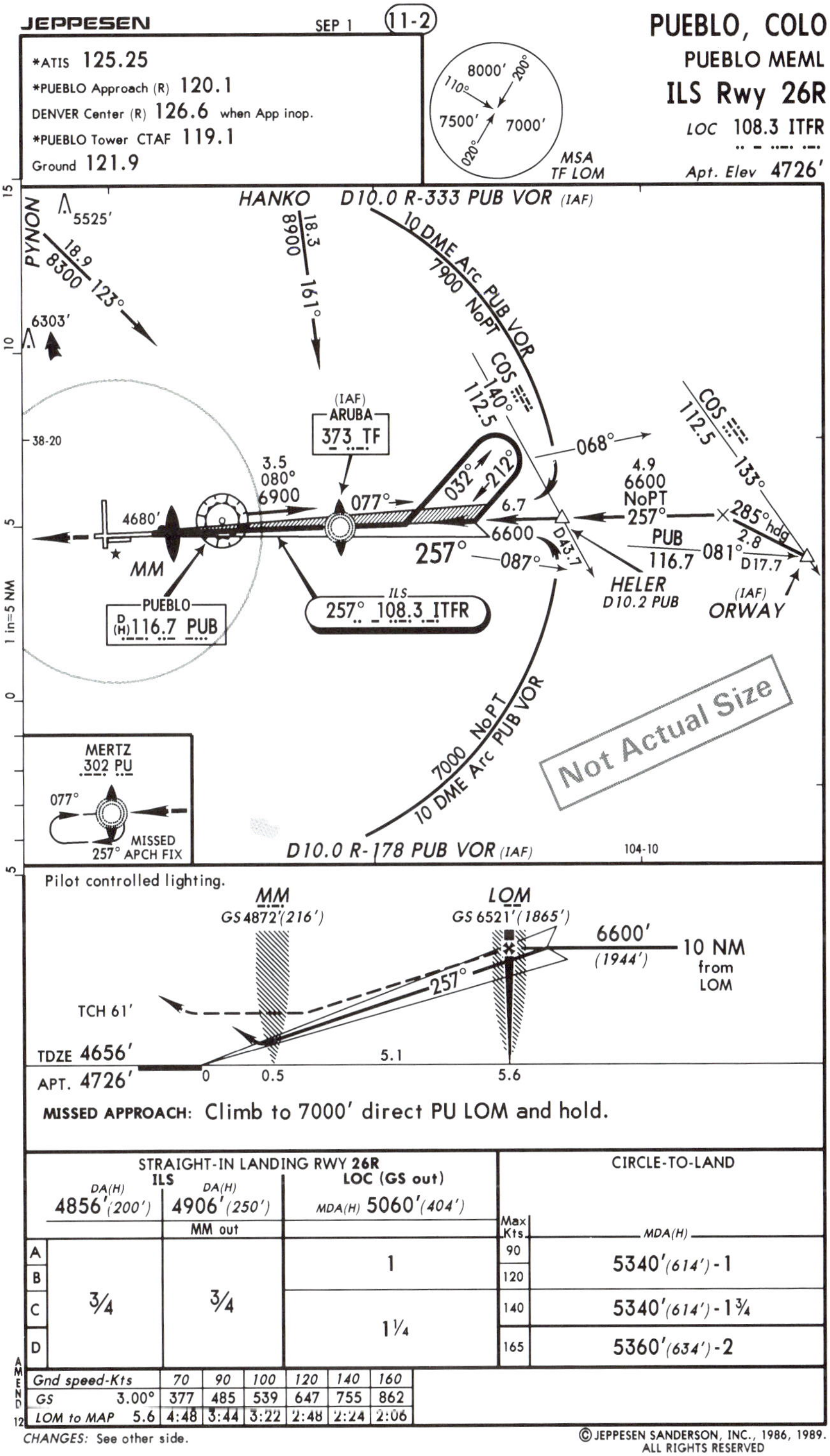

Figure 4-10. A review of the approach chart reveals a feeder route from Hanko Intersection to the Aruba LOM on the 161° bearing to the station. The minimum altitude for the feeder route is 8,900 feet MSL. During the procedure turn, your minimum altitude within 10 n.m. of the LOM is 6,600 feet MSL.

from full left deflection, the ADF bearing indicator will begin to move away from the nose position and the blue marker beacon light will flash. At this point, turn the aircraft to an outbound heading of 077°, the localizer course outbound, then report, *"Pueblo Approach, Arrow 9433P, Aruba outbound."* Approach control will then instruct you to *". . . report Aruba inbound to the tower on 119.1."*

Procedure turns are the most common method to use for course reversal.

After you are established on the approach course outbound, it will be necessary to reverse direction. The most common maneuver used to reverse course and establish your aircraft on the intermediate approach segment or the final approach course is the **procedure turn.** The outbound side of the course where the turn is made, the direction of turn, the distance within which the turn must be completed, and the minimum altitude for the turn are all specified in the procedure. However, unless otherwise restricted, the point at which you may begin the turn and the type and rate of turn are normally left to your discretion.

The FAA recommends that you begin the procedure turn as soon as practical, normally within two minutes of station passage. This should allow you to complete it within 10 n.m. of the fix. After tracking outbound from Aruba, turn the aircraft to 032° and fly for approximately one minute for the procedure turn outbound. During this segment, you should complete the prelanding checklist. After one minute, complete a 180°, standard-rate turn to 212°. Set the course selector to 257° as a reminder of the inbound course. Course interception is apparent from the CDI information, as well as a 045° relative bearing on the ADF indicator. After course interception, apply any necessary wind correction to keep the localizer needle centered and complete the approach to DH as previously described.

ILS PROCEDURES WITH RADAR

Radar is used to expedite traffic by providing vectors to the final approach course.

When radar is approved for approach control service, it is also used to provide vectors to published instrument approach procedures, such as the ILS. Radar vectors can be provided for course guidance and for expediting traffic to the final approach course of any established instrument approach procedure. Whenever ATC is providing radar vectors to the ILS approach course, you normally will be advised by the controller or through ATIS. The following is an example of a typical vector clearance. *"Mooney 782JM, descend and maintain 7,000, turn right heading 210° vector to the ILS runway 26R approach course."* As you get closer to the localizer, another vector and the approach clearance will be issued. For example, *"Mooney 782JM, position 11/2 miles east of the outer marker, turn right heading 257°, maintain 7,000 until intercepting the localizer; cleared for ILS runway 26R approach, contact the tower 119.1 at Aruba."*

After receiving an approach clearance, maintain your last assigned altitude until you are established on a published segment of a route or instrument approach procedure. After you are established, published altitudes apply to descent within each succeeding segment, unless ATC assigns a different altitude. This is important when you are receiving radar vectors or when you are operating on an unpublished route, since a premature descent could compromise your obstruction clearance. Controllers are supposed to provide you with an altitude to maintain until you are established on a published segment of a route or instrument approach, or they may withhold the approach clearance until you are established.

When cleared for an approach, you must be established on a published route or instrument approach procedure before you can initiate a descent from your last assigned altitude.

If it becomes necessary for spacing, you may also be vectored across the approach course. If you are near the approach course and have not been informed that you will be vectored across it, question the controller. You should not turn inbound on the final approach course unless you have received an approach clearance. Approach control will normally issue this clearance with the vector to intercept the final approach course, and the vector should enable you to establish your aircraft on the final approach course prior to reaching the final approach fix.

In the event you are already inbound on the final approach course, you will be issued an approach clearance before you reach the final approach fix. Radar separation will be maintained between you and other aircraft, and you will be expected to complete the approach using the ILS as the primary means of navigation. Therefore, once you are established inbound on the final approach course, you may not deviate from it unless you receive a clearance to do so from ATC. On final approach, you will be instructed to monitor the tower frequency. The radar controller will issue any advisories on the tower frequency. After passing the final approach fix inbound, you are expected to proceed direct to the airport and complete the approach, or to execute the published missed approach procedure. Radar service is automatically terminated when the landing is completed.

If ATC observes that you are deviating from the final approach course during simultaneous approaches, you will be informed on tower frequency of your aircraft's position and asked what your intentions are.

Under certain conditions, ILS approaches may be flown to parallel runways at the same time. The term "**simultaneous ILS approaches**" is used when the runways are separated by at least 4,300 feet between centerlines, and radar service is provided to ensure proper lateral separation of the aircraft. "**Parallel ILS approaches**" may be flown when the runway separation is at least 2,500 feet and aircraft maintain a minimum staggered separation. Radar monitoring is not required. According to the AIM glossary, when aircraft are established inbound toward the airport on the adjacent approach courses, they are separated by at least two miles.

A parallel ILS approach provides aircraft a minimum of two miles separation between successive aircraft on the adjacent localizer course.

ILS PROCEDURES WITH DME

Although separate ILS approach procedures share many similarities, there also are many differences due to facilities, equipment, terrain, and airport location. The important thing to remember is that every instrument approach is unique. There is no such thing as a "typical" ILS procedure, so you always need to study each approach carefully. As an example, consider the ILS DME runway 11 approach to Walker Field in Grand Junction, Colorado. [Figure 4-11]

When an ILS/LOC is equipped with DME, the facility box on Jeppesen charts has the notation "ILS/DME." On NOS charts, a DME/TACAN channel is shown in the facility box.

A review of the approach chart shows that Grand Junction is served by an approach control facility with radar. This is indicated by "(R)" following the name of the approach control. The highest MSA in the area is 11,700 feet MSL. In addition to a course from the EKR VORTAC, there is a 14 DME arc available from the Grand Junction VORTAC for transitioning to the final approach course. Also, note the outlined facility box for the localizer. The notation "ILS DME" on Jeppesen charts means that DME information is provided from the localizer transmitter on 110.3 MHZ. NOS charts include the DME/TACAN channel in the localizer frequency box. A note in the profile view reminds you to use the IGJT ILS DME when you are on the localizer course. You can see the potential for error if the DME remains tuned to the Grand Junction VORTAC during the approach.

The Grand Junction VORTAC is offset from the approach course and can be used to identify Rhone Intersection, which is the nonprecision final approach fix. You can also use a 6.2 DME indication from the IGJT ILS DME for the same purpose. Of course, the precision approach FAF is the point where an aircraft at the published glide slope intercept altitude intercepts the glide slope.

Notes in the plan view indicate availability of pilot controlled lighting and arrivals via V-187 LOMMA are not authorized. Although not all obstructions are charted, the highest charted obstruction within the plan view is noted by an arrow. This is the 8,720 foot obstruction at the top of the chart.

With all components of the ILS in operation, your DA(H) is 5,033 feet, and you need one-half statute mile visibility to complete the approach. The missed approach procedure calls for you to climb on the runway heading to 6,100 feet, then a climbing right turn to 10,000 feet and fly direct to the JNC VOR, and hold as published.

If the glide slope is inoperative or fails during your approach, the localizer (GS out) minimums apply and you may continue the approach to the applicable MDA.

If any component of the ILS such as the glide slope or approach lighting system is inoperative, you should refer to the appropriate column in the landing minimums section. For example, if the glide slope is inoperative, the approach is no longer considered a precision approach and the localizer (GS out) minimums apply. This is also true in cases where the glide slope warning flag appears during the approach, whether inside or outside the FAF. Notice that a DH is not specified in the "GS out" column and an MDA(H) of 5,180 feet applies. The visibility requirements remain the same, except for Category D aircraft, and when the approach light system is out.

Without the required visual cues, a missed approach must be started at the MAP, even if you haven't completed the descent to MDA.

The localizer (GS out) missed approach point is 1.8 IGJT ILS DME. Timing to the missed approach point is not published for this procedure because DME, which is included in the title, is a required navaid and it should be

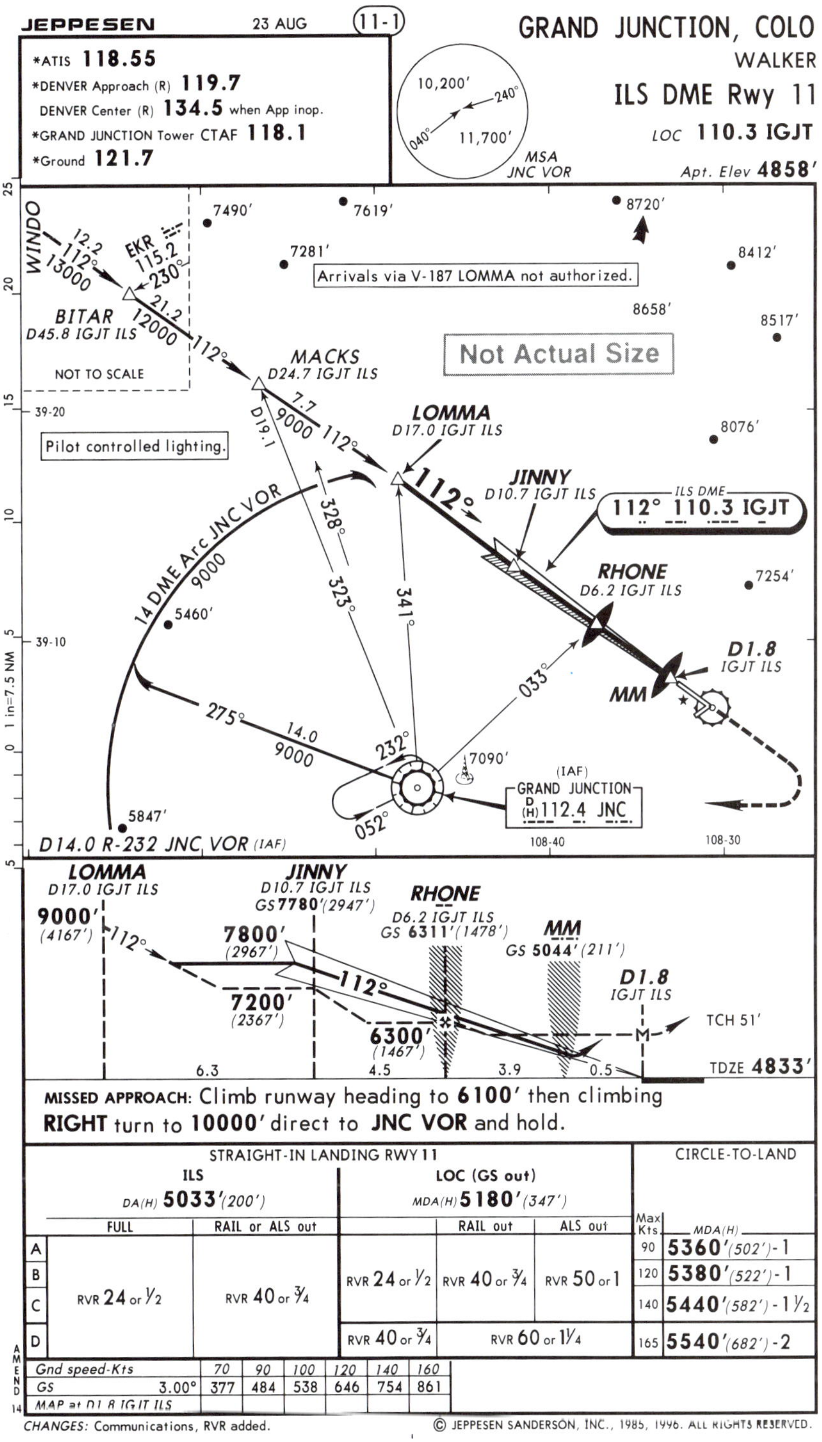

Key information on newer Jeppesen charts is displayed in big, bold type. This key information includes communication frequencies, primary NAVAID frequency and identifier, procedure headings and altitudes, airport and runway elevation, DH or MDA, and missed approach turn limit and direction, course, and altitude.

Figure 4-11. You are required to have DME equipment for the approach when specified in the approach chart title. Absence of the DME notation in the title means you are not required to have DME equipment, even if it is available at the facility. On this approach, you need DME to fly the arc and to identify key points along the approach. You may want to refer back to this approach chart while following the discussion on the opposite page.

used for determining the MAP. At the MAP you must execute the missed approach if you have not established the required visual cues. This applies even if you have not yet reached the MDA.

BACK COURSE APPROACHES

Back course approaches are common when ILS equipment is installed. Each one has its own unique features, so you need to study the approach chart carefully before flying the procedure. The LOC BC runway 13 approach to Hibbing, Minnesota, is an example of a back course procedure. [Figure 4-12]

A review of the approach chart shows two feeder routes are available for transitioning to the IAF at Kinny Intersection. One feeder is the 045° radial from the Grand Rapids VOR/DME, and the other is the 307° radial from the Hibbing VOR/DME. Both have a minimum altitude of 3,500 feet MSL. If you use Kinny as the IAF, you must make a procedure turn to reverse course. If your aircraft is DME-equipped, you can fly either of the designated 20 DME arcs instead of a procedure turn.

If your aircraft is DME-equipped and the localizer and DME transmitters are collocated, you normally receive DME information when you select the localizer frequency.

Kinny Intersection also is the final approach fix which is on the localizer at 14 DME from the Hibbing VOR/DME. Without DME, you can identify the fix with the 045° radial from the Grand Rapids VOR/DME. Notice on this procedure that DME information comes from the Hibbing VOR/DME (not the localizer). Where the localizer and DME transmitters are collocated, most equipment automatically provides DME information when you select the localizer frequency. This is not the case on this procedure, since the DME frequency is paired with the Hibbing VOR/DME frequency of 110.8 MHz.

You are required to fly at or above the minimum altitude specified on the approach procedure.

The descent profile for this approach requires that you remain at or above 3,500 feet MSL from any of the IAFs to the FAF. After the FAF, you can descend to the MDA of 2,040 feet. Further descent requires visual contact with the runway environment unless you have DME equipment. From the final approach fix inbound, several supplemental DME fixes are available to aid you in maintaining orientation with the runway. The note in the profile view reminds you to disregard any glide slope indications. At the 9.2 DME fix, there is a visual descent point, or VDP, designated for the approach. A VDP defines the point when you can begin descent from the MDA for a straight-in landing, provided you have the required visual cues. If your aircraft is not DME-equipped, you should fly the approach as though the VDP did not exist.

Another advantage to having a DME-equipped airplane for this approach is that you can descend another 260 feet to the straight-in MDA. This is shown in the DME minima table. The letters "T" and "A" in triangles below the minima tables remind you that both nonstandard takeoff and alternate minimums apply to this airport.

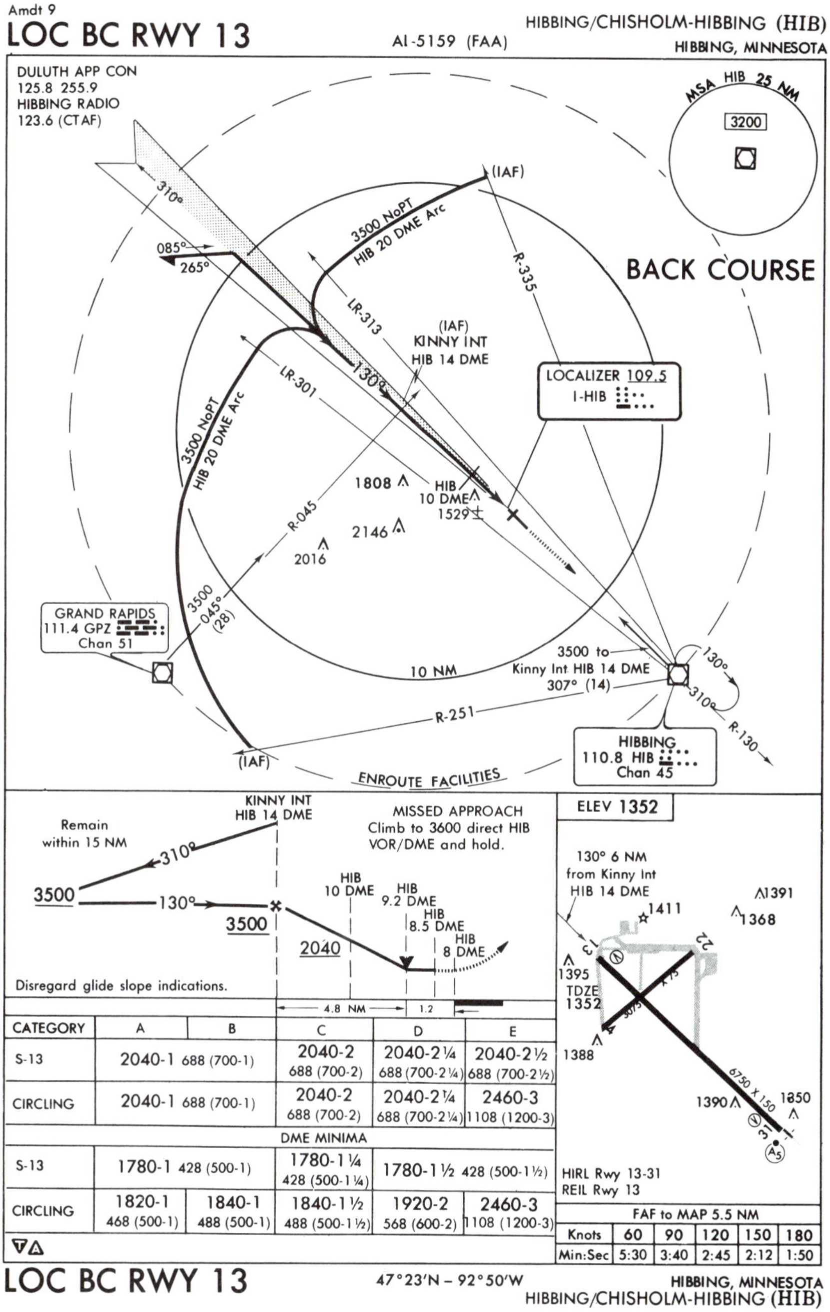

Figure 4-12. Remember, reverse sensing will occur on the localizer inbound when you are using a basic VOR. With HSI equipment, you can avoid reverse sensing if you set the front course on the course selector.

The missed approach point can be identified by timing the approach from the FAF to the MAP, or by using the 8.5 DME HIB fix. The missed approach procedure is a climb to 3,600 feet direct to the Hibbing VOR/DME and hold on the 130° radial.

CHECKLIST

After studying this section, you should have a basic understanding of:

✓ **Ground components** — What the various parts of an ILS are and how they provide precise guidance to the runway.

✓ **Marker beacons and compass locators** — How they are used in conjunction with an ILS and what equipment you need to receive them.

✓ **ILS visual aids** — What lighting and marking aids are associated with a full ILS and how equipment outages affect descent minimums.

✓ **ILS categories** — What the terms Category I, II, and III mean and what the landing minimums are for each category.

✓ **Flying the ILS approach** — What the recommended procedures are for making pitch and power adjustments to maintain the glide slope.

✓ **Approach transitions** — How DME arcs and NDBs can be used to establish an aircraft on the approach course.

✓ **ILS procedures with DME** — How you can tell which ILS approaches require DME and which ones use DME only for supplemental fixes.

✓ **ILS procedures with radar** — How radar vectors are used to abbreviate an ILS approach.

✓ **Back course approaches** — What their unique features are and how they differ from front course approaches.

SECTION B

VOR APPROACHES

VOR approaches are widely used in the national airspace system. These approaches provide for final descents as low as 250 feet above the runway. However, the MDAs for VOR approach procedures typically range from 500 to 1,000 feet.

APPROACH CLEARANCE

In many cases, you may receive an approach clearance from a terminal approach control facility at the destination airport. At locations which do not have approach control, your clearance will be issued directly from an ARTCC, if air/ground communications permit, or indirectly through flight service stations. A typical VOR approach clearance from a radar-equipped approach control facility may be issued as follows: "*Cessna 1736U, now three and one-half miles north of Delta; turn right heading 180° for vector to final approach course, maintain 7,000 until crossing Tyler VOR, cleared for the VOR runway 20 approach; contact the tower 118.3 over the VOR inbound.*" In this case, a radar vector is provided in lieu of the initial approach segment, and the published procedure turn or other initial approach segment will not be used.

When vectors are not used at terminal locations or at nonradar approach control facilities, the following clearance may be issued. "*Cessna 1736U, cleared for VOR approach, report the VOR inbound to the tower on 118.3.*" In this case, if more than one VOR approach is available, you may have your choice of procedures. When the controller wants you to fly a specific procedure, it will be stated; for example, "*Cessna 1736U, cleared for VOR runway 36 approach.*"

At nontower airports without approach control service, the ARTCC may issue a cruise clearance. An example of such a clearance is, "*Cessna 1736U, cleared to the Winnsboro Airport, cruise 5,000.*" If you have been issued a cruise clearance, you may proceed to the destination airport, execute any published approach procedure available, and land if visibility minimums are met. Normally, cruise clearances are used for relatively short flights in uncongested areas.

VOR APPROACH PROCEDURE

Nonprecision approaches are of two general types — those that use a navaid located beyond the airport boundaries and those with the navaid located on the airport. You can tell which type is used for individual procedures from the approach chart profile. [Figure 4-13]

An example of an approach with an off-airport navaid facility is the VOR approach procedure for Winnsboro, Texas. It uses the Quitman VORTAC, located 5.5 n.m. southwest of the airport. [Figure 4-14]

At Winnsboro Municipal, the final approach course is not aligned with the runway. Therefore, straight-in landing minimums are not shown, and only circling maneuvers to either runway are permitted, as indicated by the notation "VOR-A". You should use the Tyler altimeter setting for this approach. It is available from Longview Approach on 128.75 or Fort Worth Center on 132.85 MHZ. The minimum safe altitude within 25 n.m. of Quitman is 2,600 feet MSL. This provides at least 1,000 feet of obstruction clearance. The minimum procedure turn altitude is listed in the profile section as 2,000 feet MSL. Further descent is not authorized until you cross the FAF. However, some approaches permit a descent between the procedure turn and the final approach fix. This procedure has an MDA of 1,360 feet MSL if the Tyler altimeter setting is available. The time from the FAF to the missed approach for various groundspeeds is shown at the bottom of the chart. The MAP also may be identified with the 5.5 DME fix. Additionally, the missed approach procedure uses a holding pattern. Now, let's conduct an approach using this chart.

OFF-AIRPORT FACILITY

Assume you are on an IFR flight in a Category A aircraft, and you plan to use the VOR-A approach to Winnsboro Municipal Airport. You are

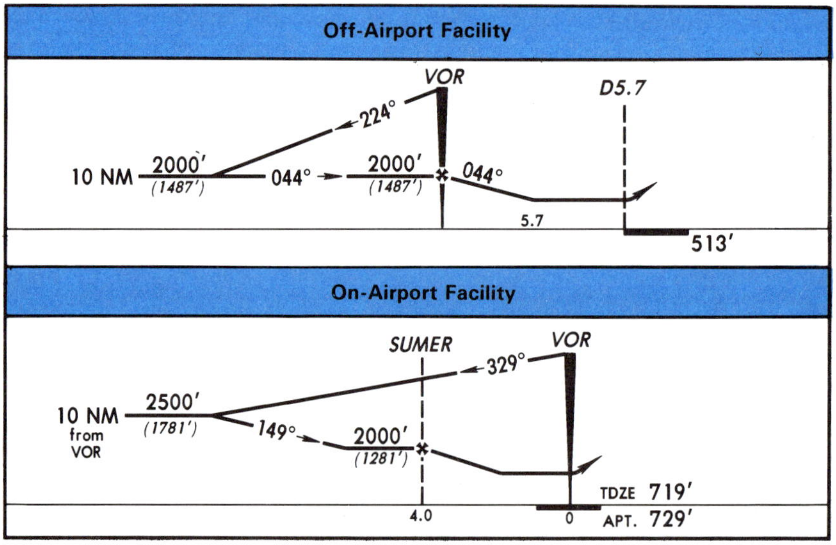

Figure 4-13. These approach chart profiles illustrate the basic differences in approaches with off-airport and on-airport facilities. When the VOR is not located on the airport, it may also be both the IAF and the FAF.

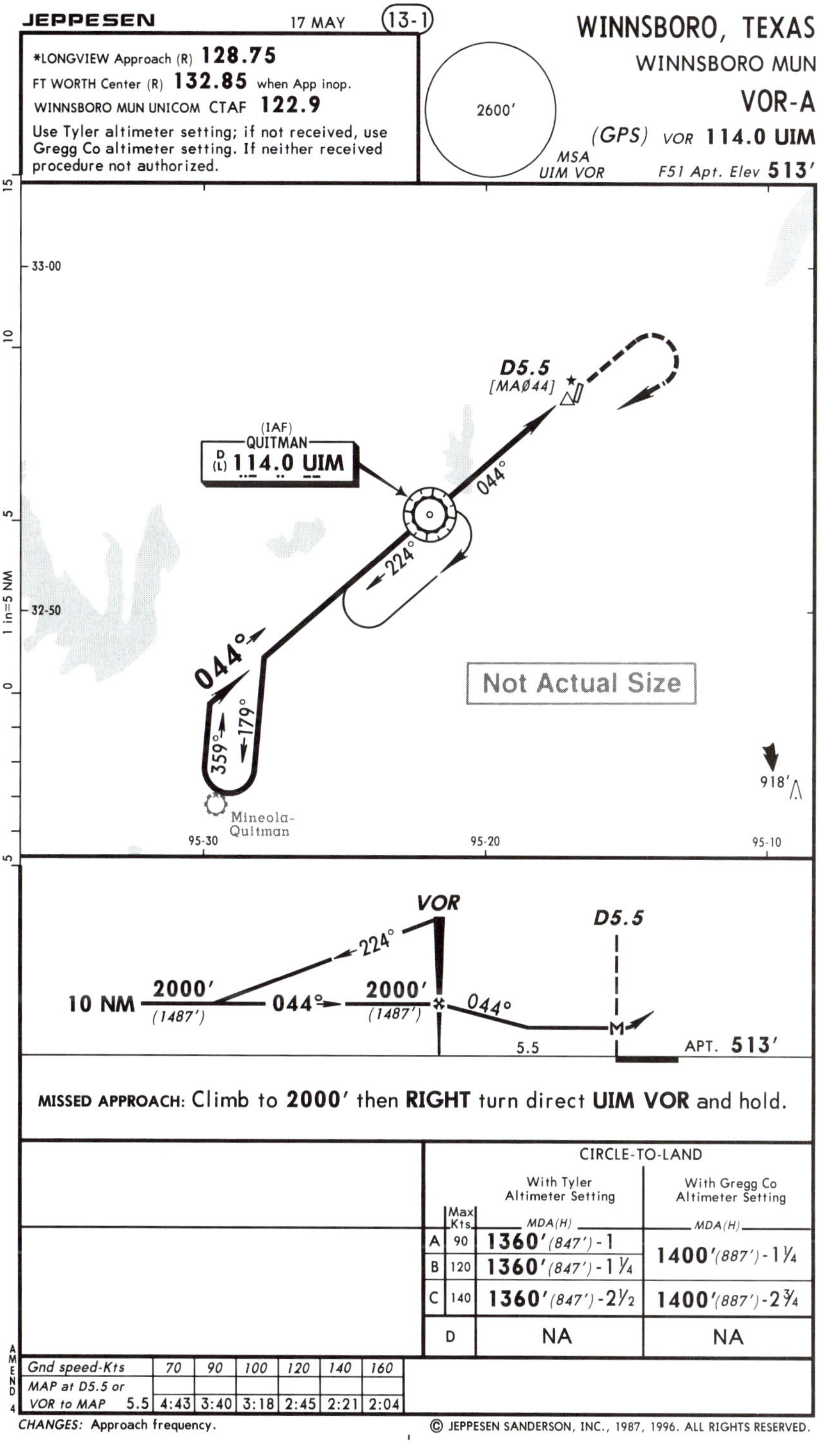

Figure 4-14. The navigation facility for this approach is also part of the enroute structure. Consequently, feeder routes, which normally provide a means of transitioning from the enroute structure, are not required.

approaching Quitman from the northwest on V-114 at an assigned altitude of 5,000 feet MSL. [Figure 4-15]

If you have at least one mile visibility at the MEA, you may request a contact approach instead of flying the published VOR procedure. If visibility is not adequate, you must comply with the published approach procedure before you descend further.

Since you will approach the Quitman VORTAC on V-114 (magnetic course of 108°), you should be prepared to turn right to an outbound intercept heading for the 224° radial and, once established, descend to 2,000 feet, the procedure turn altitude. If you turn initially to 224°, you will only parallel the course. [Figure 4-16]

As you cross the VOR, begin timing for the outbound leg and verify that your altitude is no lower than 2,000 feet. For this particular approach, outbound timing before the procedure turn depends on the speed of the aircraft, since the procedure turn should be completed within 10 n.m. of the facility. For most light, single-engine aircraft, two minutes is recommended to remain within the 10 n.m. limit. You should also adjust outbound timing for the effects of known wind. According to the procedure turn depiction, you may turn left to a magnetic heading of 179° and maintain that heading for one minute. Adjust the course selector to the inbound approach course (044°) and, after one minute, start a standard-rate turn to an inbound procedure turn heading of 359°. During the procedure turn (initial segment), you may set the airspeed and configuration for the final approach. Immediately following inbound course interception, you should determine the heading required to maintain course. Check the heading indicator against the compass and reset it, if necessary.

At the VOR inbound, you must note the time you passed the station, begin a descent, and report that you are at the FAF inbound. An FAF inbound report is not required when you are in radar contact. [Figure 4-17]

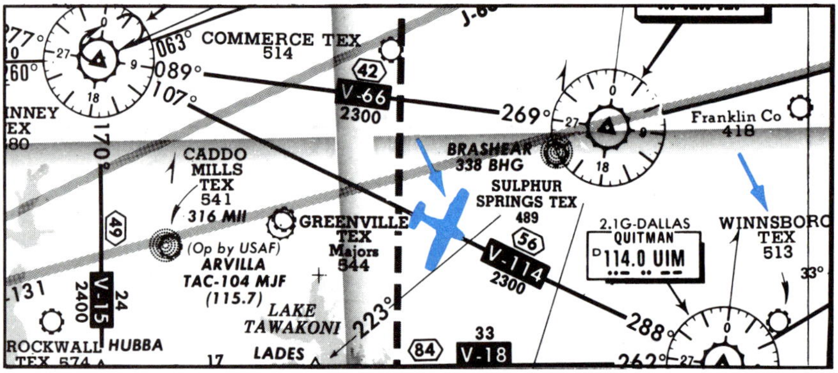

Figure 4-15. On this route, it is likely you will receive your approach clearance directly from Fort Worth Center. If Center issues a cruise clearance, you can descend first to the MEA of 2,300 feet MSL.

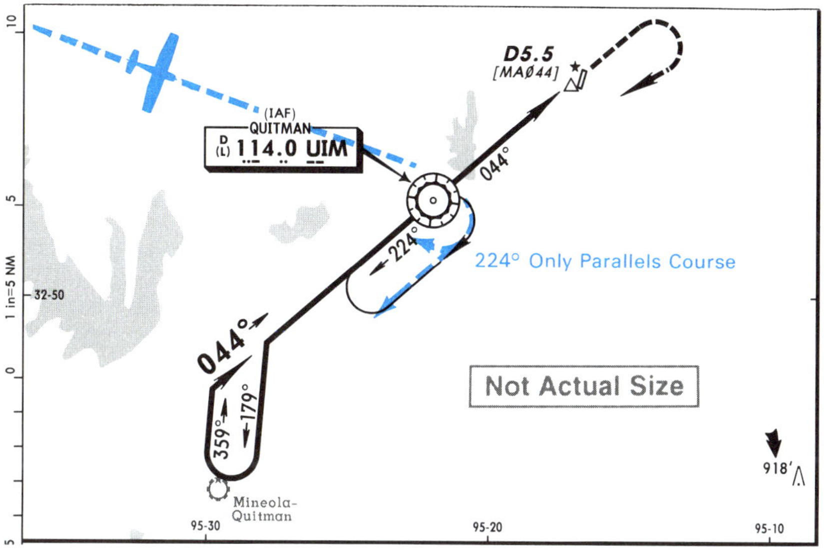

Figure 4-16. Due to the large difference in magnetic course between V-114 and the outbound approach radial, an intercept angle of 45° is appropriate. You should turn right to 269° and reset the course selector to 224°, the outbound course for the initial segment. As the CDI begins to center, begin a left turn to 224°.

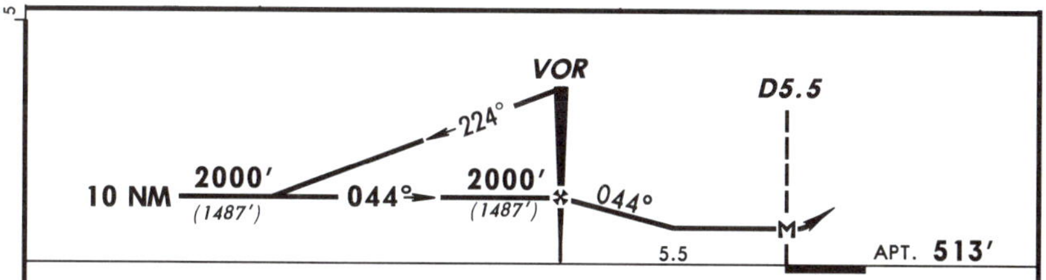

	Max Kts	CIRCLE-TO-LAND With Tyler Altimeter Setting MDA(H)	CIRCLE-TO-LAND With Gregg Co Altimeter Setting MDA(H)
A	90	1360'(847')-1	1400'(887')-1¼
B	120	1360'(847')-1¼	
C	140	1360'(847')-2½	1400'(887')-2¾
D		NA	NA

Gnd speed-Kts	70	90	100	120	140	160
MAP at D5.5 or VOR to MAP 5.5	4:43	3:40	3:18	2:45	2:21	2:04

AMEND 4

CHANGES: Approach frequency.

Figure 4-17. If your approach groundspeed is 90 knots, the conversion table at the bottom of the approach procedure shows the time from the FAF to MAP will be 3 minutes, 40 seconds. You should complete the descent to the Category A MDA of 1,360 feet MSL well before this time has elapsed. As you approach the time limit, attempt to establish visual contact with the airport. If you have not located the airport at the end of 3 minutes and 40 seconds, you must make a missed approach.

If the airport and runway environment are clearly visible before reaching the MAP and, in your judgment, you can make a safe approach and landing, you can initiate a circling maneuver to the favored runway. You must not descend below the MDA during the circling maneuver until you are in a position from which you can make the final descent to the runway. If at any time during the circling maneuver you are unable to keep the airport in sight, you must execute the missed approach procedure immediately.

You may land straight in from an approach with only circling minimums, provided certain requirements are met.

Remember, straight-in minimums are shown only on instrument approach charts when the final approach course is within 30° of the runway alignment and a normal descent can be made from the MDA or DH to the surface. If these conditions are not met, straight-in minimums are not published and the circling minimums apply. The fact that a straight-in minimum is not published doesn't mean that you cannot land straight in if you have the active runway in sight and have sufficient time to make a normal approach and landing. Of course, you must have clearance from the tower to land at a controlled field.

The missed approach procedure requires you to climb to 2,000 feet, then turn right and return to Quitman VORTAC. During the right turn, you should rotate the OBS approximately 180° until the CDI centers with a TO indication. As soon as practical after beginning the missed approach procedure, you should make a report to Fort Worth Center similar to the following: *"Fort Worth Center, Cessna 1736U, missed approach, request clearance for another approach (or clearance to the alternate)."*

If you reach Quitman before receiving additional clearance, the normal procedure is to execute a parallel entry into the holding pattern and report again: *"Fort Worth Center, Cessna 1736U, 2,000 feet, holding southwest of Quitman at 17, standing by for clearance."* Holding pattern entries are discussed in detail in Chapter 5.

ON-AIRPORT FACILITY

An example of a VOR approach with an on-airport facility is the procedure for Jefferson County Airport, Beaumont, Texas. This airport has a control tower, as well as a radar-equipped approach control facility. Beaumont Approach Control will issue your approach clearance after you receive a handoff from Center. [Figure 4-18]

The profile view shows the minimum altitude for crossing Compa Intersection and for completing the procedure turn. You may descend to a minimum of 1,600 feet MSL for the procedure turn, but you must not descend below that altitude until passing the FAF. Compa Intersection is identified by the 345° radial of the Sabine Pass VOR/DME and the 300° radial of the Beaumont VORTAC. You can make the final descent to the MDA for Category A aircraft to 480 feet MSL for a circling approach, or

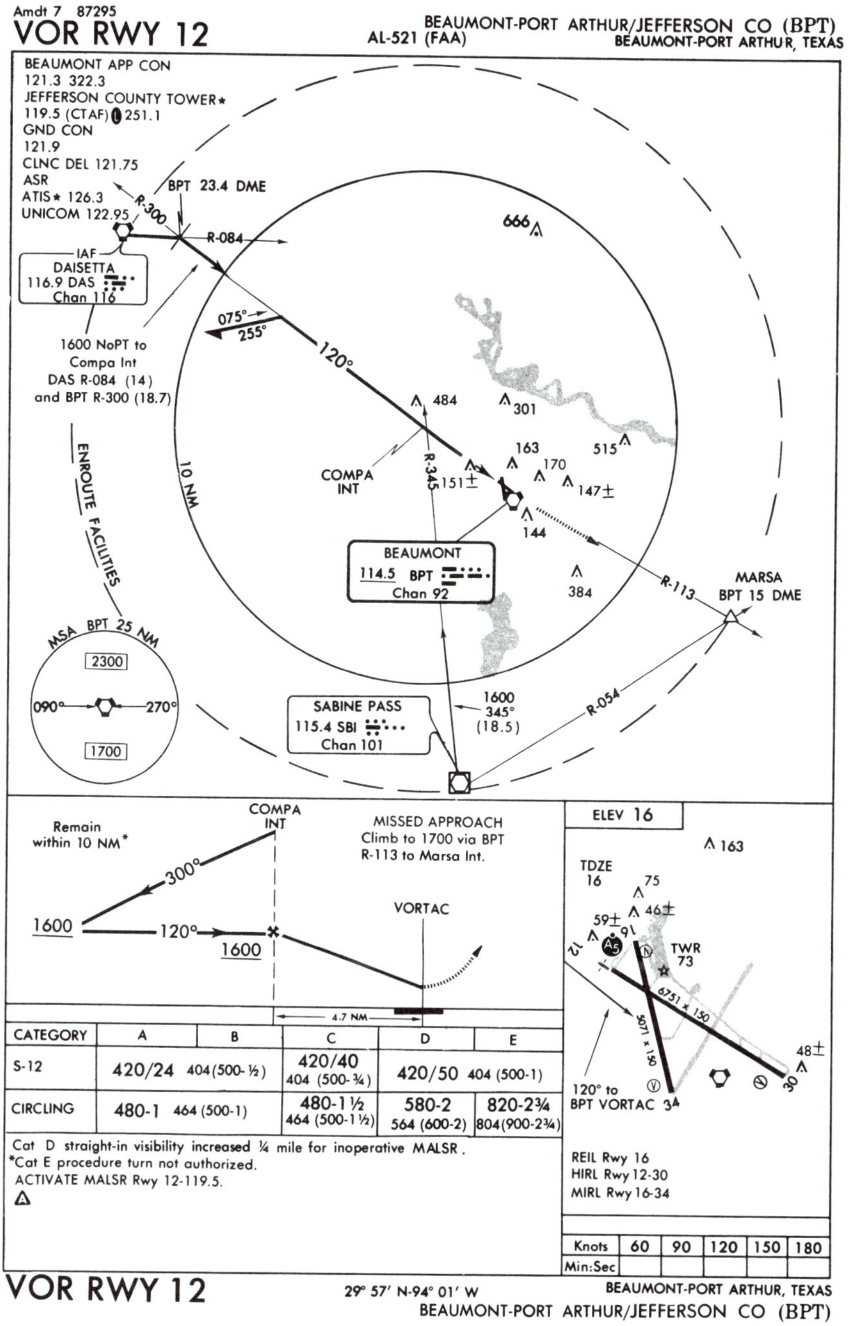

Figure 4-18. The final segment of this approach is aligned with runway 12, and both straight-in landing and circling minimums are provided. An initial approach segment is available from Daisetta VORTAC, and a feeder route is provided from Sabine Pass VOR/DME to Compa Intersection.

to 420 feet MSL for a straight-in landing on runway 12. The required visibility is one mile for a circling approach or RVR 2,400 for a straight-in landing. Since the VOR is the MAP on this procedure, timing from Compa Intersection is not required or shown on the chart. If the runway environment for runway 12 (or the airport when you are circling) is not visible at the MAP, you must execute the missed approach procedure. This is required whether or not you have reached the MDA. The missed approach procedure also applies if you cannot maintain visual contact with the runway or airport during the landing maneuver.

VOR/DME PROCEDURES

When the notation "VOR/DME" appears on an approach procedure, it means the use of DME equipment is mandatory for the approach. Although many procedures use supplemental DME fixes, the absence of the notation "DME" in the chart heading means the use of DME equipment is not mandatory for the approach. An approach procedure which requires DME navigation equipment is the VOR/DME approach to runway 31 for the Gregg County Airport in Longview, Texas. [Figure 4-19]

When cleared for an approach, you must not descend to the next lower altitude until you have reached a point where it is authorized.

According to the profile view, after passing Kapps, you can then descend to 2,000 feet MSL, which is the procedure turn altitude. After completing the procedure turn, you must remain at or above that altitude initially; at 14.0 DME (Argen Intersection) you may descend to 1,500 feet MSL. Start your next descent when crossing the FAF at the 8.0 DME fix inbound and your final descent to the MDA at the 6.0 DME stepdown fix. The limits of the DME arcs are marked by the 059° radial and the 172° radial. In many cases, these radials coincide with airways. If you were approaching on any airway within this span and had an appropriate clearance, you could intercept the 14 DME arc and proceed with the approach. When turning inbound from a DME arc, you must not descend until you are established on the approach course. When the procedure provides lead radials, you may not begin a descent to a lower altitude after crossing the lead radial unless the lower altitude is specified in the procedure. Otherwise, you must be established on course before descending.

After your flight receives a handoff from Fort Worth Center to Longview Approach Control, you will be provided with an approach clearance. The following is an example of an approach clearance using a DME arc: *"Piper 9733P, cleared for the VOR/DME runway 31 approach to the Gregg County Airport via the 14 DME arc. Contact Gregg Tower 119.2 crossing Kapps inbound."* A cross designates the nonprecision FAF in the profile section at the point where the final descent begins. A stepdown fix limits your descent to 800 feet MSL until you pass the 6.0 DME fix. The lowest MDA for a straight-in landing is 700 feet MSL.

The MAP on VOR/DME approaches is usually based on DME distance rather than time.

If a missed approach becomes necessary, you need to initiate it at the 3.9 DME indication. The missed approach procedure requires a climbing left turn to 2,500 feet via the Gregg County 172° radial to the holding pattern shown at Pipes Intersection. This requires either a teardrop or a parallel

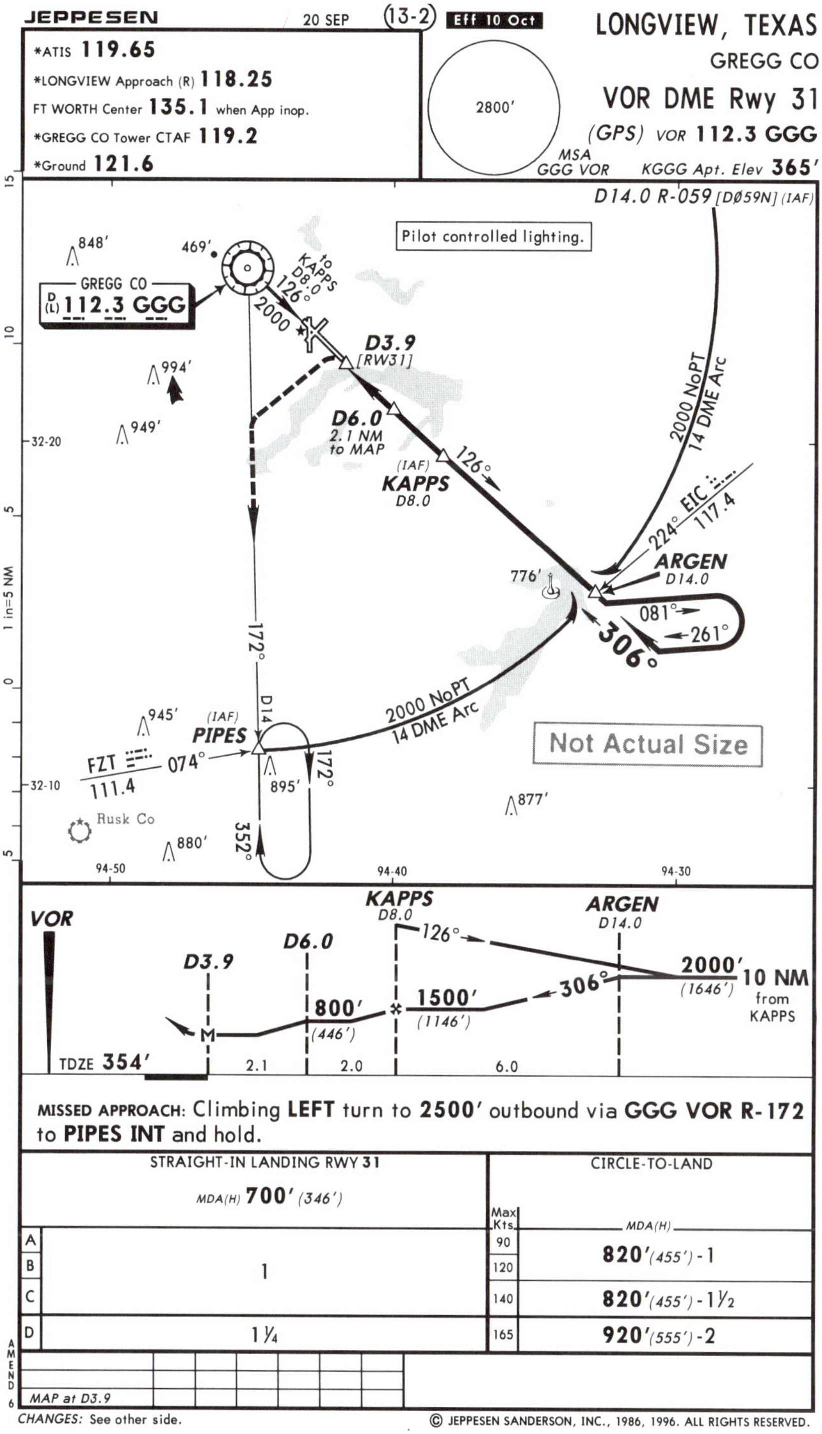

Figure 4-19. This approach provides a procedure turn and two 14-mile DME arcs for transition to the final approach course. The procedure turn is applicable when you use the feeder route from Gregg County VORTAC (126° radial) to arrive at the Kapps IAF, which is also an 8.0 DME fix.

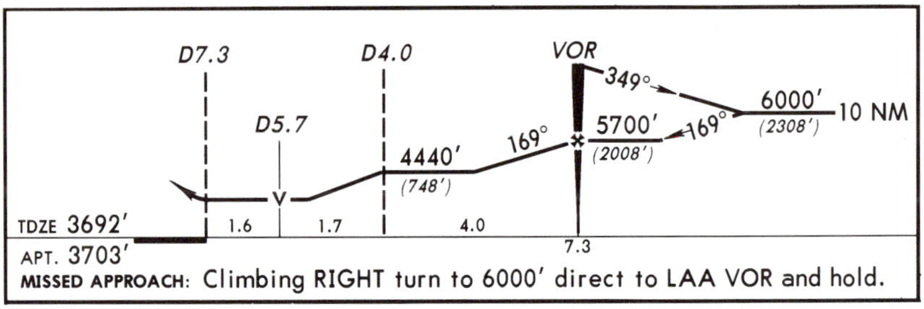

Figure 4-20. In this profile view example, you can descend to the lowest MDA if you can identify the 4.0 DME fix. After the fix, an MDA of 4,200 feet MSL is authorized. At the VDP (5.7 DME), you can begin a normal descent to the runway, provided you have the required visual cues.

holding pattern entry. If you execute the missed approach, you are normally instructed by the tower to contact approach or departure control, as appropriate. If you do not receive additional clearance before reaching the intersection, you should enter the holding pattern and report to ATC.

Since DME was required in the previous example, the minimums section reflected only one straight-in MDA. On VOR approaches that do not require DME but which still use supplemental DME fixes, two straight-in MDAs may be authorized. The lowest MDA applies only if you can identify the stepdown fix. In addition, it is not unusual to encounter a stepdown fix and a VDP on the same VOR approach. [Figure 4-20]

Keep in mind that, without DME, the stepdown fix minimum altitude (4,440 feet MSL) becomes the MDA, since you are unable to identify the stepdown fix. If this is the case, you won't be able to identify the VDP either, so you should fly the approach as though the VDP doesn't exist. Also, remember that VDPs apply to nonprecision approaches where you are making a straight-in landing; they are not applicable to circling approaches.

RNAV APPROACH PROCEDURES

A waypoint is a predetermined geographical position used to define an RNAV route or RNAV approach.

In addition to VOR and VOR/DME approaches, there are a significant number of VORTAC-based RNAV procedures throughout the U.S. Although RNAV approach charts appear different from VOR procedures from an operational viewpoint, they are actually quite similar. The primary difference between an RNAV approach and a VOR approach procedure is that the fixes on an RNAV procedure are waypoints instead of intersections. A waypoint is a predetermined geographical position used to define an RNAV route, instrument approach, or a reporting position. Although they are commonly defined by radials and distances from a VORTAC, they also may be defined by geographical coordinates. In the case of VORTAC-based RNAV, you preset radial and distance information into your aircraft's RNAV equipment. Then, you can recall the waypoints as you need them during the approach. Let's examine an RNAV approach chart to see how they are designed. [Figure 4-21]

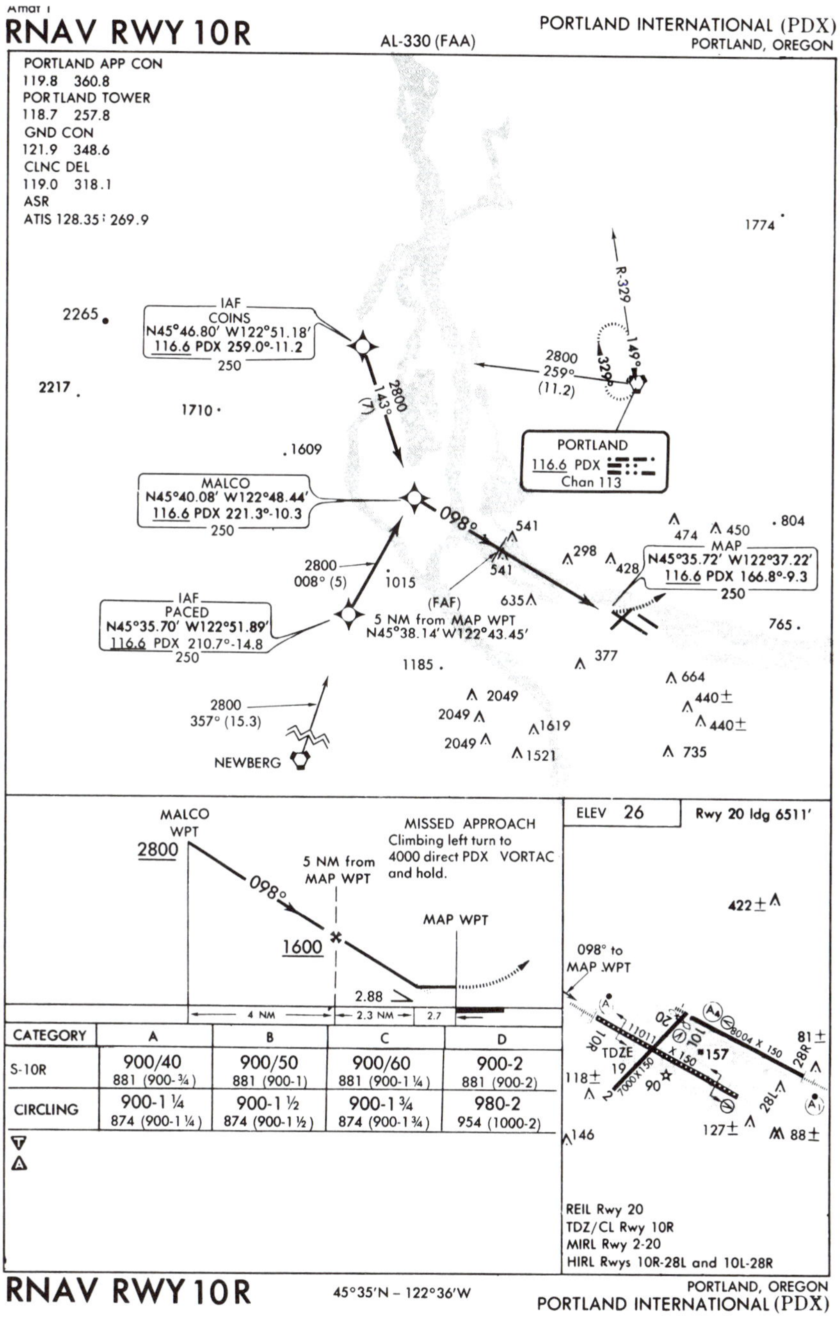

Figure 4-21. On an RNAV approach, a number of waypoints make up the fixes required to fly the procedure. Depending on your arrival route, not all of them are required.

There can be as few as two waypoints on an RNAV approach, because one waypoint may be used for more than one fix.

For each RNAV approach, the number of established waypoints will vary, although a minimum of two are required. When only two are necessary, you can expect that each one is used for more than one of the required approach fixes. The minimum number of waypoints required to fly the complete RNAV RWY 10R approach procedure at Portland, including the missed approach procedure, is three. You can use either of the IAF waypoints, plus MALCO and the MAP. The actual missed approach procedure requires you to fly direct to the PDX VORTAC, which is not a waypoint.

The waypoint data box provides the frequency of the navaid, radial/distance information, and the elevation of the VORTAC site.

The RNAV RWY 10R approach for Portland International Airport is a typical example of an RNAV procedure. Feeder routes are designated at Portland and Newberg VORTACs so you can position the aircraft on an initial approach segment. COINS and PACED are the two IAFs for this approach. Notice the distinctive waypoint data boxes. Each contains the navaid frequency, as well as the radial and distance information required to establish the waypoint. For example, the MAP is located on the 166.8° radial, 9.3 n.m. from the Portland (PDX) VORTAC on frequency 116.6 MHz. The number in the bottom line of the box indicates the elevation of the VORTAC site, which is 250 feet MSL in this case. With RNAV equipment, waypoint passage is the same as station passage with VOR; the TO/FROM indicator changes to FROM. The latitude and longitude information within these boxes is for use with other types of RNAV equipment, such as the inertial navigation system (INS).

On an RNAV approach which uses a holding pattern course reversal, you should turn inbound when the DME indicates you are at the specified distance from the waypoint.

As with other types of approaches, an RNAV procedure begins at the initial approach waypoint, which is the designated IAF. Initial approach segments based on a procedure turn are not established for RNAV procedures. When a course reversal is required, a holding pattern is used instead of a procedure turn. The legs of the holding patterns are usually based on DME distance from a waypoint. For example, if a holding pattern is shown with leg lengths of four nautical miles, you should turn inbound when your DME shows you are four nautical miles from the waypoint. You can see that the Portland procedure doesn't use a course reversal.

The intermediate segment of your approach to Portland starts when you turn to intercept the approach course after crossing the MALCO waypoint. You must cross this fix no lower than 2,800 feet MSL.

You may descend to the MDA after crossing the FAF, which is shown in both the plan and profile views.

When you cross MALCO, you can initiate a descent to 1,600 feet MSL. You can descend to the MDA after crossing the FAF, which is five nautical miles from the MAP waypoint. This is shown in both the plan view and the profile view of the approach chart.

RNAV approaches may have straight-in landing and/or circling minimums, as appropriate.

Typically, RNAV approaches meet the same requirements as nonprecision approaches. You can recognize this on the Portland approach by noting the presence of an MDA rather than a DH. This procedure has both straight-in landing and circling minimums. Assume V_{S0} for your aircraft is 77 knots, which puts it in Approach Category "B." For a straight-in landing to runway 10R, at least RVR 5,000 is required, and you can descend to the 900-foot MDA. To land on runway 20 from an approach to runway 10 means that circling minimums apply. For Category B aircraft, the circling MDA also is 900 feet MSL, but the required visibility is one and one-half statute miles.

For this approach, the MAP waypoint is designated at the approach end of runway 10R. At the MAP waypoint, if your TO/FROM indicator changes to FROM and you do not have the runway environment in sight, you are required to comply with the missed approach procedure. This procedure requires you to make a climbing left turn to 4,000 feet direct to the PDX VORTAC and hold as depicted.

If the runway environment is not in sight after you have crossed the MAP waypoint, you must fly the missed approach procedure.

Some RNAV systems incorporate a vertical guidance feature in addition to normal horizontal guidance (azimuth/distance). With vertical guidance equipment, you can select a waypoint not only at a designated surface location, but also at a desired altitude. You also can select a specific glide path angle, which means you can have glide path information on an RNAV approach. The approach chart provides the final approach angle. On the Portland procedure, for example, the notation "2.88° ⦣ " is shown between the FAF and the MAP. This means that, if you follow a 2.88° glide path from the FAF, you will reach the MDA at a distance of 2.3 n.m. from the FAF or 2.7 n.m. from the MAP waypoint. RNAV equipment with vertical guidance capability is not a requirement for standard RNAV approaches. However, they do require an FAA-approved RNAV receiver capable of two-dimensional guidance.

The final approach angle for RNAV-equipped aircraft that have vertical path guidance is shown on the approach chart.

CHECKLIST

After studying this section, you should have a basic understanding of:

- ✓ **Approach clearance** — What types of clearances you can expect for airports with and without radar facilities.
- ✓ **VOR approach using off-airport facility** — What the unique characteristics of an approach are when the primary navaid is located beyond the airport boundaries.
- ✓ **VOR approach using on-airport facility** — What advantages on-airport facilities provide and why timing to the MAP usually isn't necessary.
- ✓ **VOR/DME approaches** — How to tell when DME equipment is required for an approach and when DME provides only supplemental fixes.
- ✓ **Supplemental DME fixes** — How some VOR approaches provide for lower MDAs when supplemental DME information is used.
- ✓ **RNAV procedures** — How VORTAC-based RNAV can be used to fly an approach based on waypoints.

SECTION C

NDB APPROACHES

In actual practice, ILS and VOR approaches are the most frequently used instrument procedures. At terminal locations, NDB approaches often are referred to as "backup" procedures, because they provide alternatives if VHF navigation failure occurs. You may use NDB procedures when the ILS is shut down for maintenance. NDB approaches also are common at small airports which are remote from VHF navigation aids. At many of these locations, NDBs often provide the only means of executing instrument approaches. NDB approaches, like VOR procedures, may use either on-airport or off-airport facilities. In addition, they may provide for straight-in landings and/or circling maneuvers. The navigation facilities may be low-powered ILS compass locators or high-powered NDBs. Airports with an ILS sometimes have an NDB approach which coincides with the localizer course. The NDB straight-in minimums are higher, however, because NDB approaches are in the nonprecision approach category. Although useful as approach navaids, radio beacons are subject to disturbances that may result in erroneous bearing information. Remember, these disturbances are caused by intermittent or unpredictable signal propagation due to such factors as lightning, precipitation static, and nighttime interference from distant stations. Nearly all disturbances which affect the ADF also affect the facility's identification signal. You may hear voice, music, or an erroneous identification when a steady, false bearing is being displayed. Since ADF indicators have no warning flags, you should continuously monitor the NDB identifier during an approach.

NDB APPROACH CHARTS

The format for NDB approach charts is similar to the other approach charts. As described in Chapter 3, the main segments of the chart include heading and border data, the plan view, the profile view, and the minimums sections. An example of an NDB procedure is the NDB approach for runway 17L at Waco/TSTI Airport in Waco, Texas. [Figure 4-22]

You may transition from the Waco VORTAC to the Leroi LOM along the Waco 055° radial. If you are inbound from the Peora or Macho Intersections, initial approach segments are provided to the Tours Intersection. Your final descent begins at the LOM, as indicated by the cross in the profile section. The NDB straight-in minimums for Category A aircraft are an MDA of 920 feet MSL and one statute mile visibility, while the circling minimums are 940 feet MSL with one mile. The missed approach procedure is a climb to 2,700 feet, direct to Robinson NDB, where a holding pattern is depicted. The time from the FAF to the MAP at a groundspeed

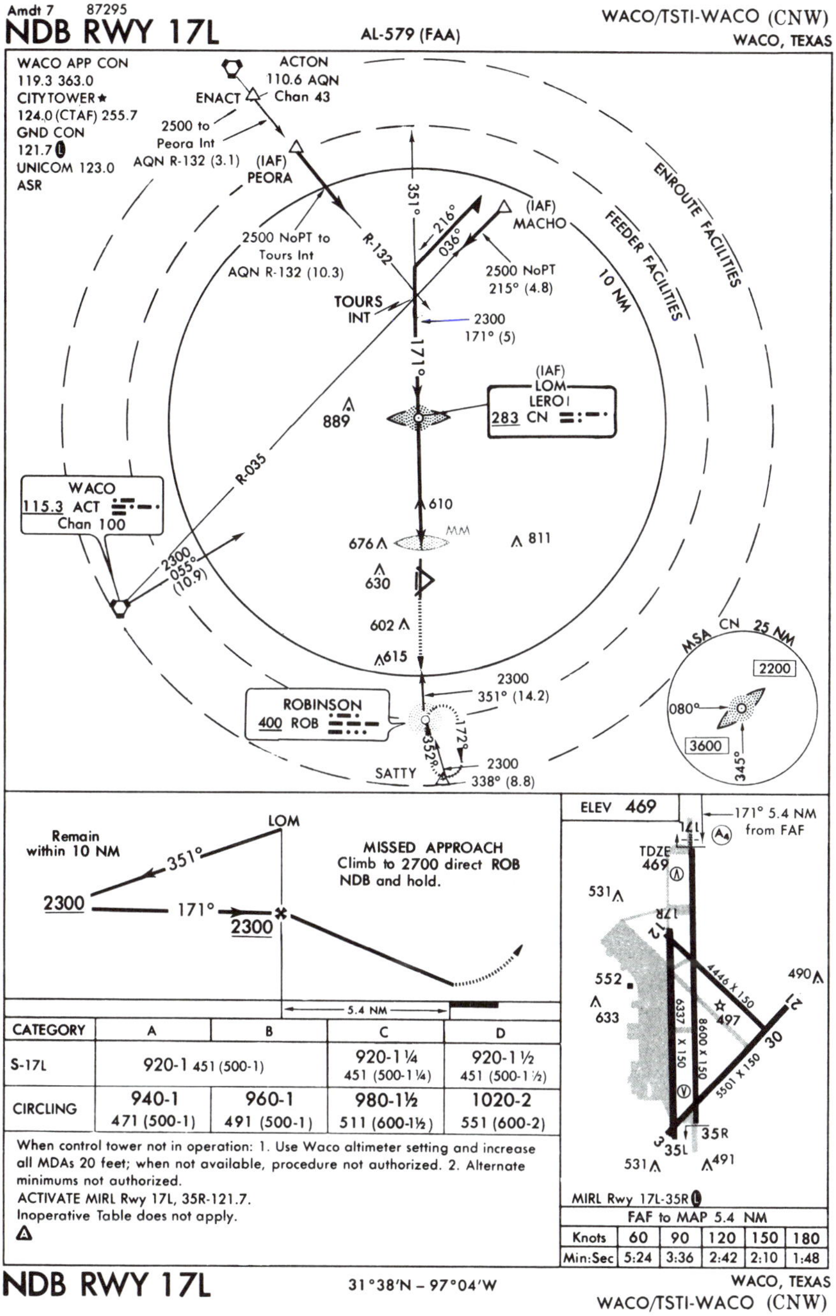

CATEGORY	A	B	C	D
S-17L	920-1 451 (500-1)		920-1¼ 451 (500-1¼)	920-1½ 451 (500-1½)
CIRCLING	940-1 471 (500-1)	960-1 491 (500-1)	980-1½ 511 (600-1½)	1020-2 551 (600-2)

When control tower not in operation: 1. Use Waco altimeter setting and increase all MDAs 20 feet; when not available, procedure not authorized. 2. Alternate minimums not authorized.
ACTIVATE MIRL Rwy 17L, 35R-121.7.
Inoperative Table does not apply.

FAF to MAP 5.4 NM

Knots	60	90	120	150	180
Min:Sec	5:24	3:36	2:42	2:10	1:48

Figure 4-22. Although radar service is available at Waco, this approach procedure also provides several feeder routes for transition from the enroute to the approach phase.

of 90 knots is 3 minutes, 36 seconds. Notes below the minimums indicate you must use the Waco altimeter setting when the tower is not in operation. In this situation, all MDAs increase by 20 feet. In addition, the airport cannot be used as an alternate when the tower is closed. Also, notice that the 17L/35R runway lights are pilot controlled, and the inoperative components table does not apply. Finally, the "A" in the triangle means nonstandard alternate minimums apply, even when the tower is operating. If you plan to list this airport as an alternate, you will need to consult the separate NOS tabulation entitled "IFR Alternate Minimums." After noting the alternate minimums, you must analyze the weather to see if the forecasts are within FAA requirements for an alternate.

NDB APPROACH PROCEDURES

Although the accuracy of an NDB approach depends on your skill using ADF navigation, this type of approach has some advantages over VOR procedures. For example, changes in bearing selection are not required and position orientation is somewhat easier because of the ADF indicator. The procedures for transitioning from enroute to the approach phase are similar to those discussed for ILS and VOR procedures. Assume you are enroute from Miles City, Montana, to Dawson Community Airport, Glendive, Montana, via V545. [Figure 4-23]

The Dawson Community Airport has a remote communications outlet (RCO) for the Miles City FSS. This means you can communicate directly with the FSS at lower altitudes in the Glendive area. The only published approach at this airport is the NDB runway 12 approach. [Figure 4-24]

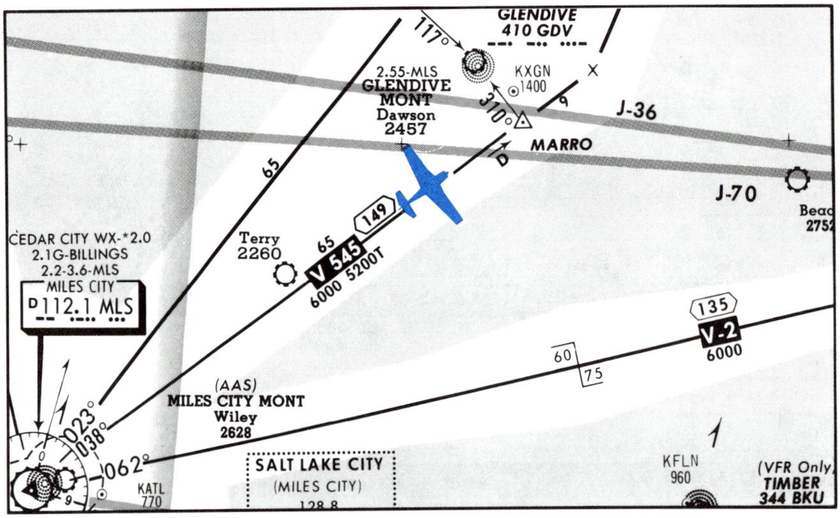

Figure 4-23. The MEA from Miles City to Marro Intersection is 6,000 feet, with a MOCA of 5,200 feet. On this route structure, Salt Lake City Center may issue a cruise clearance to an aircraft inbound to Glendive. *"Mooney 1776R, cleared to Dawson Community Airport via V545 Marro, direct Glendive, cruise 7,000, Miles City altimeter 29.87, contact Miles City radio on 122.55 for advisory."*

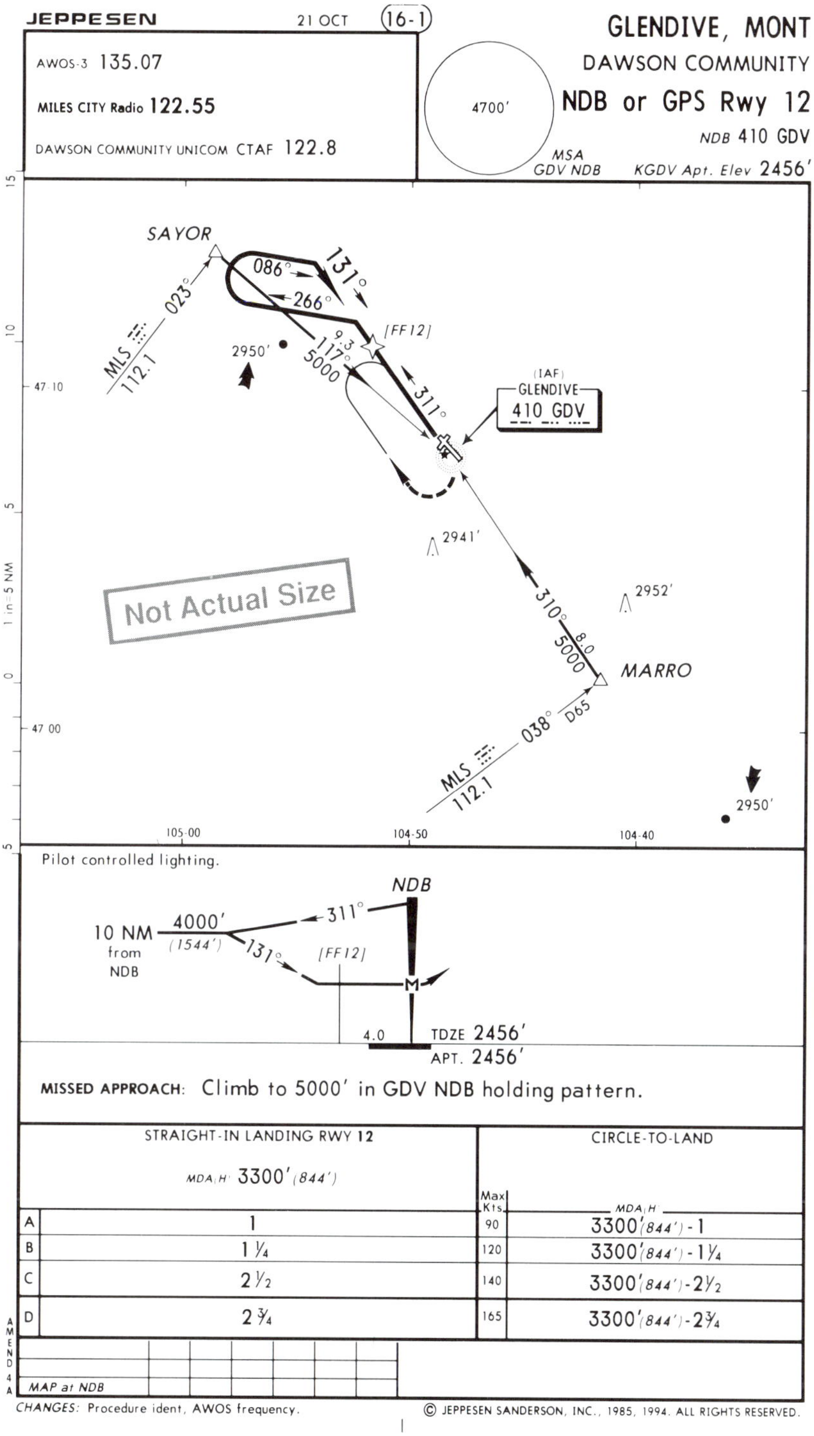

Figure 4-24. The NDB approach procedure for Glendive is representative of approaches found at smaller, remote airports. Note that the title of this approach procedure also includes GPS. Information on GPS approaches is contained in Chapter 4, Section D.

APPROACH CHART REVIEW

This procedure provides a transition via the feeder route from Marro Intersection to the Glendive NDB. The minimum altitude for this segment is 5,000 feet MSL and, after departing the airway, you must fly the entire procedure using ADF navigation. A procedure turn is mandatory, and the minimum procedure turn altitude of 4,000 feet MSL is specified to remain well above the highest reference point in the vicinity of the airport.

The consecutive course changes you must make after leaving V545 are 310° (feeder route), 311° (initial approach segment), and 131° (final approach segment). Since the Glendive NDB is an on-airport facility, an intermediate approach segment is not required. A cross does not appear in the profile section, because an FAF is not designated. When you are flying this approach, the task of determining the missed approach point is eliminated, since the MAP is the NDB. Therefore, the time to the MAP is not specified in the conversion table.

With approved weather service, the MDA for Category A aircraft is 3,300 feet MSL for both straight-in landings and circling approaches, and the visibility minimum is one statute mile. For operators without approved weather service, the minimums are increased to 3,580 feet MSL and one and one-fourth statute miles. The missed approach procedure requires you to climb in a holding pattern to 5,000 feet MSL. Notice that the depicted holding pattern is northwest of the NDB on the approach course.

FLYING THE APPROACH

You should select the Glendive NDB frequency (410 kHz) and positively identify the station. Your VOR receiver should already be set to the Miles City VORTAC frequency of 112.1 MHz. One communications radio may be set to the CTAF frequency of 122.8 MHz and the other to 122.55 MHz for Miles City Radio. For operators without approved weather service, the current Miles City altimeter setting is required for the approach, so contact the FSS for the latest weather information. Then, see if you can obtain an airport advisory from Dawson UNICOM. This may influence your decision to make a straight-in or circling approach. This also is a good time to check your heading indicator against the magnetic compass and reset it for the approach, if necessary.

TRANSITION

Your next concern is to identify Marro Intersection. You can do this easily with DME equipment. Without DME, you will need to rely on your ADF indicator. As you approach the intersection on V545, you notice that a magnetic heading of 045° is required to maintain the airway centerline. [Figure 4-25]

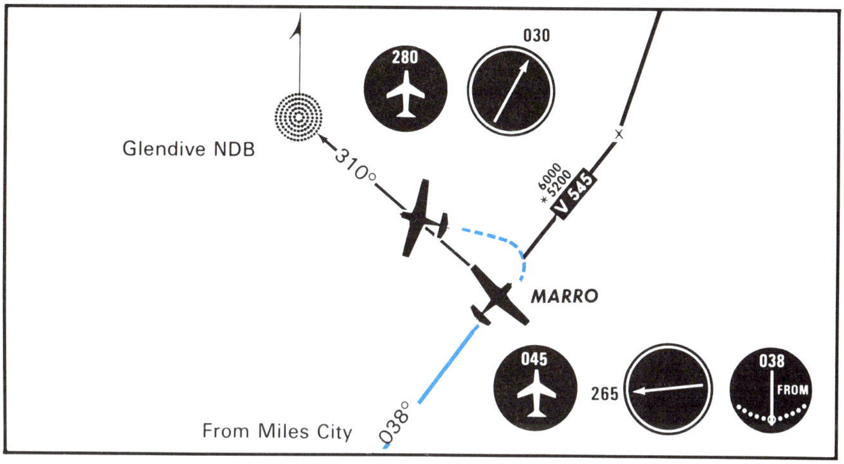

Figure 4-25. By applying the ADF formula, you conclude that your aircraft is over the intersection when the relative bearing is 265°. At that point, turn the aircraft left to a 30° intercept heading of 280°. When the ADF indicator shows 030°, you have intercepted the course, and you should turn the aircraft to a heading of 310°. When established on the feeder route, you may begin the descent to 5,000 feet MSL. Depending on local wind conditions, you may also find it necessary to apply wind correction while inbound to the NDB.

INITIAL APPROACH SEGMENT

Do not begin your descent to 4,000 feet until you reach the NDB. Then, slow the aircraft to the proper speed. If you pass directly over the station, the bearing indicator will reverse to 180°. The flight should be continued outbound on the 311° bearing from the NDB for approximately two minutes before you start the procedure turn. Depending on your groundspeed and wind conditions, you may need to adjust your timing to remain within 10 n.m. of the NDB.

To fly the procedure turn as depicted on the approach chart, turn left to 266° for one minute, followed by a right turn to 086°. You should complete the prelanding checklist at this time. Since the procedure turn course is 45° to the approach course, a 045° relative bearing will indicate course interception and completion of the procedure turn.

FINAL APPROACH SEGMENT

When inbound, give a position report to Miles City Radio, and begin descent to the MDA. Then, self-announce your position on Dawson UNICOM for the benefit of any local traffic. If you establish visual contact with the runway environment, check the windsock to confirm the wind direction/velocity. Since the straight-in and circling minimums are the same for this procedure, you may decide to circle after establishing visual contact. On most approaches, however, you must make this decision before your final descent, since circling MDAs normally are higher than straight-in MDAs.

MISSED APPROACH

If at least one of the required visual references for runway 12 cannot be clearly identified prior to the MAP (station passage) on a straight-in landing approach, you must execute the missed approach procedure. It requires you to climb in a holding pattern at the NDB to 5,000 feet MSL. After you are established in the holding pattern, you need to contact Miles City Radio to report the missed approach and request additional clearance.

CHECKLIST

After studying this section, you should have a basic understanding of:

- ✓ **NDB approaches** — How NDB procedures are charted and what similarities they have to VOR approaches.
- ✓ **Off-airport and on-airport facilities** — How the location of a navaid with respect to the airport determines the characteristics of the NDB approach.
- ✓ **ADF intersections** — How to identify intersections formed by NDB and VOR facilities.
- ✓ **ADF indicator** — How you can execute an NDB approach with ADF equipment only.

SECTION D

GLOBAL POSITIONING SYSTEM

The **global positioning system** (GPS) is a space-based navigation system which was originally developed by the Department of Defense (DOD) for the military. It provides highly accurate position, time, and velocity information around the world to an unlimited number of users. It is available 24 hours a day and is unaffected by weather.

Although the DOD operates and monitors the system for proper operation, it may be used by civilians who have the proper equipment. However, the GPS system accuracy is degraded somewhat for nonmilitary or non-government users in order to deny pinpoint accuracy to an enemy during wartime. For most civilian users, GPS provides horizontal positioning accuracy of approximately 100 meters, or 328 feet. By comparison, accuracy of VOR, VOR/DME, and TACAN is approximately one nautical mile.

COMPONENTS

GPS consists of three main components that act together to provide a very precise worldwide positioning and navigation system. These are the GPS satellites, ground stations, and the GPS receivers.

The space component of GPS includes twenty-four satellites, three of which are spares. They orbit the earth at an altitude of approximately 11,000 miles. Each satellite circles the earth twice a day, allowing at least four satellites to be in view 24 hours a day, anywhere on earth.

The control segment of GPS consists of a network of ground monitoring facilities. These stations continually monitor satellite performance in order to maintain the accuracy of the system.

The third component of the system is the airborne GPS receiver and its associated antennas. The receiver processes satellite data, determines position, displays database information, and provides various other functions depending on the manufacturer.

OPERATION

GPS uses ranging and triangulation from a group of satellites which act as precise reference points. A GPS receiver measures distance from a satellite using the travel time of a radio signal. The satellites contain extremely accurate atomic clocks which provide exact timing of the broadcast signals. A GPS receiver also contains an internal clock. The

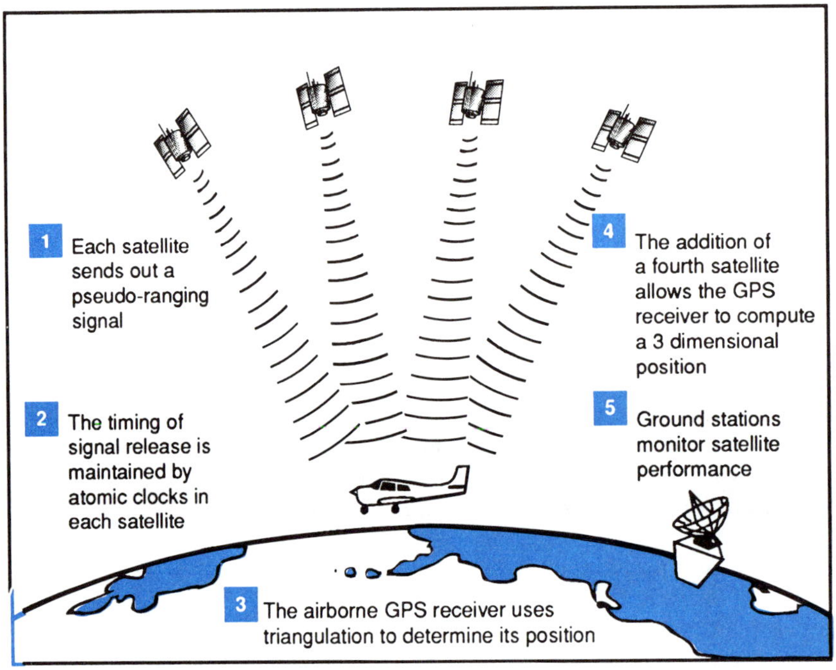

Figure 4-26. Three GPS components, including a satellite constellation, an airborne GPS receiver, and ground monitoring stations, act together to provide precise position information.

arrival time of a signal is used to compute the distance traveled by a signal, and therefore a precise position can be determined. The GPS receiver requires at least three satellites to triangulate a receiver's lateral position. A fourth satellite is needed for timing corrections to provide a three-dimensional position. [Figure 4-26]

GPS receivers use latitude and longitude coordinates to identify a position. However, receivers typically use names of fixes, waypoints, and station identifiers instead of lat/long coordinates on the GPS display. A waypoint is a predetermined geographical position, similar to the waypoints used for RNAV. Once a receiver calculates its own position, it can then determine and display the distance, bearing or direction, and estimated time enroute to the next waypoint.

Before using GPS for IFR flight, be sure the database is current.

A GPS receiver that is approved for IFR operations must have an updatable database. The waypoints, fixes, and station identifiers used for GPS navigation are loaded into the receiver, usually by means of a magnetic card containing the information. The database is updated on a regular basis, and it is important that you make sure your GPS receiver has current data. You can find additional information on GPS operations in the *Aeronautical Information Manual* and in Advisory Circular 90-94, which provides guidelines for using global positioning equipment.

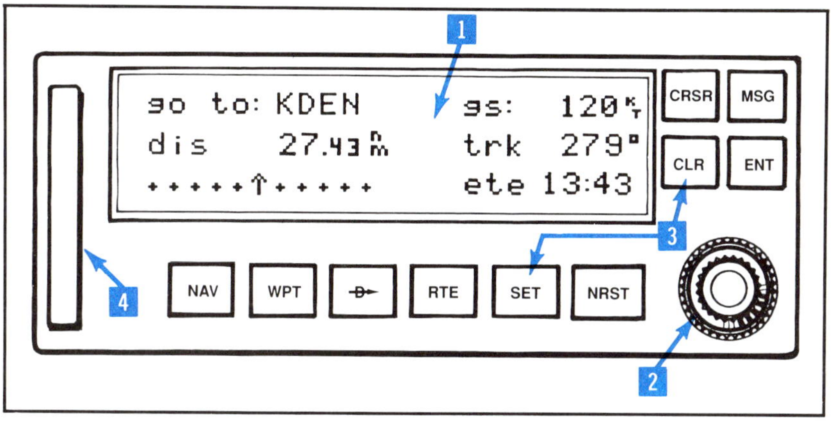

Figure 4-27. Although GPS control panels will vary from one manufacturer to another, most units contain many of the same elements. These include: a digital "page" display (item 1), knobs to control the cursor and to select information (item 2), function keys (item 3), and a slot for the database card (item 4).

Refer to the supplements section of the Airplane Flight Manual (AFM) or Pilot's Operating Handbook (POH) to determine the type and degree of authorized operations for the GPS receiver. You should be thoroughly familiar with the GPS receiver installed in your aircraft, and you must operate the equipment in accordance with the AFM or POH. Before using a GPS unit for IFR operations, you must ensure that it has been approved in accordance with Technical Standard Order (TSO) C-129, which prescribes the standards for GPS equipment. It must also be installed in accordance with AC 20-138, which provides guidance for airworthiness approval for GPS equipment. Handheld receivers are restricted to VFR flight only. Some panel mounted receivers are also restricted to VFR flight and must be placarded as such in the aircraft. Other receivers may be approved for IFR enroute and terminal operations and should be placarded if not authorized for GPS approaches. [Figure 4-27]

You must determine the operations for which your GPS unit is authorized, and operate it in accordance with the approved AFM or POH.

As with ground-based navigation aids, it is extremely important for pilots to be aware of any loss of integrity, or position accuracy, of the GPS signals. Many receiver units monitor integrity by use of an internal system referred to as Receiver Autonomous Integrity Monitoring (RAIM). If position calculations are not within certain parameters during GPS operations, the GPS unit should warn the pilot of a loss of integrity. You might think of this RAIM failure warning as similar to a traditional VOR warning flag. Some units may be connected with an annunciator panel to provide warning information.

NAVIGATION

After you have determined that you may use GPS to navigate a portion or all of your flight, you should state your navigation capability on the flight plan. GPS navigation is similar to RNAV, since you proceed direct from waypoint to waypoint. Therefore, you normally use the same equipment code, which is a "/R." Remember to ask for any NOTAMs on the GPS constellation from the FSS briefer. Also, keep in mind that the authorization and procedures discussed here apply only within the United States. Before you fly in another country, be sure to check with the civil aviation authority for that nation to determine if GPS navigation is approved.

ENROUTE AND TERMINAL

When using GPS and operating under IFR, additional navigation equipment is required for enroute and terminal operations.

GPS can be used to overfly the published Victor airways or Jet routes, SIDs, and STARs. Enroute and terminal flight can be conducted in much the same way as RNAV operations for direct flights. If you choose to fly a direct route, check the minimum enroute altitudes and special use airspace along the route. Specific equipment and operational considerations apply to all IFR flight, including both oceanic and domestic enroute as well as terminal area operations. These are outlined in the AIM and AC 90-94. Terminal operations are defined as those between the last enroute fix and the IAF for an approach. [Figure 4-28]

NONPRECISION APPROACHES

To fly an approach using GPS, the approach must be retrievable from the database.

There are currently two major categories of GPS approaches — stand alone and overlay. Stand alone approaches are designed for use with GPS navigation equipment only. They are identified on approach charts by the title, which contains only the term "GPS" and a runway number. Stand

GPS IFR ENROUTE AND TERMINAL REQUIREMENTS	
OPERATION	**GPS REQUIREMENTS**
DOMESTIC GPS IFR ENROUTE AND TERMINAL AREA	• GPS[1] • Traditional navigation equipment (VOR, DME, TACAN, and/or NDB) appropriate to the route of flight must be installed and operational. They don't need to be on and monitored unless RAIM fails. • All the compatible underlying navaids along the route of flight must be operational.
OCEANIC GPS IFR ROUTE TYPE: • Routes that require dual long range navigation systems (such as dual INS or dual OMEGA).	• GPS[1] • One other long range navigation system compatible with the route of flight.
• Routes that require only one long range navigation system.	• GPS[1] • One other means of navigation compatible with the route of flight.
[1] GPS units that are approved, in accordance with FAA TSO C129, for IFR enroute, terminal, and approach operations.	

Figure 4-28. For enroute, terminal, and some approach operations, you will need additional navigation equipment on your aircraft. The type of equipment depends on the operation. You may also need to ensure that ground-based navaids are available.

alone approach charts are being added to NOS and Jeppesen publications as they are approved. Meanwhile, the GPS overlay approach program authorizes pilots to use GPS to fly nonprecision approaches that were originally designed for use with ground based navaids such as VOR, VOR/DME, TACAN, NDB, NDB/DME, and RNAV. However, GPS is not authorized for use on localizer (LOC), localizer directional aid (LDA), or simplified directional facility (SDF) approaches. Whether an approach is a stand alone or an overlay, it must be retrievable from the database if you use GPS as the primary means of navigation. [Figure 4-29]

OVERLAY PROGRAM

The GPS approach overlay program was designed with three phases. Phase I ended in February, 1994, when the FAA declared GPS operational for civil operations. Phase II began at that time. It allows pilots to use

For a Phase II GPS approach, the navaids and avionics must be operational.

GPS IFR APPROACH REQUIREMENTS AND PHASE DESCRIPTION

APPROACH TYPE	ATC GPS CLEARANCE	GPS APPROACH NAVAID REQUIREMENTS	APPROACH CHARTS
OVERLAY PHASE II	"GPS..." not included in the clearance	• GPS[1] • The underlying ground navaids at the destination airport must be operational and the aircraft associated avionics must be installed and operational. • The route to and approach at the alternate airport must have non-GPS and non-LORAN-C aircraft compatible navaids available and operational. • No underlying ground navaids need to be monitored unless RAIM falls.	NOS: No mention of GPS JEPP: Possible (GPS) below title. Not listed in Terminal Introduction
OVERLAY PHASE III	"GPS..." only included in the clearance	• GPS[1] • The underlying ground navaids at the destination airport may be inoperative, and the aircraft associated avionics don't need to be installed. • The route to and approach at the alternate airport must have non-GPS and non-LORAN-C aircraft compatible navaids available and operational. • No underlying ground navaids need to be monitored unless RAIM falls.	NOS: "...or GPS" in title JEPP: "...or GPS" in title, or listed in Terminal Introduction
STAND ALONE	"GPS..." only included in the clearance	• Same as Phase III	NOS & JEPP: "GPS..." only in title

[1] GPS units approved, in accordance with TSO C129, for IFR enroute, terminal, and approach operations.

Figure 4-29. It is important to know whether your intended approach is a Phase II, Phase III, or a stand alone. You must understand what equipment and navaids you will need to have available. You may want to refer back to this chart during the following discussion.

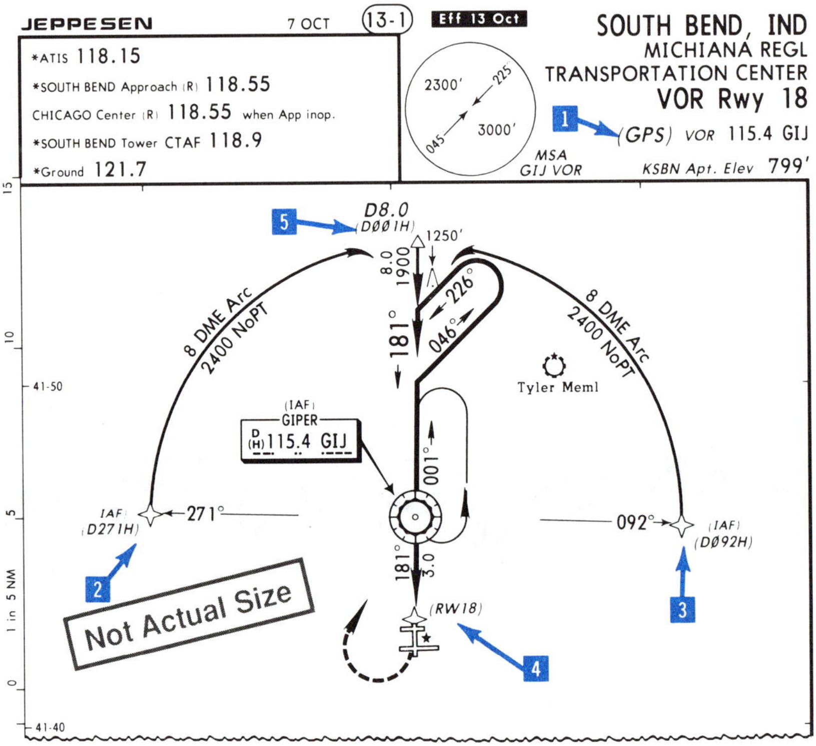

Figure 4-30. A Phase II Jeppesen chart which has been updated for GPS contains "(GPS)" under the title (item 1). Unnamed fixes shown with a waypoint star and identifier are the two IAFs (items 2 and 3) and the missed approach point (item 4). The 8 DME fix on the 001° radial (item 5), while not named, is a designated fix, as shown by the triangle. Therefore, a waypoint star is not used, but a five-letter identifier is provided.

GPS to fly an existing nonprecision approach without monitoring the navaids which define the approach, unless RAIM fails. However, the navaids required for the approach must be operational, and the associated avionics must be installed in the aircraft and be operational. If an alternate airport is required, it must have an operational non-GPS or non-LORAN-C approach which your aircraft is equipped to fly. [Figure 4-30]

APPROACH CHARTS

Phase II GPS overlay approaches will not contain "GPS" in the title of the approach. While NOS Phase II approach charts are no different, some Jeppesen charts may contain certain enhancements. On a Jeppesen chart which has been updated for GPS Phase II, the notation "(GPS)" appears beneath the title, unnamed waypoints which are not FAA-designated fixes are shown by a four-point star symbol, and waypoint identifiers are provided. If you are uncertain whether an approach is certified for GPS Phase II, check your database. Approved GPS approaches will be included in the database. Typically, all approaches for an airport are listed under the airport's four-letter identifier. As a minimum, databases are required to contain waypoints for the initial approach fix (IAF), final approach fix (FAF), missed approach point (MAP), and the missed approach holding point for each approach. Intermediate fixes (IFs) and all named fixes are also included in the database.

If an IAF or other fix has a pronounceable five-letter name, that name is used on the chart and in the database. DME fixes are typically given a five-character identifier which consists of the letter "D," a radial, and an alphabet letter which corresponds to the DME from the station. For example, the IAF waypoint D271H is a DME fix on the 271° radial at 8 DME, since "H" is the eighth letter of the alphabet. However, some DME fixes at a greater distance from the station may be coded with two letters from the VOR name followed by the radial, such as "AX141."

When a missed approach point is at a navaid, the MAP waypoint is coded using the navaid identifier. When a MAP is based on timing or a DME distance, a MAP waypoint is coded and added to the database. The identifier usually consists of either "RW" followed by the runway number, or "MA" followed by the final approach course.

Another type of coded waypoint which appears on some Jeppesen GPS approach charts is the sensor FAF. This waypoint is used on approaches that do not have a final approach fix. An example would be a VOR approach where the VOR is located on the airport, and the MAP is at the VOR. If a stepdown fix exists and is greater than 2 n.m. to the MAP, the stepdown fix is coded in the database as the sensor FAF. Otherwise, a sensor FAF is coded at least 4 n.m. from the MAP. Although sensor FAFs are coded into the database, they will not appear on Phase II or Phase III NOS charts. Sensor FAFs are shown on Jeppesen charts which have been updated for GPS.

FLYING A PHASE II APPROACH

Remember, to fly a GPS approach, you must be able to retrieve the data for that approach from your receiver's database. Although GPS displays vary according to manufacturer, all units will display the waypoints in the same sequence as they are presented on the approach chart.

It is important that you use the original approach title, such as "NDB Runway 11," and receive approval from ATC for that approach by the same name. If instead, you requested the "GPS Runway 11" approach, you would in fact be asking for a different approach. Consequently, you could possibly fly an approach for which you are not cleared, and could fly into unprotected airspace. [Figure 4-31]

For a Phase II GPS approach, be sure to use the original name for the approach, not "GPS."

With the GPS unit in normal operation, the CDI functions as if tracking a localizer.

Normal operation of the GPS unit is referred to as TO-TO navigation. The aircraft is between two waypoints and is flying TO the selected waypoint. In this mode, the CDI needle functions like it is tracking a localizer signal; that is, movement of the OBS has no effect on the CDI needle or the TO/FROM indicator. Upon passing the active waypoint, equipment will usually sequence automatically to the next waypoint for you to fly TO it.

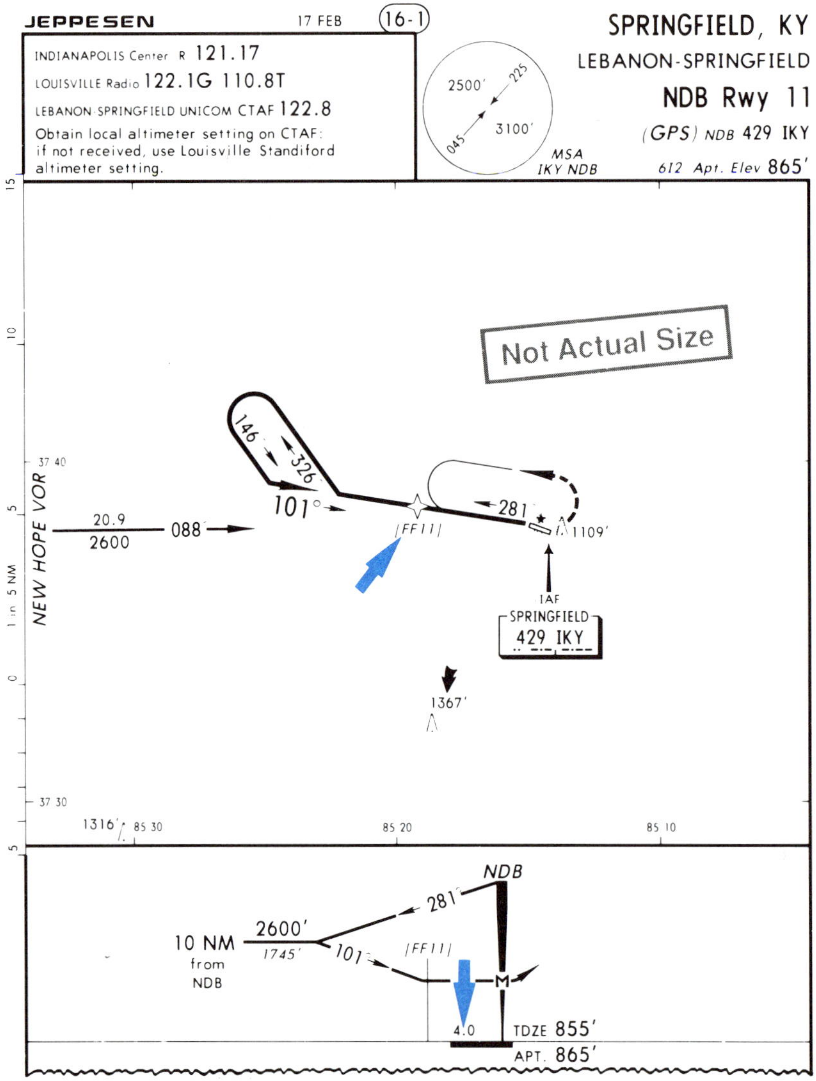

Figure 4-31. This Phase II approach was designed without a final approach fix. A sensor FAF has been added for GPS operations. It is designated by a waypoint star and coded as "FF11." In this case, it is located 4 n.m. from the MAP.

At times, you may need to select TO-FROM navigation. When you are flying to a selected waypoint, your CDI will respond as if the waypoint is a VOR station. You may fly a course either TO or FROM the

In TO-FROM navigation, the CDI functions like it is tuned to a VOR.

Although the waypoints are presented in the proper sequence for the approach, you must be aware that you cannot always fly direct from one waypoint to another. For instance, if you begin an approach at an IAF which is on a DME arc, you will need to use normal piloting techniques to maintain the published ground track along the arc. This is the case with the IAFs shown for the South Bend VOR Rwy 18 approach .

When flying a DME arc, you will need to use normal piloting techniques.

Another exception to normal operations occurs when you fly a procedure turn like the one shown for the NDB Rwy 11 at Springfield, KY, in Figure 4-31. When you cross the IAF over the NDB, the next waypoint presented is the sensor FAF, "FF11." If you did nothing with your GPS receiver, and crossed the sensor FAF while outbound on the procedure turn, the next waypoint displayed would be the MAP. However, when you complete the procedure turn and proceed inbound, you will need to have the sensor FAF displayed. Another step is required to present the waypoints in the correct order. After passing the IAF outbound and prior to the sensor FAF, you will need to place your receiver on hold (TO-FROM navigation). When you complete the procedure turn and have intercepted the inbound course, return the receiver to automatic waypoint sequencing (TO-TO navigation) at least 2 n.m. prior to the sensor FAF.

To execute a procedure turn with a sensor FAF, you will normally need to put your receiver on hold after crossing the IAF outbound.

For the same reason, you will also need to select TO-FROM navigation when you enter holding over a waypoint such as the IAF. Using the Springfield, KY, NDB approach again, let's say you are instructed to hold as published over the NDB. You will need to place your receiver on hold prior to crossing the NDB the first time. If you do not, the GPS unit will automatically select the next waypoint, which is the sensor FAF. You should leave the unit on hold until you are cleared for the approach, then select normal operation. When you cross the NDB and turn outbound, the next waypoint selected will be the sensor FAF. At that time, you will again place the unit on hold, as discussed in the preceding paragraph, until you have intercepted the inbound course.

When holding over a waypoint, you will usually need to place your GPS unit on hold until cleared for the approach.

You will recall that the CDI has greater sensitivity when using ILS signals than when using VOR. In a similar manner, the CDI sensitivity changes while flying a GPS approach. During enroute and terminal operations, full scale deflection of the CDI represents 5 n.m. from centerline. When you select the approach mode, the sensitivity smoothly transitions to one n.m. either side of centerline. At 2 n.m. from the FAF and within 30 n.m. of the airport, the sensitivity smoothly transitions to 0.3 n.m. for full scale deflection. It remains at 0.3 n.m. until the first waypoint on the missed approach is selected by the pilot. Once selected, sensitivity transitions to 1 n.m. either side of centerline. [Figure 4-32]

Sensitivity is 0.3 n.m. for full scale deflection during final approach until the pilot selects the first waypoint on the missed approach procedure.

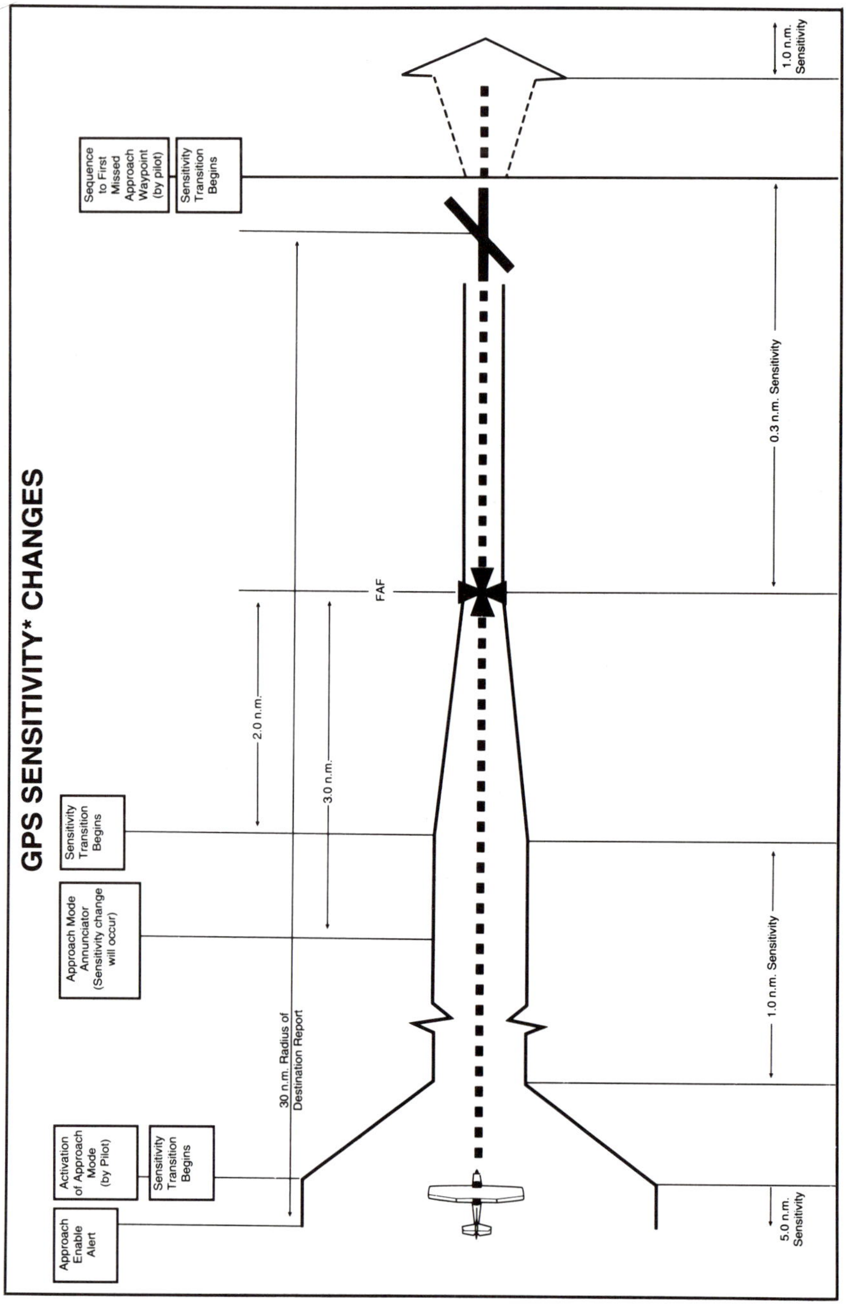

Figure 4-32. Sensitivity of the CDI refers to the distance indicated by full scale deflection. For example, with 1.0 n.m. sensitivity, full CDI deflection represents 1.0 n.m. from centerline. The sensitivity changes during an approach from 5 n.m. to 1.0 n.m., then to 0.3 n.m. during the final approach, then back to 1.0 n.m. when the first missed approach waypoint is selected.

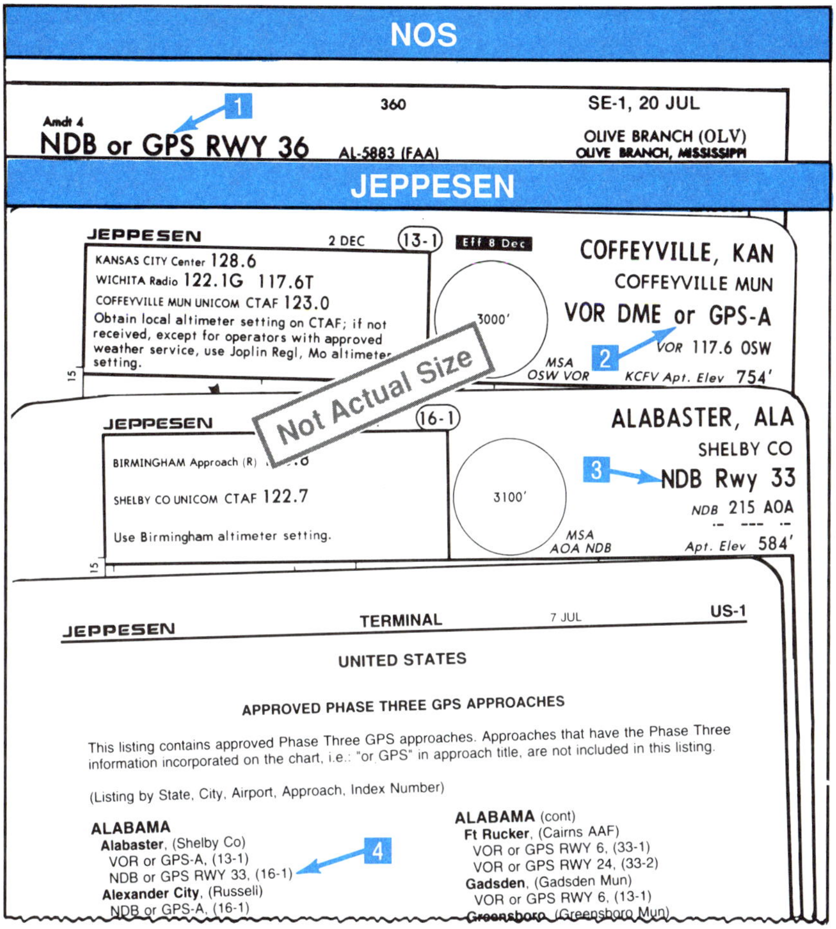

Figure 4-33. All Phase III GPS approaches on NOS charts are designated by the title "...or GPS" (item 1). Jeppesen Phase III approach charts may contain "...or GPS" in the title (item 2). Charts that do not contain a reference to GPS, such as the NDB Runway 33 at Alabaster, AL, (item 3) may or may not be approved for GPS Phase III approaches. You will need to refer to the Terminal Introduction pages of the Jeppesen Airway Manual. This NDB Runway 33 approach is an approved Phase III GPS approach (item 4).

PHASE III APPROACHES

GPS Phase III approaches are a step closer to stand alone approaches. The ground navaids at the destination airport may be inoperative, and the aircraft avionics for that approach need not be installed. However, you will need to pay careful attention during flight planning. When an alternate airport is required, the route to and approach at the alternate must have non-GPS and non-LORAN-C navaids which you are able to use. Also, during a Phase III approach, you are required to monitor the ground navaids if RAIM fails. Practically speaking, then, you will need to have ground-based navaids available. [Figure 4-33]

Ground-based navaids are not required on a Phase III approach unless RAIM fails, but you must be able to fly to an alternate without GPS or LORAN-C.

Phase III approaches are designated by the words "...or GPS" in the title on NOS charts. Jeppesen charts will either have "...or GPS" in the title, or the approach is listed in the Terminal Introduction of the Jeppesen Airway Manual.

When requesting a Phase III GPS approach, use the GPS title, not the navaid name.

Procedures for flying a Phase III approach are the same as for a Phase II approach. However, you must request a Phase III approach by its GPS name, not the underlying navaid name. For example, if you wanted to use GPS to fly the "NDB or GPS RWY 36" at Olive Branch, MS, you would need to request the "GPS Runway 36" approach. [Figure 4-34]

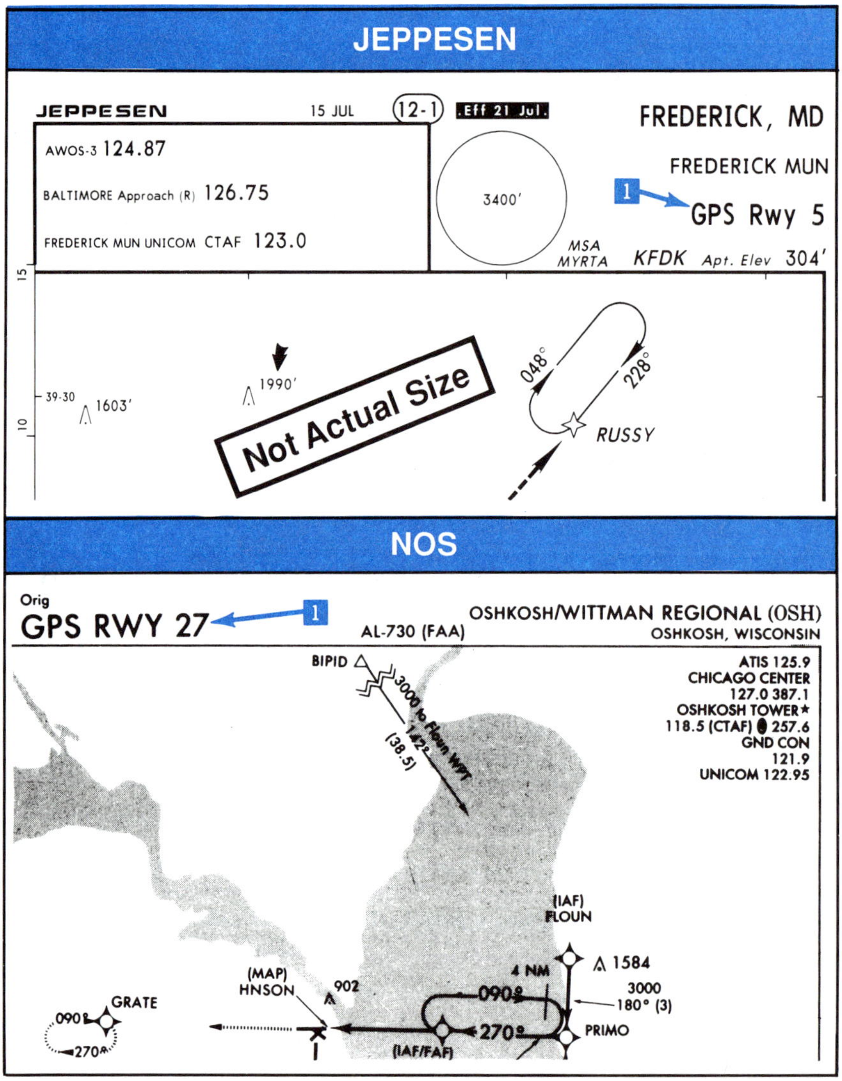

Figure 4-34. A stand alone GPS approach is titled only as a GPS approach, without reference to any ground-based navaids (item 1).

STAND ALONE GPS APPROACHES

Some GPS approaches are not overlaid on an existing approach, but are designed as stand alone approaches. The procedures for flying a stand alone approach are the same as for Phase III approaches. As is the case with a Phase III approach, it is important to note that non-GPS and non-LORAN-C navaids are still required for navigating to and flying an approach at an alternate. The significant difference in approach charts is that a stand alone approach contains only "GPS" in the title.

The navaid requirements for flying to an alternate are the same for both Phase III and stand alone approaches.

USING GPS

Before you attempt to fly IFR with GPS, you must be thoroughly familiar with all aspects of using your receiver. You also need to understand the procedures for using GPS during part or all of your flight. Since you are required to operate according to an AFM or POH, this should be your primary source of information. Taking instruction from a competent instructor is recommended.

If your circumstances permit, you should practice a number of GPS approaches in a simulator. You could also use GPS in an aircraft during visual conditions, or with an instructor or safety pilot, until you are confident in both the equipment and your abilities. An objective of practicing with this equipment is situational awareness. It is vital for instrument navigation, and especially so with GPS navigation. When available, ground-based navaids may be used to help you with orientation.

After you have gained a safe level of competence with your GPS equipment, you will need to ensure you are currently proficient before flying IFR. If it has been some time since you used GPS, you should take the time for refresher training which may be necessary to maintain a high skill level with the equipment.

FUTURE OF GPS

There will continue to be a mixture of Phase II, Phase III, and stand alone GPS approaches for some time. With the exception of LOC, LDA, and SDF, many nonprecision approaches in the U.S. are available under Phase II of the overlay program. These may eventually become Phase III approaches as they change to include "or GPS" in the titles. The FAA is also continuing to develop and authorize more stand alone approaches.

Precision approaches have not been approved for GPS, since the GPS signals received by most civil users currently do not meet the position accuracy requirements. Work is progressing on a system known as differential GPS that may provide the accuracy needed for precision approaches. The system uses a GPS receiver placed at a precisely surveyed location. This receiver computes the difference between its position calculated from the satellite data and its known position. This correction factor is then transmitted either through a satellite or directly to an airborne receiver so that it can refine its own position calculation. Using five satellites, the system has been tested with a position accuracy of 0.1 meters. With further testing, approval, and funding for receiver sites, precision approaches may be available in the future.

GPS navigation is rapidly becoming a widely used system, and is likely to become a standard for airborne navigation. As approved applications become more widespread and GPS receiver prices continue to decrease, the use of GPS is bound to become even more popular.

CHECKLIST

After studying this section, you should have a basic understanding of:

✓ **GPS components** — What they are and how signals are used for navigation.

✓ **Authorized operations** — How to determine what operations are approved and what limitations may apply to specific GPS equipment.

✓ **Additional equipment** — What non-GPS navigational equipment is required for various flight operations.

✓ **Overlay approach program** — What the phases are, how they compare to stand alone approaches, and what is required for each type of approach.

✓ **Receiver operation** — How to use the GPS equipment in your airplane.

CHAPTER 5

IFR OPERATIONAL CONSIDERATIONS

INTRODUCTION

This chapter brings together several subject areas that you have studied in previous sections. You already have a basic knowledge of the way the air traffic control system operates and the role ATC clearances play in the coordination of instrument traffic. "IFR Operational Considerations" is a chapter that describes the three phases of instrument flight — departure, enroute, and arrival. It is designed to give you an overview of IFR operations so you will know what to expect. Each phase is covered in a separate section which integrates appropriate ATC procedures, clearances, and regulations with chart interpretation and other operational factors. For example, what are the important items that you should consider before any IFR departure, and what should you be prepared for during the enroute phase? What type of arrival can you expect, and what happens if you are unable to land from an instrument approach at your destination because of weather? This chapter provides a framework for obtaining operational insight into the types of situations you may encounter when you file IFR.

SECTION

A

DEPARTURES

An IFR departure in a radar environment may be as simple as holding the headings assigned by the departure controller while monitoring your position from local navaids. In other cases, you may be adhering to a detailed standard instrument departure. At remote locations, you may fly the entire departure without the benefit of radar vectors or a SID. This section covers important considerations for IFR departures, beginning with takeoff minimums.

IFR takeoff minimums are specified by regulations. Although these minimums do not specifically apply to private aircraft operating IFR under FAR Part 91, good judgment should dictate your compliance. Therefore, the following discussion is intended to help you determine takeoff minimums and apply them to the IFR departure.

TAKEOFF MINIMUMS

FAR Part 97 prescribes the standard instrument approaches for airports in the United States. Whenever an approach is established, it usually specifies the weather minimums that are required for commercial operators to take off. Examples include FAR Part 121 air carriers, FAR Part 135 air taxi operators, and operators of large aircraft, among others. In cases where takeoff minimums for a particular airport are not published but an instrument approach procedure is prescribed, the following standard minimums apply — one statute mile visibility for single- and twin-engine airplanes and one-half statute mile visibility for aircraft with more than two engines.

In most cases, takeoff weather minimums are based only on visibility. In some cases, the required visibility may be greater than standard due to terrain, obstructions, or departure procedures. In addition, some airports may require both a minimum ceiling and visibility before you can take off. Climb performance may also be a factor. When a ceiling is required, it is generally due to obstructions that you must avoid during the departure. However, the ceiling and visibility requirements may be waived in some cases where a minimum climb gradient is published and you can comply with it. The climb gradient specified ensures obstruction clearance, so standard visibility minimums are adequate. If you note nonstandard visibility or ceiling requirements for takeoff minimums, it should alert you to the fact that some type of operational limitation exists and you will need additional information before departure. [Figure 5-1]

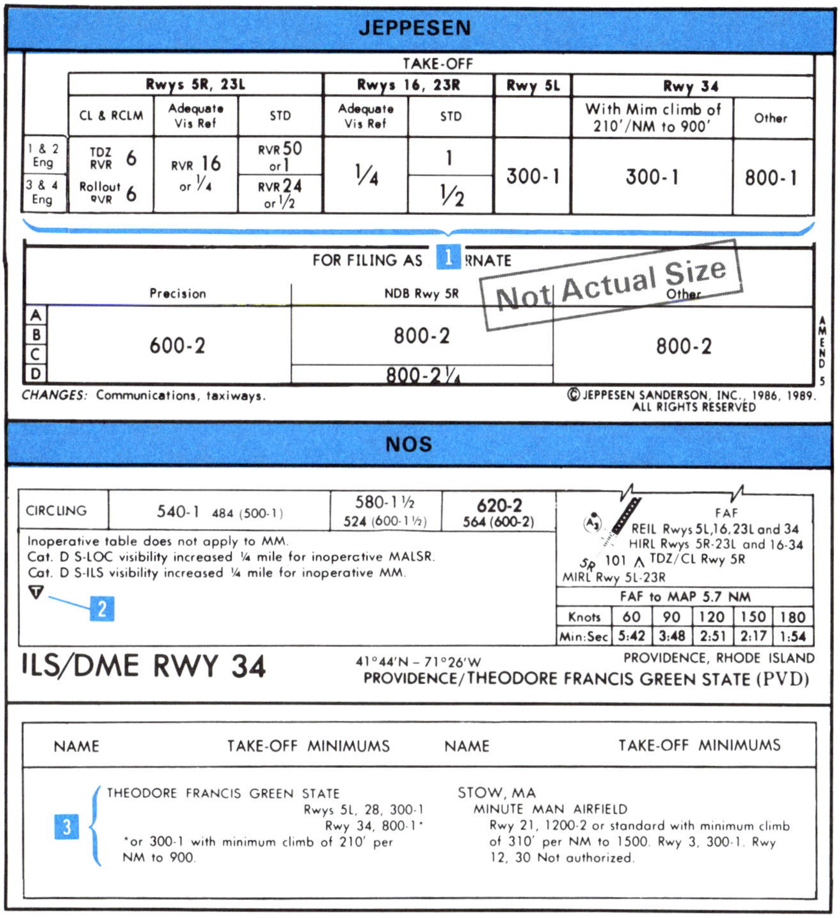

Figure 5-1. Takeoff minimums are printed on Jeppesen airport charts so you can read them directly (item 1). On NOS charts, the triangle symbol enclosing a "T" within the information box indicates that special minimums have been established for that airport (item 2). You must then consult the separate listing in the front of each regional approach chart book to determine the appropriate takeoff minimums (item 3).

VISIBILITY

Prevailing visibility is reported in statute miles or fractions of miles.

As mentioned before, takeoff minimums usually are based on visibility, which may be expressed in terms of prevailing visibility, tower visibility, runway visibility value, or runway visual range. **Prevailing visibility** and **tower visibility** are the greatest distance a weather observer or tower personnel can see throughout one-half the horizon, and they need not be continuous. They are reported in statute miles or fractions of miles and recorded on the aviation routine weather report (METAR). When visibility is critical at an airport which has a weather observing station and a control tower, both take observations. When the tower observation is less than four miles, the lowest of the two (METAR or tower) is the prevailing visibility. The other observation may be included in remarks. Except for these circumstances, the reported prevailing visibility normally reflects a surface observation.

Runway visibility value (RVV) is the distance down the runway you can see unlighted objects or unfocused lights of moderate intensity; it is reported in statute miles or fractions of miles. RVV usually is derived from a transmissometer for a particular runway. The current METAR format does not include reporting of RVV.

RVR values represent the distance, in hundreds of feet, you can expect to see down the runway from the approach end.

Runway visual range (RVR) is always a transmissometer value. It represents the distance you can expect to see down the runway from a moving aircraft. RVR is based on sighting high intensity runway lights or on the visual contrast of other targets, whichever yields the greater visual range. It is expressed in hundreds of feet and, when available, is used instead of prevailing visibility in determining minimums for particular runways. RVR is horizontal visual range, not slant visual range. It is based on the measurement of a transmissometer located near the touchdown point of the instrument runway. The primary instrument runways at major airports may have as many as three transmissometers providing RVR readings. [Figure 5-2]

It is important to remember that the prevailing visibility or RVR shown in the surface aviation weather report should normally be used only for informational purposes. The current visibility at the actual time of departure is the value you should use for determining compliance with takeoff minimums. Control towers report RVR when the prevailing visibility is one mile or less and/or the RVR is 6,000 feet or less.

Definitions	Conversions RVR (ft.)	Visibility (s.m.)
Touchdown RVR — The RVR visibility readout values obtained from RVR equipment serving the runway touchdown zone.	1,600	¼
	2,400	½
Mid-RVR — The RVR readout values obtained from RVR equipment located midfield of the runway.	3,200	⅝
	4,000	¾
	4,500	⅞
Roll-out RVR — The RVR readout values obtained from RVR equipment located nearest the roll-out end of the runway.	5,000	1
	6,000	1¼

Figure 5-2. When takeoff minimums are charted as touchdown, mid-, and roll-out RVR, they apply to commercial operators who are approved for lower than standard takeoff minimums. You should use the more conservative standard minimums as your guide. Remember to convert RVR values to visibility in statute miles when RVR is out of service.

DEPARTURE CONSIDERATIONS

IFR departures are designed according to the criteria established in the ***U.S. Standard for Terminal Instrument Procedures*** (TERPs). The following information is a review of TERPs criteria that apply to obstacle clearance requirements for departures, departure routes, and SIDs. All published departures, including SIDs, are constructed to allow specific clearance from obstacles at a given distance from the runway. As you know from Chapter 3, the required obstacle clearance is based on an aircraft climbing at least 200 feet per nautical mile. A slope of 152 feet per nautical mile, beginning no higher than 35 feet above the departure end of the runway, is evaluated for obstacles. If no obstacles penetrate the slope, no departure procedures are published. If obstacles do penetrate the slope, a minimum ceiling and/or climb gradient may be required. In some cases, the aircraft may have to be maneuvered to avoid obstacles. Some departures may require a combination of these restrictions to ensure a safe departure.

STANDARD INSTRUMENT DEPARTURE (SID)

SIDs are coded departure routes established to expedite departures at airports with a large volume of traffic. In general, they are intended to simplify clearance delivery and departure procedures for both you and air traffic control. If you are operating from an airport where SIDs are published, you can expect to have a SID routing included in your ATC clearance. Your acceptance of a SID is not mandatory, but you must advise ATC if you do not wish to use one. The recommended procedure is to file "NO SID" in the remarks section of your flight plan. The least desirable method is to advise ATC verbally when the SID is assigned. If you wish to fly a SID, you must have either the charted procedure or at least the textual description of the appropriate SID in your possession. This is necessary because many SIDs are lengthy and complex. A detailed explanation of the procedure by ATC would defeat the original purpose of a SID.

To fly a SID, you must have the charted procedure or at least the textual description in your possession; otherwise, you should file "NO SID" in your flight plan.

COMPLIANCE WITH A SID

Before you file for a SID or accept one in a clearance, it is very important that you determine whether your airplane's performance will allow you to meet its requirements. When necessary for obstruction clearance, SIDs specify a climb gradient in feet per nautical mile. To convert this figure to rate of climb per minute, divide the groundspeed by 60 and multiply by the climb gradient. For example, if a SID requires an altitude gain of 200 feet per nautical mile and your planned groundspeed is 150 knots, the required rate of climb is 500 feet per minute ($150 \div 60 \times 200 = 500$).

To convert climb gradient to climb rate, divide groundspeed by 60 and multiply by the climb gradient figure.

You should also recognize that some SIDs require you to maintain a climb gradient to altitudes in excess of 10,000 feet. Therefore, the calculated continuous climb performance must be valid to the altitude required in the SID. This could be a significant factor if you encounter

REQUIRED CLIMB RATE (ft. per NM)	GROUND SPEED (KNOTS)						
	30	60	80	90	100	120	140
200	100	200	267	300	333	400	467
250	125	250	333	375	417	500	583
300	150	300	400	450	500	600	700
350	175	350	467	525	583	700	816
400	200	400	533	600	667	800	933
450	225	450	600	675	750	900	1050
500	250	500	667	750	833	1000	1167
550	275	550	733	825	917	1100	1283
600	300	600	800	900	1000	1200	1400
650	325	650	867	975	1083	1300	1516
700	350	700	933	1050	1167	1400	1633

Figure 5-3. Tables also are available for converting a minimum climb gradient to a minimum climb rate, based on your groundspeed. Using this NOS table, a gradient of 400 feet per nautical mile (item 1) at a groundspeed of 100 knots (item 2) requires a 667 f.p.m. rate of climb (item 3). Jeppesen normally provides a climb gradient/climb rate table on the SID itself.

adverse weather conditions, such as high density altitude, turbulence, and/or icing during the climb to cruising altitude. Another consideration is the average winds aloft during your departure. Since climb gradient is based on groundspeed, a tailwind will require an even greater rate of climb. If you feel that you cannot comply with these requirements, you cannot accept the SID, and you should notify ATC well in advance of departure. [Figure 5-3]

IFR DEPARTURE PROCEDURES

In addition to SIDs, IFR departure procedures also are established when necessary for airports that have published instrument approaches. When you are using NOS charts, the IFR departure procedures are tabulated in the front of each regional approach chart book, as well as each SID book. If you are using Jeppesen approach charts, the procedure is printed on the airport chart, which is usually located on the reverse side of the first approach chart for each airport. At larger terminals, the airport chart may be printed on a separate sheet. Regardless of the source, IFR departure procedures are designed to help you avoid obstructions in the departure path during your transition to the enroute phase. As mentioned earlier, when an aircraft cannot maintain obstacle clearance at 200 feet per nautical mile in accordance with TERPs criteria, a departure procedure must be established. The published procedure assures obstacle clearance from the departure end of the appropriate runway to the minimum enroute altitude. When using NOS charts be sure you refer to the IFR takeoff minimums and departure procedures listing to determine the requirements that will apply. [Figure 5-4]

INSTRUMENT APPROACH PROCEDURES (CHARTS)

IFR TAKE-OFF MINIMUMS AND DEPARTURE PROCEDURES

Civil Airports and Selected Military Airports

CIVIL USERS: FAR 91 prescribes take-off rules and establishes take-off minimums as follows:
(1) Aircraft having two engines or less – one statute mile. (2) Aircraft having more than two engines – one-half statute mile.
MILITARY USERS: Special IFR departure procedures, not published as Standard Instrument Departure (SIDs), and civil take-off minima are included below and are established to assist pilots in obstruction avoidance. Refer to appropriate service directives for take-off minimums.
Airports with IFR take-off minimums other than standard are listed below. Departure procedures and/or ceiling visibility minimums are established to assist pilots conducting IFR flight in avoiding obstructions during climb to the minimum enroute altitude. Take-off minimums and departures apply to all runways unless otherwise specified. Altitudes, unless otherwise indicated, are minimum altitudes in feet MSL.

NAME	TAKE-OFF MINIMUMS	NAME	TAKE-OFF MINIMUMS

ALBUQUERQUE INTL, NM
Rwy 8, 2400-2*
1 IFR DEPARTURE PROCEDURE: Comply with SID or radar vectors; or: Rwy 8 turn left or right as cleared. All aircraft climb direct ABQ VORTAC. Departures on R-147 CW 023 climb on course. All others climb westbound to cross ABQ VORTAC at or above 10,000.
*or standard (FAR 135: RVR 2400) with minimum climb of 424′ per NM to 8000.

AUSTIN, TX
AUSTIN EXECUTIVE AIRPARK
Rwy 36, 400-2 or standard with minimum climb of 250′ per NM to 1100.

LAKEWAY AIRPARK
2 IFR DEPARTURE PROCEDURE: Rwy 16, climb runway heading to 2000 prior to turning westbound or comply with radar vectors.

Figure 5-4. Both of these examples show that, in addition to higher than standard takeoff minimums, special IFR departure procedures are established. Item 1 indicates steeply rising terrain in the Albuquerque area, and item 2 requires a straight climb to 2,000 feet before turning westbound on course during departure from Lakeway Airpark.

COMPLIANCE WITH IFR DEPARTURE PROCEDURES

An IFR departure procedure, unlike a SID, is not assigned as a portion of your IFR clearance unless it is required for separation purposes. In general, it is your responsibility to determine if one has been established, then comply with it. In IFR conditions, the departure procedure is the only method of ensuring terrain and obstacle clearance. However, if the weather is VFR, you can expedite your departure by requesting a clearance to climb in VFR conditions from the appropriate ATC facility. This relieves you from flying the entire IFR departure procedure, although you are then responsible for your own terrain clearance and traffic separation during the climb. When a clearance is issued to climb in VFR conditions, the controller may abbreviate the term by saying, "*. . . climb VFR . . . ,*" and if required, "*. . . climb VFR between 4,000 and 10,000.*" If there is reason to believe that flight in VFR conditions may become impractical, your clearance also may provide for that possibility.

RADAR DEPARTURES

Radar departures are often assigned at radar-equipped approach control facilities and require close coordination with the tower. If your flight is to be radar vectored immediately after takeoff, the tower will advise you of the heading to be flown, but not necessarily the purpose of the heading.

During the IFR departure, you should not contact departure control until advised to do so by the tower.

This type of advisory will be issued to you either in your initial IFR routing clearance or by the tower just before takeoff. Once you have received your takeoff clearance, you should understand that coordination of your flight is the responsibility of the tower controller. After you are airborne, you can expect a handoff to the departure controller. However, it is important that you wait until you are instructed by the tower to "*. . . contact departure control.*" Leaving the tower frequency prematurely can complicate the handoff between the tower controller and departure control and may lead to confusion with respect to your position.

On your initial call to departure control, you should give only your aircraft or flight number, the altitude you are climbing through, and the altitude to which you are climbing. The controller wants to verify that your reported altitude agrees with the altitude being displayed by your Mode C transponder equipment. If your reported altitude does not agree with the altitude shown on the radar display, the controller cannot use your Mode C readout for traffic separation purposes. In this situation, you will be asked to verify your altitude, as necessary, so it can be correlated. In the event you omit the altitude report from your initial callup, the same problem occurs. If your response and the radar display correlate, your Mode C altitude readout can be used for separation.

The term "*radar contact*" is used by ATC to advise you that your aircraft has been identified and radar flight following will be provided until radar identification has been terminated.

After verifying position, altitude, and altitude assignment, the departure controller will state, "*radar contact.*" This means your aircraft is identified on the radar display and radar flight following will be provided until service is terminated. You can expect radar services appropriate to the airspace in which you are operating. However, the controller will not provide terrain and obstruction clearance just because your flight is in radar contact. During departure, terrain and obstruction clearance remains your responsibility until the controller begins to provide navigational guidance in the form of radar vectors.

"*Resume own navigation*" is a phrase used by ATC to advise you to assume responsibility for your own navigation.

During the departure, you may expect to be vectored out of the terminal area to a point where you can intercept your assigned route. Once the final vector is provided, the controller usually will state something similar to, "*. . . intercept V-81 and resume own navigation.*" This simply indicates that the controller is no longer providing radar vectors and expects you to intercept and continue on course using your own navigation equipment. Keep in mind that the phrase "*. . . resume your own navigation*" does not negate radar separation service.

In certain circumstances, you may be issued a vector which takes your flight off a previously assigned route. When this occurs, you will be advised briefly what the vector is to achieve. Generally, it is issued for weather avoidance, terrain clearance, or traffic separation. Radar service will be provided until your aircraft has been reestablished on course and you have been advised of your position. In some cases, a handoff may be made to another radar controller with continuing radar surveillance capabilities. If you feel that any vector is given in error, you are encouraged to question the controller and verify the purpose of the vector. In addition, you should maintain orientation to your current position at all times so you are prepared to resume your own navigation, if necessary.

DEPARTURE RESTRICTION

When you are departing an airport which is not served by a control tower, ATC may find it necessary to issue a departure restriction with your clearance. An IFR departure is based upon reserving a specific block of airspace for one aircraft at a specific time. This way, ATC can provide traffic separation between IFR aircraft in the area. To accomplish separation, ATC may occasionally find it necessary to include void times such as, "*. . . clearance void if not airborne by 1530Z.*" This avoids reservation of large blocks of airspace for indefinite periods of time. To avoid possible search and rescue operations, you must advise ATC as soon as possible but no later than 30 minutes after the void time if you do not depart.

OBTAINING THE IFR CLEARANCE

When you are departing an airport that has a control tower, you should plan on receiving your IFR clearance from ground control or, if appropriate, clearance delivery. At airports with only an FSS, you should request your clearance when you are ready to taxi. The FSS will request your clearance from the controlling ARTCC and relay it to you when it is available. If you are planning to depart an uncontrolled airport when IFR conditions prevail, you must contact the ARTCC directly by radio or telephone the nearest FSS to receive your clearance. When VFR conditions prevail at an uncontrolled airport, you may receive your clearance from ARTCC after takeoff. However, you should always maintain VFR conditions until you have obtained your IFR clearance and have ATC approval to proceed on course in accordance with the clearance.

NONRADAR IFR DEPARTURE PROCEDURE

IFR departures in a nonradar environment require procedures similar to those discussed previously in this section. For example, you should plan to comply with any published departure procedure and initiate the same report with regard to your present altitude and the altitude to which you are climbing. You can expect the controller to request additional reports from you in order to monitor your flight progress. These requested reports may include crossing a particular navigation fix, reaching an altitude, or intercepting and proceeding on course. By using this technique, the handoff to center is not only expedited, but the controller is aware of the airspace you have vacated and can make it available to another IFR flight. After you have been handed off to the center controller, you will be identified on radar and placed in "*radar contact.*"

IFR CLIMB CONSIDERATIONS

You should climb at an optimum rate to within 1,000 feet of your assigned altitude.

ATC expects you to maintain a continuous rate of climb of at least 500 f.p.m. to your assigned cruising altitude. If you are unable to maintain this climb rate, you should notify ATC of your reduced rate of climb. Unless ATC advises "*At pilot's discretion*," you are expected to climb at an optimum rate consistent with your airplane's performance to within 1,000 feet of your assigned altitude. Then attempt to climb at a rate of between 500 and 1,500 f.p.m. for the last 1,000 feet of climb.

While climbing on an airway, you are required by regulation to maintain the centerline except when maneuvering in VFR conditions to detect and/or avoid other air traffic.

After you have made the IFR departure and are established on the airway, another climb consideration becomes important. FAR Part 91 specifies that, when flying IFR on a Federal airway, you must fly the centerline of that airway during climb, cruise, and descent. However, the regulation further provides that you are not prohibited from maneuvering the aircraft to pass well clear of other aircraft in VFR conditions. In addition, the FAA recommends that, while climbing in VFR conditions, you make gentle turns in each direction so you can continuously scan the area around you. As previously discussed, whenever you are operating on an IFR flight plan in VFR conditions, you are responsible for collision avoidance.

CHECKLIST

After studying this section, you should have a basic understanding of:

✓ **Takeoff minimums** — What their significance is, who must comply, and where you can find them.

✓ **Visibility** — What the various types are and how to apply them when determining takeoff minimums.

✓ **SIDs** — When you can expect one on an IFR departure and why you must have the charted procedure or textual description to fly it.

✓ **IFR departure procedures** — How you can tell when one has been established for an airport and why you should comply with its provisions.

✓ **Radar departures** — What information the controller needs and when terrain and obstruction clearance is provided.

✓ **Nonradar departures** — How to obtain the initial clearance and comply with any established IFR departure procedures.

✓ **IFR climb considerations** — How ATC expects you to climb to your assigned altitude and what climb rates to use.

SECTION B

ENROUTE OPERATIONS

Once you reach the IFR enroute phase of flight, you must consider additional operational requirements. These include communications, compulsory reporting procedures, position monitoring, orientation, and correct response to ATC clearances. This section describes typical situations you can expect during IFR enroute operations, including holding patterns.

ENROUTE RADAR PROCEDURES

Air route traffic control centers receive the majority of IFR traffic from radar-equipped departure control facilities. As your flight transitions to the enroute phase, you can expect a handoff from departure control to the ARTCC. The radar handoff between the two facilities is a quick and simple procedure, but correct communications techniques are important.

COMMUNICATIONS

As your flight leaves the departure controller's airspace, either by radar vector or your own navigation, you will be instructed to contact the center. The instructions issued by the controller include the name of the facility, the appropriate frequency, and any pertinent remarks. For example, the controller may simply advise you to "*. . . contact Kansas City Center on 119.65,*" or "*. . . contact Kansas City Center on 119.65 leaving 6,000.*" Another example is "*. . . contact Kansas City Center on 119.65 and advise him of your heading.*" When you are instructed to change to another frequency, you should acknowledge receipt of the frequency change by repeating it back to the controller. This readback also verifies that you understand the instruction and have received the correct frequency.

During a handoff, ATC instructions include the name of the facility to contact, the appropriate frequency, and any pertinent remarks.

Your initial callup procedure to the center is basically the same as the one used when you contacted departure control. Include the facility identification, your aircraft identification, altitude, and assigned altitude; for example, "*Chicago Center, Cheyenne 131PT, 7,000.*" When you are climbing to an assigned altitude, you should state your present altitude and the assigned altitude. As an example of this type of communications technique, your initial callup should be: "*Fort Worth Center, Cessna 325, 6,500, climbing to 9,000.*" When your transmission is received, the controller verifies your position and compares your reported altitude to that shown by your Mode C equipment. Once verified, the controller acknowledges your transmission and states, "*radar contact.*" You can now expect

"*Radar contact*" means your aircraft has been identified on the controller's radar display, and flight following will be provided until radar identification is terminated.

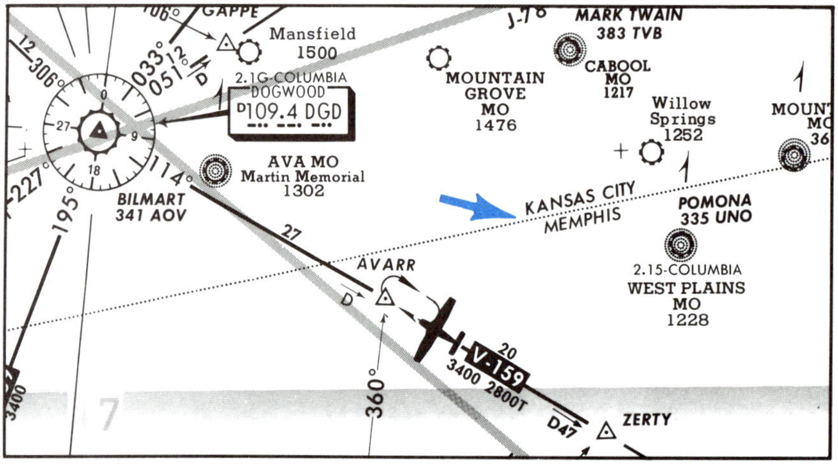

Figure 5-5. An enroute chart shows you where you can anticipate a handoff from one center to another. For example, assuming you are northwestbound on V-159, you can expect a handoff from Memphis Center to the Kansas City Center in the vicinity of Avarr Intersection.

radar flight following and radar services until you are advised, *"radar contact lost"* or *"radar service terminated."* [Figure 5-5]

FACILITY RADIO FAILURE

If you cannot establish communications on the newly assigned frequency, return to the previous frequency for further instructions.

When you are operating as pilot in command under IFR in controlled airspace, FARs require you to maintain a continuous watch on the appropriate frequency. If you make a required frequency change and center does not acknowledge your callup, return to the previously assigned frequency. Although it is a rare occurrence, ARTCC facilities are subject to transmitter/receiver failures, and radio contact may be momentarily impossible. Each ARTCC frequency has at least one backup transmitter and receiver which can be put into service quickly with little or no disruption of ARTCC service. Technical problems of this type may cause a delay, but the switch-over process rarely takes more than 60 seconds. Therefore, you should wait at least one minute before deciding that the center has experienced a radio failure. If you cannot establish contact using the newly assigned frequency, return to the one previously assigned and request an alternate frequency. If you are still unable to establish radio contact, try again on any ARTCC frequency. Failing that, contact the nearest FSS in the area for further instructions.

COMPULSORY REPORTING PROCEDURES

The fixes used in the flight plan to define an off-airway route become compulsory reporting points in a non-radar environment.

Compulsory reporting is not only regulatory, but also is an important pilot responsibility that assists ATC in maintaining adequate aircraft separation standards. In addition, you should understand that certain reporting procedures apply to both radar and nonradar environments, even though radar contact negates certain reports. In very broad terms, you may consider any change in altitude, speed, aircraft performance, or aircraft navigation capability to warrant a compulsory report. In a nonradar environment, for example, you must make position reports over the designated VOR stations and intersections along your route of flight. They are depicted on IFR enroute charts by solid triangles, rather than open triangles. Compulsory reporting points are also applicable when operating on an IFR flight plan in accordance with a VFR-on-top clearance. If you are on a direct course that is not on an established airway, you must report over the fixes used in your flight plan that define the route. Whether your route is on airway or off airway, position reports are mandatory in a nonradar environment, and they must include specific information relating to position, time, altitude, the ETA at and name of the next fix, and the name only of the succeeding fix. [Figure 5-6]

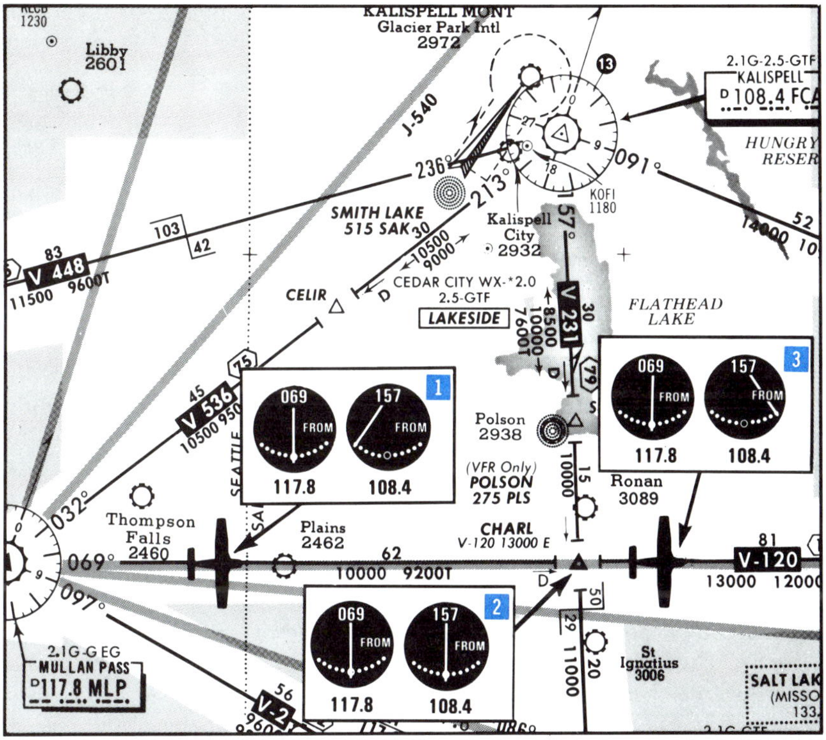

Figure 5-6. Assume that your route of flight is eastbound on V-120. When you cross the Mullan Pass VOR, ATC advises, *"radar contact lost."* This statement by ATC indicates you are now operating in a nonradar environment and must initiate position reporting. As you proceed outbound on the 069° radial of the Mullan Pass VOR, the first fix is the Charl Intersection which is identified by DME or the 157° radial of the Kalispell VOR. Also, note that Charl Intersection is a compulsory reporting point since the intersection symbol is a solid triangle. Note that the number two VOR receiver is tuned to the Kalispell VOR (108.4 MHz) and 157° is set on the OBS. The course needle deflects toward the station before you reach Charl (item 1). When you reach Charl, the needle centers (item 2), and when you pass the intersection, the course needle deflects away from the selected VOR station (item 3). Incidentally, this applies regardless of your direction of travel on the airway.

You must notify ATC when you leave an assigned altitude and whenever your true airspeed changes by 5% or 10 knots, whichever is greater.

There are several other compulsory reports. For example, when you are cleared from one altitude to a newly assigned altitude, you must report leaving the previous altitude. However, when you reach the newly assigned altitude, you are not required to report unless ATC requests you to do so. A change in airspeed also is a mandatory report. Specifically, if you elect to change your true airspeed at cruising altitude by 5% or 10 knots (whichever is greater), you must advise ATC.

ATC must be notified if you are unable to climb or descend at a rate of at least 500 feet per minute.

Other compulsory reports relate to airplane performance and airplane navigational capability. For example, in the event your airplane is unable to maintain at least a 500-foot per minute rate of climb or descent, you must notify ATC. Also, if any navigation receiver (such as VOR, ILS, or ADF) should fail in controlled airspace, you must report the failure to ATC.

Other situations which require a report include the time and altitude reaching a holding fix or point to which cleared, leaving any assigned holding fix or point, encountering unforecast weather conditions, and any other information relating to safety of flight. You should also report the FAF inbound on a nonprecision approach and when leaving the outer marker or fix used in lieu of it on a precision approach. Keep in mind that all of these reports are required when your flight is in radar contact, with the exception of the position report, revised ETA, and FAF inbound. In the event the controller advises *"radar service terminated,"* you should resume normal position reporting and revise your ETA, if appropriate. The remainder of the reports must be made regardless of radar coverage and without a specific request from ATC. [Figure 5-7]

SPECIAL USE AIRSPACE

As you know, special use airspace is designated to confine activities that may be hazardous to civil aircraft. Therefore, limitations are imposed upon aircraft operations that are not a part of those activities. In general, ATC will not issue an IFR route clearance that penetrates an active restricted area. However, if the restricted area is not active and has been released to the controlling agency (FAA), ATC will allow you to operate within it without issuing a specific clearance to do so. If your filed route of flight takes you through a restricted area that is active, ATC normally issues you a revised routing clearance which ensures your flight avoids the affected airspace. In contrast, nonparticipating IFR traffic may be cleared through an active MOA if ATC can provide separation. With the exception of controlled firing areas, special use airspace areas are depicted on IFR enroute charts and should be considered during preflight planning. Refer to the AIM for information regarding IFR operations in special use airspace.

COMPULSORY IFR REPORTS — RADAR/NONRADAR	
Vacating one assigned altitude or flight level for another	*"Saratoga 6758L, leaving 7,000, climbing to 10,000."*
VFR on top change in altitude	*"Saratoga 6758L, VFR on top, climbing to 10,500."*
Unable to climb/descend at 500 feet per minute	*"Saratoga 6758L, maximum climb rate 400 feet per minute."*
Missed approach	*"Saratoga 6758L, missed approach, request clearance to Omaha."*
TAS variation from filed of 5%, or 10 knots, whichever is greater	*"Saratoga 6758L, advises TAS decrease to 150 knots."*
Time and altitude or flight level reaching a holding fix or point to which cleared	*"Saratoga 6758L, Fargo Intersection at 05, 10,000, holding east."*
Leaving any assigned holding fix or point	*"Saratoga 6758L, leaving Fargo Intersection."*
Loss of nav/com capability	*"Saratoga 6758L, ILS receiver inoperative."*
Unforecast weather conditions or any other information relating to safety of flight	*"Saratoga 6758L, experiencing moderate turbulence at 10,000."*

STANDARD POSITION REPORT — NONRADAR	
Identification	*"Saratoga 6758L,*
Position	*Shreveport,*
Time	*15,*
Altitude/flight level	*11,000,*
IFR or VFR for report to FSS only	*IFR,*
ETA over the next reporting fix	*Quitman 40,*
Succeeding reporting points	*Scurry next."*
Pertinent remarks	*(Infrequently used)*

ADDITIONAL REPORTS — NONRADAR	
Leaving FAF or OM inbound on final approach	*"Saratoga 6758L, outer marker inbound, leaving 2,000."*
Revised ETA of more than three minutes	*"Saratoga 6758L, revising Scurry estimate to 55."*

Figure 5-7. This listing shows the IFR compulsory reports along with sample phraseology. A standard position report and the additional reports which are required in a nonradar environment are also shown.

HOLDING PATTERNS AND PROCEDURES

Holding patterns are used to delay aircraft for various reasons. During the enroute phase of flight, ARTCCs may assign holding patterns to regulate the flow of traffic and maintain separation. Approach control facilities often use a holding "stack" to sequence aircraft for the active instrument approach. In other cases, you may find it necessary to climb in a holding pattern to the MEA when enroute obstruction clearance dictates. Holding patterns also are used when you reach a clearance limit or may be required following a missed approach. In addition, you may request a holding pattern if you need to wait for the weather to improve before starting an approach.

STANDARD HOLDING PATTERNS

A **standard holding pattern** is a racetrack-shaped maneuver. The turns are performed to the right at a standard rate of three degrees per second, provided the angle of bank does not exceed 30°. If a flight director system is used, the bank limit is 25°. A **nonstandard holding pattern** utilizes the same procedure, but employs left turns. The length of the holding pattern legs varies with the groundspeed and altitude of the aircraft. In a no-wind condition, you fly each leg for one minute at and below 14,000 feet MSL or for one and one-half minutes above 14,000 feet MSL. Therefore, if the groundspeed of your aircraft is 120 knots, the resulting leg lengths will be two miles for holding patterns at or below 14,000 feet MSL. [Figure 5-8]

OUTBOUND AND INBOUND TIMING

After you enter a holding pattern, you should fly the initial outbound leg for one or one and one-half minutes, whichever is appropriate for your

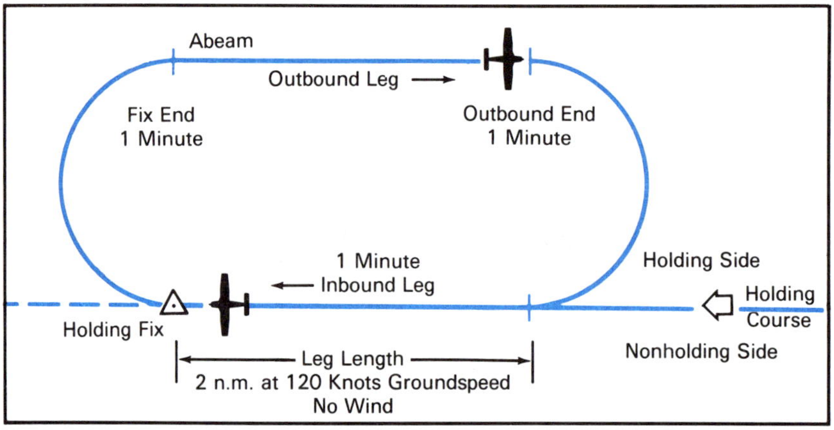

Figure 5-8. A standard holding pattern uses right-hand turns at a standard rate of three degrees per second. The side of the holding course where the holding pattern is executed is called the "holding side." Take a few minutes to study the labels for the other segments, and refer back to this illustration as needed in the accompanying discussion.

operating altitude. The timing for the outbound leg usually begins over/abeam the holding fix, whichever occurs later. If the abeam position cannot be identified, you should start timing when you complete the turn outbound. Adjust the timing for subsequent outbound legs to achieve proper inbound timing (one minute or one and one-half minutes). This procedure is required, because the wind may cause significant differences between the groundspeed of the inbound and outbound legs. In the remainder of this discussion, we will use one-minute legs for simplicity.

Outbound leg timing for the holding pattern begins over/abeam the fix, whichever occurs later, or, if the abeam position cannot be identified, at the completion of the turn outbound.

If you find that the inbound leg of a holding pattern requires 45 seconds instead of one minute, you should lengthen the outbound leg. The normal procedure is to lengthen the outbound leg by the same amount that the wind has shortened the inbound leg (in this case, 15 seconds). Although this procedure results in slightly unequal time for the two legs, it will suffice unless winds are extremely high in relation to the true airspeed of the aircraft. You should continue to check the inbound timing and make outbound adjustments as necessary during subsequent holding pattern circuits.

When DME is used to define a holding pattern, the same holding procedures apply, except that DME distances are used in lieu of timing. The holding fix is on a radial at a specified DME. The leg length may be given by ATC, or the turn inbound may be at a designated DME.

CROSSWIND CORRECTION

Crosswind conditions can require pronounced heading corrections. During inbound tracking, you can determine the wind correction angle (WCA) required to maintain the inbound course. Once you have determined it, plan to triple the correction for the outbound leg. If you do not apply this technique, you may penetrate unprotected airspace. In addition, it may be difficult to become established on the inbound course before you reach the fix, resulting in more pattern irregularity. [Figure 5-9]

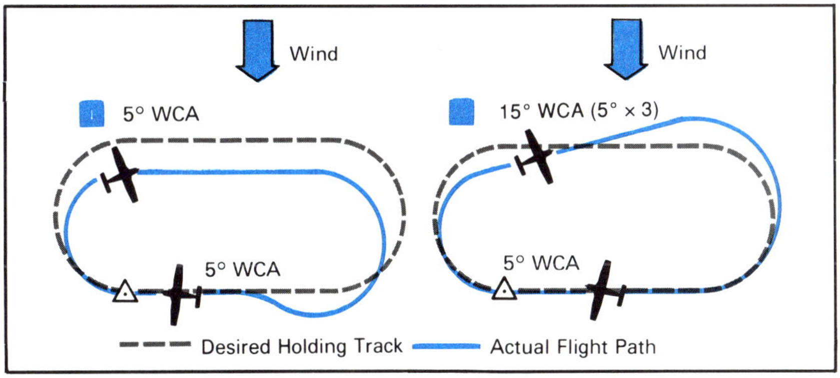

Figure 5-9. If you do not use sufficient wind correction on the outbound leg (item 1), the aircraft will overshoot the course following the turn inbound. The correct procedure recommended in the AIM is to triple the inbound correction and apply it to the outbound heading (item 2).

To reduce the size of the holding pattern and, subsequently, the amount of protected airspace, aircraft speed in the holding pattern is limited according to altitude and aircraft type.

MAXIMUM HOLDING SPEED

Obviously, as the aircraft's speed increases, the holding pattern size increases. In order to reduce the amount of airspace that must be protected by ATC, maximum holding speeds have been designated for specific altitude ranges. However, if the holding speed is less than you feel is necessary due to turbulence or other operational requirements, you should advise ATC of your revised holding speed. Also, if the indicated airspeed of your aircraft exceeds the applicable maximum holding speed, ATC expects you to reduce speed within three minutes of your ETA at the holding fix. Since holding is a delaying maneuver, you may want to further reduce your airspeed for greater fuel savings and endurance. [Figure 5-10]

HOLDING PATTERN ENTRIES

Recommended holding pattern entry procedures are designed to establish the aircraft on the holding radial or course without excessive maneuvering. The three types of holding pattern entries are direct, teardrop, and parallel. The type of entry you use depends on your heading as you approach the holding fix. [Figure 5-11]

Your aircraft's magnetic heading when you arrive at the fix determines the type of entry to use.

Once you establish the entry sectors, you can determine your position in relation to the holding course. For example, if the holding course is the 090° radial of a VOR station, the sectors are established by the 020° radial and the 200° radial. If you are approaching the VOR between the 020° and 200° radials, use a direct entry. If you are approaching from the other side of the station and your heading is between 020° and 090°, use a teardrop entry. If your heading is between 090° and 200°, a parallel entry is correct. You may allow 5° variance in heading when determining the correct entry. Therefore, if you are approaching a fix on a sector boundary, you may use either of two entries in most cases.

MAXIMUM HOLDING AIRSPEEDS			
Propeller Driven, Including Turboprop		**Civil Turbojet**	
All Altitudes	175 KIAS	Above 14,000 FT. MSL	265 KIAS
		14,000 FT. MSL – Above 6,000 FT. MSL	230 KIAS 210 KIAS where published
		6,000 FT. MSL – Minimum Holding Altitude	200 KIAS

Figure 5-10. This diagram shows the maximum holding airspeeds (MHA) in knots indicated airspeed for propeller-driven and civil turbojet aircraft for each altitude range.

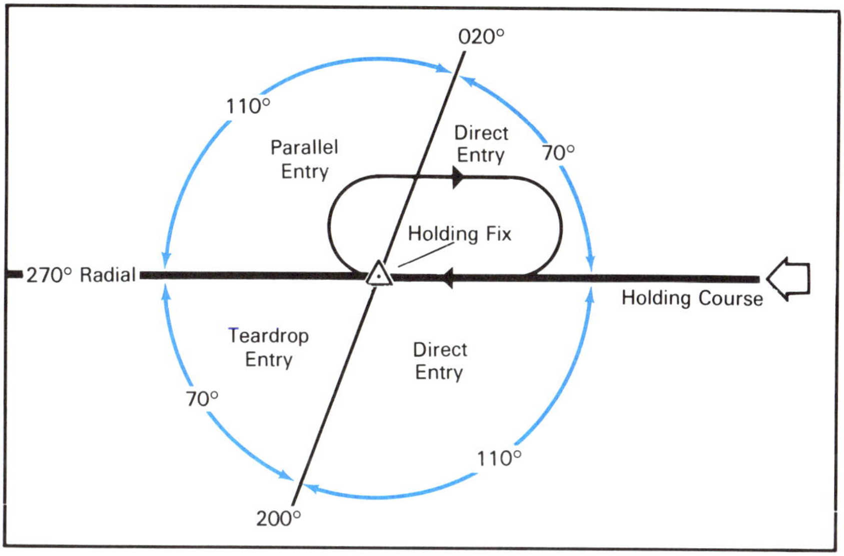

Figure 5-11. Entry sectors are established by forming angles of 70° on the holding pattern side of the holding course and 110° on the nonholding side. It will be helpful to refer to this illustration during the following discussion.

The **direct entry** procedure is the least complicated. It is also the one most often used, because it can be applied throughout 180° of azimuth in relation to the holding fix. When you use the direct entry, you simply fly across the fix, turn right to the outbound heading, and fly the pattern. [Figure 5-12]

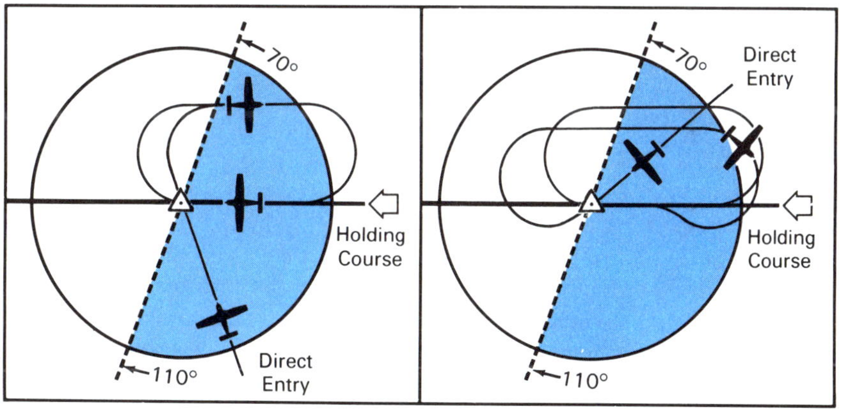

Figure 5-12. In this example, assume the holding course is on the 090° radial. After you have passed the station, simply turn right to a heading of 090° and time for one minute, then turn right again and intercept the radial inbound. In this example, since you are tracking inbound on the 090° radial from the station, you should set 270° on the course selector as you turn inbound on the radial. It may take several circuits before you establish the right amount of wind correction to maintain the holding course precisely.

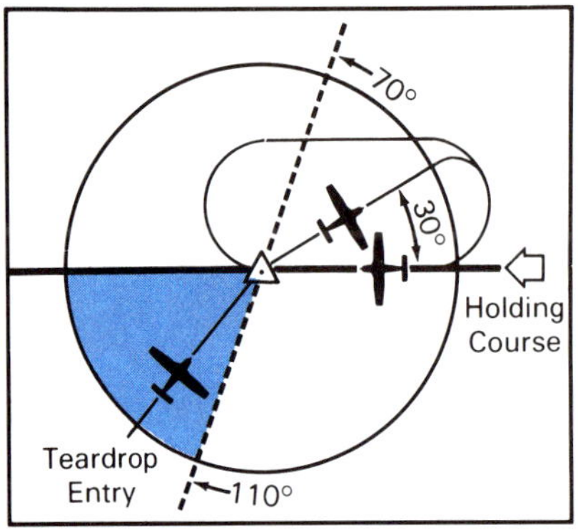

Figure 5-13. This example shows that the initial heading of 060° positions the airplane away from the holding course on the holding side. You should set 270° on the course selector and turn right after one minute to intercept the inbound holding course.

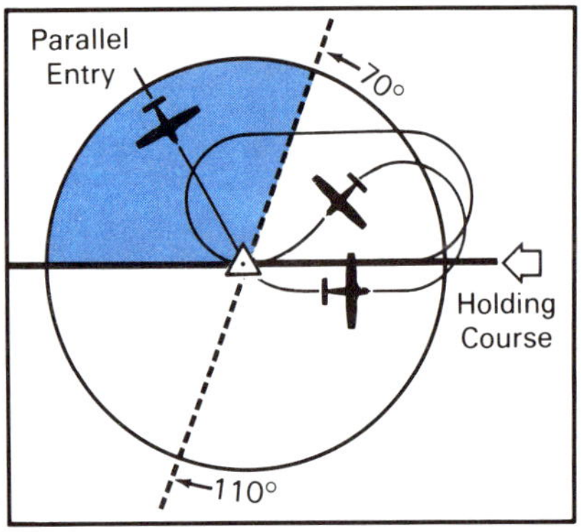

Figure 5-14. In this example, make the first turn parallel to the outbound course using a heading of 090°. After one minute, initiate a left turn and return to the holding fix or intercept the holding course.

During the **teardrop entry**, you must maneuver the aircraft on the holding side of the holding course before you establish the aircraft inbound. After crossing the VOR, turn right to a heading which is approximately 30° away from the holding course, but on the holding side of the pattern. Once you are established on this heading, begin timing for approximately one minute, turn right to intercept the holding course inbound, and return to the fix. [Figure 5-13]

The **parallel entry** procedure is used throughout 110° of azimuth and involves paralleling the holding course on the nonholding side. After crossing the holding fix, turn the airplane to a heading that parallels the outbound course and begin timing for one minute. Then, initiate a left turn to pass through the holding course and reintercept it from the holding side or proceed to the holding fix. [Figure 5-14]

VISUALIZING ENTRY PROCEDURES

Pilots have improvised various ways to visualize entry procedures prior to arrival over the fix. These include drawing holding patterns on aeronautical charts, using the wind side of a flight computer, employing the aircraft heading indicator, referring to a holding pattern inscription on a plotter, and using specially designed pattern entry computers. The value of any of these methods depends on how well they help you mentally visualize the holding pattern and the appropriate entry.

There is an alternate method which does not require separate "paraphernalia," and it may be the most advantageous. In many cases, holding patterns are assigned without advance notice, leaving you little time to determine the correct entry procedure with separate tools or devices. This method requires you to visualize your arrival at the fix. If the holding course is behind the aircraft when you arrive at the fix, make a direct entry. When the holding course is ahead and to the right of the aircraft, use a teardrop entry. If the holding course is ahead and to the left of the aircraft as it crosses the fix, use a parallel entry. You may also apply this method to nonstandard holding patterns. The direct entry sector is still behind the aircraft; however, holding courses ahead and to the right of the aircraft require a parallel entry. Conversely, holding courses ahead and to the left of the aircraft require a teardrop entry.

ATC HOLDING INSTRUCTIONS

If ATC does not specify the direction of turn in the holding clearance, all turns must be made to the right.

When controllers anticipate a delay at a clearance limit or fix, you will usually be issued a holding clearance at least five minutes before you arrive at the fix. If the holding pattern assigned by ATC is depicted on the appropriate aeronautical chart, you are expected to hold as published, unless advised otherwise by ATC. In this situation, the controller will issue a holding clearance which includes the name of the fix, directs you to hold as published, and includes an expect further clearance (EFC) time. An example of such a clearance is: "*Cessna 21T, hold east of Byson Intersection as published, expect further clearance at 1521.*" When ATC issues a clearance requiring you to hold at a fix where a holding pattern is not charted, you will be issued complete holding instructions. This information includes the direction from the fix, name of the fix, course, leg length (if appropriate), direction of turns (if left turns required), and the EFC. You are required to maintain your last assigned altitude unless a new altitude is specifically included in the holding clearance, and you should fly right turns unless left turns are specified. Also note that all holding instructions should include an expect further clearance time. If you lose two-way radio communications, the EFC allows you to depart the holding fix at a specific time. [Figure 5-15]

ELEMENTS OF HOLDING CLEARANCES	
Holding Pattern Published — Sample Clearance	
1. The direction to hold from the holding fix	*". . . Hold northwest*
2. Holding fix	*of Drako Intersection as published.*
3. Expect further clearance time	*Expect further clearance at 2102."*
Holding Pattern Not Charted — Sample Clearance	
1. The direction to hold from the holding fix	*". . . Hold west*
2. Holding fix	*of Green Intersection*
3. The specified radial, course, magnetic bearing, airway number, or route	*on V-8,*
4. The outbound leg length in minutes or nautical miles when DME is used	*five-mile legs,*
5. Nonstandard pattern, if used	*left turns.*
6. Expect further clearance time	*Expect further clearance at 1528."*

Figure 5-15. These samples show the two types of holding clearances you may expect to receive, depending on whether the holding pattern is published on the chart.

If you are approaching your clearance limit and have not received holding instructions from ATC, you are expected to follow certain procedures. First, you request further clearance before you reach the fix. If you cannot obtain further clearance, you are expected to hold at the fix in compliance with the published holding pattern. If a holding pattern is not charted at the fix, you are expected to hold on the inbound course using right turns. This procedure ensures that ATC will provide adequate separation. [Figure 5-16]

IFR CRUISING ALTITUDES

The hemispheric rule for IFR flight specified in FAR 91.179 applies only to flight in uncontrolled airspace. IFR altitudes for flight in controlled airspace are specified by ATC. Since most pilots file IFR altitudes which agree with the hemispheric rule, ATC usually assigns even thousand-foot altitudes for westbound flights below 18,000 feet MSL and odd-thousand foot altitudes for eastbound flights. In controlled airspace, however, you may request and/or receive altitude assignments that do not comply with east/west rules.

As another example, assume you are IFR below 18,000 feet and you are operating in accordance with a VFR-on-top clearance. You may select

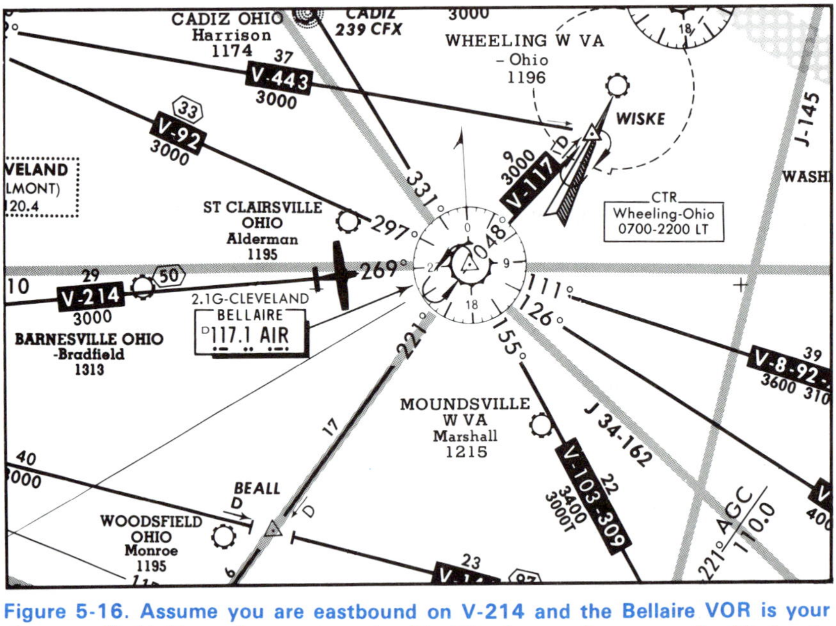

Figure 5-16. Assume you are eastbound on V-214 and the Bellaire VOR is your clearance limit. If you have not received holding instructions, you should plan to hold southwest on the 221° radial using left-hand turns, as depicted. If this holding pattern was not charted, you would hold west of the VOR on V-214 using right-hand turns.

any VFR cruising altitude appropriate to your direction of flight between 18,000 feet MSL and the MEA. Of course, you must still report any change in altitude to ATC and comply with all other IFR compulsory reports that apply. VFR on top is not authorized above 18,000 feet MSL in Class A airspace.

You can select any appropriate VFR cruising altitude between the MEA and 18,000 feet MSL when you have a clearance to operate VFR on top.

Below 18,000 feet MSL, you are expected to keep your altimeter adjusted to the current setting, as reported by a station within 100 n.m. of the aircraft. In areas where stations are more than 100 n.m. from the aircraft, you may use the closest appropriate station's altimeter setting. During IFR flight, ATC advises you periodically of the current altimeter setting.

At or above 18,000 feet MSL, you are required to set your altimeter to 29.92 in. Hg. This requirement results in some problems when actual altimeter settings along the route of flight are lower than 29.92. For example, an aircraft flying at FL 180 using an altimeter setting of 29.92 (assuming standard temperature) could be at the same true altitude as an aircraft in the low altitude sector. This could happen if the aircraft in the low altitude sector is using an altimeter setting of 28.92 and indicating 17,000 feet MSL. Consequently, FAR 91.81 specifies the **lowest usable flight levels** for given altimeter setting ranges. As local altimeter settings fall below 29.92, a pilot in the high altitude sector must cruise at progressively higher indicated altitudes to ensure adequate separation from other aircraft operating in the low altitude structure.

DESCENDING FROM THE ENROUTE SEGMENT

As you approach your destination, ATC will issue a descent clearance so you arrive in approach control's airspace at an appropriate altitude. However, there are several factors regarding a descent clearance that you should understand in order to comply with the clearance. For example, there are two basic descent clearances that ATC may issue. The first is a simple clearance to descend to and maintain a specific altitude. Generally, this ATC clearance is for enroute traffic separation purposes and your prompt compliance is not only expected but also required. You should descend at the optimum rate consistent with the operating characteristics of your aircraft until you are 1,000 feet above the assigned altitude. Remember, you must report vacating any previously assigned altitude for a newly assigned altitude and you should make your last 1,000 feet of descent at a rate of 500 to 1,500 f.p.m. The second type of descent uses the phrase, "*at pilot's discretion.*" When you are issued a pilot's discretion descent clearance, you may begin the descent whenever you choose. In addition, you are authorized to level off temporarily at any intermediate altitude during the descent. However, once you vacate an altitude during the descent, you may not return to that altitude.

During a descent at pilot's discretion, you may temporarily level off at any intermediate altitude.

A clearance for a descent at pilot's discretion may also be included as part of a normal descent clearance, such as ". . . *cross the Joliet VOR at or above 12,000, descend and maintain 5,000."* This clearance authorizes you to descend from the assigned altitude at your discretion, so long as you cross the Joliet VOR at or above 12,000 feet MSL. After that, you should descend at normal rates until you reach the assigned altitude of 5,000 feet.

A descent clearance which specifies a crossing altitude authorizes you to descend at your discretion for that portion of the flight to which the crossing altitude restriction applies. Any other time that an authorization to descend at pilot's discretion is intended, it must be specifically stated by the controller. Let's consider another example, using an enroute chart excerpt. [Figure 5-17]

Clearances to descend at pilot's discretion are not only an ATC option but also may be requested to achieve optimum operating efficiency. For example, if you are enroute above an overcast layer, you may request a descent at your discretion to allow you to stay above the overcast for as long as possible. This is particularly important when the outside air temperature is conducive to icing conditions. With this request, you can stay at your cruising altitude longer in order to conserve fuel and avoid prolonged periods of IFR flight in icing conditions. You can also use this type of descent to minimize turbulence by leveling off at an intermediate altitude where the air is smoother.

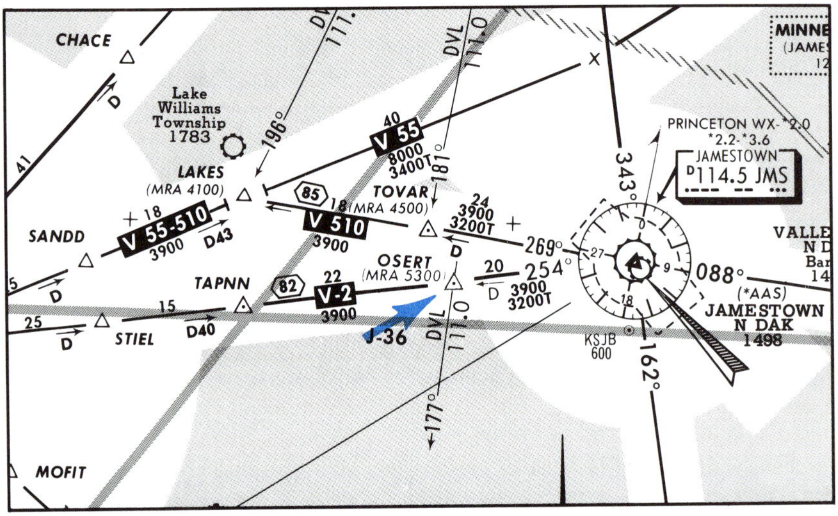

Figure 5-17. Assume that you are westbound on V-2 at 14,000 feet over the Osert Intersection, and ATC issues the following clearance: *". . . descend now to 10,000, cross Stiel Intersection at or above 6,000 feet, descend and maintain 4,000."* In this case, you are expected to descend promptly to 10,000 feet upon receipt of the clearance. You may then descend at your discretion, as long as you cross Stiel Intersection at or above 6,000 feet. However, once you reach Stiel, the descent is no longer at your discretion and you must descend promptly to 4,000 feet.

CHECKLIST

After studying this section, you should have a basic understanding of:

✓ **Radar environment** — How radar simplifies reporting procedures, what the phrase *"radar contact"* means, and how a radar handoff is accomplished.

✓ **Nonradar procedures** — What it means when the controller advises *"radar service terminated"* or *"radar contact lost."*

✓ **Compulsory IFR reports** — What they are and which ones are exempt in a radar environment.

✓ **Special use airspace** — What happens if your filed route of flight penetrates an active restricted area or an MOA.

✓ **Holding patterns** — Why they are used and how you should enter them to remain in protected airspace.

✓ **Holding clearance** — What the controller will say when a full holding clearance is issued and how it can be abbreviated when a holding pattern is published.

✓ **IFR cruising altitudes** — What you should know about cruising altitudes in controlled and uncontrolled airspace and what the significance is of the lowest usable flight level.

✓ **Descents** — What ATC expects during a descent from cruising altitude and how to interpret a clearance to *"descend at pilot's discretion."*

SECTION C

ARRIVALS AND APPROACHES

You should start preparing for your arrival and approach well before you descend from the enroute phase of flight. In this section, you will be introduced to a number of practical operating procedures that apply to the arrival phase. Some examples are standard terminal arrivals and radar and nonradar approaches, including straight-in, circling, and side-step maneuvers. The importance of correct missed approach procedures also is discussed. This section, like others in this chapter, is operationally oriented and stresses compliance with ATC clearances and regulations. The subject areas are summarized from appropriate source publications, including the following: *U.S. Standard for Terminal Instrument Procedures (TERPs), FAA Order 7110.65 — Air Traffic Control, Airman's Information Manual, AC 61-27 — Instrument Flying Handbook,* and *FAR Part 91.*

STANDARD TERMINAL ARRIVALS

ATC may assign a STAR procedure unless you specify "NO STAR" in the remarks section of your flight plan.

As you approach the destination airport, ATC can assign your flight a STAR procedure. You should understand that ATC may do this even if you have not requested the procedure. If you accept the clearance, you must have the charted procedure, or at least an approved textual description of the STAR, in your possession. If you do not wish to fly the standard arrival procedure, you should specify "NO STAR" in the remarks section of your IFR flight plan. You also can use the less desirable method of refusing the STAR by verbally notifying ATC when the clearance is issued.

As a general rule, STAR procedures are established to simplify the clearances issued to arriving aircraft. As you know, STARs often include navigation fixes that are used to provide transition and arrival routes from the enroute structure to the final approach course or to a fix where radar vectors will be provided. When ATC issues a STAR clearance, it will include the name of the STAR, the specific transition route, and an altitude assignment. A typical clearance issued by ARTCC would be, ". . . *cleared via the Proud Four Arrival, Woodstown Transition, maintain 8,000.*" [Figure 5-18]

After you receive a STAR clearance, you should complete certain tasks pertinent to your arrival in the terminal area. An important task to complete early is to listen to ATIS to obtain current weather information and

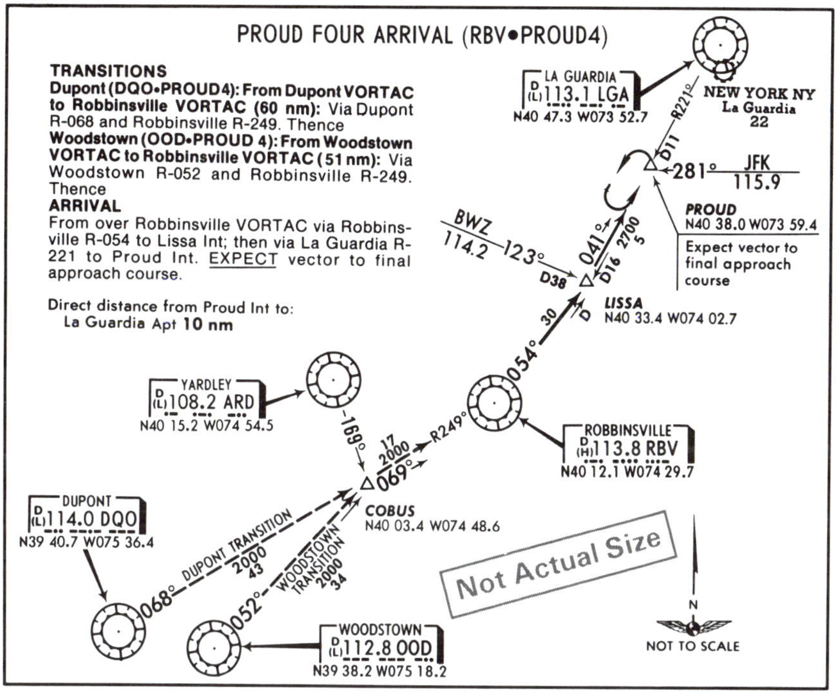

Figure 5-18. Note that there are two transition routes associated with the Dupont and Woodstown VORTACs that allow you to navigate from the enroute structure to the Robbinsville VORTAC. In this example, the transition routes end and the arrival route begins at Robbinsville. This also is reflected in the narrative description for each transition. The arrival description uses the terminology "from over the Robbinsville VORTAC," and therefore identifies the fix where the arrival route begins. Also, note that the bold line defining the arrival route terminates at Proud Intersection. The note adjacent to the intersection indicates you can expect radar vectors to the final approach course.

airport data concerning landing runways, as well as any NOTAMs. By monitoring ATIS in the arrival stage, you also receive advance information on instrument approaches. If you are arriving at an airport that does not have ATIS and two or more instrument approach procedures are available, ATC will advise you of the type of approach to expect during IFR weather conditions. However, this information will not be furnished where visibility is three miles or more and the ceiling is at or above the highest initial approach altitude for the airport. Once you have determined what approach you will use, review the chart thoroughly before you enter the terminal area. This review should include radio frequencies, inbound course, MDA or DH, the missed approach point, and the missed approach procedure. When you have completed the approach chart review, consider completing the descent and approach checklists as appropriate to your aircraft. This way, you will have completed important prelanding items early in the arrival stage, so you may concentrate on precise maneuvering and navigation during the approach.

Altitude and airspeed control are additional requirements as you proceed inbound. During your arrival in the terminal area, ATC may issue a descent clearance which includes a crossing altitude restriction. To comply with the crossing altitude, you must first determine the distance available to descend and then estimate the rate of descent necessary to comply with the crossing altitude restriction. [Figure 5-19]

During the arrival, expect to make adjustments in indicated airspeed at the controller's request. With a high-performance airplane, it is very common for ATC to request an airspeed adjustment in order to achieve traffic sequencing and separation. Speed adjustments reduce the amount of vectoring required in the terminal area.

An airspeed assignment may be issued in several ways, depending on the controller's separation needs. Controllers may use the following

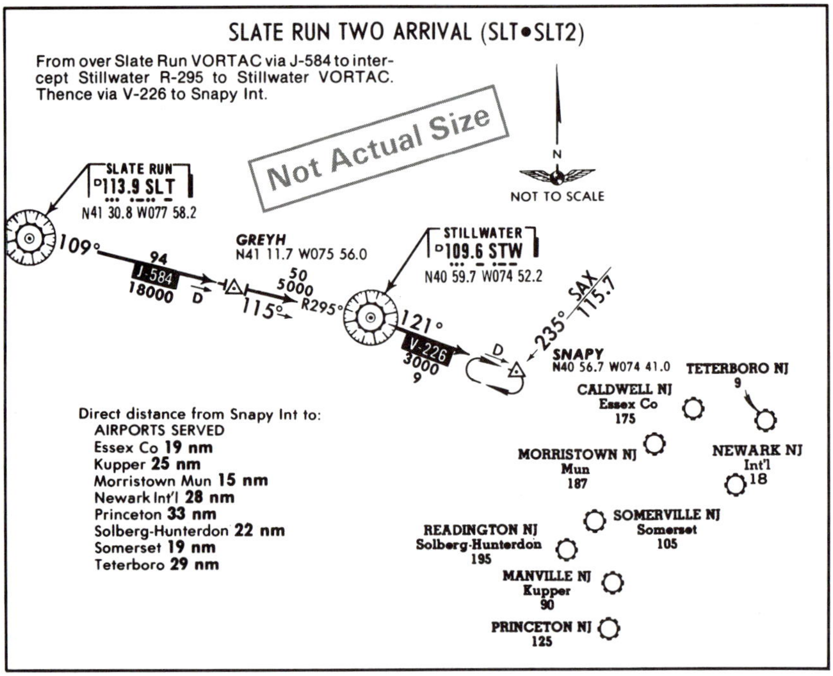

Figure 5-19. Assume that you are inbound to the Stillwater VORTAC on the Slate Run Two Arrival. Your position is 45 DME from the VORTAC, your cruising altitude is 10,000 feet, and ATC issues the following clearance, *". . . cross Snapy Intersection at 4,000, maintain 4,000."* This means you are authorized to descend at your discretion, as long as you cross the intersection at the specified altitude. Assuming your planned rate of descent is 500 f.p.m. and your groundspeed is approximately 180 knots (3 n.m. per minute), the total distance required for this descent is 36 n.m. ($6{,}000 \div 500 \times 3 = 36$). Since the Snapy Intersection is 9 n.m. southeast of the Stillwater VORTAC, you should begin the descent no less than 27 DME northwest of Stillwater to reach 4,000 feet MSL at Snapy Intersection.

examples when issuing speed restrictions, "*. . . reduce speed to 160 knots,*" "*. . . increase speed to 160 knots,*" "*. . . slow to 150 knots after crossing Aggie Intersection,*" "*. . . do not exceed 180 knots,*" "*. . . if practical, increase speed 20 knots,*" and so on.

When you are operating a reciprocating-engine or turboprop aircraft within 20 miles of the airport of intended landing, 150 knots usually is the lowest speed you will be assigned. If your aircraft cannot maintain the assigned speed, you should advise ATC. In this event, the controller may ask you to maintain the same speed as the preceding or succeeding aircraft, if practical. ATC expects you to maintain the specified speed adjustment within ±10 knots. In other cases, you may be asked to increase or decrease your speed by 10 knots, or multiples thereof. When a speed adjustment is no longer needed, ATC will advise you to "*. . . resume normal speed.*"

ATC normally will not assign an airspeed lower than 150 knots when you are flying a reciprocating-engine or turboprop aircraft within 20 miles of the airport where you intend to land.

Keep in mind that the maximum speeds specified in FAR Part 91.117 still apply during speed adjustments. It is your responsibility to advise ATC if an assigned speed adjustment would cause you to exceed these limits. However, for operations in Class C or Class D airspace at or below 2,500 feet AGL within 4 nautical miles of the primary airport, ATC has the authority to request or approve a higher speed than prescribed in FAR Part 91.117.

USE OF ATC RADAR FOR APPROACHES

You may remember from previous discussions, that where radar is approved for approach control service, it also can be used in conjunction with published instrument approach procedures to provide guidance to the final approach course or to the traffic pattern for a visual approach. In addition, approach control radar is used for ASR and PAR approaches and to monitor nonradar approaches.

Approach control radar is used to provide guidance to the final approach course, for ASR and PAR approaches, and for monitoring nonradar approaches.

During an arrival, you will normally be cleared to the airport or an outer fix that is appropriate to your arrival route. Once you have been handed off to approach control, you will continue inbound to the airport or to the fix in accordance with your last route clearance. ATC will advise you to expect radar vectors to the final approach course for a specific approach procedure unless that information is included in ATIS. Radar vectors and altitude assignments will be issued as needed for spacing and aircraft separation. Therefore, it is very important that you not deviate from the headings and altitudes assigned by approach control.

MINIMUM VECTORING ALTITUDES

During the process of radar vectoring, the controller is responsible for assigning altitudes that are at or above the minimum vectoring altitude. These altitudes are established in terminal areas to provide terrain and obstruction clearance.

The MVA provides 1,000 feet of obstruction clearance in nonmountainous areas and, in most cases, 2,000 feet in designated mountainous areas. Minimum vectoring altitudes also must provide a margin of at least 300 feet above the floor of controlled airspace. Therefore, the altitudes you are assigned by ATC when you are being radar vectored provide terrain and obstruction clearance, and you are always vectored within controlled airspace. The MVA in a given sector may be lower than the nonradar MEA, MOCA, or other minimum altitude shown on instrument charts.

VECTORS TO THE FINAL APPROACH COURSE

Radar vectors to the final approach course provide a method of intercepting and proceeding inbound on the published instrument approach procedure. Although radar vectors are frequently used in conjunction with ILS approaches, they also can be used with nonprecision approaches such as localizer back course, NDB, and VOR. Regardless of the type of approach in use, ATC will attempt to provide an intercept angle no greater than 30°. When ATC issues the intercept heading, it normally is qualified with additional instructions. For example, the controller may state, ". . . *turn right heading 320°, maintain 7,000 until intercepting the localizer* . . ."

If it becomes apparent the assigned heading will cause you to pass through the final approach course, you should maintain that heading and question the controller.

If the controller does not clear you for the approach, you must maintain the assigned heading even though it will cause you to cross through the final approach course. If this appears imminent and you have not been advised that you will be vectored through it, question the controller. Of course, the controller may intentionally vector you through the final approach course to achieve traffic separation. If this is the case, the controller should advise you and state the reason such as, ". . . *expect vector across final approach course for spacing.*" In any event, you should not turn inbound on the final approach course until you are cleared for the approach.

APPROACH CLEARANCE

When you are cleared for an approach while being radar vectored, you must maintain your last assigned altitude until established on a segment of the published approach.

When you are being vectored for an approach, the assigned headings normally permit you to establish the aircraft on the final approach course prior to the FAF. As a general rule, the approach clearance will be issued in conjunction with the final vector. If you are being radar vectored and are not established on a segment of a published route or an instrument approach procedure, the controller will stipulate a minimum altitude for you to maintain until you intercept the final approach course.

Once you are on the final approach course, make any further descent according to the minimum altitudes published on the approach chart. As an example, a typical clearance would be, ". . . *six miles northwest of Garys, turn left heading 150, maintain 3,200 until established on the*

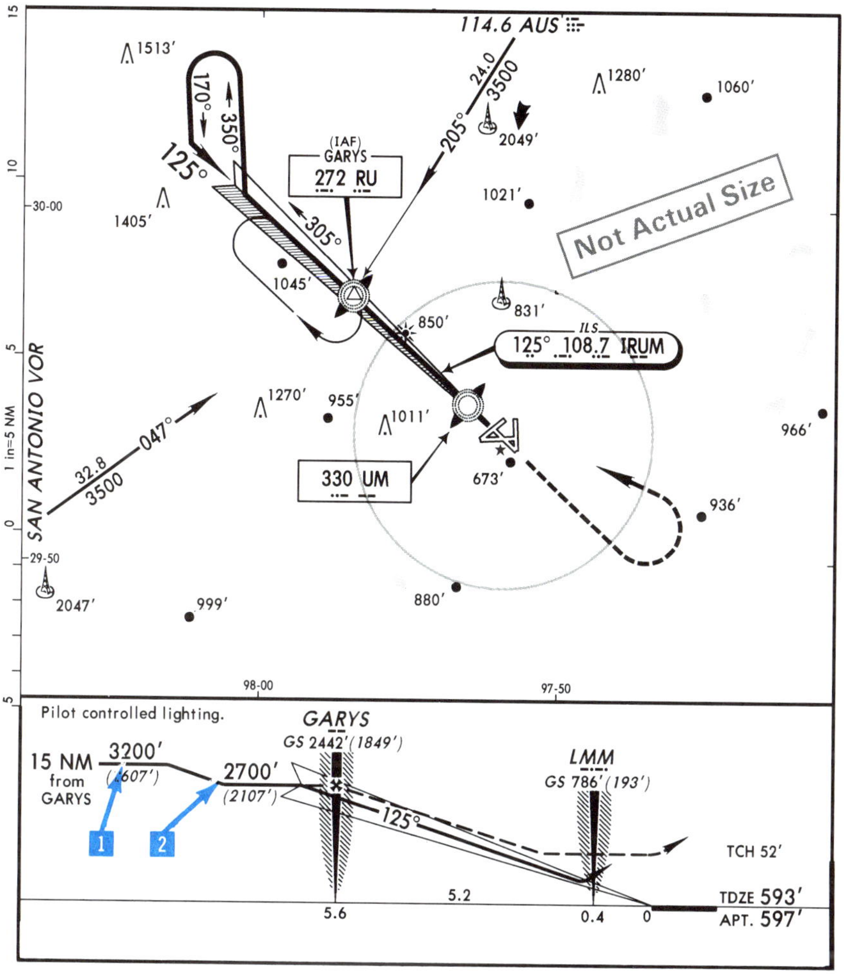

Figure 5-20. As you approach the localizer course, you must maintain the last assigned heading of 150° and altitude of 3,200 feet (item 1) until you turn inbound and become established on the final approach course. Then, you can refer to the profile view to determine the minimum altitude for further descent, which is 2,700 feet MSL (item 2), until you intercept the glide slope. After you acknowledge the clearance, you should change to the CTAF frequency and broadcast your position and intentions.

localizer, cleared for ILS runway 12 approach, change to advisory frequency approved." [Figure 5-20]

NONRADAR APPROACHES

The previous discussion was devoted to procedures that use approach control radar to establish an aircraft on the final approach course inbound for a straight-in approach. However, you also can make straight-in approaches without being radar vectored. Generally, you begin these

A straight-in approach does not require nor authorize a procedure turn or course reversal.

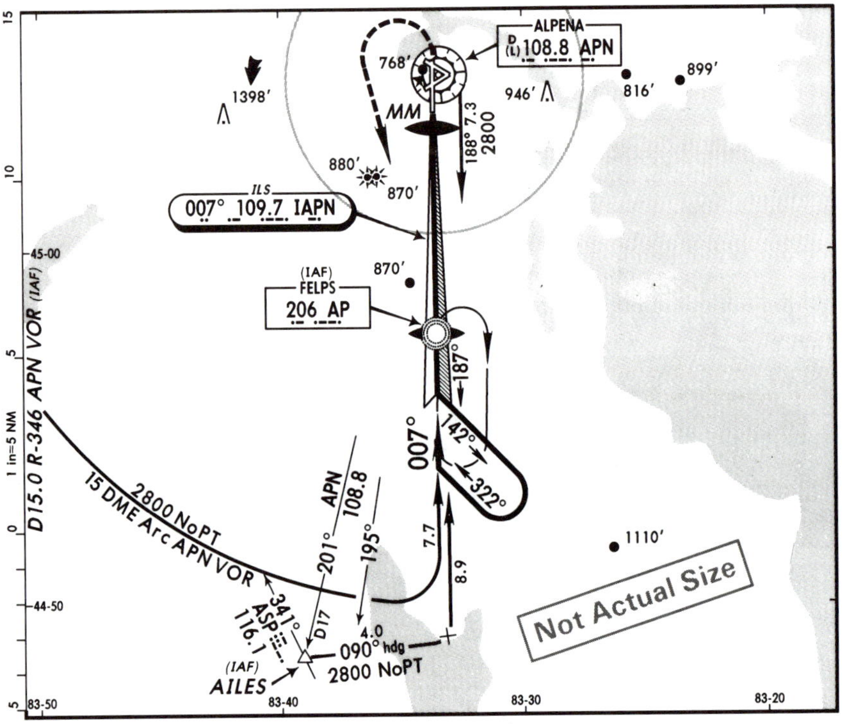

Figure 5-21. This approach chart excerpt shows two methods you can use to establish your flight on a straight-in approach. The first is to use the 15-mile DME arc to intercept the localizer. The second requires a 090° heading from the Ailes Intersection (lower left) to intercept the final approach course. Note that both of these routes specify "NoPT," which means a procedure turn is not authorized.

approaches at an outlying IAF, then fly the initial and intermediate segments, which places you on the final approach segment. [Figure 5-21]

Another situation where a straight-in approach without radar vectors is required is where airways provide a means of navigating directly to the final approach course. When this occurs, the approach chart may show a NoPT arrival sector formed by airways leading to an enroute navaid which also serves as an initial approach fix. This arrangement allows flights inbound on Victor airways within the sector to proceed straight-in on the final approach course. Keep in mind that your arrival must be on an airway within the sector. [Figure 5-22]

COURSE REVERSAL

Some approach procedures do not permit straight-in approaches unless you are being radar vectored. In these situations, you will be required to complete a procedure turn or other course reversal, generally within 10 miles of the fix, to establish your aircraft inbound on the intermediate or final approach segments. During a procedure turn, the maximum speed

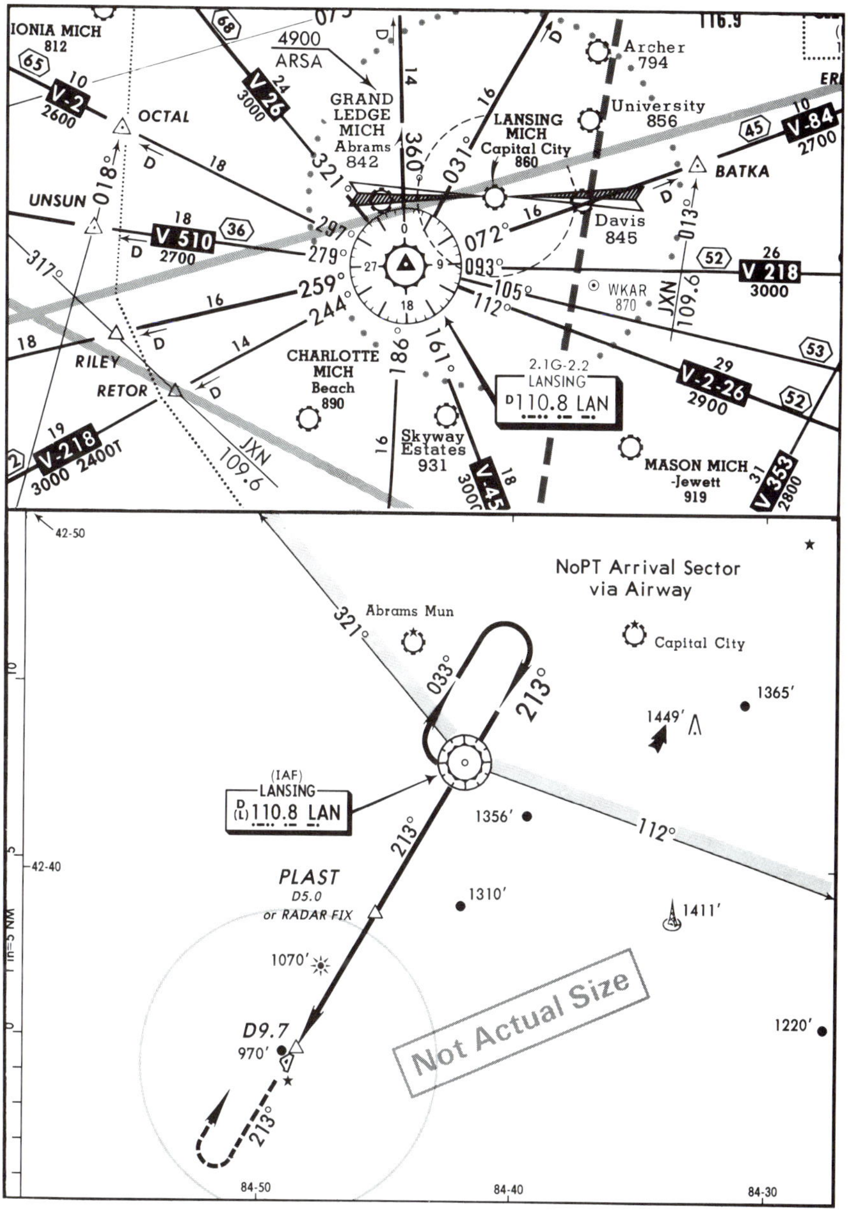

Figure 5-22. The upper portion of this illustration shows the Victor airway structure at the Lansing VOR. The lower portion of the illustration shows an excerpt of the VOR runway 20 approach at Charlotte, Michigan, which is based on the Lansing VOR. The NoPT arrival sector via airway coincides with V26 (321° radial) through V-2-26 (112° radial). According to the note, all inbound flights on airways within the arrival sector must proceed straight-in on the VOR approach. The following is a typical approach clearance for this type of procedure, *". . . cleared for straight-in VOR runway 20 approach to the Charlotte Airport, circle to land runway 2, change to advisory frequency approved."*

is 250 knots IAS. Unless a holding pattern or teardrop procedure is published, the point where you begin the turn and the type and rate of turn are optional. Keep in mind, you must remain within the airspace designated for the course reversal. If you are above the procedure turn altitude, you may begin descent as soon as you cross the IAF.

When a holding pattern is published in lieu of a procedure turn, the holding pattern must be flown using the published time or distance as depicted in the profile view.

The 45°-type procedure, the racetrack pattern (holding pattern), the teardrop procedure, or the 80°/260° procedure are mentioned in the AIM as acceptable variations for course reversal. When a holding pattern is published in place of a procedure turn, you must make the standard entry and follow the depicted pattern to establish your aircraft on the inbound course. Additional circuits in the holding pattern are not necessary or expected by ATC if you are cleared for the approach prior to returning to the fix. In the event you need additional time to lose altitude or become better established on course, you should advise ATC and obtain approval for any additional turns. When a teardrop is depicted and a course reversal is required, you also must fly the procedural track as published.

Approach charts provide headings, altitudes, and distances for a course reversal. Published altitudes are "minimum" altitudes, and you must complete the maneuver within the distance specified on the profile view (normally within 10 n.m.). You also are required to maneuver your aircraft on the procedure turn side of the final approach course. These requirements are necessary to stay within the protected airspace and maintain adequate obstacle clearance. [Figure 5-23]

TIMED APPROACHES FROM A HOLDING FIX

Timed approaches from a holding fix are generally conducted at airports where the radar service for traffic sequencing is out of service or is not

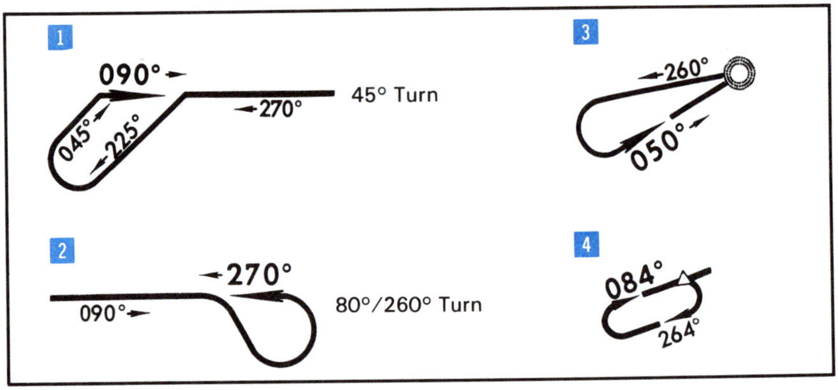

Figure 5-23. This illustration shows the different methods of course reversal you may encounter when establishing your aircraft on the approach course inbound. Item 1 represents the standard 45° procedure turn and item 2 is an 80°/260° course reversal. Item 3 depicts a teardrop pattern, while item 4 is a holding, or racetrack, pattern. When a teardrop or holding pattern is used in lieu of a procedure turn, you must fly the pattern as charted unless you are being radar vectored.

available and numerous aircraft are waiting for their approach clearance. You should be aware that this procedure is not utilized at all airports; certain conditions must prevail before ATC initiates timed approaches. The following list of conditions indicates when timed approaches may be conducted.

Timed approaches from a holding fix are only conducted at airports which have operating control towers.

1. A control tower is in operation at the airport where the approaches are conducted.
2. Direct communications are maintained between you and the approach controller until you are instructed to contact the tower.
3. If more than one missed approach procedure is available, none require course reversal.
4. If only one missed approach procedure is available, course reversal is not required, and ceiling and visibility are equal to or greater than circling minimums at that airport.
5. When cleared for the approach, you shall not execute a procedure turn.

Controllers may not specifically state that timed approaches are in progress. However, if you are issued a time to depart the holding fix inbound, it means that timed approaches are being used. The holding fix may be the FAF on a nonprecision approach. On a precision approach, it may be the outer marker or a fix used in lieu of the outer marker. When timed approaches are in progress, you will be given advance notice of the time you should leave the holding fix. When you are given a time to leave the holding fix, you should adjust your flight path within the holding airspace in order to leave the holding fix as closely as possible to the designated time. [Figure 5-24]

When making a timed approach from a holding pattern at the outer marker, adjust the holding pattern so you will leave the outer marker inbound at the assigned time.

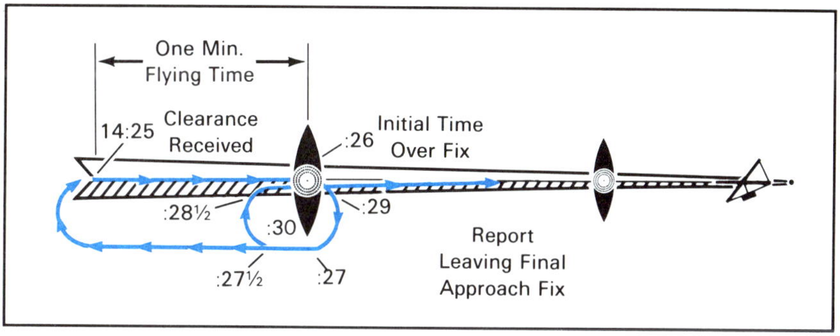

Figure 5-24. Assume that you are turning inbound in the holding pattern at 1425Z when ATC advises you to depart the holding fix inbound on the approach at 1429Z. In this example, you will have to adjust the holding pattern track to take up the remaining four minutes. Since it will require one minute to reach the holding fix, you must adjust one circuit in the holding pattern to last three minutes. You do this by making the outbound turn in the holding pattern (one minute), flying outbound for 30 seconds, turning inbound (one minute), and flying inbound for 30 seconds. By making this adjustment, you should arrive over the final approach fix inbound at or near 1429Z.

NONPRECISION APPROACH PLANNING

As you learned in previous sections, the advantage of an ILS is that you can simply descend on the electronic glide slope to the decision height (DH). However, nonprecision approaches require you to plan your descent from the final approach fix to the MDA so the aircraft is in a position to land. This is especially important when you are planning a straight-in landing, since it prevents you from arriving at the MDA and the MAP at the same time. [Figure 5-25]

Remember that FARs prohibit you from continuing below the DH or MDA unless you have established the required visual cues. FARs do provide for a descent below the DH or MDA if you have sighted the approach lights or other visual references. However, if you are using the approach lights for reference, you may not descend lower than 100 feet above the touchdown zone elevation unless the red terminating bars or the red side row bars are also distinctly visible and identifiable. Red terminating bars and side row bars are used in ALSF-1 and ALSF-2 approach lighting systems. The red terminating bars located near the

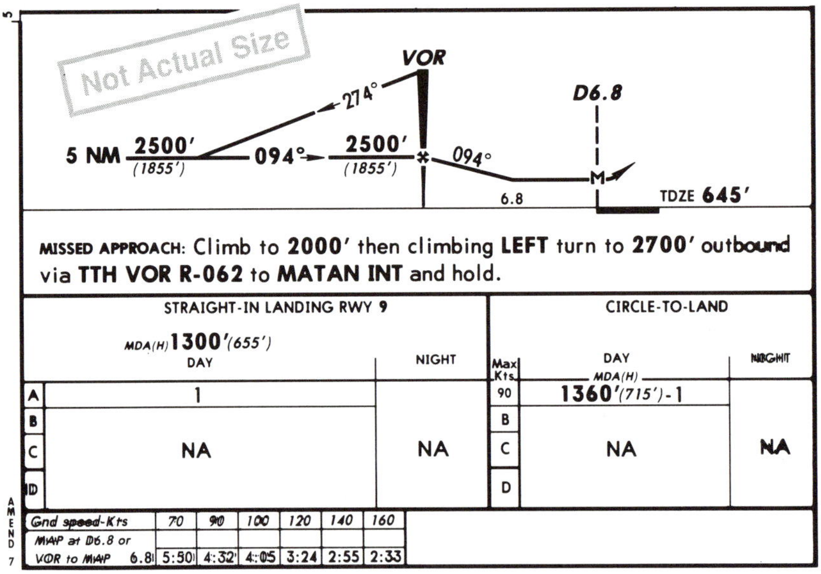

Figure 5-25. This chart excerpt shows the profile view, minimums box, and conversion table for a typical nonprecision approach. Note that the procedure turn altitude and the minimum altitude prior to the FAF is 2,500 feet and the MDA is 1,300 feet. Also, note that the length of the final approach segment is 6.8 miles and the missed approach point is 6.8 DME from the VORTAC which coincides with the runway threshold. In addition, the conversion table shows the time required to fly from the FAF to the missed approach point at various groundspeeds. Normally, you should descend at a rate that allows you to reach the MDA prior to the MAP. If you do this, you will be in a position to establish a normal rate of descent from the MDA to the runway, using normal maneuvers.

threshold pertain to ALSF-1 systems, and the red side row bars are included in the ALSF-2 system. Refer to FAR 91.175 for additional information on specific requirements to descend below the DH or MDA.

CIRCLING APPROACH CONSIDERATIONS

On occasion, you will find that unfavorable winds or a runway closure will make a straight-in landing impractical. If this is the case, you must plan to make a circling approach to the runway. Circling minimums appropriate to each aircraft category are established in accordance with TERPs criteria. Each circling approach is confined to a protected area which is defined by TERPs and also published in the AIM. You should understand that the size of this area varies with aircraft category. Only if you remain within the protected area are you assured obstacle clearance at the MDA during circling maneuvers. In addition, you must remain at or above the circling MDA unless the aircraft is continuously in a position from which a descent to a landing on the intended runway can be made, using a normal rate of descent and normal maneuvering.

While circling, however, it is important to realize that your protected area is limited to 1.3 n.m. from the runway ends for Category A aircraft and extends to 4.5 n.m. for Category E. Generally, the higher the category, the higher the MDA. If you are operating a Category A aircraft at the appropriate MDA, you must remain within 1.3 n.m. from the ends of the runways during the maneuver. This applies even if the visibility minimum for the approach is greater than 1.3 n.m. Higher visibility minimums do not constitute authorization to leave the protected area while circling to land. [Figure 5-26]

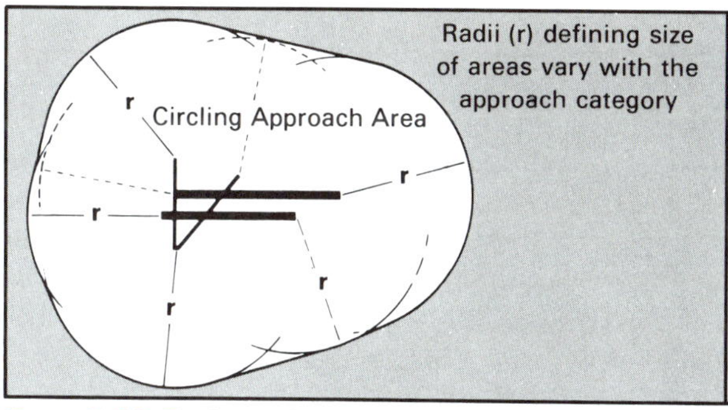

Approach Category	Radius (n.m.)
A	1.3
B	1.5
C	1.7
D	2.3
E	4.5

Figure 5-26. During a circling approach, you are provided obstacle clearance as long as you maneuver within the protected area. The left portion of this illustration shows how the protected area is established by the connection of arcs drawn from each runway end. The arc radii differ according to the approach category table in the right portion of this illustration. As approach speeds increase the turn radii increase. This often results in higher circling MDAs for aircraft in the higher categories. If obstacles are present near the airport within the protected area, a procedural note is added which prohibits circling within that area. As an example, a procedural note might say, "Circling NA E of Rwy 17-35" (circling not authorized east of runways 17/35).

When you are planning a circling approach, make sure you know your aircraft category so you can apply the appropriate MDA. As you know, categories are based on 1.3 times the stall speed in the landing configuration. However, if you use a higher than normal approach speed, you should use the minimums of the next higher category. For example, if you are circling in a Category B aircraft, but you are operating above the Category B speed limit, you must use the minimum descent altitude and visibility appropriate to the next higher category.

CIRCLING APPROACH PROCEDURES

In simple terms, the circling approach procedure involves flying the approach, establishing visual contact with the airport, and positioning the aircraft on final approach to the runway of intended landing. However, the circling approach is not a simple maneuver, since you are required to fly at a low altitude at a fairly slow airspeed while remaining within a specifically defined area. Remember that you fly the circling approach at or above the MDA, and you cannot descend from the MDA until the airplane is properly positioned to make a normal descent to the landing runway. To position your airplane properly on final approach, several options are available to you as you enter the traffic pattern, including an upwind, base, or downwind entry. [Figure 5-27]

You should be very cautious when making circling approaches because they can be extremely hazardous when combined with such factors as low visibility, hilly or mountainous terrain, and/or night operations. First of all, you must remain within the protected area for your aircraft category while circling. Secondly, you must realize that once you descend below the circling MDA, obstacle clearance is your own responsibility. If

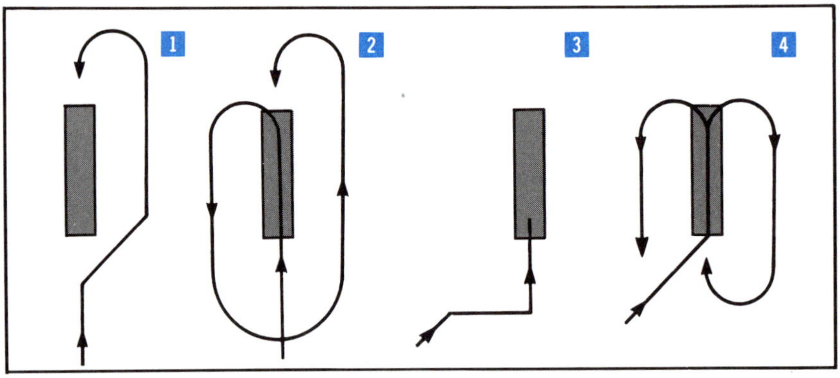

Figure 5-27. This illustration shows four possible options from the FAA *Instrument Flying Handbook* for the circling approach. While you are maneuvering at the MDA, use the pattern on the left (item 1) when you sight the runway in time to enter a downwind leg. If you do not see the runway in time for the downwind leg, fly the pattern indicated by item 2. When your final approach course intersects the runway centerline at less than 90°, and you sight the runway early enough to establish a base leg, use the third pattern (item 3). The pattern on the right (item 4) applies if you do not see the runway in time for a base leg entry. In this case, you can circle in either direction.

conditions prevent you from seeing well enough to guarantee obstruction clearance during the final descent, you should not continue the approach. Many accidents have occurred during circling approaches because the pilot could see lights in the vicinity but could not see hilly or mountainous terrain between the aircraft and the airport.

MISSED APPROACHES

A published missed approach procedure is carefully designed and flight tested so you will have adequate obstacle clearance throughout the missed approach segment. Each procedure is unique to the airport and to the particular approach. Depending on obstacles and surrounding terrain, a missed approach segment may designate a straight climb, a climbing turn, or a climb to a specified altitude, followed by a turn to a specified heading, navaid, or navigation fix. This is why you should always review the missed approach procedure before making an approach.

1) Fly
max power
Climb
MAP procedure
2) Tinker
Radio call
Non-Life Threating
stuff.

The most common reason for a missed approach is low visibility conditions that do not permit you to establish required visual cues. Additional reasons include improper alignment on the final approach course, wind shear which causes a deviation from your desired airspeed, sudden runway closure due to a disabled aircraft, inoperative airborne or ground navigation components, or an ATC request because of inadequate aircraft separation on final approach.

Regardless of the reason for a missed approach, it is very important that you can maneuver the aircraft safely throughout the missed approach segment. This is generally not difficult when you begin the missed approach at the missed approach point. In this situation, you simply fly the procedure as described and depicted on the approach chart. On occasion, however, you will be required to initiate a missed approach from a position that is not at the missed approach point and may not be on the missed approach segment.

For example, assume you decide to make a missed approach before reaching the MAP. In this situation, remember that no consideration for an abnormally early turn is given in the procedure design. If the procedure you are flying requires a turn, you must fly the approach to the MAP before turning. There is no prohibition against climbing early, but you must delay the turn until you are over the MAP.

Another good example is when you are executing a circling maneuver and suddenly lose sight of the runway. According to the AIM, you should make an initial climbing turn toward the landing runway to become established on the missed approach course. Remember, the airspace over the airport affords you the greatest obstacle clearance protection. Since the missed approach point for a nonprecision approach usually is the runway threshold, a turn toward the runway will keep you

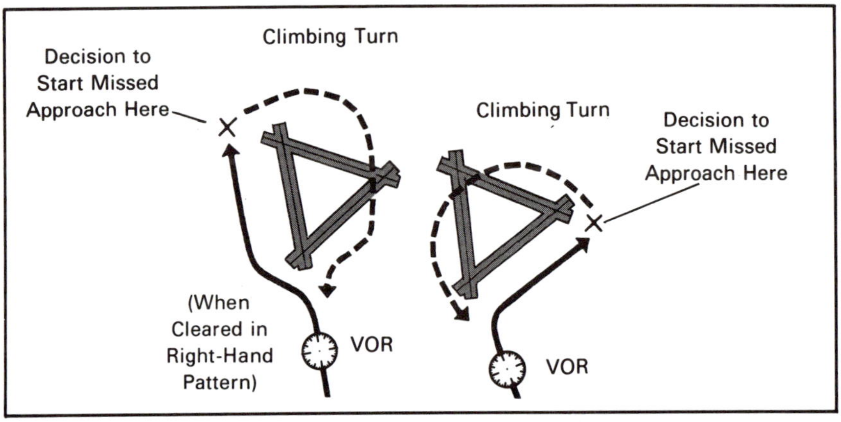

Figure 5-28. Since a circling maneuver may be accomplished in more than one direction, different patterns are required to become established on the missed approach course. The one to use depends on the position of the aircraft at the time visual reference is lost. These patterns are shown in the *Airman's Information Manual* and are intended to assure that an aircraft will remain within the circling and missed approach obstruction clearance areas.

over the airport and may position you very close to the actual missed approach point. [Figure 5-28]

VISUAL AND CONTACT APPROACHES

Contact and visual approaches simplify instrument approach procedures. They also permit you to shift to visual references when the destination airport is VFR or near VFR. Both of these procedures relieve you of the requirement to fly the complete instrument approach procedure.

VISUAL APPROACHES

When visual approaches are initiated by ATC, it is done to expedite traffic since they shorten the flight path to the airport.

Although you may request visual approaches, they are often initiated by ATC to expedite the flow of traffic to an airport. At a controlled airport, the reported weather must be at least a 1,000-foot ceiling and 3 miles visibility. Typically, the clearance is initiated by the controller when you report the airport or the preceding aircraft in sight. In the event you have the airport in sight, but do not see the aircraft you are following on the approach, the controller can still clear you for a visual approach. However, in this situation the controller retains the responsibility to maintain aircraft and wake turbulence separation. If you do report the preceding aircraft in sight, it becomes your responsibility to maintain your own separation from that aircraft and avoid associated wake turbulence. At airports without an operating control tower and no weather reporting facility, ATC will authorize a visual approach only if there is a reasonable assurance that the weather will be at least 1,000 and 3. It is the pilot's responsibility to remain clear of clouds. Also, keep in mind that acceptance of a visual approach does not relieve you of the responsibility to close your IFR flight plan when you arrive at an uncontrolled airport.

CONTACT APPROACHES

Contact approaches are issued only for airports with published approach procedures (public or private). ATC issues clearances for contact approaches upon pilot request when the reported ground visibility at the destination is one statute mile or greater. ATC cannot initiate a contact approach.

When operating under a clearance for a contact approach, you must remain clear of clouds and maintain one mile visibility to the airport (special VFR minimums). During a contact approach you are responsible for your own obstruction clearance, but ATC provides separation from other IFR or special VFR traffic. Separation from normal VFR traffic is not provided.

A contact approach may be to your advantage for several reasons. For example, it requires less time than the published instrument procedure, allows you to retain your IFR clearance, and provides separation from IFR and special VFR traffic. On the other hand, obstruction clearance and VFR traffic avoidance become your responsibility.

USE OF VASI LIGHTS

Whether you are making a straight-in landing or a circle-to-land maneuver, you can use VASI lights to your advantage once you encounter visual meteorological conditions. Not only do they provide a visual glide path that enhances the safety of your descent, their use is regulatory under some conditions. Keep in mind that at airports with operating control towers, you must maintain a glide path at or above the VASI while approaching a runway served by a VASI installation. In addition, you should use the VASI installation to supplement the electronic glide slope during an ILS approach once you have established visual contact with the runway. If a glide slope malfunction occurs when you are in instrument meteorological conditions, you must apply the localizer-only minimums (MDA) and report the malfunction to ATC. Once you have established visual references, you may continue the descent at or above the VASI glide path for the remainder of the approach.

SIDESTEP MANEUVER

Under certain conditions, your approach clearance may include a **sidestep maneuver**. At some airports where there are two parallel runways that are 1,200 feet or less apart, you may be cleared to execute an approach to one runway followed by a straight-in landing on the adjacent runway. An example of a sidestep clearance would be, "*. . . cleared for ILS 26L approach, sidestep to runway 26R.*" With this type of clearance, you are

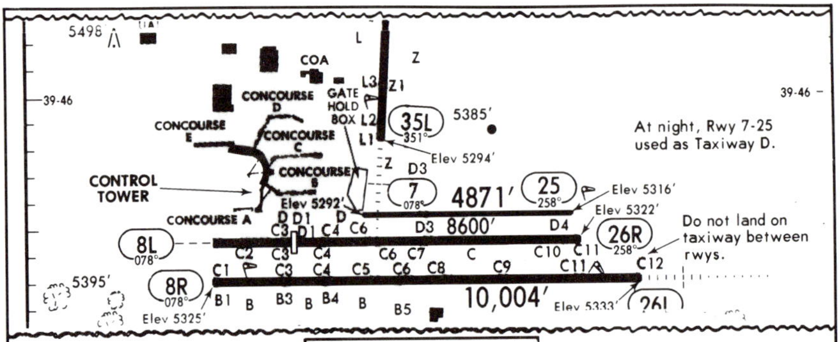

Not Actual Size

	STRAIGHT-IN LANDING RWY 26L ILS DA(H) 5533'(200')	ILS DA(H) 5533'(200')	DA(H) 5583'(250')	LOC (GS out) MDA(H) 5800'(467')			SIDESTEP LANDING RWY 26R MDA(H) 5800'(477')	CIRCLE-TO-LAND Max Kts	MDA(H)
	FULL	RAIL or ALS out	MM out		RAIL out	ALS out			
A	RVR 24 or 1/2	RVR 40 or 3/4	RVR 24 or 1/2	RVR 24 or 1/2	RVR 40 or 3/4	RVR 50 or 1	1	90	5880'(547')-1
B								120	
C				RVR 40 or 3/4	RVR 60 or 1 1/4		1 1/2	140	5880'(547')-1 1/2
D			RVR 40 or 3/4	RVR 50 or 1	1 1/2		2	165	5900'(567')-2

Figure 5-29. The upper portion of this illustration is an excerpt which shows the parallel runway configuration for runways 26L/26R. The lower part shows the minimums box for the ILS runway 26L approach. Note that the MDA for the sidestep landing to runway 26R (item 1) is lower than the published circling minimums (item 2).

expected to fly the approach to the primary runway and begin the approach to a landing on the parallel runway as soon as possible after you have it in sight. [Figure 5-29]

RUNWAY VISUAL ILLUSIONS

Once you have reached visual conditions on the approach, you should be aware that certain visual illusions may be apparent as you approach the runway. They are the product of various runway features, terrain features, and atmospheric conditions which can create the sensation of incorrect height above the runway or incorrect distance from the runway threshold. It is important to recognize these illusions so you can avoid a lower than normal approach. [Figure 5-30]

CLOSING YOUR IFR FLIGHT PLAN

The requirement to close your flight plan is established in FARs 91.153 and 91.169. You can cancel an IFR flight plan anytime you are operating in VFR conditions below 18,000 feet MSL. If your destination airport has an operating control tower, your IFR flight plan is automatically closed when you land.

Situation	Illusion	Result
Upsloping Runway or Terrain	Greater Height	Lower Approaches
Narrower-Than-Usual Runway	Greater Height	Lower Approaches
Featureless Terrain	Greater Height	Lower Approaches
Rain on Windscreen	Greater Height	Lower Approaches
Haze	Greater Height	Lower Approaches
Downsloping Runway or Terrain	Less Height	Higher Approaches
Wider-Than-Usual Runway	Less Height	Higher Approaches
Bright Runway and Approach Lights	Less Distance	Higher Approaches
Penetration of Fog	Pitching Up	Steeper Approaches

Figure 5-30. This table shows various situations that may create specific illusions. It includes the likely result of each illusion. You should pay particular attention to those which lead to a lower than normal final approach profile.

If you are flying to an airport that does not have an operating control tower, you are responsible for closing your own IFR flight plan. You can do this after landing through an FSS or by other direct communications with ATC. According to the AIM, you also can cancel in VFR conditions while you are still airborne and able to communicate with ATC by radio. This is appropriate in cases where an FSS is not available and air/ground communications with ATC are not possible at low altitudes. This saves you the time and expense of cancelling by telephone. It also releases the airspace to other aircraft.

When you land at an airport without an operating tower, close your IFR flight plan with the nearest FSS or ATC facility.

CHECKLIST

After studying this section, you should have a basic understanding of:

✓ **Arrival procedures** — What methods are available for transitioning from the enroute phase to the approach phase.

✓ **STARs** — When you can expect one, and what you should know about them from an operational standpoint.

✓ **Crossing restrictions** — How to descend at "pilot's discretion" so you will cross a fix or navaid at the altitude specified by ATC.

✓ **Speed adjustments** — When you can expect speed adjustments and how the controller will issue them.

- ✓ **Radar arrival —** How radar vectors are used for transition to the final approach course and how they reduce the time required for an approach.
- ✓ **NoPT arrival sector —** How you can transition directly to final without making a procedure turn and without being radar vectored.
- ✓ **Course reversals —** What types you may encounter and which ones require you to maintain the procedural track.
- ✓ **Straight-in approach —** What it means from an operational viewpoint and what ATC expects you to do.
- ✓ **Circling approach —** What circumstances require a circle-to-land maneuver, what the options are, and what governs the protected airspace for the circling area.
- ✓ **Missed approach procedures —** When they are mandatory and how they can vary, depending on your position when you start them.
- ✓ **Visual illusion —** What conditions produce certain illusions and what results you can expect if you fail to compensate.

CHAPTER 6

METEOROLOGY

INTRODUCTION

This chapter provides an analysis of aviation weather which is appropriate to both VFR and IFR flight operations. Although some areas provide a necessary review of basic weather concepts, this presentation generally assumes private level knowledge as a starting point. The first section considers a wide range of general weather factors and includes a review of weather theory. The next section analyzes various weather hazards and provides guidance for avoiding them during flight. The next two sections cover printed reports and forecasts, as well as graphic weather products, with an emphasis on instrument meteorological conditions (IMC). The final section considers important high altitude aspects of meteorology.

SECTION A

WEATHER FACTORS

Based on previous training and your current level of flight experience, you probably have acquired a good working knowledge of weather theory. This first section is a review of selected weather theory concepts.

THE ATMOSPHERE

You will recall that the atmosphere is divided into a series of layers, or spheres, based on common characteristics within each layer. The **troposphere** is the one closest to the earth. The **tropopause** is a thin layer of the atmosphere at the top of the troposphere. It acts as a lid to confine most of the water vapor, as well as most of the weather, below the tropopause. Above the tropopause are three more atmospheric layers — **stratosphere**, **mesosphere**, and **thermosphere**. [Figure 6-1]

The average temperature lapse rate changes abruptly at the tropopause.

Of the various atmospheric layers, the troposphere contains the vast majority of earth's weather. Within this layer, air temperature decreases with increasing altitude. At the tropopause, the rate of temperature decrease changes abruptly. Above this point in the stratosphere, temperature continues to decrease, but at a slower rate. Near the top of the stratosphere, temperature begins to increase with increasing altitude. Although most of the earth's weather remains in the troposphere, certain types of weather, such as severe thunderstorms, may extend into the stratosphere.

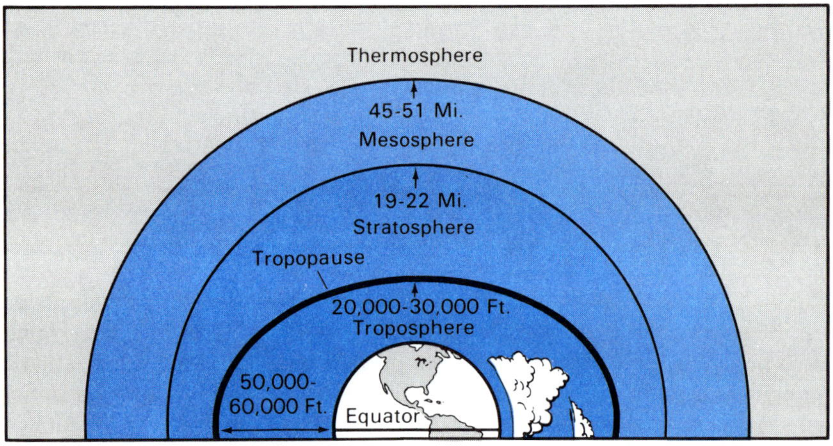

Figure 6-1. The height of the troposphere varies from about 20,000 feet at the poles to 60,000 feet at the equator. For a given latitude, it is higher in the summer than it is in the winter. In the mid-latitudes, it averages about 37,000 feet.

ATMOSPHERIC CIRCULATION

The primary cause of weather is uneven heating of the earth's surface by the sun; solar radiation is the driving force that sets the atmosphere in motion. Uneven heating modifies air density and creates circulation patterns, resulting in changes in pressure. This is one of the main reasons for differences in altimeter settings between weather reporting stations. Meteorologists plot pressure readings on weather maps and connect points of equal pressure with lines called **isobars**. The resulting pattern reveals the **pressure gradient**, or change in pressure over distance. When isobars are far apart, the gradient is considered to be weak, while closely spaced isobars indicate a strong gradient. Isobars also help to identify pressure systems, which are classified as highs, lows, ridges, troughs, and cols. A **col** is a neutral area between two highs or two lows. [Figure 6-2]

Uneven heating of the earth's surface is the major cause of weather.

Air flows from the cool, dense air of highs into the warm, less dense air of lows. The speed of the resulting wind depends on the strength of the pressure gradient. A strong gradient tends to produce strong winds, while a weak gradient results in lighter winds. The force behind this

Wind is caused by differences in air pressure.

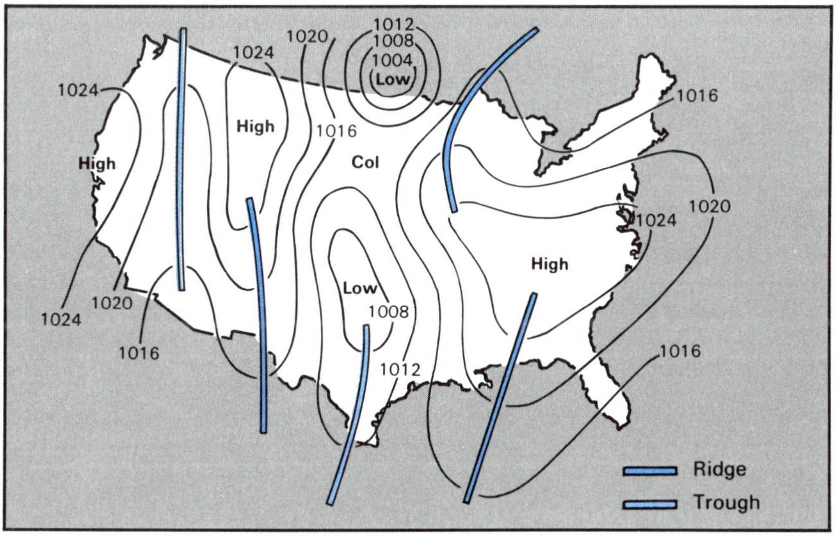

Figure 6-2. In high pressure systems, air is moving outward, depleting the quantity of air. As a result, a high or ridge is characterized by descending air which favors dissipation of cloudiness. This is why a high or a ridge is generally associated with good visibility, calm or light winds, and few clouds. When air converges into a low, it cannot go outward against the pressure gradient so it must ascend. This is why a low or trough is an area of rising air. Rising air is conducive to cloudiness and precipitation, which is why low pressure is generally associated with bad weather. This may include low clouds, poor visibility, precipitation, gusty winds, and turbulence. Weather may be very violent in the area of a trough. A col can designate either a neutral area between two highs and two lows, or the intersection of a ridge and a trough.

movement is caused by the pressure gradient and is referred to as pressure gradient force.

Coriolis force deflects air to the right in the northern hemisphere.

If the earth did not rotate, pressure gradient force would propel wind directly from highs to lows. Instead, the earth's rotation introduces another force, called **Coriolis**, that deflects the flow of air to the right in the northern hemisphere and to the left in the southern hemisphere. The amount of deflection produced by Coriolis force varies with latitude. It is negligible at the equator and increases toward the poles. It also is proportional to the speed of the airmass. The faster the air moves, the greater the deflection.

Wind flows clockwise around a high and counterclockwise around a low. Air flows outward and downward from a high and inward and upward toward a low.

Pressure gradient and Coriolis forces work in combination to create wind. Pressure gradient force causes air to move from high pressure areas to low pressure areas. As the air begins to move, Coriolis force deflects it to the right in the northern hemisphere, resulting in a clockwise flow around a high. The deflection continues until pressure gradient force and Coriolis force are in balance and the wind flows roughly parallel to the isobars. As the air flows into a low pressure area, it moves counterclockwise around the low. [Figure 6-3]

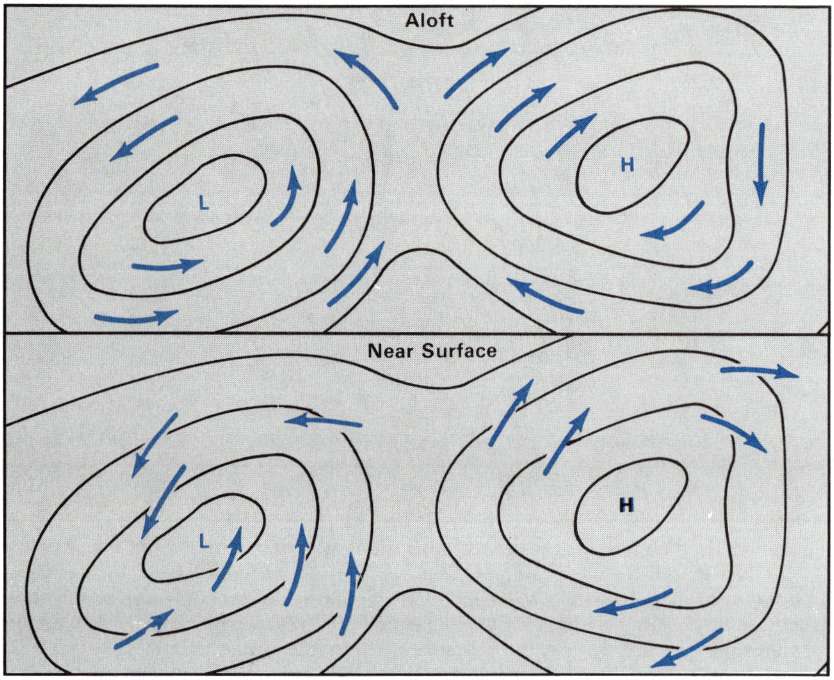

Figure 6-3. When pressure gradient force and Coriolis are balanced, airflow circulation aloft is parallel to the isobars. Within about 2,000 feet of the ground, surface friction slows the wind, and Coriolis force is weakened. Pressure gradient force then predominates, causing the wind to flow at an angle to the isobars. The wind angles toward the low pressure center and away from the high pressure center.

MOISTURE

Weather is very dependent upon the moisture content of the air. If the air is dry, the weather is usually good. If the air is very moist, poor or even severe weather can occur. Water is present in the atmosphere in three states: solid, liquid, and gas. Changes in state occur through one of five processes. Water vapor (a gas) is added to the atmosphere by the process of **evaporation** or sublimation. It is removed from the atmosphere by **condensation** or sublimation. **Sublimation** is the changing of ice (solid water) directly to water vapor, or of water vapor to ice. In sublimation, the liquid state is bypassed. Of course, melting is the change of ice to water, and freezing is the reverse process.

The atmosphere accumulates moisture through evaporation and sublimation.

As water changes from one physical state to another, an exchange of heat takes place. In fact, every physical process of weather is accompanied by this heat exchange. When water evaporates, heat is absorbed. The heat absorption is known as the latent heat of evaporation. When water vapor condenses, the heat absorbed by water vapor during evaporation is released. The heat released is referred to as the latent heat of condensation. It is an important factor in cloud development. The heat exchange between melting and freezing is small and has relatively little effect on weather.

Every physical process of weather is accompanied by or is the result of a heat exchange.

As temperature decreases, the amount of moisture the air can hold also decreases. When the **dewpoint** is reached, the air contains all the moisture it can hold at that temperature, and it is said to be **saturated**. In this case, the relative humidity is 100%. You can also say that as the temperature/dewpoint spread decreases, relative humidity increases.

The amount of moisture the air can hold depends on air temperature. Dewpoint is the temperature at which air becomes saturated.

As air cools to its saturation point, the processes of condensation and sublimation change invisible water vapor into visible states. Clouds, fog, or dew will always form when water vapor condenses. Most commonly, moisture takes the form of clouds or fog.

Clouds, fog, or dew will always form when water vapor condenses.

Clouds are composed of very small droplets of water or ice crystals. When they form near the surface, they are referred to as fog. You can anticipate the formation of fog or very low clouds by monitoring the **temperature/dewpoint spread**. When the spread reaches 4°F (2°C) and continues to decrease, the air is nearing the saturation point.

A small and decreasing temperature/dewpoint spread indicates conditions are favorable for fog formation.

When these water droplets grow to a size where the atmosphere can no longer support their weight, they fall as precipitation. Water droplets that remain liquid fall as drizzle or rain. Rain that remains liquid even though its temperature is below freezing is considered to be **supercooled**, and is referred to as freezing rain. When supercooled water strikes an object, such as an airplane in flight or the earth's surface, it turns to ice. Ice pellets result if the rain freezes as it falls. This always indicates freezing rain at some higher altitude and the existence of a layer of warmer air aloft.

Ice pellets usually are evidence of freezing rain and a warmer layer of air at higher altitudes.

In clouds with strong vertical currents where temperatures are cold, water droplets may freeze. As they rise and fall, they increase in size from collisions with other freezing water droplets. Eventually they become too large for air currents to support, and they fall as hail.

If wet snow is encountered at your flight altitude, the temperature is above freezing.

Precipitation that forms by sublimation falls as snow, if the temperature of the air remains below freezing. As precipitation falls, it may change its state due to the temperature it encounters. Falling snow may melt to form rain. Melting snow means the temperature is above freezing at your altitude. With low relative humidity, rain may evaporate before it reaches the surface. When this occurs, it is called **virga**, and usually appears as streamers of precipitation trailing from clouds.

Frost forms when a surface is at or below the dewpoint of the surrounding air and the dewpoint is below freezing.

On cool, still nights, surface features and objects may cool to a temperature below the dewpoint of the surrounding air. Moisture then condenses out of the air in the form of dew. If the temperature of the aircraft is at or below the dewpoint and the dewpoint is below freezing, moisture sublimates out of the air and forms frost.

ATMOSPHERIC STABILITY

Stability is the atmosphere's resistance to vertical motion. The stability of a parcel of air actually determines whether it will rise or descend in relation to the air around it. Stable air resists vertical movement, while unstable air has a tendency to move vertically. The combined effects of temperature and moisture determine the stability of the air and, to a large extent, the type of weather produced. The greatest instability occurs when the air is both warm and moist. Tropical weather, with its almost daily thunderstorm activity, is a perfect example of weather that occurs in very unstable air. Air that is both cool and dry resists vertical movement and is very stable. A good example of this can be found in polar regions in winter, where stable conditions often result in very cold temperatures with generally clear weather.

Air that moves upward expands due to lower atmospheric pressure. When air moves downward, it is compressed by the increased pressure at lower altitudes. As the pressure of a given portion of air changes, so does its temperature. The temperature change is caused by a process known as adiabatic heating or cooling, which is a change in the temperature during expansion or compression when no energy is added to or removed from the air. As the molecules of air rise and expand, the temperature of the air decreases. When air descends, the opposite is true.

You can use the actual or ambient lapse rate to determine the stability of the atmosphere.

The rate at which temperature decreases with an increase in altitude is referred to as its **lapse rate**. As you ascend through the atmosphere, the temperature decreases at an average rate of 2°C (3.5°F) per 1,000 feet. The adiabatic lapse rate of a parcel of air that is lifted depends on the amount of moisture present in the air. The dry air adiabatic lapse rate (unsaturated air) is 3°C (5.4°F) per 1,000 feet. The moist air adiabatic lapse rate varies from 1.1°C to 2.8°C (2°F to 5°F) per 1,000 feet; however, it is always less than the rate for dry air. The ambient lapse rates can be used to indicate the stability of the air. For example, moist air has less stability and a lower lapse rate than dry air.

When you compare the adiabatic lapse rates, you can see that they can be used to indicate the stability of the air. For example, moist air is less stable than dry air, because it cools at a slower rate. This means that moist air must rise higher before its temperature cools to that of the air around it.

When warm, moist air begins to rise in a convective current, cumulus clouds often form at the altitude where the temperature and dewpoint reach the same value. When lifted, unsaturated air cools at about 5.4°F per 1,000 feet, and the dewpoint temperature decreases at about 1°F per 1,000 feet. Therefore, the temperature and dewpoint converge at about 4.5°F per 1,000 feet.

You can use these values to estimate cloud bases. For example, if the surface temperature is 80°F and the surface dewpoint is 62°F, the spread is 18°F. This difference, divided by the rate that the temperature approaches the dewpoint (4.5°F), will help you judge the approximate height of the base of the clouds in thousands of feet (18 ÷ 4.5 = 4 or 4,000 feet AGL).

Although temperature usually decreases with an increase in altitude, the reverse is sometimes true. When temperature increases with altitude, a **temperature inversion** exists. Inversions are usually confined to fairly shallow layers and may occur near the surface or at higher altitudes. They act as a lid for weather and pollutants. Visibility is often restricted by fog, haze, smoke, and low clouds. Temperature inversions often occur in stable air with little or no wind and turbulence. One of the most familiar types of ground- or surface-based inversions forms from radiation cooling just above the ground on clear, cool nights. An inversion can also occur when cool air is forced under warm air, or when warm air spreads over cold. Both of these are called frontal inversions.

A stable layer of air and a temperature increase with altitude are features of a temperature inversion.

Stable and unstable air have predictable characteristics in terms of cloud types, precipitation, visibility, turbulence, and icing. For example, if stable air is forced up a slope (orographic lifting), stratus-type clouds may form. If the same air is unstable and sufficient moisture is present, clouds with extensive vertical development may develop. In unstable air, any upward current enhances precipitation. By observing these features,

When air is forced aloft from orographic lifting, the stability of the air before it is lifted determines the type of clouds that will form.

you should be able to distinguish between stable and unstable air. [Figure 6-4]

CLOUDS

Clouds are classified into four **families**, based on their characteristics and the altitudes where they occur.

Low — Bases range from the surface to 6,500 feet AGL. Clouds may be cumulus, stratocumulus, or stratus.

Middle — Bases range from 6,500 feet AGL to 23,000 feet AGL. Clouds may be altocumulus, altostratus, or nimbostratus.

High — Bases usually range from 16,500 feet AGL to 45,000 feet AGL. Clouds may be cirrus, cirrocumulus, or cirrostratus.

Extensive Vertical Development — Bases range from 1,000 feet AGL or less to 10,000 feet AGL or more; tops sometimes exceed 60,000 feet MSL. Cloud types may be towering cumulus or cumulonimbus.

The words "alto" and "cirro" are used to denote cumulus and stratus clouds from the middle and high families, respectively. The prefix "nimbo" and the suffix "nimbus" are used to name clouds that produce rain.

	Stable Air	Unstable Air
Clouds	Wide areas of layered clouds or fog; gray at low altitude, thin white at high altitude	Extensive vertical development; bright white to black; billowy
Precipitation	Small droplets in fog and low-level clouds; large droplets in thick stratified clouds; widespread and lengthy periods of rain or snow	Large drops in heavy rain showers; showers usually brief; hail possible
Visibility	Restricted for long periods	Poor in showers or thunder-showers, good otherwise
Turbulence	Usually light or nonexistent	Moderate to heavy
Icing	Moderate in mid-altitudes; freezing rain, rime, or clear ice	Moderate to heavy clear ice
Other	Frost, dew, temperature inversions	High or gusty surface winds, lightning, tornadoes

Figure 6-4. Stable air is generally smooth, with layered or stratiform clouds. Visibility usually is restricted, with widespread areas of clouds and steady rain or drizzle. Unstable air is usually turbulent, with good surface visibility outside of scattered rain showers.

Stratus clouds are layered clouds that form in stable air near the surface due to cooling from below. These clouds frequently produce low ceilings and visibilities, but usually have little turbulence. Icing conditions are possible if temperatures are at or near freezing. Stratus clouds may form when stable air is lifted up sloping terrain. Stratus clouds also may form along with fog when rain falls through cooler air. In this case, the rain raises the humidity of the cooler air to the saturation point. [Figure 6-5]

Stratus clouds form in stable air.

Cumulus clouds form in convective currents resulting from the uneven heating of the earth's surface. They usually have flat bottoms and dome-shaped tops. Widely spaced cumulus clouds that form in fairly clear skies are called fair weather cumulus and indicate a shallow layer of instability. You can expect turbulence at and below the cloud level, but little icing or precipitation.

Turbulence can be expected at and below the bases of fair weather cumulus.

Stratocumulus clouds are white, puffy clouds that form as stable air is lifted. They often form as a stratus layer breaks up or as cumulus clouds spread out. [Figure 6-6]

Altostratus clouds are flat, dense clouds that cover a wide area. They are a uniform gray or gray-white in color. Although they produce minimal turbulence, they contain moderate icing.

Altocumulus clouds are gray or white, patchy clouds of uniform appearance that often form when altostratus clouds start to break up. They may produce light turbulence and icing.

Nimbostratus clouds are gray or black clouds that can be several thousand feet thick and contain large quantities of moisture. If temperatures are near or below freezing, they may create heavy icing.

Cirrus clouds are thin, wispy clouds composed mostly of ice crystals that usually form above 30,000 feet. White or light gray in color, they

Figure 6-5. Stratus clouds have a gray, uniform appearance and generally cover a wide area.

Figure 6-6. These are mainly stratocumulus clouds. A few cumulus clouds indicate an area with slightly more convective currents than the surrounding areas.

often exist in patches or narrow bands that cross the sky. They are sometimes blown from the tops of thunderstorms or towering cumulus clouds.

Towering cumulus clouds are similar to cumulus clouds, except they have more vertical development. Their bases begin at altitudes associated with low to middle clouds, while their tops may extend into the altitudes associated with high clouds. They look like large mounds of cotton with billowing cauliflower tops. Their color may vary from brilliant white at the top to gray near the bottom. Towering cumulus clouds indicate a fairly deep layer of unstable air. They contain moderate to heavy convective turbulence with icing and often develop into thunderstorms. [Figure 6-7]

Embedded thunderstorms are frequently obscured by massive cloud layers and usually cannot be seen.

Frequently, these cloud types are obscured by other cloud formations. When this happens, they are said to be **embedded**. This can be a particularly hazardous situation. Instrument flight in these conditions is unwise unless your aircraft is equipped with a weather avoidance system such as radar. Night operations also make it more difficult to see hazardous cloud formations. A discussion of weather avoidance procedures is provided in the next section.

With moist, unstable air and very warm surface temperatures, expect strong updrafts and cumulonimbus clouds.

Cumulonimbus clouds, which are more commonly called thunderstorms, are large, vertically developed clouds that form in moist, unstable air. They are gray-white to black in color and contain large amounts of moisture, turbulence, icing, and lightning. You can think of them as severe versions of the towering cumulus cloud. Thunderstorms are discussed in more detail later in Section B.

Figure 6-7. Towering cumulus clouds develop in unstable air and may contain convective turbulence and icing conditions serious enough to warrant a diversion.

AIRMASSES

An airmass is a large body of air with fairly uniform temperature and moisture content. It usually forms where air remains stationary or nearly stationary for at least several days. During this time, the airmass takes on the temperature and moisture properties of the underlying surface. The area where an airmass acquires the properties of temperature and moisture that determine its stability is called its source region.

Airmasses generally have uniform properties of temperature and moisture.

As an airmass moves over a warmer surface, its lower layers are heated, and vertical movement of the air develops. Depending on temperature and moisture levels, this can result in extreme instability, characterized by cumuliform clouds, turbulence, and good visibility. When an airmass flows over a cooler surface, its lower layers are cooled and vertical movement is inhibited. As a result, the stability of the air is increased. If the air is cooled to its dewpoint, low clouds or fog may form. This cooling from below creates a temperature inversion and may result in low ceilings and visibility for long periods of time.

Moist, unstable air typically has cumulus clouds and showery precipitation. Warming from below decreases airmass stability, while cooling from below increases stability.

FRONTS

When an airmass moves out of its source region, it comes in contact with other airmasses that have different moisture and temperature characteristics. The boundary between airmasses is called a **front**. Since the weather along a front often presents a serious hazard to flying, you need to have a thorough understanding of this weather.

Fronts are boundaries between airmasses.

When you cross a front, you move from one airmass into another with different properties. The changes between the two may be very abrupt, indicating a narrow frontal zone. On the other hand, the changes may occur gradually, indicating a wide and, perhaps, diffused frontal zone. These changes can give you important cues to the location and intensity of the front.

A change in the temperature is one of the easiest ways to recognize the passage of a front. At the surface, the temperature change usually is very noticeable and may be quite abrupt in a fast-moving front. With a slow-moving front, it is usually less pronounced. When you are flying through a front, you can observe the temperature change on the outside air temperature gauge. However, the change may be less abrupt at middle and high altitudes than it is at the surface.

The most reliable indications that you are crossing a front are a change in wind direction and, less frequently, wind speed. Although the exact new direction of the wind is difficult to predict, the wind always shifts to the right in the northern hemisphere. When you fly through a front at low to middle altitudes, you always need to correct to the right in order to maintain your original ground track.

When you fly through a front, you will notice a change in wind direction. Wind speed may also change.

As a front approaches, atmospheric pressure usually decreases, with the area of lowest pressure lying directly over the front. Pressure changes on the warm side of the front generally occur more slowly than on the cold side. The important thing to remember is that you should update your altimeter setting as soon as possible after crossing a front.

The type and intensity of frontal weather depend on several factors. Some of these factors are the availability of moisture, the stability of the air being lifted, and the speed of the frontal movement. Other factors include the slope of the front and the moisture and temperature variations between the two fronts. Although some frontal weather can be very severe and hazardous, other fronts produce relatively calm weather.

TYPES OF FRONTS

A **cold front** separates an advancing mass of cold, dense air from an area of warm, lighter air. Because of its greater density, the cold air moves along the surface and forces the less dense, warm air upward. The speed of a cold front usually dictates the type of weather associated with the front. However, there are some general weather characteristics that are found in most cold fronts. These include:

1. Cumulus clouds
2. Turbulence
3. Showery precipitation
4. Strong, gusty winds
5. Clearing skies and good visibility after the front passes

Fast-moving cold fronts are pushed along by intense high pressure systems located well behind the front. Surface friction acts to slow the movement of the front, causing the leading edge of the front to bulge out and to steepen the front's slope. Because of the steep slope and wide differences in moisture and temperature between the two airmasses, these fronts are particularly hazardous. [Figure 6-8]

The leading edge of a **slow-moving cold front** is much shallower than that of a fast-moving front. This produces clouds which extend far behind the surface front. A slow-moving cold front meeting stable air usually causes a broad area of stratus clouds to form behind the front. When a slow-moving cold front meets unstable air, large numbers of vertical clouds often form at and just behind the front. Fair weather cumulus clouds are often present in the cold air, well behind the surface front. [Figure 6-9]

Steady precipitation with little turbulence usually precedes a warm front.

Warm fronts occur when warm air overtakes and replaces cooler air. They usually move at much slower speeds than cold fronts. The slope of a warm front is very gradual, and the warm air may extend up over the

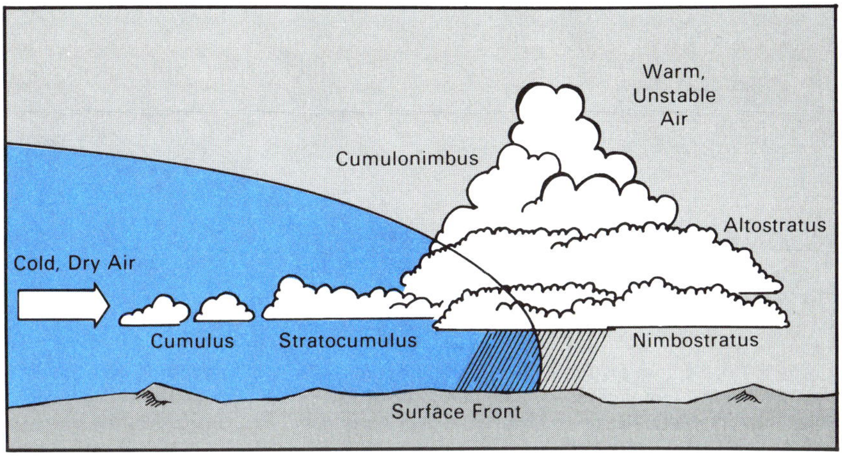

Figure 6-8. Fast-moving cold fronts force warmer air to rise. This causes widespread vertical cloud development along a narrow frontal zone if sufficient moisture is present. An area of severe weather often forms well ahead of the front. The weather usually clears quickly behind a cold front. You will often notice reduced cloud cover, improved visibility, lower temperatures, and gusty surface winds following the passage of a fast-moving cold front.

cool air for several hundred miles ahead of the front. Some of the common weather patterns found in a typical warm front include:

1. Stratus clouds, if the air is moist and stable
2. Little turbulence, except in an unstable airmass
3. Steady precipitation ahead of the front
4. Poor visibility with haze or fog
5. Wide area of precipitation

The stability and moisture content of the air in a warm front determines what type of clouds will form. If the air is warm, moist, and stable,

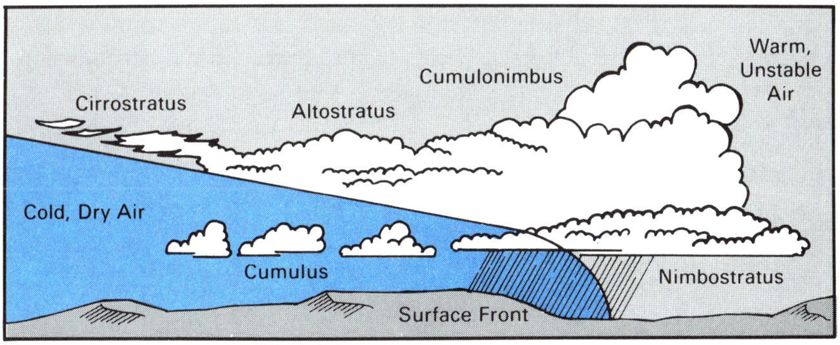

Figure 6-9. As a slow-moving cold front meets unstable air, cumulonimbus and nimbostratus clouds may develop near the surface front, creating hazards from icing and turbulence.

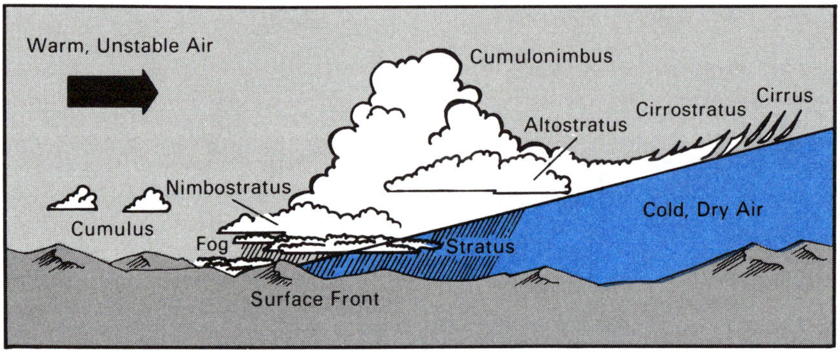

Figure 6-10. Stratus clouds usually extend out ahead of a slow-moving warm front. Vertical clouds sometimes develop along and ahead of the surface front, depending on the stability and moisture content of the warm air.

stratus clouds will develop. If the air is warm, moist, and unstable, cumulus clouds will develop. [Figure 6-10]

Stationary fronts have qualities of both warm and cold fronts.

When the opposing forces of two airmasses are relatively balanced, the front that separates them may remain stationary and influence local flying conditions for several days. The weather in a **stationary front** is usually a mixture of that found in both warm and cold fronts.

A **frontal occlusion** occurs when a fast-moving cold front catches up to a slow-moving warm front. The difference in temperature within each frontal system is a major factor that influences the type of front that develops. A **cold front occlusion** develops when the fast-moving cold front is colder than the air ahead of the slow-moving warm front. In this case, the cold air replaces the cool air at the surface and forces the warm front aloft. [Figure 6-11]

A **warm front occlusion** takes place when the air ahead of the slow-moving warm front is colder than the air within the fast-moving cold front. In this case, the cold front rides up over the warm front, forcing the cold front aloft. [Figure 6-12]

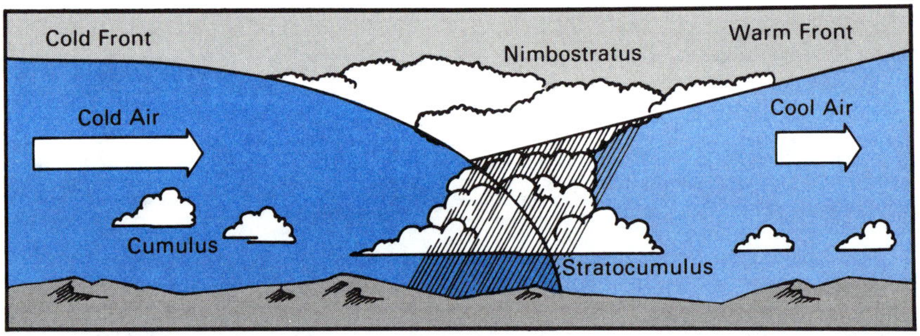

Figure 6-11. When the air being lifted by a cold front occlusion is moist and stable, the weather will be a mixture of that found in both a warm and cold front.

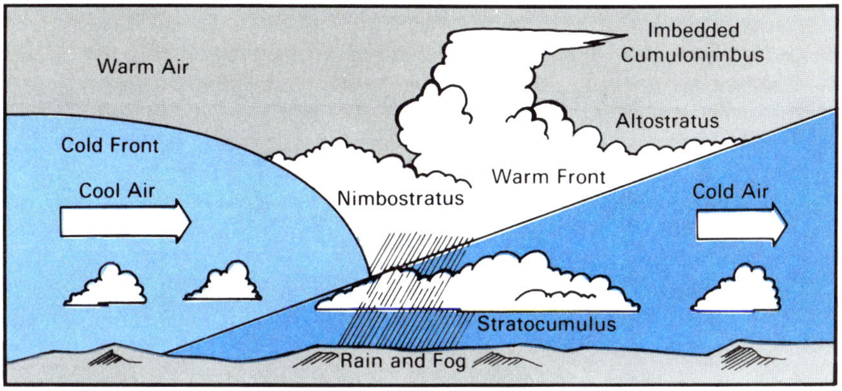

Figure 6-12. When the air being lifted by a warm front occlusion is moist and unstable, the weather will be more severe than that found in a cold front occlusion.

A **frontal wave** is another phenomenon which results primarily from the interaction of two contrasting airmasses. The wave usually begins as a disturbance, or bend, along a slow-moving cold front or stationary front. If the disturbance continues, the wave increases in size and a counterclockwise, or cyclonic, circulation forms. One section of the front begins to move as a warm front, and the other section moves as a cold front. This deformation is the frontal wave.

Pressure at the peak of the wave falls, creating a low pressure center. Then, as the cyclonic circulation intensifies, the winds begin to move the fronts. Since a cold front moves faster than a warm front, the cold front catches up with the warm front, and the two close together, or occlude. The result is an occlusion, or an occluded front. Weather conditions change rapidly in occlusions and usually are most severe during the initial stages of development.

Occlusions combine the weather of both a warm and cold front into one extensive system. You can expect a line of showers and thunderstorms typical of a cold front to merge with the low ceilings of the warm front. Precipitation and low visibilities are widespread over a large area on either side of the surface position of the occlusions. In addition, look for strong winds around an intense low at the northern end of the occlusion.

CHECKLIST

After studying this section, you should have a basic understanding of:

✓ **Atmospheric pressure —** What atmospheric pressure is, why it varies from one location to another, and how it affects wind.

✓ **Moisture in the atmosphere —** How it is measured, under what conditions it increases, and how it affects weather.

- ✓ **Change of state** — What the physical states of moisture in the atmosphere are and the processes involved in change of state.
- ✓ **Clouds and fog** — How clouds and fog are formed and how temperature/dewpoint spread can be used to predict cloud formation.
- ✓ **Stability** — What it is and what determines whether air is stable or unstable.
- ✓ **Characteristics of stable and unstable air** — What cloud types, precipitation, visibility, and turbulence you can expect to encounter in stable and unstable air.
- ✓ **Clouds** — How clouds are classified and the characteristics associated with each class.
- ✓ **Airmasses** — What they are, where they come from, and how they influence weather.
- ✓ **Fronts** — What the types of fronts are and their associated weather.

WEATHER HAZARDS

This section covers various weather hazards that you need to be aware of to conduct VFR or IFR flight operations safely. Instrument operations, in particular, require a sound knowledge of hazardous weather phenomena. When flying in instrument meteorological conditions, you are often unable to observe weather hazards directly. Without a knowledge of the conditions that produce hazardous weather, your ability to avoid dangerous conditions is extremely limited.

THUNDERSTORMS

Thunderstorm formation requires an unstable lapse rate, a lifting force, and high moisture levels.

Thunderstorms produce some of the most dangerous weather elements in aviation, and you should avoid penetrating them. Remember, there are three conditions necessary to create a thunderstorm — air that has a tendency toward instability, some type of lifting action, and relatively high moisture content. [Figure 6-13]

The lifting action may be provided by several factors, such as rising terrain (orographic lifting), fronts, or the heating of the earth's surface

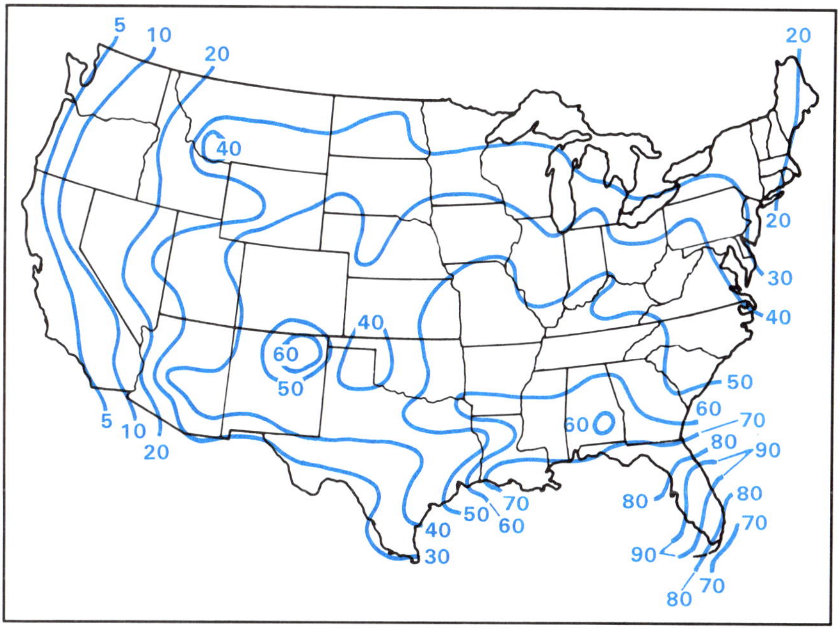

Figure 6-13. This map show the average number of thunderstorms that occur each year in different parts of the country. As you can see, you should be particularly alert for storms in the southeast.

(convection). Thunderstorms progress through three definite stages — cumulus, mature, and dissipating. You can anticipate the development of thunderstorms and the associated hazards by becoming familiar with the characteristics of each stage. Remember that other weather phenomena may prevent you from seeing their characteristic shapes. For example, a cumulonimbus cloud may be embedded, or obscured, by massive cloud layers.

The cumulus stage is characterized by continuous updrafts.

In the **cumulus stage**, a lifting action initiates the vertical movement of air. As the air rises and cools to its dewpoint, water vapor condenses into small water droplets or ice crystals. If sufficient moisture is present, heat released by the condensing vapor provides energy for the continued vertical growth of the cloud. Because of strong updrafts, precipitation usually does not fall. Instead, the water drops or ice crystals rise and fall within the cloud, growing larger with each cycle. Updrafts as great as 3,000 f.p.m. may begin near the surface and extend well above the cloud top.

Thunderstorms reach the greatest intensity during the mature stage, which is signaled by the beginning of precipitation at the surface.

As the drops in the cloud grow too large to be supported by the updrafts, precipitation begins to fall to the surface. This creates a downward motion in the surrounding air and signals the beginning of the **mature stage**. The resulting downdraft may reach a velocity of 2,500 f.p.m. The down-rushing air spreads outward at the surface, producing a sharp drop in temperature, a rise in pressure, strong gusty surface winds, and wind shear conditions. The leading edge of this wind is referred to as a **gust front**, or the **first gust**. As the thunderstorm advances, a rolling, turbulent, circular-shaped cloud may form at the lower leading edge of the cloud. This is called the **roll cloud**. [Figure 6-14]

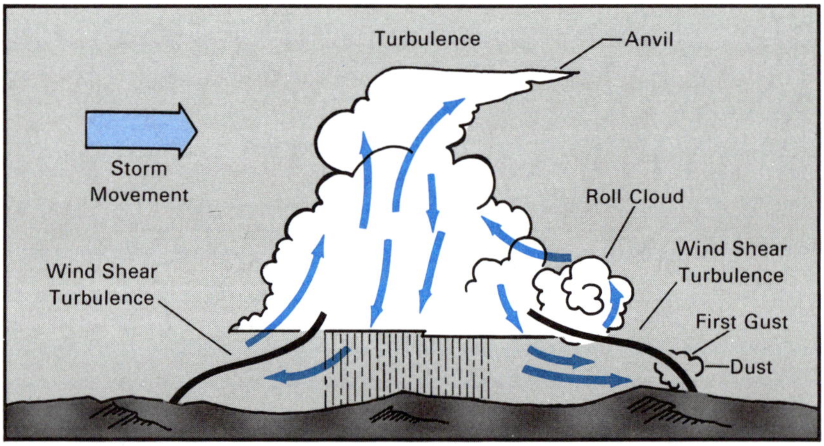

Figure 6-14. Wind shear areas can be found on all sides of a thunderstorm, as well as directly under it. Early in the mature stage, the updrafts continue to increase up to speeds of 6,000 f.p.m. The adjacent updrafts and downdrafts cause severe turbulence. The most violent weather occurs during this phase of the life cycle.

As the mature stage progresses, more and more air aloft is disturbed by the falling drops. Eventually, the downdrafts begin to spread out within the cell, taking the place of the weakening updrafts. Because upward movement is necessary for condensation and the release of the latent energy, the entire thunderstorm begins to weaken. When the cell becomes an area of predominant downdrafts, it is considered to be in the **dissipating stage**. During this stage, the upper level winds often blow the top of the cloud downwind, creating the familiar anvil shape. The anvil, however, does not necessarily signal the storm's dissipation; severe weather can still occur well after its appearance.

A dissipating thunderstorm is characterized by predominant downdrafts.

Thunderstorms usually have similar physical features, but their intensity, degree of development, and associated weather do differ. They are generally classified as airmass or frontal storms. **Airmass thunderstorms** generally form in a warm, moist airmass and are isolated or scattered over a large area. They are usually caused by solar heating of the land, which results in convection currents that lift unstable air. These thunderstorms are most common during hot summer afternoons when winds are light or in coastal areas at night. Airmass storms can also be caused by orographic lifting. Although they are usually scattered along individual mountain peaks, they may cover large areas. They also may be embedded in other clouds, making them difficult to identify when approached from the windward side of a mountain. Nocturnal thunderstorms can occur in late spring and summer during the late night or early morning hours when relatively moist air exists aloft. Usually found from the Mississippi Valley westward, nocturnal storms cover many square miles, and their effects may continue for hours at a given location.

Airmass storms are usually caused by convection or orographic lifting.

Frontal thunderstorms can be associated with any type of front. Those which occur with a warm front are often obscured by stratiform clouds. You should expect thunderstorms when there is showery precipitation near a warm front. In a cold front, the cumulonimbus clouds are often visible in a continuous line parallel to the frontal surface. Occlusions can also spawn storms. A **squall line** is a narrow band of active thunderstorms which normally contains very severe weather. While it often forms 50 to 200 miles ahead of a fast-moving cold front, the existence of a front is not necessary for a squall line to form.

A squall line is a nonfrontal band of thunderstorms that contains the most severe types of weather-related hazards.

Thunderstorms typically contain many severe weather hazards, and may include lightning, hail, turbulence, gusty surface winds, or even tornadoes. These hazards are not confined to the cloud itself. For example, you can encounter turbulence in VFR conditions as far as 20 miles from the storm. It may help to think of a cumulonimbus cloud as the visible part of a widespread system of turbulence and other weather hazards. In fact, the cumulonimbus cloud is the most turbulent of all clouds. Indications of severe turbulence within the storm system include the cumulonimbus cloud itself, very frequent lightning, and roll clouds.

Severe turbulence is indicated by frequent lightning and roll clouds associated with cumulonimbus clouds.

Lightning is always associated with thunderstorms.

Lightning is one of the hazards which is always associated with thunderstorms and is found throughout the cloud. While it rarely causes personal injury or substantial damage to the aircraft structure in flight, it can cause temporary loss of vision, puncture the aircraft skin, or damage electronic navigation and communications equipment. Studies have shown that aircraft have been struck from 14,000 to 38,000 feet in temperature ranges from +5°C to -50°C. Your aircraft can also be struck by lightning when you are in the clear but still in the vicinity of a thunderstorm.

Hail is another thunderstorm hazard. You can encounter it in flight, even when no hail is reaching the surface. In addition, large hailstones have been encountered in clear air several miles downwind from a thunderstorm. Hail can cause extensive damage to your aircraft in a very short period of time.

The cumulonimbus is the type of cloud that produces the most severe turbulence.

Thunderstorm turbulence is another serious hazard which develops when air currents change direction or velocity rapidly over a short distance. The magnitude of the turbulence depends on the differences between the two air currents. Within the thunderstorm cloud, the strongest turbulence occurs in the shear between the updrafts and downdrafts. Near the surface, there is an area of low-level turbulence which develops as the downdrafts spread out at the surface. These create a **shear zone** between the surrounding air and the cooler air of the downdraft. The resulting area of gusty winds and turbulence can extend outward for many miles from the center of the storm. We will discuss turbulence in greater detail later in this section.

Funnel clouds are violent, spinning columns of air which descend from the base of a cloud. Wind speeds within them may exceed 200 knots. If a funnel cloud reaches the earth's surface, it is referred to as a **tornado**. If it touches down over water, it is called a **waterspout**.

THUNDERSTORM AVOIDANCE

When conditions permit, you should circumnavigate thunderstorms. Remember that hail and severe turbulence may exist well outside the storm cloud. During night operations, lightning may provide you a clue as to a storm's location; however, lightning may not yet have developed in younger storms and may have ceased in older ones. Even without visible lightning, storms may still contain destructive turbulence or hail. The best approach during night operations or during flight under instrument conditions is to avoid areas where thunderstorms exist or are likely to develop.

If the aircraft you are flying is equipped with a weather avoidance system, such as weather radar, you can use it to avoid thunderstorms. With radar, you should avoid intense thunderstorm echoes by at least 20 miles. You should not fly between intense radar echoes unless they are at least 40 miles apart. Keep in mind that this is only a general recommendation.

Avoid intense radar echoes by at least 20 miles and do not fly between them if they are less than 40 miles apart.

Airborne weather radar is designed for avoiding severe weather, not for penetrating it. Weather radar detects drops of precipitation; it does not detect minute cloud droplets. Therefore, it should not be relied on to avoid instrument weather associated with clouds and fog. Also, be sure you are familiar with the operation of the systems and the manufacturer's recommendations appropriate to your system.

Airborne weather radar provides no assurance of avoiding IFR weather conditions.

TURBULENCE

We have already discussed turbulence briefly with regard to thunderstorms. To avoid or minimize the effects of turbulence, you must understand its causes and know where it is likely to be found. If you enter turbulence unexpectedly, unintentionally enter a thunderstorm, or expect that you may encounter turbulence, reduce power to slow the airplane to maneuvering speed (V_A) or less, or the airspeed recommended for rough air, and attempt to maintain a level flight attitude. Accept variations in airspeed and altitude, since this will help to avoid the high structural loads that you may impose on the aircraft if you try to maintain them precisely. If you encounter turbulence during a landing approach, increase your airspeed to slightly above normal approach speed to maintain more positive control of the aircraft.

If you encounter turbulence, establish maneuvering speed and attempt to maintain a level flight attitude.

Turbulence normally is linked to **wind shear**, which is a sudden, drastic shift in wind speed and/or direction that may occur at any altitude in a vertical or horizontal plane. It can subject your aircraft to sudden updrafts, downdrafts, or extreme horizontal wind components, causing loss of lift or violent changes in vertical speeds or altitude. In fact, when landing during gusty conditions, you should use a power-on approach and landing to minimize the effect of the wind and to retain positive control of the aircraft. Wind shear is associated with temperature inversions, the jet stream, thunderstorms, and frontal inversions. With a temperature inversion, it occurs when cold, calm air is near the surface and warmer air aloft is in motion. This can be a hazard during climbouts or approaches when your airspeed is relatively slow. Be alert for a sudden change in airspeed and carry an extra margin of speed if you suspect an inversion.

Wind shear can exist at any altitude and may occur in a vertical or horizontal direction. When landing in gusty conditions, fly a power-on approach and landing.

With fronts, the most critical period is either just before or just after frontal passage. With a cold front, wind shear occurs just after the front passes and for a short time afterward. Studies indicate the amount of wind shear in a warm front is generally much greater than in a cold front. The wind shear below 5,000 feet AGL in a warm front may last for approximately six hours. The most critical period is before the front passes; the problem ceases to exist after it passes.

Look for wind shear before a warm front passes and after a cold front passes.

A horizontal wind shear can result in a sudden change in indicated airspeed. The amount of change is directly related to how fast the wind speed or direction changes. For example, a sudden 20-knot decrease in headwind results in a 20-knot loss in indicated airspeed. This can be extremely hazardous if you are flying near the surface.

A **low-level wind shear alert system** (LLWAS) is installed at some airports. It continuously compares the wind around the airport perimeter with wind at a center field location. If the system indicates a significant difference, tower controllers will advise you of the wind velocities at two or more of the sensors; for example, *"Bonanza 1776R, center field wind 270° at 10; south boundary wind 140° at 30."* Consult the *Airport/Facility Directory* listings to determine if an airport has LLWAS.

Microbursts are intense, localized downdrafts seldom lasting longer than 15 minutes from the time the burst first strikes the ground until dissipation. Maximum downdrafts may reach 6,000 feet per minute.

A **microburst** is an intense, localized downdraft of brief duration which spreads out in all directions when it reaches the surface. This creates severe horizontal and vertical wind shears which pose serious hazards to aircraft, particularly those near the surface. An individual microburst typically covers an area of less than two and a half miles in diameter at the surface and usually lasts no longer than 15 minutes. Peak winds last two to four minutes and attendant downdrafts can be as strong as 6,000 feet per minute. [Figure 6-15]

Any convective cloud can produce this phenomenon. Although microbursts are commonly associated with heavy precipitation in thunderstorms, they often occur in virga. If there is no precipitation, your only cue may be a ring of dust at the surface. If you suspect the presence of microbursts in your local area, delay your takeoff or landing.

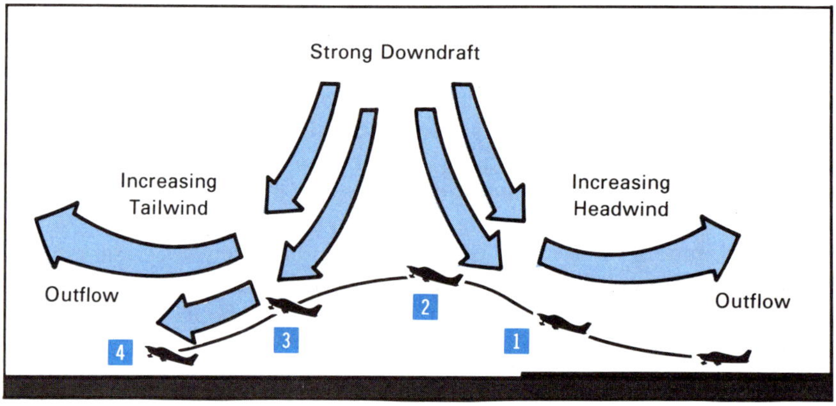

Figure 6-15. As this aircraft takes off into a microburst, it first experiences a headwind which increases performance without a change in pitch and power (position 1). A headwind of 45 knots may result in a total of a 90-knot windshear within the microburst. This is followed by a decreasing headwind and performance, and a strong downdraft (position 2). Performance continues to deteriorate as the wind shears to a tailwind in the downdraft (position 3). The most severe downdraft will be encountered between positions 2 and 3, which may result in terrain impact or operation dangerously close to the ground (position 4).

The following discussion reflects Federal Aviation Administration recommendations that apply to wind shear encounters during a stabilized landing approach. You should not attempt approaches and departures during known wind shear conditions.

On an approach, you should monitor the power and vertical velocity required to maintain the glide path. If you encounter an unexpected wind shear during an approach, it may be difficult to stay on the glide path at normal power and descent rates. For example, if a headwind shifts to a tailwind or calm, the indicated airspeed decreases, the pitch attitude decreases, and your aircraft tends to sink below the glide path. Increase power to regain airspeed and fly up to the glide slope, but be ready to reduce power once you have recovered. Your groundspeed is greater than when you had the headwind, so your rate of descent must be greater than it was before you encountered the wind shear.

If the power and descent rates required to maintain the glide path are abnormal, you should be alert for wind shear. *

When a tailwind shears to calm conditions or a headwind, the opposite situation occurs. Your indicated airspeed and pitch attitude both increase, and the tendency is to go above the correct glide path. You should reduce power to slow your airspeed and descend to the glide path, but be ready to increase your power again. You now need more power and a slower rate of descent than you did when you had a tailwind. If there is ever any doubt that you can regain a reasonable rate of descent and land without abnormal maneuvers, you should apply full power and make a go-around or missed approach.

As you know, convective turbulence occurs frequently during the summer months. When the air is moist, the convective currents are marked by the development of cumulus clouds. Some of these clouds may grow into towering cumulus clouds and develop into thunderstorms. Another source of turbulence includes buildings or rough terrain, which interfere with the normal wind flow. This phenomenon, referred to as mechanical turbulence, is often experienced in the traffic pattern when the wind blows around hangars and other structures on the airport.

Mechanical turbulence also occurs when strong winds flow nearly perpendicular to mountain ridges. When you fly at low altitudes over hilly terrain, ridges, or mountain ranges, the greatest danger from turbulence usually is encountered on the lee side flying into the wind. As the unstable air spills down the leeward side, violent downdrafts may develop. These downdrafts can exceed the climb capability of an aircraft. The situation is somewhat different when stable air crosses a mountain barrier. In this case, the airflow usually is smooth on the windward side of the ridge. Wind flow across the barrier is laminar — that is, it tends to flow in layers. The barrier can generate mountain waves that extend 100

The greatest turbulence normally occurs as you approach the lee side of mountain ranges, ridges, or hilly terrain in strong headwinds.

miles or more downwind, and the crests may reach well above the highest mountain peaks. [Figure 6-16]

Clear air turbulence (CAT) is commonly thought of as a high altitude phenomenon, although it can take place at any altitude. According to the AIM, CAT can be a serious operational problem for jet aircraft flying above 15,000 feet. Its presence carries no visual warning, since it is a form of turbulence not associated with cumuliform clouds or thunderstorms. CAT may be caused by wind shear, convective currents, or obstructions to normal wind flow. It often develops in or near the **jet stream**, which is a narrow band of high altitude winds near the tropopause. Later in this chapter, the jet stream is discussed more fully in the section on high-altitude considerations.

REPORTING TURBULENCE

You are encouraged to report encounters with turbulence, including the frequency and intensity. Turbulence is considered to be **occasional** when it occurs less than one-third of a given time span, **intermittent** when it covers one-third to two-thirds of the time, and **continuous** when it occurs more than two-thirds of the time. You can classify the intensity using the following guidelines:

Light — Slight erratic changes in altitude or attitude; slight strain against seat belts. **Light chop** is slight, rapid bumpiness without appreciable changes in altitude or attitude.

Moderate — Changes in altitude or attitude, but aircraft in positive control at all times; usually changes in indicated airspeed; definite strains against seat belts. **Moderate chop** is rapid bumps or jolts without appreciable changes in altitude or attitude.

Severe — Abrupt changes in altitude or attitude; usually large variations in indicated airspeed; aircraft may be momentarily out of control; occupants forced violently against seat belts.

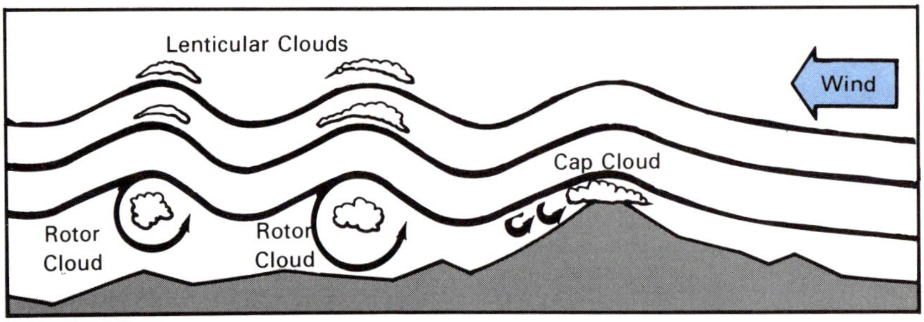

Figure 6-16. The crests of the waves may be marked by lenticular clouds. Their official name is standing lenticular altocumulus (ACSL). These clouds form in the updrafts and dissipate in the downdrafts so they have a stationary appearance as the wind blows through them. Rotor clouds may also form, and cap clouds may obscure the mountain peaks. Rotor clouds signify possible severe or extreme turbulence.

Extreme — Aircraft practically impossible to control; may cause structural damage.

WAKE TURBULENCE

Broadly defined, **wake turbulence** refers to the phenomena that result from the passage of an aircraft through the atmosphere. Whenever an airplane generates lift, air spills over the wingtips causing **wingtip vortices**. The greatest wake turbulence danger is produced by large, heavy aircraft operating at low speeds and high angles of attack. Since these conditions usually exist on takeoff and landing, you should be alert for wake turbulence near airports used by large airplanes. In fact, wingtip vortices from commercial jets can induce uncontrollable roll rates in smaller aircraft. The probability of induced roll is greatest when your heading is aligned with the generating aircraft's flight path. The vortices from large aircraft in flight sink at a rate of about 400 to 500 feet per minute and level off at a distance of about 900 feet below the generating aircraft. Vortex strength diminishes with time and distance behind the generating aircraft. On the ground in a no-wind condition, the vortices tend to remain on the ground and move outward at about two or three knots. Therefore, a crosswind of one to five knots tends to cause the upwind vortex to remain on the runway and the downwind vortex to drift to a parallel runway. The most hazardous situation is a light, quartering tailwind, since it can push the vortex forward into the touchdown zone. [Figure 6-17]

Wingtip vortices can exceed the roll rate of an aircraft, especially when flying in the same direction as the generating aircraft.

If you are in a small aircraft approaching to land behind a large aircraft, controllers must ensure a separation of four miles (six miles behind a heavy jet). However, if you accept a clearance to follow an aircraft you have in sight, the responsibility for wake turbulence avoidance is transferred from the controller to you. On takeoff, controllers provide a two-minute interval behind departing heavy jets (three minutes for intersection takeoffs or takeoffs in the opposite direction on the same runway). You may waive these time intervals if you wish, but this is not a wise procedure.

Situation	Proper Action
Landing after large airplane lands	Stay above large airplane's glide path and touch down beyond its touchdown point.
Landing after large airplane takes off	Touch down well before large airplane's liftoff point.
Taking off after large airplane takes off	Lift off before large airplane's rotation point and climb out above or upwind of its flight path.
Taking off after large airplane lands	Lift off beyond its touchdown point.

Figure 6-17. This table summarizes wake turbulence avoidance procedures.

Jet engine blast is another form of wake turbulence. It can damage or even overturn a small airplane if it is encountered at close range. To avoid exhaust velocities in excess of 20 knots, you must stay several hundred feet behind a jet with its engines operating, even when it is at idle thrust. At takeoff thrust, you need to be as far as 2,100 feet from a heavy jet.

LOW VISIBILITY

One of the most common aviation weather hazards is low visibility. As a pilot, you usually are concerned with two types of visibility — prevailing visibility and flight visibility. As you know from Chapter 5, prevailing visibility represents the horizontal surface visibility which is recorded in surface aviation weather reports. In FARs, **flight visibility** is defined as the average forward horizontal distance, from the cockpit of an aircraft in flight, at which prominent unlighted objects may be seen and identified by day and prominent lighted objects may be seen and identified by night. However, your most practical concern in flight operations often is slant-range visibility. The slant-range visibility may be greater or less than the surface horizontal visibility, depending on the depth of the surface condition. Slant-range visibility is important in the approach zone when you are landing from an instrument approach. Horizontal visibility at the surface is most important during IFR takeoff operations. Poor visibility creates the greatest hazard when it exists together with a low cloud ceiling.

RESTRICTIONS TO VISIBILITY

Industrial areas typically produce more fog since they have more condensation nuclei.

Let's now take a look at some ground-based weather hazards which can restrict visibility. These include fog, haze, and smog. Fog requires both sufficient moisture and condensation nuclei on which the water vapor can condense. It is more prevalent in industrial areas, due to an abundance of condensation nuclei. One of the most hazardous characteristics of fog is its ability to form rapidly. It can completely obscure a runway in a matter of minutes.

Radiation fog forms in moist air on clear, calm nights.

Fog is classified according to the way it forms. **Radiation fog**, also known as **ground fog**, is very common. It forms over fairly level land areas on clear, calm, humid nights. As the surface cools by radiation, the adjacent air is also cooled to its dewpoint. Radiation fog usually occurs in stable air associated with a high pressure system. As early morning temperatures increase, the fog begins to lift and usually "burns off" by mid-morning. If higher cloud layers form over the fog, visibility will improve more slowly.

Advection fog can form day or night over extensive areas and is usually more persistent than radiation fog.

Advection fog is caused when a low layer of warm, moist air moves over a cooler surface, which may be either land or water. It is most common under cloudy skies along coastlines, where sea breezes transport air from the warm water to cooler land. Winds up to 15 knots will intensify the fog. Above 15 knots, turbulence creates a mixing of the air, and the fog usually lifts sufficiently to form low stratus clouds. Advection fog is usually more persistent and extensive than radiation fog and can form rapidly during day or night. It commonly forms in winter when airmasses move inland from the coasts. **Upslope fog** forms when moist, stable air is forced up a sloping land mass. Like advection fog, upslope fog can form in moderate to strong winds and under cloudy skies.

Advection fog and upslope fog are both dependent upon wind for their formation.

Precipitation-induced fog may form when warm rain or drizzle falls through a layer of cooler air near the surface. Evaporation from the falling precipitation saturates the cool air, causing fog to form. This fog can be very dense, and usually does not clear until the rain moves out of the area. **Steam fog** occurs as cool air moves over warmer water. It rises upward from the water's surface and resembles rising smoke.

Haze, smoke, smog, and blowing dust or snow can also restrict your visibility. **Haze** is caused by a concentration of very fine salt or dust particles suspended in the air. It occurs in stable atmospheric conditions with relatively light winds. Haze makes contrasting colors less distinct, so objects are harder to see. It also creates the illusion of being higher than actual above the runway, and can cause pilots to fly a lower approach. **Smoke** is usually much more localized; it is generally found in industrial areas and is a hazard only when it drifts across your intended landing field. On the other hand, **smog**, which is a combination of fog and smoke, can produce very poor visibility over a large area. Blowing dust and blowing snow present similar problems. They occur in moderate to high winds and can extend to an altitude of several thousand feet.

Restrictions to visibility, such as haze, create the illusion of being a greater distance above the runway, causing pilots to fly a lower approach.

ICING

As you know, there are two general types of icing with which you must be familiar — induction and structural. **Induction icing** includes carburetor icing as well as air intake icing. Carburetor icing forms in the carburetor venturi. It is most likely to occur when the outside air temperature is between approximately -7°C (20°F) and 21°C (70°F) and relative humidity is above 80%. Some aircraft may have a carburetor air temperature gauge to help detect potential carburetor icing conditions. A typical gauge is marked with a yellow arc between -15° and +5° Celsius. The yellow arc indicates the carburetor temperature range where carburetor icing can occur. However, it is dangerous for the indicator to be in the yellow arc only if the moisture content of the air, as well as the air temperature, is conducive to ice formation. Carburetor heat should be used as recommended in the pilot's operating handbook.

Carburetor icing is most likely with high humidity and temperatures between -7°C (20°F) and 21°C (70°F).

Air intake icing, like airframe icing, usually requires the aircraft surface temperature to be 0°C or colder with visible moisture present. However, it also can form in clear air when relative humidity is high and temperatures are 10°C (50°F) or colder.

If you encounter rain that freezes on impact, the temperatures are above freezing at some higher altitude.

Structural icing builds up on any exposed surface of an aircraft, causing a loss of lift, an increase in weight, and control problems. There are two general types — rime and clear. Mixed icing is a combination of the two. **Rime ice** normally is encountered in stratus clouds and results from instantaneous freezing of tiny water droplets striking the aircraft surface. It has an opaque appearance caused by air being trapped in the water droplets as they freeze. The major hazard is its ability to change the shape of an airfoil and destroy its lift. Since rime ice freezes instantly, it builds up on the leading edge of airfoils, but it does not flow back over the wing and tail surfaces. **Clear ice** may develop in areas of large water droplets which are found in cumulus clouds or in freezing rain beneath a warm front inversion. Freezing rain means there is warmer air at higher altitudes. When the droplets flow over the aircraft structure and slowly freeze, they can glaze the aircraft's surfaces. Clear ice is the most serious of the various forms of ice because it has the fastest rate of accumulation, adheres tenaciously to the aircraft, and is more difficult to remove than rime ice.

Ice, snow, or frost having the thickness and roughness of sandpaper can reduce lift by 30% and increase drag by 40%.

The effects of ice buildup on aircraft are cumulative. Ice increases drag and weight, and decreases lift and thrust. Tests have shown that ice, snow, or frost with a thickness and roughness similar to medium or coarse sandpaper on the leading edge and upper surface of a wing can reduce lift by as much as 30% and increase drag by 40%. These changes in lift and drag will significantly increase the stalling speed. If you encounter icing, you must quickly alter your course or altitude to maintain safe flight. In extreme cases, two to three inches of ice can form on the leading edge of an aircraft in less than five minutes.

Two conditions are necessary for a substantial accumulation of ice on an aircraft. First, the aircraft must be flying through visible moisture, such as rain or clouds. Second, the temperature of water or of the aircraft must be 0°C or lower. Keep in mind that aerodynamic cooling can lower the temperature of an airfoil to 0°C, even though the ambient temperature is a few degrees warmer.

High clouds are least likely to contribute to aircraft structural icing.

When water droplets are cooled below the freezing temperature, they are in a supercooled state. They turn to ice quickly when disturbed by an aircraft passing through them. Clear icing is most heavily concentrated in cumuliform clouds in the range of temperature from 0°C to -10°C, usually from altitudes near the freezing level to 5,000 feet above the freezing level. However, you can encounter clear icing in cumulonimbus clouds with temperatures as low as -25°C. In addition, supercooled water and icing have been encountered in thunderstorms as high as 40,000 feet, with temperatures of -40°C. Small supercooled water droplets also can cause rime icing in stratiform clouds, although rime does not usually accumulate as fast as clear ice. Continuous icing can be expected in stratiform clouds in the temperature range from 0°C to -20°C. You are least likely to encounter icing in the high cloud family, since these clouds are composed mainly of ice crystals. Icing may also occur as a mixture of both rime and clear. This commonly is the case with frontal systems which can produce a wide variety of icing conditions. [Figure 6-18]

ESTIMATING FREEZING LEVEL

Knowing the location of the freezing level is important when you select a cruising altitude for an IFR flight. You can estimate the freezing level by using temperature lapse rates. The standard, or average, temperature lapse rate is approximately 2°C (3.5°F) per 1,000 feet. If the surface temperature is 60°F, subtract 32°F and divide by 3.5°F to determine the freezing level (60 - 32 ÷ 3.5 = 8). This means the freezing level is approximately 8,000 feet AGL, assuming the actual lapse rate is close to 3.5°F.

AVOIDING ICE ENCOUNTERS

Obviously, you should avoid areas where icing is forecast or expected. If you unintentionally get into an icing situation without ice removal equipment, there usually are only limited courses of action available. Also, remember that you must obtain approval from ATC if you need to make a diversion while operating on an IFR flight plan. When you encounter icing while flying in stratiform clouds, you can climb or descend out of the visible moisture or to an altitude where the temperature is above freezing or lower than -10°C.

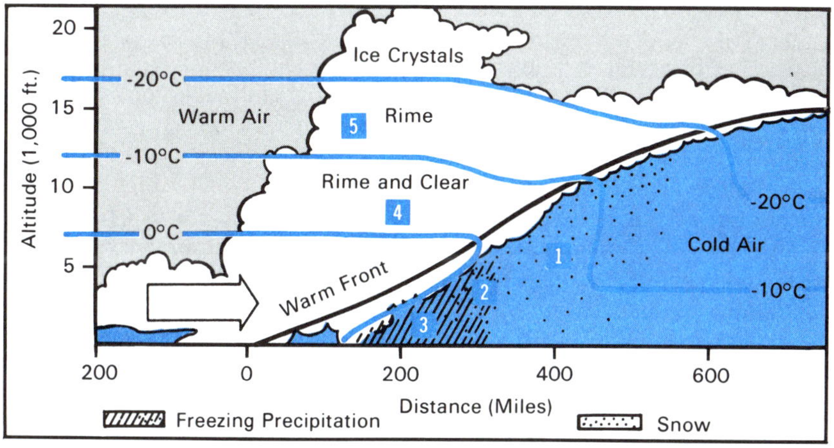

Figure 6-18. This profile of a warm front shows the typical icing areas. At position 1, rime ice is possible when you are flying in snow falling from stratus clouds. Wet snow also could be encountered below the freezing level. As you proceed under the warm front at position 2, you may encounter ice pellets formed by raindrops which have frozen while falling through colder air. Ice pellets normally do not adhere to an aircraft; however, they always indicate there is freezing rain at a higher altitude. The level of the freezing rain will be lower as you fly under the frontal zone. You should anticipate icing from freezing rain or drizzle at position 3. The area of freezing rain may be quite narrow, or it may extend for several miles. You should expect a mixture of rime and clear ice at position 4. If you are flying in the upper level of clouds associated with this warm front where the temperatures are far below freezing, rime ice is the predominant type of icing you will encounter at position 5.

If you encounter cumuliform clouds and suspect icing conditions, a simple change of course may be the best action. If this is impractical, descend to a lower altitude to avoid the ice. Overflying developed cumuliform clouds may be beyond the performance capability of your aircraft. Your remaining option is a 180° turn; in some cases, this is the best choice.

If you encounter freezing rain, you must take immediate action. One option is to climb into the warmer air above. If you choose to do this, start the climb at the earliest possible moment. Otherwise, you may not have enough power to reach a higher altitude, because ice buildup quickly degrades your aircraft's performance. Also, if the frontal surface slopes up in the direction of your flight, you may not be able to climb fast enough to reach the warm air. Other options include making a 180° turn and reversing course or descending to a lower altitude. You may choose to make a descent if the free air temperature at lower altitudes is above freezing or cold enough to change the precipitation into ice pellets (sleet). A 180° turn or a descent is probably the best course of action if the altitude of warmer air aloft is unknown and your aircraft's performance will not permit a rapid and extended climb. If enough ice accumulates to make level flight impossible, descent is inevitable, so trade altitude for airspeed to maintain control.

In weather forecasts or pilot reports, aircraft structural icing is normally classified as trace, light, moderate, or severe depending on the accumulation rate. A **trace** means ice is perceptible, but accumulation is nearly balanced by its rate of sublimation. De-icing equipment is unnecessary, unless icing is encountered for an extended period of time. **Light** ice accumulation can be a problem during prolonged exposure (over one hour) if you do not have adequate de-icing/anti-icing equipment. In **moderate** icing conditions, even short encounters become potentially hazardous unless you use de-icing/anti-icing equipment. **Severe** icing produces a rate of accumulation greater than the reduction or control capabilities of the de-icing/anti-icing equipment.

Most small general aviation aircraft are not approved for flight into icing conditions. Those that are have special de-ice and anti-ice equipment to protect aircraft surfaces, as well as the induction system. They have also been flight tested to demonstrate their ability to fly in icing conditions. However, even these aircraft cannot fly in severe icing, and prolonged flight in moderate icing can become hazardous due to ice accumulation on unprotected surfaces. Since ice protection systems vary widely in both operation and effectiveness, be sure to consult the POH for detailed information on system operation. A general discussion of de-ice and anti-ice systems is included in Chapter 9.

Frost is a related element which poses a serious hazard during takeoffs. It interferes with smooth airflow over the wings and can cause early airflow separation, resulting in a loss of lift. This means the wing stalls at a lower than normal angle of attack. Although frost increases drag, the aircraft should be able to reach takeoff speed. The danger is that it may stall shortly after liftoff. Always remove all frost from the aircraft surfaces before flight.

If frost is not removed from the wings before flight, it may cause an early airflow separation which decreases lift and increases drag. This will cause the aircraft to stall at a lower-than-normal angle of attack.

HYDROPLANING

Hydroplaning is caused by a thin layer of standing water that separates the tires from the runway. It causes a substantial reduction in friction between the aircraft tires and the runway surface and results in poor or nil braking action at high speeds. If severe enough, it may result in your aircraft skidding off the side or end of the runway. Hydroplaning is most likely at high speeds on wet, slushy, or snow-covered runways which have smooth textures.

High aircraft speed, standing water, slush, and a smooth runway texture are factors conducive to hydroplaning.

The best remedy for hydroplaning is to prevent its occurrence. You should study hydroplaning speeds and recommended procedures for your aircraft. When selecting a runway, pick one which is longer than required and allows you to use only light braking pressures to stop the aircraft safely. Since it may take several seconds for the wheels to reach their rotational speed after landing, do not apply the brakes too quickly. More detailed information on hydroplaning is available in Chapter 8.

COLD WEATHER OPERATIONS

Prior to a flight in cold weather, there are additional precautions which you must observe. As mentioned above, you must remove any frost on the aircraft. You should also inspect all control surfaces and their associated control rods and cables for snow or ice which may interfere with their operation. Be sure to check the crankcase breather lines, since vapor from the engine can condense and freeze, preventing the release of air from the crankcase.

Be sure to check for snow or ice which may interfere with the operation of the aircraft controls.

Another important consideration is whether or not to preheat the aircraft prior to flight. If temperatures are so low that you will experience difficulty starting the engine, you should preheat it. You should also consider preheating the cabin, as well as the engine compartment. This is recommended for proper operation of instruments, many of which are adversely affected by cold temperatures.

Preheating the cabin is just as important as preheating the engine.

A final consideration before you try to start the engine is priming. Here, you should go by the manufacturer's recommended procedures in the POH. Overpriming may result in poor compression and difficulty in starting the engine. Another cold starting problem is caused by icing on sparkplug electrodes. This can occur when the engine fires for only a few revolutions and then quits. Heating the engine is the only remedy for "frosted plugs," short of removing them from the engine and heating them individually.

CHECKLIST

After studying this section, you should have a basic understanding of:

- ✓ **Thunderstorms —** What the conditions are for their formation and what the types, characteristics, and hazards of thunderstorms are.
- ✓ **Turbulence —** What causes it, how it is classified, and what you can do to minimize its effects.
- ✓ **Low visibility —** What the difference is between prevailing and flight visibility, and how slant-range visibility is often your most important limitation.
- ✓ **Structural icing —** What the types of structural icing are, how each type forms, the hazards associated with each, and how to escape icing conditions.
- ✓ **Cold weather operations —** What procedures and precautions are required during cold weather operations.

SECTION C

PRINTED REPORTS AND FORECASTS

Weather information is available through a vast network of government agencies, including the National Weather Service (NWS), the Federal Aviation Administration (FAA), and the Department of Defense (DOD). Private companies and contracted individuals also contribute to the network, as do several foreign countries. One of the most important facilities within the NWS is the National Meteorological Center (NMC), located in Washington, D.C. This facility is the central data processing center for the collection and distribution of basic weather observations. The NMC also prepares the nation's winds and temperatures aloft forecasts, which you will review later in this section.

Aviation forecasts are prepared by over 50 Weather Service Forecast Offices (WSFOs). They prepare terminal forecasts, route forecasts, and recorded weather information for the entire nation. Another important aviation weather facility is the National Aviation Weather Advisory Unit (NAWAU) in Kansas City, MO, which issues all area forecasts, as well as severe weather watches and various types of in-flight advisories. Aviation weather is distributed to FSSs through the Weather Message Switching Center (WMSC) in Atlanta, Georgia, and Salt Lake City, Utah. It is transmitted by high speed communications and automated data processing equipment.

You know that as pilot in command of an aircraft, you are required to familiarize yourself with all available information before beginning a flight. For a flight under IFR or any flight not in the vicinity of an airport, that information must include the latest or most current weather reports and forecasts. Normally, you will receive your formal weather briefings from either of two weather service outlets — an FSS or a Weather Forecast Office (WFO). Weather Forecast Offices provide pilot weather briefings in areas not served by an FSS; they also supply backup service to the FSS network. While personal visits to an FSS or WFO to interpret weather reports and forecasts have become less frequent, keep in mind that a variety of private weather briefing services also are available. These systems make weather products available to you directly, so you will have a continuing need to understand and interpret the various reports, forecasts, and charts. Even when you obtain your preflight weather briefing by telephone from an FSS or WFO outlet, a working knowledge of weather products will help you question the briefer when you need additional information or clarification.

AVIATION ROUTINE WEATHER REPORTS

During preflight planning, the **aviation routine weather report (METAR)** allows you to assess existing conditions and evaluate the accuracy of forecasts. You can also review several previous reports to get a general idea of the weather trend. A current report also indicates whether your destination is presently above VFR or IFR minimums, as appropriate. If it is not and the forecast calls for only slight improvement by your ETA, the destination weather may not permit you to land. The METAR also is important at airports without a control tower or flight service station, since the last surface observation is the latest official weather available at that location.

There are two types of aviation weather reports—the regularly scheduled METAR, which is taken every hour, and the **aviation selected special weather report (SPECI)**. The Special observation is taken when a significant change occurs in one or more elements of the weather report. The type of report is shown in the first block of the printed report. [Figure 6-19]

Appearing next is the four-letter International Civil Aviation Organization (ICAO) identifier for the reporting station. After the station identifier the date (day of the month) and UTC or Zulu time of the observation is shown.

The first weather element in the METAR is the wind direction and speed. This information can provide important clues to local weather developments, such as thunderstorms or frontal passage. Under certain conditions with high winds, you might be aware of blowing dust or snow. On the other hand, calm winds (00000KT) with a temperature/dewpoint spread of 2°C or less can mean that the formation of fog is likely. Gusty winds are shown by a "G," and may also influence your flight planning.

The prevailing visibility, reported in statute miles (SM), is a critical factor in determining if conditions are VFR or IFR. When runway visual range (RVR) is shown, you should pay careful attention to the minimums required for an instrument approach.

When the prevailing visibility is less than one mile and/or the RVR is 6,000 feet or less, you may see RVR values in the visibility section of the METAR. When RVR is reported, it is denoted by the letter "R," followed by the actual runway number, a "/," and the recorded visual range in feet (meters outside the U.S.) When visibility is variable, the letter "V" follows the reported RVR, and the tendency is shown by the first letter of Up, Down, or No change. When runway visual range is greater than (Plus) 6,000 feet, it is shown as P60. An RVR less than the minimum value that the transmissometer can detect is preceded by the letter "M" for minus. For example, an RVR of less than 1,000 feet may be reported as "M10."

1 METAR 2 KTPA 122150Z 3 AUTO 4 08020G38KT 5 1/2SM R36L/2400FT

6 +TSRA 7 SCT008 OVC012CB 8 20/18 9 A2995 10 RMK A02 TSB24RAB24 SLP134

1 Type of Report

METAR Hourly Observation
SPECI Special, unscheduled report

2 Station Designator and Date/Time

Four-letter ICAO location identifier
Date and Time of report in Zulu (UTC)

3 Modifier

Auto = Automated
COR = Corrected

4 Wind Information

First three digits = Direction, or VRB = Variable
Next two digits = Speed in knots (KT), or three digits if speed >99KT. G = highest gust, followed by speed. 00000KT = calm

5 Visibility

Prevailing visibility, statute miles (SM). Runway visual range: R, runway number, / , visual range in feet (FT).

6 Present Weather

Intensity: -light, +heavy, no sign for moderate.

Proximity: VC = weather 5 to 10 miles from airport center

Descriptor: for precipitation or obstructions to visibility:

TS Thunderstorm	DR low drifting
SH Shower(s)	MI Shallow
FZ Freezing	BC Patches
BL Blowing	PR Partial

Precipitation types:

RA Rain	GR Hail (1/4" increments)
DZ Drizzle	GS Small hail/snow pellets
SN Snow	PE Ice Pellets
SG Snow grains	IC Ice Crystals
	UP Unknown Precipitation (Automated stations only)

Obscurations to visibility:

FG Fog (vsby<5/8SM)	PY Spray
BR Mist (vsby 5/8 to 6SM)	SA Sand
FU Smoke	DU Widespread Dust
HZ Haze	VA Volcanic Ash

Other Phenomena:

SQ Squalls	SS Sandstorm
DS Duststorm	PO Dust/Sand swirls
FC Funnel cloud	+FC Tornado or Waterspout

7 Sky Condition

Amount of sky cover:
SKC Clear (no clouds)
FEW (Less than 1/8 to 2/8 sky cover)
SCT Scattered (3/8 to 4/8 sky cover)
BKN Broken (5/8 to 7/8 sky cover)
OVC Overcast (8/8 sky cover)

Height: three digits in hundreds of feet AGL

Type: towering cumulus (TCU) or cumulonimbus (CB) clouds reported after the height of their base.
Vertical visibility (VV): height into a total obscuration in hundreds of feet.

8 Temperature/Dew Point

Degrees Celsius, two-digit form. Prefixed "M" = minus (below zero)

9 Altimeter

Inches of mercury, prefixed by "A".

10 Remarks

Prefixed by "RMK"
A01 or A02 = Station type (automated)
PK WND 28045/30 = Peak wind from 280°, 45 knots, at 30 minutes past the hour.
WSHFT 30 FROPA = Wind shift accompanied by frontal passage beginning at 30 minutes after the hour.
B (Began) or E (Ended) followed by time (minutes after the hour).
TS SE MOV NE = Thunderstorms southeast moving northeast
SLP134 = Sea level pressure in hectoPascals (1013.4 hPa)

DECODED REPORT: Routine observation for Tampa, FL, on the 12th day of the month at 2150 UTC. Automated Station. Wind from 080° at 20 knots with gusts to 38 knots. Prevailing visibility 1/2 statute mile, runway 36 Left visual range 2,400 feet. Thunderstorm with heavy rain. Scattered clouds at 800 feet AGL, overcast cumulonimbus clouds with bases of 1,200 feet AGL. Temperature 20°C, dewpoint 18°C. Altimeter setting 29.95 inches of mercury. Remarks: Automated station, precipitation discriminator indicated by A02, thunderstorm began 24 minutes past the hour, rain began 24 minutes past the hour, sea level pressure 1013.4 hectoPascals (Note: 1 hPa = 1 millibar).

Figure 6-19. A METAR usually contains most of these elements. An item may be omitted if it is not occurring at the observation time or is not pertinent. Visibility and sky condition are especially important for deciding if a VFR or IFR flight plan is appropriate.

If the wind direction varies 60° or more, the variability follows the wind group, with the extremes in direction separated by a "V." For example, 190V260 indicates the wind direction varies from 190° to 260°.

Some typical examples of weather and descriptors are: +SN for heavy snow, -RA FG for light rain and fog, BR HZ for mist and haze (visibility >5/8 SM), FZDZ for freezing drizzle, BLDU for blowing dust, and VCSHRA for a rain shower in the vicinity.

Sky cover reported as SCT025TCU BKN080 is interpreted as scattered towering cumulus clouds with bases at 2,500 feet AGL, and broken clouds at 8,000 feet.

Fog obscuring the sky might be reported as 1/8SM FG VV006. The horizontal visibility is one eighth mile in fog, and vertical visibility is 600 feet. On automated reports, visibility less than 1/4 statute mile is reported as M1/4.

A temperature of eight degrees below zero Celsius is reported as M08.

A METAR containing the RMK RAE42SNB42 means that rain ended at 42 minutes past the hour and snow began at that time.

The next section provides you with weather and obstructions to vision, allowing you to further evaluate the weather at the airport. For example, if a thunderstorm is in progress, the cloud cover and visibility may be only a local problem. If fog is indicated but temperatures are forecast to rise, you can expect visibilities to improve. Intensity levels are shown with a "-" for light, a "+" for heavy, and no sign indicates moderate intensity. The letters "VC" are used for weather in the vicinity, that is, between 5 and 10 miles of the airport. You should realize that a thunderstorm in the vicinity could possibly be over the airport at the time of your arrival.

Either an obscuration or the lowest layer of clouds classified as broken or overcast constitutes a ceiling.

The other key element for determining if conditions are VFR or IFR is the sky condition. Although not specifically designated in a METAR, a ceiling is the AGL height of the lowest layer of clouds that is reported as broken or overcast or the vertical visibility into an obscuration, such as fog or blowing snow. Since a broken layer covers 5/8 to 7/8 of the sky, a ceiling exists whenever more than half of the sky is covered.

When conditions are below VFR, you will need to check the published IFR takeoff, landing, and alternate minimums. For example, if the published takeoff minimums are "600-2," the ceiling must be at least 600 feet AGL and visibility must be at least two statute miles. A reported sky condition of "002 SCT 006 BKN" with a visibility of "3SM" is above minimums, since the lowest ceiling is 600 feet and the visibility is greater than two miles. The lower scattered layer at 200 feet does not constitute a ceiling and is, therefore, not restrictive to an IFR departure.

The temperature and dewpoint are reported in Celsius. When the temperature/dewpoint spread is 2°C or less and fog has not yet developed, it should alert you to seek additional information. You would want to know if the temperature will increase or decrease, or if wind would prevent the formation of fog.

The altimeter setting in inches of mercury can provide another clue to the weather conditions. A low setting indicates a low pressure system, and

The remark, T00081016, is temperature and dewpoint in tenths °C. The first zero indicates a plus value for temperature (+.8 °C) and the first one indicates a minus value for dewpoint (-1.6 °C).

Coded Data	Explanations
A02	Automated station with precipitation discriminator
PK WND 20032/25	Peak wind from 200° at 32 knots, 25 minutes past the hour
VIS 3/4V1 1/2	Prevailing visibility variable 3/4 to 1 and 1/2 miles
FRQ LTG NE	Frequent lightning to the northeast
FZDZB45	Freezing drizzle began at 45 past the hour
RAE42SNB42	Rain ended and snow began at 42 past the hour
PRESFR	Pressure falling rapidly
SLP045	Sea level pressure in hectoPascals and tenths, 1004.5 hPa
T00081016	Temperature/dewpoint in tenths °C, .8 °C/–1.6 °C

Figure 6-20. Examples of coded remarks are shown on the left, followed by the explanations.

you should be aware of the associated weather. A rising or falling altimeter in recent hours is another indicator of approaching weather.

Weather events which are considered significant to aircraft operations are included in the remarks section. Beginning and ending times of precipitation or thunderstorms, as well as other potentially hazardous weather phenomena, are typically printed in the remarks. In general, more remarks are included when the weather is bad and the airport is approved for IFR operations. [Figure 6-20]

ADDITIONAL REPORTS

Other reports or messages may be available to help you form a more complete picture of the weather conditions. Examples are freezing level reports, notices to airmen, pilot reports, and radar weather reports.

In addition to making surface observations, certain stations also record freezing level data. The report is identified by the letters "**RADAT**" (freezing level data) which is followed by a group of coded data.

RADATs contain information on the highest relative humidity at any freezing level (if more than one is present) and the MSL heights of as many as three freezing levels. Multiple freezing levels can occur with inversions and frontal activity, although one is most common.

For example, "RADAT 90045," means the relative humidity was 90% and the only crossing of the freezing level (0° isotherm) was at 4,500 feet MSL. "RADAT 87L024105" is an example with two freezing levels. It means the temperature was 0°C or less below 2,400 feet MSL, above 0°C from 2,400 to 10,500 feet MSL, and below 0°C again above 10,500 feet MSL. The "L" indicates that the relative humidity of 87% refers to the lower freezing level. The letters "M" or "H" are used to indicate relative humidity at the middle or upper levels, respectively. RADAT reports are usually issued twice each day. Freezing level information also is available on graphic weather charts.

Notices to Airmen (NOTAMs) are normally transmitted as separate lines of information. Pilot weather reports (PIREPs) also may be transmitted separately. PIREPs can help you determine cloud tops, actual icing conditions, and turbulence levels which have been noted by other pilots. They are your best source for observed weather between reporting points. When you encounter unexpected weather conditions, you are encouraged to make a pilot report. If the ceiling is at or below 5,000 feet, or visibility is at or below five miles, air traffic control facilities are required to solicit PIREPs. If you make a PIREP, the ATC facility or the FSS can add your report to the distribution system, and it can be used to brief other pilots before flight or to provide in-flight advisories. [Figure 6-21]

ELEMENTS	EXPLANATIONS
UA or UUA — Type of Report	UA is routine PIREP; UUA is urgent PIREP
/OV — Location	In relation to VOR or route segment (station identifier, radial, DME)
/TM — Time	Coordinated Universal Time (UTC)
/FL — Altitude	Above mean sea level (MSL)
/TP — Type of aircraft	Example, C172
/SK — Sky cover	Cloud height and coverage (sky clear, few, scattered, broken, or overcast)
/WX — Weather	Precipitation, visibility, restrictions to vision
/TA — Temperature	Degrees Celsius
/WV — Wind	Direction in degrees true north and speed in knots
/TB — Turbulence	Light, moderate, severe, as appropriate
/IC — Icing	Trace, light, moderate, severe, as appropriate
/RM — Remarks	To clarify the report or for additional information

CODED PIREP:

UA/OV OKC 063064/TM 1522/FL 080/TP C172/SK 020 BKN 045/[illegible] OVC 070/TA -04/WV 245040/TB LGT/RM IN CLR

DECODED PIREP:

Routine pilot report . . . 64 n.m. on the 63° radial from Oklahoma City VOR . . . at 1522 UTC . . . flight altitude 8,000 ft . . . type of aircraft is a Cessna 172 . . . base of broken layer at 2,000 feet with tops at 4,500 feet. Base of overcast layer at 6,000 feet tops at 7,000 feet . . . outside air temperature is minus four degrees Celsius . . . wind is from 245° at 40 kts . . . light turbulence and clear skies.

Figure 6-21. PIREPs are made up of several elements, as indicated in the upper portion of this figure. The lower part provides a sample PIREP and its plain-language interpretation. Notice that MSL altitudes are used. Although PIREPs should be complete and concise, you should not be overly concerned with strict format or terminology. The important thing is to forward the report so other pilots can benefit from it.

You can determine the thickness of cloud layers by using a METAR and a PIREP which includes cloud tops.

Bases and tops of clouds may be included in pilot reports, with the heights shown in feet MSL. This information, in conjunction with the station ceiling and field elevation, can be used in some cases to determine the thickness of cloud layers. Assume you see the following PIREP, ". . . UA . . . /SK OVC 045," which means a pilot reported the top of the overcast as 4,500 feet MSL. If the field elevation is 1,500 feet MSL and the ceiling is reported as 500 feet, the cloud layer may be 2,500 feet thick (4,500 – 2,000 = 2,500). Notice that you first need to convert the ceiling from an AGL value to an MSL value in order to solve the problem.

RADAR WEATHER REPORTS

Radar weather reports describe areas of precipitation, along with information on the type, intensity, and trend.

General areas of precipitation, especially thunderstorms, are observed by radar on a routine basis. Most radar stations issue **radar weather reports** (SDs) each hour at 35 minutes after the hour, with intervening special reports as required. These reports are routinely transmitted on weather service circuits, and some are included in FSS weather broadcasts. You can use these reports to determine the location of precipitation, its type, severity, trend, and direction of movement. [Figure 6-22]

1: OKC 1934 · 2: LN · 3: 8 · 4: TRW ++/+ · 5: 86/40 164/60 199/115 · 6: 15W · 7: C2425

8: MT 570 AT 159/65 · 9: 2 INCH HAIL RPRTD THIS ECHO

1. **Location identifier and time of radar observation (UTC)**
2. **Echo pattern** — Line in this example
3. **Coverage in tenths** — 8/10 in this example
4. **Type, intensity, and trend of weather** — Thunderstorm (T), very heavy rainshowers (RW++), increasing in intensity (/+)
5. **Azimuth (reference true N) and range in nautical miles (n.m.) of points defining the echo pattern**
6. **Dimensions of echo pattern** — 15 n.m. wide
7. **Pattern movement** — From 240° at 25 knots; "C" indicates movement of individual storms or "cells"
8. **Maximum tops and location** — 57,000 feet, 159° radial, 65 n.m.

Remarks — Self-explanatory using plain language contractions

ECHO PATTERN		INTENSITY		INTENSITY TREND	
Symbol	**Pattern**	**Symbol**	**Intensity**	**Symbol**	**Trend**
CELL	Single Cell	--	Light	+	Increasing
LN	Line	(NONE)	Moderate		
FINE LN	Fine line	+	Heavy	-	Decreasing
AREA	area	++	Very heavy		
SPRL BAND	Spiral band	X	Intense	NC	No change
AREA	area	XX	Extreme		
		U	Unknown	NEW	New echo

Figure 6-22. Much of the symbology used in SDs is explained in this sample report for Oklahoma City, Oklahoma.

TERMINAL AERODROME FORECASTS

A terminal aerodrome forecast (TAF) is an essential report for any VFR flight and an indispensable one for an instrument flight. You always want to know what weather to expect upon arrival at your destination. A TAF allows you to select the most favorable approach, based on the forecast ceiling, visibility, and winds. You also need this forecast to determine if

The terminal aerodrome forecast is the primary source of weather information for the destination airport at your ETA.

an alternate airport is required. If one is required, you need the TAF at the alternate to see if it qualifies. [Figure 6-23]

TAFs generally are issued four times a day, each valid for a 24-hour period, beginning at 0000Z, 0600Z, 1200Z, and 1800Z. Each forecast is valid for a 24-hour period. The report header will indicate either TAF for a routine forecast, or TAF AMD for an amended forecast. A corrected (COR) or delayed (RTD) TAF will be identified only in the communications header preceding the actual forecast.

"P6SM" indicates the visibility is greater than six statute miles.

The basic format of the TAF provides the forecast wind, visibility, weather if significant to aviation, and the sky condition. The codes used in the TAF are nearly all the same as those used for METAR. One of the exceptions is for good visibility. When the visibility is greater than six statute miles, it is reported as "P6SM."

When you evaluate a forecast during your preflight planning, the main body will help you determine whether conditions will be VFR or IFR. In addition, you should check for temporary conditions and forecast change groups around your expected arrival time. Also, pay careful attention to the probability forecasts for thunderstorms or other precipitation.

For example, a forecast for "051212 11010KT P6SM SCT050 FM1800 31010G20KT 3SM BKN015 PROB40 1921 1SM TSRA" would require careful evaluation. On the fifth day of the month, beginning at 1200Z, the weather is forecast to be VFR. However, beginning at 1800Z (FM1800), the wind will shift, visibility will drop, and the ceiling will be 1500 feet AGL. Another important factor is the 40% probability of a thunderstorm with rain between the hours of 1900 and 2100Z. This information will help you decide what type of flight plan you will need to file and whether an alternate is required. You may also decide to change your time of arrival to avoid deteriorating weather conditions.

AREA FORECASTS

Aviation area forecasts (FAs) are issued for six regions in the contiguous United States three times each day by the NAWAU in Kansas City, MO. FAs for Alaska are issued by WSFOs, located in Anchorage, Fairbanks, and Juneau, for their respective areas three times daily. The WSFO in Honolulu issues FAs four times a day for the Hawaiian Islands. In addition, a specialized FA for the Gulf of Mexico is issued twice a day by the National Hurricane Center in Miami. This area includes the coastal plains and coastal waters (approximately 100 to 180 miles from the coast) extending from the

```
1 TAF
2 KOKC 051130Z 051212 14008KT 5SM BR BKN030 TEMPO 1316 1 1/2SM BR
     FM1600 16010KT P6SM SKC BECMG 2224 20013G20KT 4SM SHRA OVC020
     PROB40 0006 2SM TSRA OVC008CB BECMG 0608 21015KT P6SM NSW=
```

(3 = 051130Z; 4 = 051212; 5 = 14008KT; 6 = 5SM; 7 = BR; 8 = BKN030; 9 = TEMPO 1316 1 1/2SM BR; 10 = FM1600 16010KT P6SM SKC BECMG 2224 20013G20KT 4SM SHRA OVC020; 11 = PROB40 0006 2SM TSRA OVC008CB; 12 = =)

1 **Type of Report** – Routine terminal aerodrome forecast (TAF)

2 **Location Identifier** – ICAO four-letter identifier

3 **Issuance Data and Time** – Two-digit day of the month and the time in Zulu (Z)

4 **Valid Period** – TAF valid time, two-digit date followed by the beginning and ending hours in Zulu

Forecast Elements: Wind/Visibility/Weather/Sky Condition

5 **Wind** – Three-digit direction, two-or three-digit speed in knots (KT)

6 **Visibility** – Expected prevailing visibility in statute miles. Forecast visibility greater than 6 miles is coded as P6SM (plus 6 SM).

7 **Weather** – Significant weather and obstructions to visibility

8 **Sky Condition** – Amount, height (CB) or vertical visibility
Cumulonimbus (CB) is the only type of cloud included in the TAF
North America TAFs will also include nonconvective low-level wind shear (WS), up to 2000 feet. When forecast, WS will appear after the cloud forecast group

9 **Temporary Conditions** – (TEMPO) Fluctuations expected to last less than an hour in each instance and, in aggregate, to cover less than half of the period, two-digit beginning time, in hours, and two-digit ending time, also in hours

10 **Forecast Change Group** – Significant, permanent changes in conditions; from (FM), four-digit hour and minute time the change is expected to begin; or becoming (BECMG) with a four-digit time group to indicate beginning hour and ending hour times

11 **Probability Forecast** – Probability of thunderstorms or precipitation (in percent), along with associated weather conditions (wind, visibility, and sky condition); the four-digit beginning hour and ending hour follows the abbreviation PROB40

12 **End of Report Separator** – (=)

DECODED REPORT:
Routine TAF for Oklahoma City, OK...on the 5th day of the month, at 1130Z...valid for 24 hours from 1200Z on the 5th to 1200Z on the 6th...Wind from 140° at 8 knots...visibility 5 statute miles in mist...broken clouds at 3,000 feet...Temporary condition, between 1300Z and 1600Z, 1 1/2 statue miles visibility in mist...From 1600Z, wind from 160° at 10 knots...visibility greater than 6 miles and sky clear...Becoming, between 2200Z and 2400Z, wind from 200° at 13 knots, gusting to 20 knots, visibility 4 miles with moderate rain showers, sky overcast at 2000 feet...Between midnight (0000Z) and 0600Z, a 40% probability of 2 statute miles visibility with thunderstorms and moderate rain, sky overcast with the bases of cumulonimbus clouds at 800 feet AGL...Becoming, 0600Z to 0800Z, wind 210° at 15 knots, visibility greater than 6 statute miles and no significant weather.

Figure 6-23. Codes used in a TAF are nearly identical to those shown for METARs. Some exceptions are for visibility greater than six miles (P6SM), and the use of Probability (PROB), Becoming (BECMG), and Temporary (TEMPO).

A calm wind is indicated on a TAF by "00000KT." This specifically implies the wind is expected to be three knots or less.

The code "WS005/27050KT" on a TAF indicates a wind shear at 500 feet AGL with forecast winds from 270° at 50 knots. When forecast, the wind shear follows the sky condition group.

Weather conditions expected to occur in the vicinity of an airport are denoted by the letters "VC." This implies within geographical proximity of between a five and ten statute mile radius of the airport, but not at the airport itself.

western Florida panhandle to Brownsville, Texas. The Gulf FA contains both marine and aviation information, and is intended to support offshore helicopter operations.

Use the FA to determine forecast weather between reporting stations and at airports where TAFs are not issued.

The area forecast is used to determine enroute weather and is your principal source of weather at airports that do not have terminal forecasts. This forecast is a good single reference for information on frontal movement, turbulence, icing, and other flight hazards for a specific area. Following the communications and product headers are three precautionary statements. The main body of the FA is divided into two main sections, "Synopsis" and " VFR Clouds and Weather."

The synopsis is an 18-hour forecast of the location and movement of frontal systems and circulation patterns. References to low ceilings and/or visibilities, strong winds, or other phenomena may also be included. The VFR Clouds/Weather section is a forecast of clouds and weather which cover an area greater than 3,000 square miles and is significant to VFR flight operations. It contains a specific 12-hour forecast followed by a 6-hour (18-hour in Alaska) outlook, for a total period of 18 hours (30 hours in Alaska). Location identifiers (airports and VORs) are used to the extent possible to outline affected areas. Two letter identifiers are used for states, and initials are used for the Great Lakes, such as LS for Lake Superior and LH for Lake Huron. Clearly defined geographical areas may also be used where helpful. FAs are amended or corrected when there is a significant change, and only the affected section needs to be amended. AIRMETs and SIGMETs also serve to amend the FA. An amended or corrected FA is identified by "AMD" or "COR," as applicable. [Figure 6-24]

WINDS AND TEMPERATURES ALOFT FORECASTS

Winds aloft are forecast in true direction and in knots.

A winds and temperatures aloft forecast (FD) provides an estimate of wind direction in relation to true north, wind speed in knots, and the temperature in degrees Celsius for selected stations and altitudes. These forecasts are important preflight considerations. For example, assume you want to know how much the forecast temperature deviates from that in the standard atmosphere (ISA) at your planned cruising altitude. You may need this to determine an accurate true airspeed from the pilot's operating handbook. First, determine the standard temperature for the selected altitude by using a flight computer, and then subtract the corresponding forecast temperature. This will give you a value that can be expressed as plus or minus ISA, such as +10 ISA, which means the temperature is forecast to be 10°C warmer than standard for that particular altitude. You

```
1   SLCC FA 141045
    SYNOPSIS AND VFR CLDS/WX
    SYNOPSIS VALID UNTIL 150500
    CLDS/WX VALID UNTIL 142300...OTLK VALID 142300-150500
    ID MT NV UT WY CO AZ NM

2   SEE AIRMET SIERRA FOR IFR CONDS AND MTN OBSCN.
    TSTMS IMPLY PSBL SVR OR GTR TURBC SVR ICG LLWS
    AND IFR CONDS.

    NON MSL HGTS ARE DENOTED BY AGL OR CIG.

3   SYNOPSIS...HIGH PRES OVER NERN MT CONTG EWD
    GRDLY. LOW PRES OVR AZ NM AND WRN TX RMNG
    GENLY STNRY. ALF...TROF EXTDS FROM WRN MT INTO
    SRN AZ RMNG STNRY..

4   .
    ID MT
    FROM YXH TO SHR TO 30SE BZN TO 60SE PIH TO LKT TO
    YXC TO YXH.
    70-90 SCT-BKN 120-150. WDLY SCT RW-. TOPS SHWRS 180.
    OTLK...VFR
    RMNDR AREA...100-120. ISOLD RW- MNLY ERN PTNS AREA.
    OTLK...VFR

    .
    UT NV NM AZ
    80 SCT-BKN 150-200. WDLY SCT RW-/TRW-. CB TOPS 450.
    OTLK...VFR

    .
    WY CO
    FROM BZN TO GCC TO LBL TO DVC TO RKS TO BZN.
    70-90 BKN-OVC 200. OCNL VSBY 3R-F. AFT 20Z WDLY SCT
    TRW-. CB TOPS 450. OTLK...MVFR CIG RW.
```

1 Heading Section
2 Precautionary Statements
3 Synopsis
4 VFR Clouds and Weather

The categorical outlook, identified by "OTLK," is included in each area breakdown. An outlook of IFR or MVFR can be due to ceilings only (CIG), restrictions to visibility TRW, F, etc.), or a combination of both. The abbreviation, "WND," is included in the outlook if winds are expected to be 20 knots or more.

Figure 6-24. This FA was issued for the Salt Lake City region. It was published on the 14th day of the month at 1045Z. The Synopsis is valid for 18 hours, until 0500Z on the 15th. The Clouds/Weather section covers a 12-hour period and the Outlook is for an additional six hours. The precautionary statements are standard for all FAs, and refer you to Airmet Sierra for details on IFR conditions and a mountain obscuration. The Synopsis is a forecast of general weather systems. The VFR Clouds and Weather section is a specific forecast by states or geographical areas.

FT	3000	6000	9000	12000	18000	24000
DEN			2315+04	2519+00	2519-16	2523-27
HOU	2619	2728-01	2635-05	2641-09	2654-21	2665-32
JOT	2526	2736-02	2642-06	2649-11	2662-22	2674-33

Figure 6-25. Depending on station elevation, FDs normally forecast winds for nine levels, from 3,000 through 39,000 feet. An FD does not forecast winds for levels within 1,500 feet of the station elevation, and temperatures are not forecast for the 3,000-foot level or for a level within 2,500 feet of the station elevation. At Denver (DEN), for example, the station elevation is over 5,000 feet MSL, so the lower two levels are omitted.

can also determine the approximate freezing level by noting the temperatures forecast for the various altitudes and applying the average temperature lapse rate of 2°C per 1,000 feet. [Figure 6-25]

When wind speed is 100 knots or more, 50 is added to the direction code and 100 subtracted from the speed.

The wind information included in the body of the FD is presented in a format similar to other reports. The first two numbers indicate the true direction from which the wind is blowing. For example, "2736–02" indicates the wind is from 270° at 36 knots, and the temperature is –2°C. Wind speeds between 100 and 199 knots are shown with 50 added to the wind direction code and 100 subtracted from the speed. For example, "7701" shows a wind from 270° (77 – 50 = 27 or 270°) at 101 knots (01 + 100 = 101). Speeds over 199 knots are coded as 199. A code of "9900" indicates light and variable winds (less than five knots). Above 24,000 feet, all temperatures are below 0°C, and the minus sign is omitted. As you know, it is best to interpolate when your planned flight altitude is between forecast levels.

SEVERE WEATHER REPORTS AND FORECASTS

Although most weather gathering activity is concerned with routine reports and forecasts, considerable effort is devoted to monitoring and reporting severe weather conditions. Except for the Pacific area, the National Hurricane Center in Miami, FL, issues hurricane advisories, and the National Severe Storms Forecast Center in Kansas City, MO, issues special forecasts and reports for severe weather conditions. These include convective outlook forecasts and severe weather watch bulletins.

Convective outlooks forecast general as well as severe thunderstorm activity for the next 24-hour period.

The **convective outlook** (AC) forecasts general thunderstorm activity for the next 24-hour period. Areas with a high, moderate, or slight risk of severe thunderstorms are included, as well as areas where thunderstorms may approach severe limits. "Approaching severe limits" means winds are greater than or equal to 35 knots but less than 50 knots, and/or hail is greater than or equal to one-half inch in diameter. ACs are used by meteorologists to prepare graphic charts. You may use the data for planning flights within the forecast period.

A **severe weather watch bulletin** (WW) is a weather report that defines areas of possible severe thunderstorms or tornadoes. WWs are issued on an unscheduled basis and are updated as required. Satellite photos have improved the accuracy of forecasting severe weather phenomena. Normally, you will know about severe weather conditions well in advance of the actual occurrence. Since severe weather forecasts and reports may affect the general public as well as pilots, they are widely disseminated through all available media.

WWs are issued on an unscheduled basis and updated as required.

IN-FLIGHT WEATHER SERVICES

As you already know, forecasting is still an inexact science, and weather conditions can change rapidly in the course of a few hours. During flight, you need to update your weather information on a continual basis. You may do this by contacting an FSS. In other situations, you can receive in-flight weather broadcasts by monitoring appropriate FSS radio frequencies. TWEBs are also available from selected NDBs and VORs. TWEB broadcasts normally provide route-oriented weather information, although the format can vary somewhat, depending on location.

TWEB route-oriented data includes forecast sky cover, cloud tops, visibility, and obstructions to vision.

Enroute flight advisory service (EFAS) is an FSS service specifically designed to provide timely weather information when pilots request it. You can usually contact an EFAS specialist on 122.0 MHz from 6 A.M. to 10 P.M. anywhere in the conterminous U.S. at a minimum altitude of 5,000 feet AGL. At 18,000 feet MSL and above, different frequencies are allocated for each ARTCC area. You should call the FSS using the name of the ARTCC followed by the term "**Flight Watch**," your aircraft identification, and the nearest VOR. The A/FD contains charts depicting the locations of flight watch control stations (parent facilities) and the outlets they use. These charts also list the high altitude EFAS frequencies for the respective centers. EFAS frequencies are not intended to be used for flight plan filing, position reports, or complete weather briefings.

The frequency for EFAS below FL180 is 122.0 MHz. Pilots requesting information beyond the scope of flight watch will be advised to use other appropriate FSS frequencies.

The weather advisories you receive are tailored to the type of flight you are conducting and are appropriate to your route and cruising altitude. In addition, EFAS is a central collection and distribution point for pilot reports (PIREPs), so pilot participation is essential for success of the program. You normally receive very current or real-time weather information. This includes any thunderstorm activity along your route.

EFAS is designed to provide weather information for the cruise portion of a flight. Active pilot participation is necessary for the program to be successful.

In-flight advisories are issued to notify you of the possibility of encountering hazardous flying conditions which may not have been forecast at the time of your preflight briefing. Whether or not the conditions may be hazardous to your flight is something that you must evaluate on the basis of your experience and the operational limitations of your airplane. The first two advisories discussed are AIRMETs and SIGMETs. AIRMET is an acronym for "airman's meteorological information," and SIGMET means "significant meteorological information." Remember that AIRMETs and SIGMETs, along with PIREPs, provide the most accurate information on icing conditions (current and forecast).

You will find the most accurate information on icing conditions in the AIRMETs, SIGMETs, and PIREPs.

AIRMETs are issued on a scheduled basis every six hours.

AIRMETs (WAs) are issued on a scheduled basis every six hours for weather phenomena which are of operational interest to all aircraft. These weather conditions are potentially hazardous to aircraft having limited capability because of lack of equipment, instrumentation, or pilot qualifications. If these conditions were adequately forecast in the FA, an AIRMET will not be issued. AIRMETs are issued for the following conditions: moderate icing, moderate turbulence, sustained winds of 30 knots or more at the surface, widespread areas of ceilings less than 1,000 feet and/or visibility of less than three miles, and extensive mountain obscurement.

SIGMETs warn of weather hazards which concern all aircraft.

SIGMETs (WSs) are issued for hazardous weather (other than convective activity) which is considered significant to all aircraft. SIGMET criteria include the following: severe or extreme turbulence, clear air turbulence or severe icing, none of which are associated with thunderstorms; duststorms, sandstorms, or volcanic ash lowering surface and/or in-flight visibilities to less than three miles; and volcanic eruptions. SIGMETs are issued whether or not the conditions were included in the area forecast. [Figure 6-26]

Convective SIGMETs include tornadoes, thunderstorms, and hail.

Convective SIGMETs (WSTs) are issued for hazardous convective weather (existing or forecast) which is significant to the safety of all aircraft. They always imply severe or greater turbulence, severe icing, and low-level wind shear, so these items are not specified in the advisory. WSTs are unscheduled and may be issued for any convective situation which the forecaster considers hazardous to all categories of aircraft. They include any of the following phenomena: tornadoes, lines of thunderstorms, thunderstorms over a wide area, hail greater than or equal to three-fourths of an inch in diameter, and/or wind gusts to 50 knots or greater. Convective SIGMETs are issued for the Eastern (E), Central (C), and Western (W) United States. Individual convective SIGMETs are numbered sequentially for each area (01-99) each day. [Figure 6-27]

SIGMET	AIRMET
DFWP **UWS** 051710 SIGMET **PAPA** 1 VALID UNTIL 052110 AR LA MS FROM MEM TO 30N MEI TO BTR TO MLU TO MEM OCNL SVR ICING ABV FRZLVL EXPCD. FRZLVL 080 E TO 120 W. CONDS CONTG BYD 2100Z.	MIAT WA 151900 AMD AIRMET **TANGO** UPDT 2 FOR TURBC VALID UNTIL 160100. AIRMET TURBC...GA FL FROM SAV TO JAX TO CTY TO TLH TO SAV MDT TURBC BLO 100 EXPCD. CONDS IPVG AFT 160000Z.

Figure 6-26. The first issuance of a SIGMET, as shown on the left, is labeled UWS (Urgent Weather SIGMET). PAPA 1 means it is the first issuance for a SIGMET phenomenon; PAPA 2 is the second issuance for the same phenomenon. In the example on the right, AMD means this is an amended AIRMET for the phenomenon (moderate turbulence) identified as TANGO. The alphanumeric designator stays with the phenomenon even when it moves across the country. Note, the maximum forecast period is four hours for a SIGMET and six hours for an AIRMET.

Consecutive Convective SIGMETs	
MKCC WST 221655 CONVECTIVE SIGMET 17C KS OK TX VCNTY GLD—CDS LINE NO SGFNT TSTMS RPRTD FCST TO 1855Z LINE TSTMS DVLPG BY 1755Z WILL MOV EWD 30-35 KT THRU 1855Z HAIL TO 1 1/2 IN PSBL.	MKCC WST 221655 CONVECTIVE SIGMET 18C SD NE IA FROM FSD TO DSM TO GRI TO BFF TO FSD AREA TSTMS WITH FEW EMBDD CELLS MOVG FROM 2725 TOPS 300 FCST TO 1855Z DSPTG AREA WILL MOV EWD 25 KT.

Figure 6-27. "WST" in the header identifies these reports as convective SIGMETs. The designators 17C and 18C indicate they are consecutive issuances for the central U.S. One forecasts a line of thunderstorms with possible hail, while the other forecasts embedded thunderstorms over a large area.

As mentioned earlier, a **severe weather watch bulletin** (WW) is a weather report that defines areas of possible severe thunderstorms or tornadoes. In order to alert forecasters, briefers, and pilots that a severe weather watch bulletin is being prepared, a preliminary message called a **severe weather forecast alert** (AWW) is broadcast before the main bulletin is issued. These preliminary messages define areas of possible severe thunderstorms or tornado activity. [Figure 6-28]

A **center weather advisory** (CWA) is another unscheduled weather advisory which is issued by an ARTCC to alert you to existing or anticipated adverse weather conditions within the next two hours. A CWA may supplement an existing in-flight advisory and is intended primarily for IFR traffic. It may also alert you when a SIGMET has not been issued, but conditions meet those criteria based on current pilot reports. CWAs are one of the most current sources for existing or forecast hazardous weather.

Air Route Traffic Control Centers broadcast AIRMETs, SIGMETs, Convective SIGMETs, AWWs, and CWAs once on all frequencies, except emergency, when any part of the area described is within 150 miles of the airspace under their jurisdiction. These broadcasts contain SIGMET or CWA identification and a brief description of the weather activity and general area affected.

Severe Weather Forecast Alert
MKC AWW 241817 WW 256 TORNADO NE 241900Z - 250000Z AXIS. . 70 STATUTE MILES EAST AND WEST OF LINE. . 55SSW EAR/KEARNEY NE/ — 65NNW OFK/NORFOLK NE/ HAIL SURFACE AND ALOFT. . 2 1/2 INCHES. WIND GUSTS. . 70 KT. MAX TOPS TO 550. MEAN WIND VECTOR 260/35.

Figure 6-28. "AWW" in the header identifies this report as a severe weather forecast alert. This message warns of possible tornado activity in Nebraska and defines the watch area. Hail and high winds also are possible. A detailed severe weather watch bulletin (WW) immediately follows the alert message.

Beginning with the second hour, a summarized alert notice is broadcast. It simply states that the particular advisory is current and briefly describes the weather phenomenon and the area affected. If you hear an alert notice and have not received the advisory, you should contact the nearest FSS for more detailed information, since it might be pertinent to your flight. Affected ARTCCs and terminal control facilities also make a single broadcast on all normal frequencies when a SIGMET, AWW, or CWA is issued. In terminal areas, local control and approach control may limit these broadcasts to weather occurring within 50 n.m. of the airspace under their jurisdiction.

HAZARDOUS IN-FLIGHT WEATHER ADVISORY SERVICE

HIWAS continuously broadcasts summaries of in-flight advisories, including AIRMETs, SIGMETs, convective SIGMETs, AWWs, and CWAs.

A program for broadcasting hazardous weather information on a continuous basis over selected VORs is called **hazardous in-flight weather advisory service** (HIWAS). The broadcasts include summarized AIRMETs, SIGMETs, convective SIGMETs, AWWs, CWAs, and urgent PIREPs. In areas where HIWAS is implemented, you should be aware that ARTCC, terminal ATC, and FSS facilities have discontinued their normal broadcasts of in-flight advisories. However, FSSs and ARTCCs announce updates to HIWAS information on all normal frequencies. While HIWAS is an additional source of hazardous weather information, it is not a replacement for preflight or in-flight briefings or real-time weather updates from flight watch. VORs used for HIWAS broadcasts can be found in the A/FD. Since HIWAS is a developing program, check the *Aeronautical Information Manual* and NOTAMs publication for the latest information.

AUTOMATED WEATHER OBSERVING SYSTEM

Another source of weather information is an automated unit which can make surface weather observations and transmit them directly to you during flight. This unit is the **automated weather observing system** (AWOS), which uses various sensors, a voice synthesizer, and a radio transmitter to provide real-time weather data. AWOS transmissions are receivable within 25 n.m. of the site, at or above 3,000 feet AGL. In many cases, AWOS signals may be received on the surface of the airport. Most units transmit a 20- to 30-second weather message once each minute. Information on AWOS types and message formats is published in the *Aeronautical Information Manual.* At locations where AWOS is installed, frequencies and telephone numbers are included in the A/FD, and where applicable, on published instrument approach procedures.

AUTOMATED SURFACE OBSERVATION SYSTEM

The primary surface weather observing system in the United States is the **automated surface observation system** (ASOS). It is designed to support aviation operations and weather forecast activities. Using an array of sensors, computers, and digitized voice communications, ASOS provides continuous minute-by-minute observations. These weather reports are broadcast over VHF frequencies, VORs, and/or telephone lines, as indicated in the A/FD. An ASOS report which has not been modified by a weather observer is noted by the word "AUTO." In this case, the report only contains information that can be reported automatically, such as cloud height, visibility, precipitation identification, pressure sensors, temperature/dewpoint, wind direction and speed, and rainfall accumulation. In addition, some automatic ASOS stations can determine the difference between liquid precipitation and frozen or freezing precipitation. The stations where this distinction can be made are identified with an "A02." An "A01" station does not have this capability.

CHECKLIST

After studying this section, you should have a basic understanding of:

✓ **Aviation routine weather reports (METARs)** — What the basic types are, the main information elements they provide, and how they are interpreted.

✓ **Pilot reports (PIREPs)** — What they are and the types of information typically included.

✓ **Radar weather reports (SDs)** — What type, intensity, and trend information they provide concerning areas of precipitation.

✓ **Terminal aerodrome forecasts (TAFs)** — What they are, where they apply, and what information they contain.

✓ **Area forecasts (FAs)** — What information the sections contain and how to interpret the information.

✓ **Winds and temperatures aloft forecasts (FDs)** — What the basic format is and how to interpret it.

✓ **Severe weather reports and forecasts** — What their significance is and what types are available.

✓ **In-flight weather services** — What the various sources of in-flight weather are, including EFAS, and how to contact them.

✓ **In-flight advisories** — What the criteria are for issuance of AIRMETs, SIGMETs, convective SIGMETs, AWWs, and CWAs and how to receive them in flight.

✓ **HIWAS, AWOS, and ASOS** — What these services are and how to access them for information.

SECTION

D

GRAPHIC WEATHER PRODUCTS

Graphic weather products allow you to quickly assess large-scale weather patterns and trends. With weather charts, you can identify areas of IFR weather at a glance and also see where IFR conditions are forecast within the next 12 to 24 hours. This is valuable information for any flight, whether VFR or IFR. This section is designed to review the charts you are already familiar with and introduce you to some you may not have used before.

SURFACE ANALYSIS CHART

The surface analysis chart, also referred to as a surface weather chart, shows weather conditions as they existed at the observation (valid) time shown on the chart. Although it is computer prepared, it takes time to collect and transmit the information, so it normally is at least two hours old by the time you see it. Some items included are actual frontal positions, sea level pressure patterns, highs and lows, temperature and dewpoint, wind direction and speed, local weather, and obstructions to vision at the valid time of the chart. [Figure 6-29]

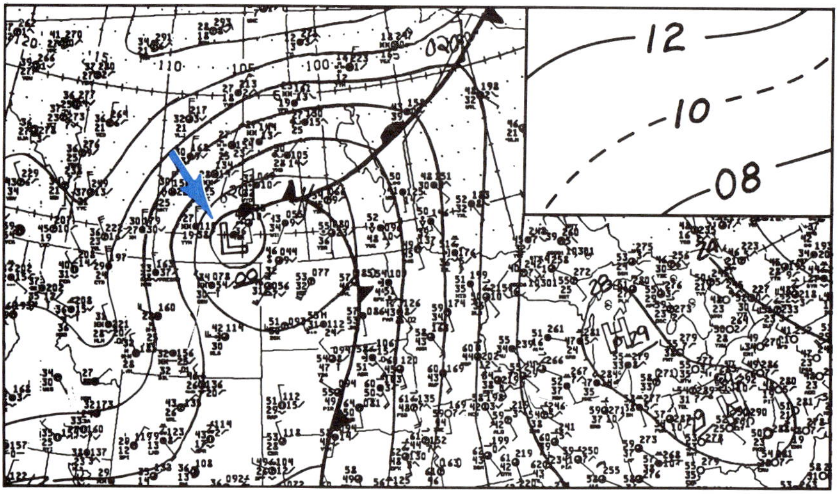

Figure 6-29. This excerpt shows a prominent low pressure center which has a cold front trailing south and a stationary front extending to the northeast. Isobars connecting points of equal pressure are usually drawn at four-millibar intervals. When they are closely spaced, it means a high pressure gradient and strong winds. The inset shows how isobars are numbered; the number "12" means the line connects points where the pressure is 1012 millibars. When the pressure gradient is weak, dashed isobars are sometimes inserted at two-millibar intervals to define the pressure pattern more clearly.

The surface analysis chart provides surface weather observations for a large number of reporting points throughout the United States. Each of these reporting points is represented on the chart by a **station model**. A round symbol represents stations where an observer makes the weather report, and a square symbol is used for automated stations. [Figure 6-30]

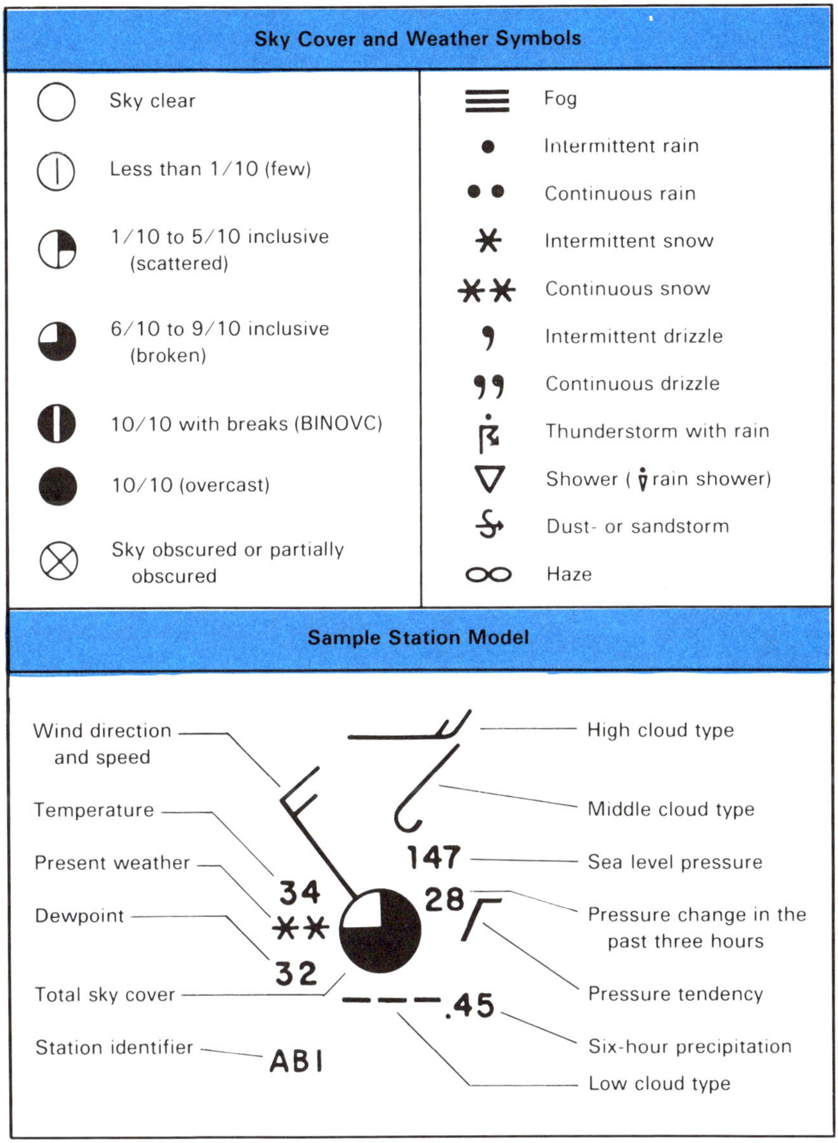

Figure 6-30. The station model uses a number of symbols like those in the upper part of this illustration. Some symbols pertain to sky cover and others represent weather conditions. According to the sample station model, sky cover was broken and continuous snowfall was in progress at the time of the observation. You may need to refer back to this illustration.

Wind direction is shown relative to true north, and wind velocity is given in knots.

Symbols extending out from the station circle give wind information. The symbol shows the general true direction the surface wind is blowing from and gives the velocity in knots. True wind direction is shown by the orientation of the wind pointer. Velocity is indicated by barbs and/or pennants attached to the wind pointer. One short barb is five knots, a longer barb is 10 knots, and a pennant is 50 knots. The wind pointer in the sample station model shows the wind is from the northwest at 15 knots. If there is no wind symbol and a double circle is shown around the station, conditions are calm.

Low cloud symbols are placed below the station model, while middle and high cloud symbols are placed immediately above it. A typical station model may include only one cloud type; seldom are more than two included. Decoding information for cloud types and a variety of other symbols is available in FAA weather publications and at flight service stations.

Temperatures are given in degrees Fahrenheit, while pressure is given in millibars.

On U.S. charts, temperatures are shown in degrees Fahrenheit; pressure is given in millibars. For example, the temperature at the sample station is 34°F, and the atmospheric pressure is 1014.7 millibars. Just below the station pressure, the number "28" indicates the pressure at this station has increased 2.8 millibars in the past three hours and is now steady. The ".45" below the pressure change tells that 45-hundredths of an inch of precipitation fell within the past six hours.

WEATHER DEPICTION CHART

The weather depiction chart provides a graphic display of both VFR and IFR weather areas.

You can think of the weather depiction chart as an abbreviated version of the surface analysis chart. It provides a simplified station model and a graphic depiction of VFR, marginal VFR, and IFR weather conditions. This makes it an excellent source to help you determine general weather conditions when you are flight planning. Information plotted on this chart is derived from surface aviation observations. Unlike the surface chart, pressure patterns and wind information are not provided. [Figure 6-31]

If total sky cover is scattered, the base of the lowest layer of clouds is shown as the cloud height in the station model.

The sky cover symbols used in the station model are nearly the same as those used for the surface analysis chart. The one exception is that automated stations are depicted with a bracket (]) to the right of the station circle. The number directly below the station is the ceiling or cloud height at that station in hundreds of feet (the same as the coded METAR). The sample station model is showing sky obscured, ceiling 400 feet, and visibility of one mile in continuous rain. In this case, the "X" in the sky coverage symbol means a total obscuration. A totally obscured sky always has a height entry for a ceiling, which represents the vertical visibility into the obscuration. A partial obscuration, which uses the same symbol, doesn't have a ceiling entry if there are no clouds above the obscuring phenomenon. If there are clouds, the height of the lowest cloud layer will

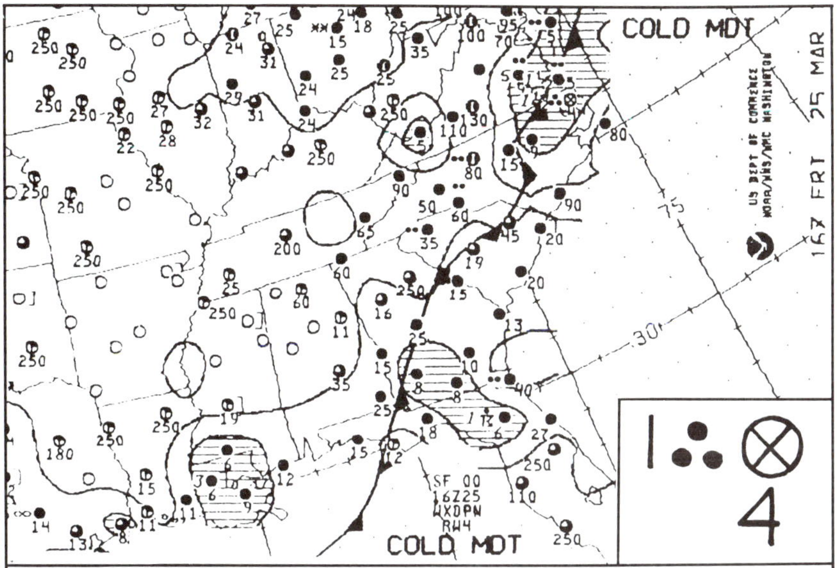

SHADED AREAS — IFR WITH CIG LESS THAN 1000 FT AND/OR VSBY LESS THAN 3 MI

CONTOURED WITHOUT SHADING — MVFR AREAS WITH CIG GREATER THAN OR EQUAL TO 1000 TO LESS THAN OR EQUAL TO 3000 FT AND/OR VSBY GREATER THAN OR EQUAL TO 3 TO LESS THAN OR EQUAL TO 5 MI

NO CONTOURS — VFR AREAS WITH NO CIG OR GREATER THAN 3000 FT AND VSBY GREATER THAN 5 MI

Figure 6-31. The inset at the lower right of this weather depiction chart excerpt shows a sample of the station model; the information at the bottom is the legend. You may want to refer back to this illustration during the following discussion.

be shown. In some cases, it may be necessary to look at a METAR to tell if a height entry is a cloud layer above a partial obscuration or vertical visibility into a total obscuration. Also, when the total sky cover is scattered and there is no obscuration, the cloud height entered is the base of the lowest layer. When the visibility is six miles or less, it is shown to the left of the station in statute miles or fractions of a mile. The reason for the obstruction to vision also may be given. In this case, it is continuous rain of moderate intensity at the time of observation, and it is not freezing.

Areas of IFR are represented by enclosed, shaded areas, meaning the ceiling is less than 1,000 feet AGL and/or the visibility is less than three statute miles. You can determine the reason for IFR conditions by referring to the station models within the IFR areas. Marginal VFR (MVFR) is indicated by enclosed areas without shading. Marginal VFR means the ceiling is 1,000 to 3,000 feet AGL inclusive and/or the visibility is three to five miles. In areas that are not enclosed, there is no ceiling, or the ceiling is greater than 3,000 feet AGL, and the visibility is greater than five miles.

RADAR SUMMARY CHART

The radar summary chart is unique because it shows thunderstorm cells and lines.

The radar summary chart provides a graphic depiction of certain types of weather phenomena which can be detected by special weather radar systems. These systems primarily detect particles of precipitation within a cloud or falling from a cloud. Because of this, the radar scope can show areas of precipitation, individual thunderstorm cells, lines of cells, and areas of thunderstorm activity. Computers prepare radar summary charts from these radar observations. The chart shows size, shape, and intensity of returns, as well as the intensity trend and direction of movement. Keep in mind that the intensity of the radar returns increases as storms become more severe.

In addition, the chart provides echo heights of the tops and bases of associated precipitation areas. However, the absence of echoes does not guarantee clear weather. Radar detects only precipitation, either in frozen or liquid form; it does not detect all cloud formations. For instance, fog is not displayed, and actual cloud tops may be higher or lower than the precipitation returns indicate. [Figure 6-32]

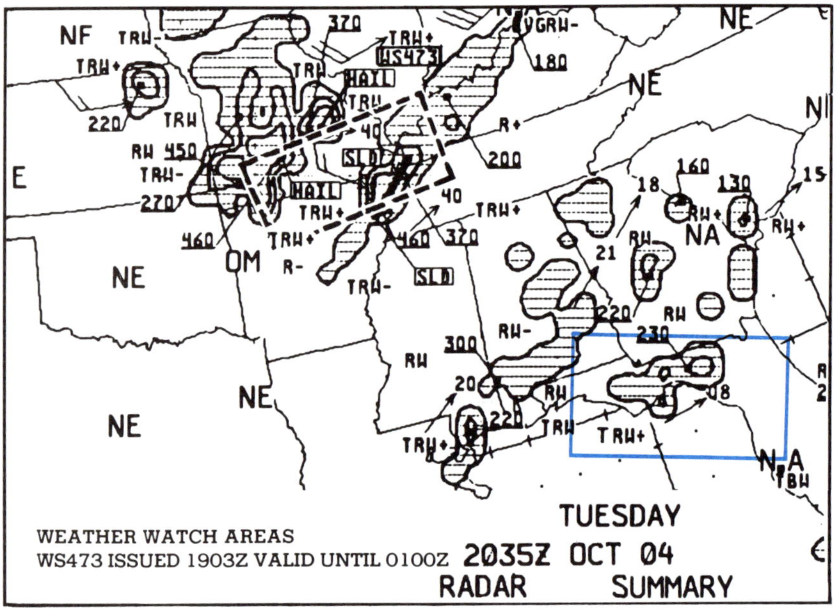

Figure 6-32. This is an excerpt of a radar summary chart. The highlighted section represents one area of precipitation. The notation TRW+ indicates that a thunderstorm and rain showers are in progress and intensity is increasing. The second contour represents strong and very strong returns, while the number "230" in the upper left indicates the top of a single cell within this area to be 23,000 feet MSL. The line from the number into the return area marks the location of the cell. The arrow symbol indicates the area of echoes is moving to the northeast at about 18 knots. You may need to refer back to this illustration during the remainder of the discussion.

Echo intensity is shown on the chart by **contours**. The six levels displayed on a radar display are combined into three contours which are printed on the radar summary chart. The first contour represents levels one and two, weak and moderate returns (light to moderate precipitation); the second shows levels three and four, or strong and very strong returns (heavy to very heavy precipitation); and the third outlines levels five and six, which represent intense and extreme returns (intense to extreme precipitation).

The radar summary chart displays precipitation in one of three contour levels representing intensity.

There are several other symbols you need to be familiar with when interpreting a radar summary chart. For example, a heavy dashed line is used to outline a severe weather watch area, as shown near the center of the illustration. In this case, the notation "WS473" appears above the upper right corner of the box. The letters "WS" identify this as a severe thunderstorm watch area. The numbers "473" mean this is the 473rd severe weather watch to be issued for the year. [Figure 6-33]

Symbol	Meaning	Symbol	Meaning
	Echo intensity level 1-2 Weak and moderate	240/80	Echo top 24,000′ MSL Echo base 8,000′ MSL
		+	Intensity increasing or new echo
	Echo intensity level 3-4 (second contour) Strong and very strong	-	Intensity decreasing
	Echo Intensity level 5-6 (third contour) Intense and extreme	SLD	Solid, over 8/10 coverage
		LEWP	Line echo wave pattern
	Dashed lines define areas of severe weather	HOOK	Hook echo
		HAIL	Hail
	Area of echoes	WS999	Severe thunderstorm watch
	Line of echoes	WT999	Tornado watch
		NE	No echos
20	Cell moving east at 20 knots	NA	Observation not available
		OM	Equipment out for maintenance
	Line or area is moving east at 20 knots (10-knot barbs)	STC	STC on — All precipitation may not be seen

Figure 6-33. Here are some of the symbols which may appear on the radar summary chart, along with brief explanations of their meanings.

A line of echoes is often shown on the chart. If there is at least 8/10 coverage, the line is labeled "SLD" at both ends. A **hook-shaped echo** (HOOK) may be associated with a tornado. In a **line echo wave pattern** (LEWP), one portion of a squall line bulges out ahead of the rest, producing strong, gusty winds. **Sensitivity time control** (STC) is a radar feature which diminishes nearby echoes to enhance reception of more distant returns. This may mask some echoes or distort their relative intensities.

Thunderstorms develop rapidly, so be sure to check other reports and forecasts for the latest weather conditions.

While the radar summary chart is a valuable preflight briefing aid, it has certain limitations. Keep in mind that it is an observation of conditions that existed at the valid time. By comparing it with the weather depiction chart, you can get a three-dimensional picture of clouds and precipitation. However, since thunderstorms develop rapidly, you should examine other weather sources for current and forecast conditions that might affect your flight.

CONSTANT PRESSURE CHARTS

Constant pressure charts are issued for five different pressure levels.

The constant pressure analysis chart is an upper air weather map on which all information is referenced to a specified pressure level. It is issued twice daily, with a valid time of 1200Z and 0000Z for each of five pressure altitude levels from 850 millibars (5,000 feet) to 200 millibars (39,000 feet). The metric equivalent of millibar is hectoPascal (1 mb = 1 hPa). The abbreviation "mb/hPa" is commonly used in reference to constant pressure levels.

The observed data for each reporting location (at the specified altitude) are plotted on the chart. The information includes the observed temperature and temperature/dewpoint spread (°C), wind direction (true north) and wind speed (knots), height of the pressure surface (meters), and changes in height over the previous 12 hours. Although the station model allows you to determine specific information at various points along your route, the constant pressure chart is most useful for quickly determining winds and temperatures aloft for your flight.

Constant pressure charts also depict highs, lows, troughs, and ridges aloft with height contour patterns that resemble isobars on a surface map. Where the height of the pressure surface is high, there is warm air over the underlying stations. Low heights indicate cold, dense air. On an upper air chart, you speak of "high height centers" and "low height centers." Frontal positions are shown in the same way as on surface charts, provided the fronts extend up to the pressure level of the chart. Fronts are observed most frequently on the 850-millibar and 700-millibar charts which correspond to approximately 5,000 and 10,000 feet, respectively. Highs and lows are also marked on the constant pressure charts. Cold upper lows are producers of bad weather, since the injection of cold air from aloft creates instability. Thus, the area underneath cold upper lows will generally have poor flying conditions.

The 850-millibar constant pressure chart portrays conditions for low-level flight at approximately 5,000 feet MSL. This is the general level of clouds associated with bad weather. This chart is valuable for forecasting thunderstorms, rain, snow, overcast, and heavy cloudiness.

The 700-millibar chart shows weather data in the vicinity of the 10,000-foot level. It shows wind conditions associated with heavy clouds and rain, but only well-developed fronts appear on this type of chart. The symbols used are the same as those used for the 850-millibar chart.

The 500-millibar chart portrays weather at about the 18,000-foot level and represents average troposphere conditions. These charts are useful in determining average wind and temperature conditions for long-range flights at or near FL180.

The 300- and 200-millibar constant pressure charts show conditions in the upper atmosphere that are significant to high altitude flight. The 300-millibar chart reflects weather at about the 30,000-foot level; the 200-millibar chart applies at the 39,000-foot level. Wind velocities on these charts are depicted with wind arrows in the same manner as on the other constant pressure charts. Height of the pressure level, wind, and temperature can easily be determined from a constant pressure chart. However, to decode a constant pressure chart, you need to know the level for which the chart was issued. [Figure 6-34]

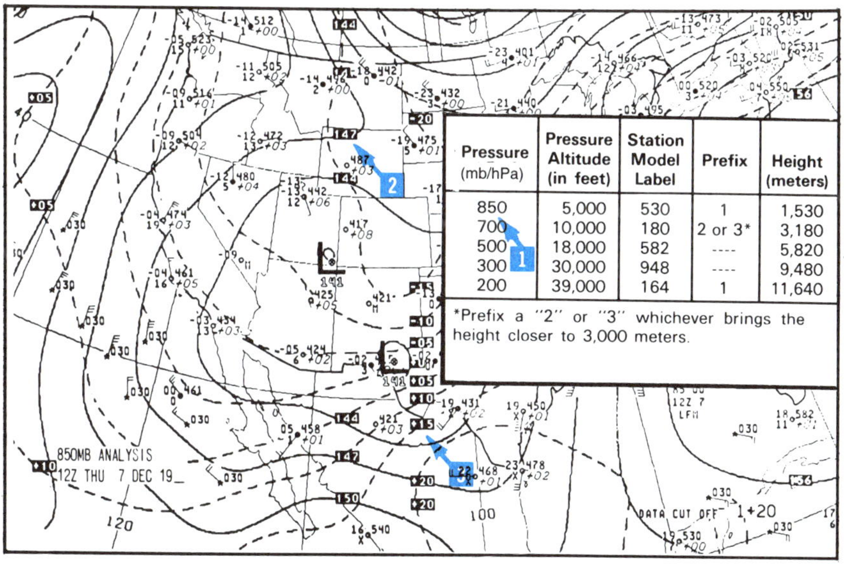

Pressure (mb/hPa)	Pressure Altitude (in feet)	Station Model Label	Prefix	Height (meters)
850	5,000	530	1	1,530
700	10,000	180	2 or 3*	3,180
500	18,000	582	----	5,820
300	30,000	948	----	9,480
200	39,000	164	1	11,640

*Prefix a "2" or "3" whichever brings the height closer to 3,000 meters.

Figure 6-34. This 850-millibar/hectoPascal chart represents a pressure altitude of 5,000 feet as shown in the table on the right (item 1). The solid line with the notation " 147" (item 2) means the height of the 850-millibar level at all points along this line is 1,470 meters (4,823 feet). An isotherm labeled "+15" (item 3) indicates a line of equal temperatures at +15°C.

Isotachs, or lines of equal wind velocity, are drawn only for the 300- and 200-millibar charts. Cross-hatching is used on these charts to denote wind speeds of 70 to 110 knots. A clear area within a hatched area indicates winds of 110 to 150 knots. Another hatched area would denote an area of 150 to 190 knots.

FREEZING LEVEL CHART

The freezing level chart is one panel of the composite moisture stability chart. It depicts the freezing level, as reported from upper air observations.

The freezing level chart is one panel of the four-panel composite moisture stability chart. It shows freezing level data from upper air observations. The chart shows the altitude of the freezing level (0° Celsius isotherm) for each observing station. A dashed line, labeled 32°F, is plotted to indicate the location of the freezing level. It also outlines areas where stations are reporting "BF" (below freezing) at the surface. Contour lines are drawn at 4,000-foot intervals, indicating the altitude of the freezing level above mean sea level (MSL) when it is not at the surface.

LOW-LEVEL SIGNIFICANT WEATHER PROG

Prognostic charts provide a forecast of expected weather conditions.

The charts we have discussed so far are all based on weather observations of the conditions that existed at a particular time. The next one we will cover is a forecast chart instead of an observation chart. The U.S. low-level significant weather prognostic, or prog, chart is valid from the surface up to the 400-millibar pressure level (24,000 feet). It is designed for use in planning flights below this altitude. [Figure 6-35]

The low-level significant weather prog chart provides a 12- and 24-hour forecast.

Low-level prog charts are issued four times each day. The valid time is printed on the lower margin of each panel. Since the two panels on the left forecast the weather 12 hours from the issue time and the two panels on the right forecast 24 hours, you can compare the two sets of panels and note expected changes between the two time frames.

SIGNIFICANT WEATHER PANELS

The symbols used to mark areas of IFR and MVFR differ from those on the weather depiction chart.

The upper panels show areas of IFR and marginal VFR weather, turbulence, and their freezing levels. Smooth lines enclose areas of forecast IFR weather; scalloped lines represent marginal VFR. As shown by the legend information, IFR areas have ceilings less than 1,000 feet and/or visibility less than three statute miles. Scalloped lines enclose areas of marginal VFR weather with ceilings from 1,000 to 3,000 feet and/or visibility of three to five miles inclusive. The symbols for portraying IFR and MVFR areas are different than those used for the weather depiction chart.

Low-level progs identify areas and altitudes where significant turbulence and icing are forecast.

Low-level progs provide turbulence and icing forecasts. Long, dashed lines enclose areas of moderate or greater turbulence. Numbers within these areas give the height of the turbulence in hundreds of feet MSL. Figures below a line show the expected base, while figures above a line represent the top of the turbulence. For example, 120 (underlined) indicates turbulence from the surface to 12,000 feet MSL. Since thunderstorms always imply moderate or greater turbulence, areas of possible thunderstorm turbulence are not outlined. Normally, the freezing level height contours

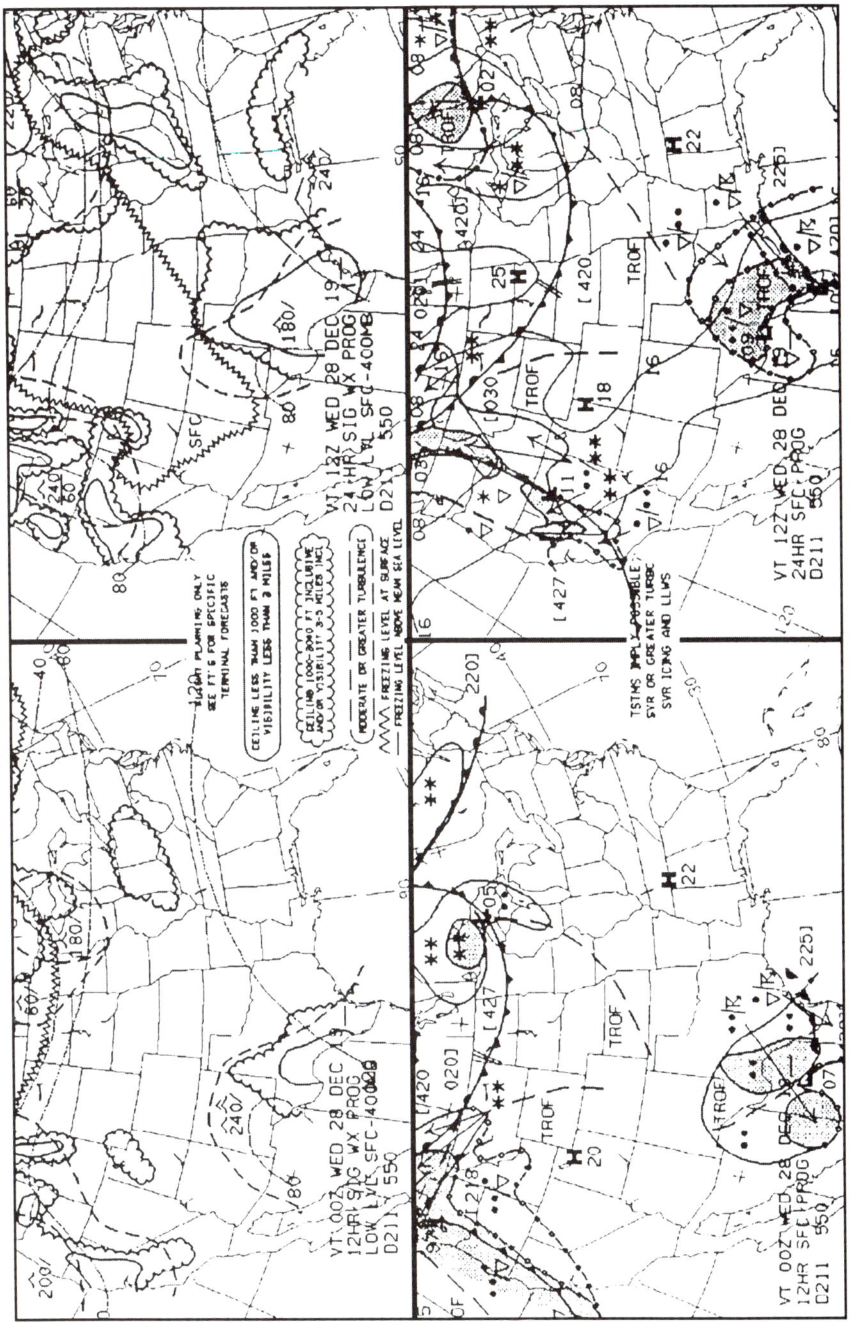

Figure 6-35. A low-level prog chart consists of four panels. The two lower panels are 12- and 24-hour forecasts of surface weather conditions, while the two upper panels are 12- and 24-hour forecasts of weather between the surface and 24,000 feet. Legend information is included between the two upper panels.

are drawn at 4,000-foot intervals with zig-zag lines. These contours are labeled in hundreds of feet MSL. A dotted line labeled "32°F," or "SFC," shows the surface location of the freezing level.

Since graphic weather charts cover large areas of the United States, they are most useful as aids during the planning phase of a cross-country flight. They should also be used as a starting point when you obtain a preflight weather briefing, since they allow you to get the "big picture" of the weather likely to affect your flight.

SURFACE PROG PANELS

The two lower panels are the surface prog panels. They contain standard symbols for fronts and pressure centers. Direction of pressure center movement is shown by an arrow; the speed is listed in knots. Areas of forecast precipitation, as well as thunderstorms, are outlined. Prog charts also have several unique symbols that are used to forecast weather conditions. [Figure 6-36]

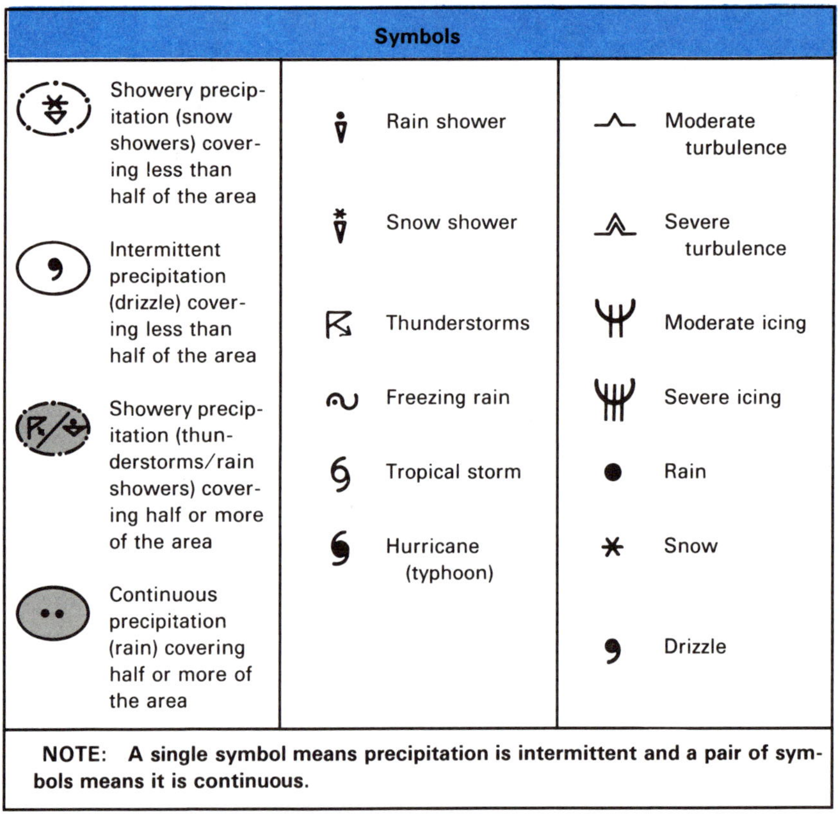

Figure 6-36. An area which is expected to have continuous or intermittent (stable) precipitation is enclosed by a solid line. If only showers are expected, the area is enclosed with a dot-dash line. When precipitation covers one-half or more of the area, it is shaded.

SEVERE WEATHER OUTLOOK CHART

The severe weather outlook chart provides a 48-hour outlook for thunderstorm activity. It is divided into two panels. The left-hand panel covers the first 24-hours beginning at 1200Z. It depicts areas of general thunderstorm activity and severe thunderstorms. The second panel covers the following day beginning at 1200Z, providing an outlook for severe thunderstorms only. The chart is issued daily and is available in the morning. [Figure 6-37]

The severe weather outlook chart provides a forecast of thunderstorm activity.

A line with an arrowhead indicates an area of probable general thunderstorm activity. Expect activity to the right of the line when you are following the line's direction of travel. The sample chart contains a line showing that much of the southern U.S. can expect thunderstorm activity in the next 24 hours. An area labeled "APCHG" indicates the thunderstorm activity may approach severe intensity.

An enclosed area shaded with diagonal lines indicates possible severe thunderstorms. A notation is included that indicates the probable percentage of coverage. "SLGT" represents a slight risk, meaning a possible coverage of 2 to 5%. A moderate risk, "MDT," indicates possible coverage of

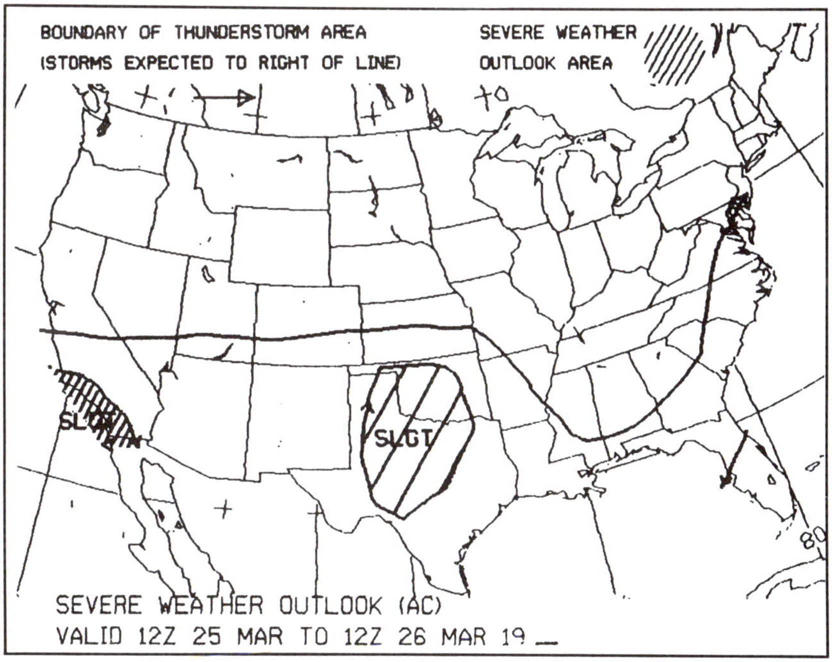

Figure 6-37. This is an example of the first 24-hour panel of a severe weather outlook chart for the time period of 1200Z on March 25 to 1200Z the following day. The legend information at the top of the chart explains some of the chart symbology.

6 to 10%. A high risk area has a possible coverage of more than 10%. Tornado watch areas also are included if they are in effect at the valid chart time.

You should use the severe weather outlook chart only for advanced planning. Since thunderstorms can develop quickly, more current information is essential to develop a true picture of the severe weather along a route.

CHECKLIST

After studying this section, you should have a basic understanding of:

✓ **Surface analysis charts** — How they represent the locations of fronts, surface pressure patterns, and weather at reporting stations across the country.

✓ **Weather depiction charts** — Why you should consult these charts to determine areas of IFR and marginal VFR weather.

✓ **Radar summary charts** — How these charts show thunderstorm activity and intensity based on radar returns from precipitation.

✓ **Constant pressure charts** — How to equate millibar levels with flight altitudes and determine winds and temperatures aloft.

✓ **Freezing level chart** — How you find the freezing level along your route of flight based on observed temperatures.

✓ **Low-level significant weather progs** — How you can interpret forecast weather conditions 12 to 24 hours in the future.

✓ **Severe weather outlook charts** — How you can determine areas of probable thunderstorm activity extending 24 or 48 hours into the future.

SECTION E

HIGH-ALTITUDE CONSIDERATIONS

So far in our discussion of meteorology, we have covered a number of subjects, including weather theory, hazards, reports, forecasts, in-flight advisories, and charts. This section is designed to provide you with a basic working knowledge of high-altitude meteorology. It begins with a discussion of the tropopause and associated phenomena, including the jet stream. Weather charts appropriate to high-altitude operations near the tropopause also are presented. Since high-altitude flight requires supplemental oxygen or pressurization systems, you may wish to review the discussion of these subjects in Chapter 9.

JET STREAM

A **jet stream** can be described as a narrow band of high velocity wind of 50 knots or more which meanders vertically and horizontally around the earth in wave-like patterns. Although jet streams occur in both the northern and southern hemisphere, this discussion is limited to the northern hemisphere.

The jet stream usually is stronger in the winter than in the summer. This is because its mean position shifts south in the winter. As the jet stream moves south, its core rises to a higher altitude, and its average speed usually increases. The core of strongest winds generally is found between 25,000 and 40,000 feet, depending on the latitude and season. [Figure 6-38]

The jet stream shifts south in the winter and generally intensifies in velocity.

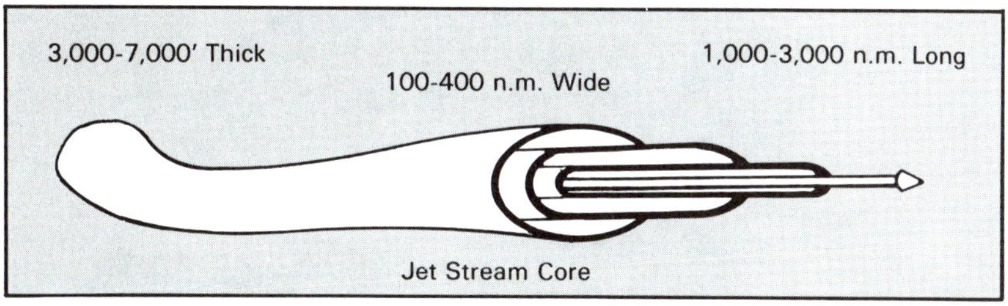

Figure 6-38. Since the jet stream is stronger in some places than in others, it rarely encircles the entire hemisphere as a continuous river of wind. A given segment normally is several thousand miles in length, several hundred miles wide, and a mile or so in depth. Winds in the jet may reach 300 knots, but generally are between 100 and 150 knots.

The boundary layer between the troposphere and the stratosphere is called the tropopause.

At the beginning of this chapter, the troposphere was described as the inner layer of the atmosphere which is adjacent to the earth. It varies in depth from an average of 60,000 feet over the equator to about 20,000 feet over the poles. Its depth is greater in the summer than in the winter. It is characterized by a decrease in temperature with an increase in altitude. The top of the troposphere is called the tropopause, which serves as the boundary between the troposphere and the stratosphere. The location of the tropopause is usually characterized by a pronounced change in temperature lapse rate. If ISA conditions prevail, the tropopause location is about 36,000 feet. Between this altitude and about 66,000 feet in the standard atmosphere, the temperature remains constant at -57°C. The tropopause acts like a "lid," since it resists the exchange of air between the troposphere and the stratosphere above. However, in the northern hemisphere, there are generally two breaks in the tropopause. One is between the polar and subtropical airmasses and the other is between the subtropical and tropical airmasses. [Figure 6-39]

JET STREAM TURBULENCE

The jet stream guides weather-producing fronts as well as high and low pressure systems. Although the jet's high winds can benefit aircraft flying at high altitudes, severe turbulence is a hazard which you must consider. Clear air turbulence (CAT) is generally described as turbulence that is not associated with any convective activity or cloud formation. It is often associated with a jet stream that is interacting with a large mountain range or deep low pressure system. These jets can sometimes be identified by long streams of cirrus cloud formations or high, "wind-swept" looking cirrus clouds. The turbulence associated with a jet

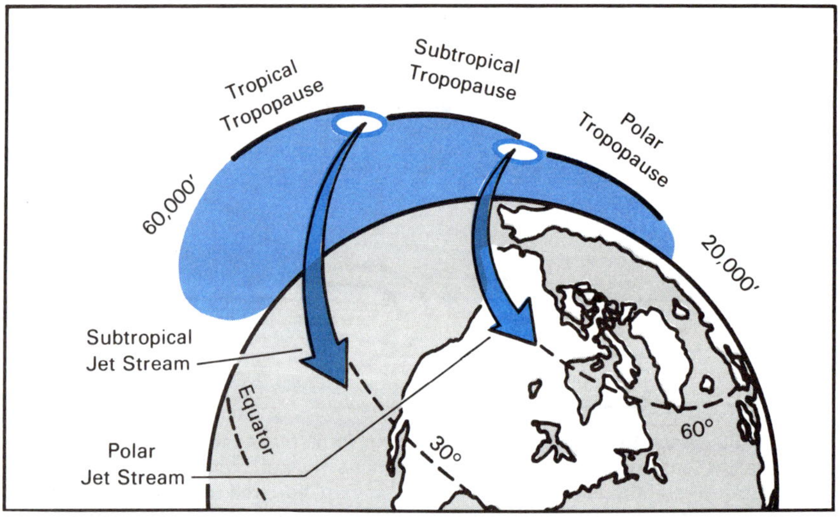

Figure 6-39. The breaks in the tropopause define the location of the jet stream. The positions of the polar and subtropical jet streams vary, depending on the seasonal migration of airmass boundaries.

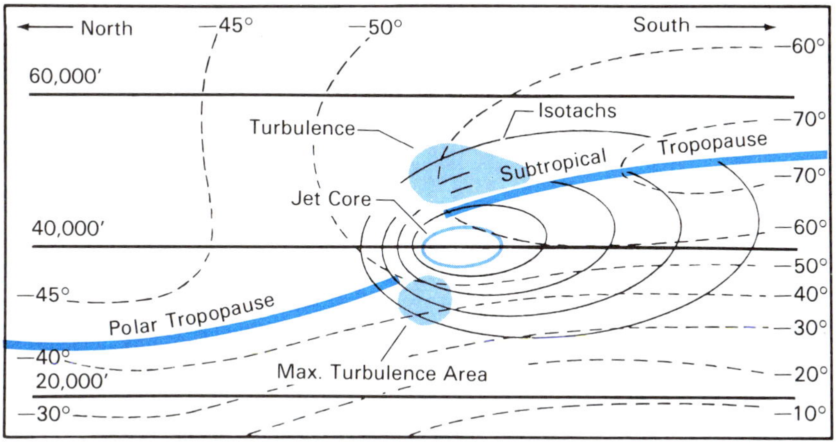

Figure 6-40. This is a cross section of a polar jet stream core. Note the wind speed gradient, shown by the spacing of the isotachs, is much greater on the polar side of the jet. For this reason, wind shear or CAT is usually greater in an upper trough on the polar side. Precise analysis of the jet stream core is not possible, so you should anticipate CAT whenever you are near an intense jet.

stream can be very strong, and because it often occurs in clear air, it is difficult to forecast accurately. As a rule of thumb, clear air turbulence can be expected when a curving jet is found north of a deep low pressure system. It can be particularly violent on the low pressure side of the jet and when the wind speed at the core is 110 knots or greater. [Figure 6-40]

OBSERVED WINDS AND TEMPERATURES ALOFT CHART

You are already familiar with winds and temperatures aloft forecasts, but you may not know that you can also review actual observations of winds and temperatures aloft. The observations for eight selected levels are sent twice daily on four-panel charts valid at 1200Z and 0000Z. One four-panel chart covers 6,000, 9,000, 12,000, and 18,000 feet MSL. The second chart covers 24,000, 30,000, 34,000, and 39,000 feet MSL. You can use these charts to select the most favorable cruising altitude for a proposed flight and to determine the temperature. Keep in mind that the observations on these charts are five to eight hours old by the time you see them.

Wind direction and speed at each observing station are shown by arrows. A calm or light wind is shown by the letters "LV," and when wind information is missing, it is shown by the letter "M." These abbreviations are plotted below the station model, which is filled in when the temperature/dewpoint spread is 5°C or less. Observed temperatures are included on the upper two panels, which are for 24,000 and 34,000 feet. A dotted bracket around the temperature indicates a calculated temperature. Wind

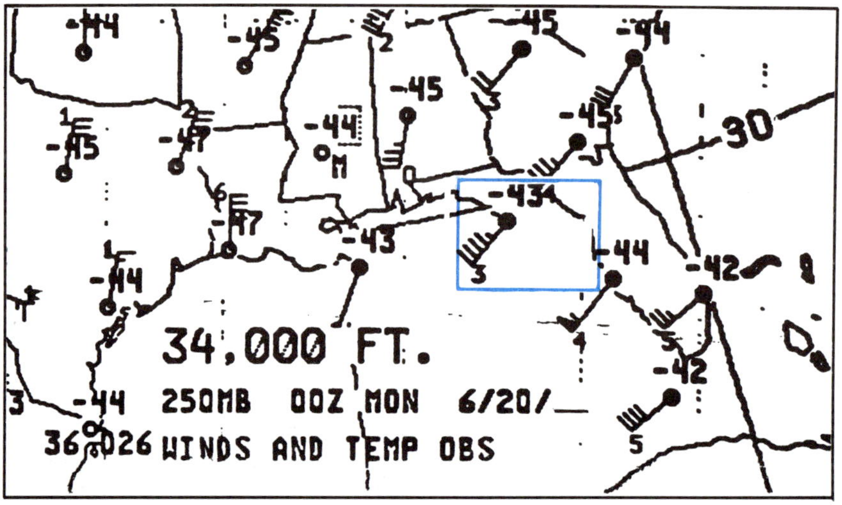

Figure 6-41. This is an excerpt of the observed winds and temperatures aloft chart for 34,000 feet. The station enclosed by the box indicates a wind speed of 45 knots. To determine wind direction, first consider the general wind direction by noting the wind direction indicator. This station indicates a wind from the southwest. The "3" at the end of the direction indicator is the second digit of the wind direction rounded to the nearest 10°. In this example, you determine the wind to be from 230°. The number "-43" indicates the observed temperature in degrees Celsius. Since the station model is filled in, you know the temperature/dewpoint spread is 5°C or less.

speed is shown by a one-half flag for each five knots, a full flag for each 10 knots, and a pennant for each 50 knots. [Figure 6-41]

TROPOPAUSE DATA CHART

Areas of missing information on the tropopause pressure, winds, and temperature chart usually indicate areas of strongest wind flow.

The tropopause data chart has several panels. The observed data panel provides pressure, wind, and temperature information relative to the tropopause for each of the upper air observing stations. Since maximum wind occurs near the tropopause, this chart is essentially a map of observed maximum winds. The letter "M" is used to indicate missing data. This usually indicates that strong winds carried radiosonde instruments too far from the observing station to obtain reliable wind data. The areas of missing wind data are actually areas of strongest wind. [Figure 6-42]

TROPOPAUSE HEIGHT/VERTICAL WIND SHEAR PROG

The letter "F" indicates a pressure altitude given in hundreds of feet.

The tropopause height/vertical wind shear prog panel shows the predicted height of the tropopause in terms of pressure altitude. These heights are preceded by the letter "F" and are given in hundreds of feet. Solid lines connect points at which the height of the tropopause intersects with standard constant pressure surfaces. Vertical wind shear in knots per 1,000 feet is also shown on this chart. Dashed lines connect points of equal vertical wind shear. [Figure 6-43]

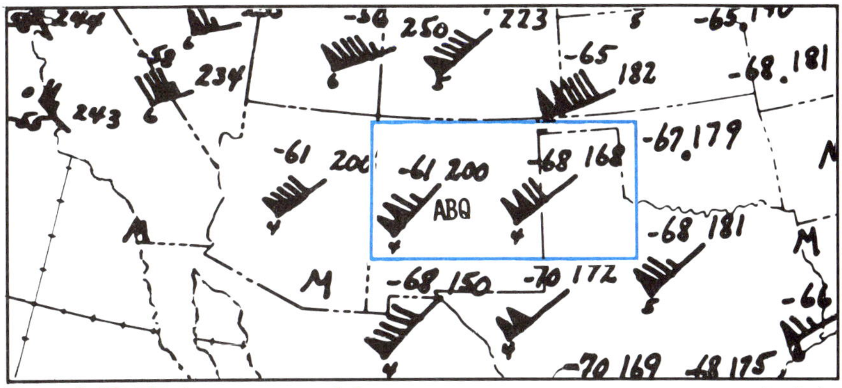

Figure 6-42. This is an excerpt of the tropopause pressure, temperature and winds chart. The plotted information for Albuquerque (ABQ) indicates the tropopause was at the 200-millibar level at the valid time of the chart. At that level, the winds were from 240° at 115 knots and the temperature was -61°C. The station to the east indicates the tropopause was at the 168-millibar level, with winds from 240° at 120 knots. The observed temperature was -68°C.

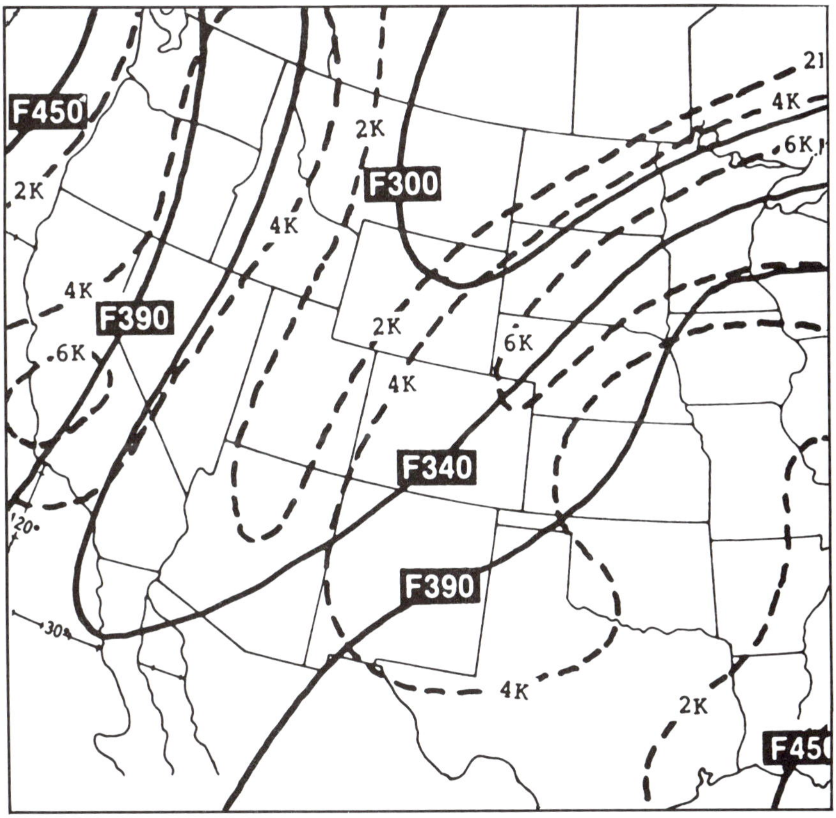

Figure 6-43. According to this chart, the predicted height of the tropopause over southern California is "F340," meaning it intersects a standard pressure surface of 34,000 feet. The vertical wind shear is predicted to be four knots per 1,000 feet.

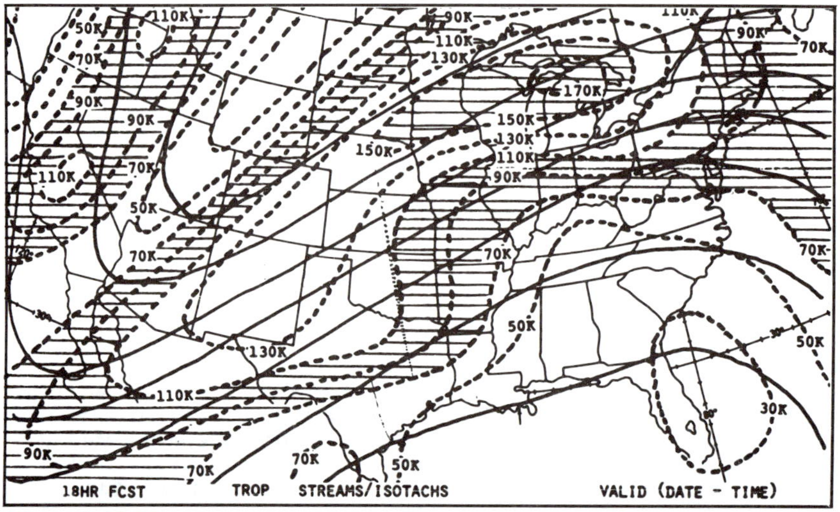

Figure 6-44. The winds parallel the solid streamlines, while wind speed is shown by dashed-line isotachs at 20-knot intervals. Areas of wind speeds between 70 and 110 knots and between 150 and 190 knots are cross-hatched. The clear area in between indicates winds of 110 to 150 knots. A high or low may be encircled by a closed streamline. You can determine whether it is a high or a low from the direction of wind circulation.

A vertical wind shear of six knots per 1,000 feet or more indicates moderate or greater turbulence.

Vertical wind shear is depicted at two-knot intervals and is averaged through a layer that extends from about 8,000 feet below to 4,000 feet above the tropopause. As a general rule, you should expect moderate or greater turbulence when the vertical wind shear is six knots or more.

TROPOPAUSE WINDS

The tropopause winds prog panel also is referred to as a maximum wind prog. It uses solid lines called **streamlines** to indicate wind direction. These streamlines usually show the wind from west to east in the middle latitudes. [Figure 6-44]

HIGH-LEVEL SIGNIFICANT WEATHER PROG

The low-level significant weather prognostic panel, which is prepared manually by the National Meteorological Center, is valid for use up to an altitude of 24,000 feet. Above that altitude, you use the high-level prognostic chart. This chart presents a forecast of thunderstorms, tropical cyclones, severe squalls, moderate or greater turbulence, widespread duststorms and sandstorms, tropopause heights, and the location of jet streams. [Figure 6-45]

Another feature of the high-level prog is that scalloped lines are used to enclose areas that have sandstorms, duststorms, and cumulonimbus clouds. Enclosed areas of cumulonimbus clouds (CB) also imply the presence of moderate or greater turbulence and icing conditions. [Figure 6-46]

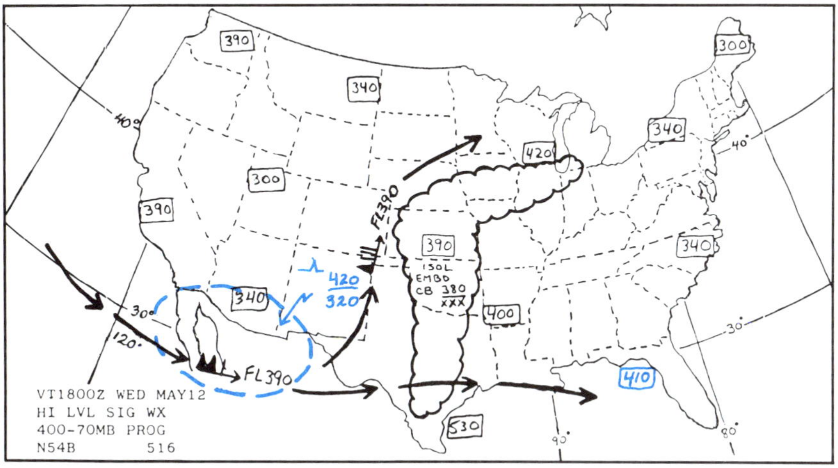

Figure 6-45. The high-level significant prog chart shown here is valid at 1800Z on Wednesday, May 12. It covers the altitude range of 400 millibars (24,000 feet) to 70 millibars (63,000 feet). However, the three-digit numbers contained in boxes represent the forecast height of the tropopause in feet MSL. For example, near the Florida coast, the tropopause is expected to be at 41,000 feet MSL. A dashed line is used to enclose areas of forecast turbulence, as shown by the area over northern Mexico. It is expected to have moderate turbulence between 32,000 feet and 42,000 feet.

Symbol	Meaning	Symbol	Meaning
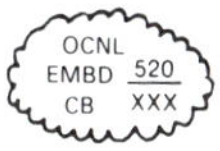	Embedded cumulonimbus, 1/8 to 4/8 coverage, tops 52,000 feet bases below 24,000 feet	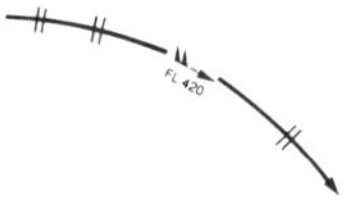	Jet stream with maximum speeds of 100 knots at FL 420 at one location and 90 knots at FL 370 at another location
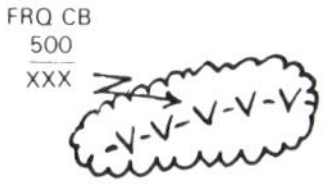	Forecast severe squall line, CB, coverage 5/8 to 8/8, bases below FL 240 and tops to FL 500		Forecast surface position, speed (knots), and direction of movement of frontal system
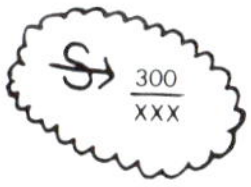	Widespread sandstorm or duststorm, bases below FL 240 (i.e., at the surface), tops FL 300		Thunderstorm area (5/8 to 8/8 area coverage, bases below FL 240, tops FL 500) associated with a tropical cyclone
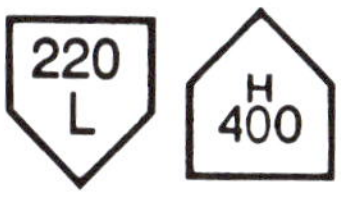	High (H) and low (L) tropopause height in hundreds of feet MSL		Volcanic activity with date and time of first eruption. The dot is the volcano's location

Figure 6-46. Some high-level prog symbols are shown here. Three "Xs" indicate that the phenomenon extends below the lower limit of the chart (24,000 feet). Area coverage terms are: ISOL (isolated, less than 1/8); OCNL (occasional, 1/8 to 4/8); and FRQ (frequent, 5/8 to 8/8). Isolated or occasional CBs are not included unless they are embedded in cloud layers or concealed by haze or dust.

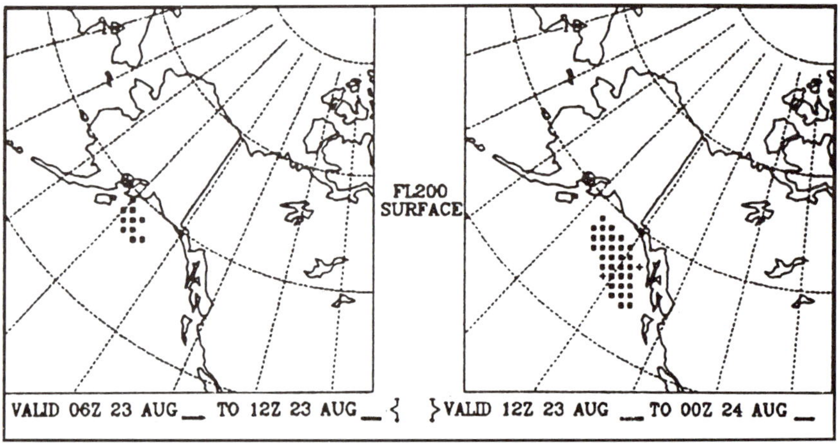

Figure 6-47. These excerpts show the forecast coverage of a volcanic ash cloud from the surface to FL200 for a 6-hour period and for the following 12 hours.

VOLCANIC ASH FORECAST CHART

In the event of a volcanic eruption, you should avoid flying in or near the ash cloud. The volcanic ash forecast transport and dispersion chart is a series of panels showing the expected location of the ash cloud at various flight levels. Each level is shown with valid times beginning 6, 12, 24, and 36 hours following the eruption. [Figure 6-47]

CHECKLIST

After studying this section, you should have a basic understanding of:

- ✓ **The jet stream** — Where the jet stream forms, how it changes position and strength, and where you can expect the most turbulence.
- ✓ **Winds and temperature aloft chart** — What information is contained on this chart and how to read the station model.
- ✓ **Tropopause pressure, wind, and temperature chart** — How to interpret the observations displayed on this chart.
- ✓ **Tropopause height/vertical wind shear chart** — How this chart predicts probable turbulence based on areas of vertical wind shear.
- ✓ **High-level significant weather prognostic chart** — What hazards are portrayed on this chart, and how to interpret the data.
- ✓ **Volcanic ash forecast transport and dispersion chart** — How to use this chart to determine the location of an ash cloud over time.

CHAPTER 7

IFR FLIGHT OPERATIONS

INTRODUCTION

This chapter brings together many different aspects of instrument flight. Section A covers flight planning from the viewpoint of an instrument pilot. Section B looks at some emergencies unique to instrument flight. Section C presents a discussion on IFR decision making and other flight considerations.

SECTION

A

IFR FLIGHT PLANNING

IFR flight planning is an extension of VFR planning. However, it requires more emphasis on selecting your route, gathering communications and navigation data, and obtaining weather. This includes reviewing IFR charts, flight publications, and regulations.

Regulations require you to perform certain pre-flight actions.

For IFR operations, FARs require you to be familiar with all available information concerning the flight. This includes weather reports and forecasts, fuel requirements, runway lengths at each airport you intend to use, takeoff and landing distance requirements, known traffic delays, and alternatives if the flight cannot be completed as planned.

To illustrate, let's plan a flight between Childress Municipal Airport in Childress, Texas, and Meacham Field at Fort Worth, Texas. Your planned departure time is 1600Z and the trip should take about two hours. Your aircraft is a Piper Warrior, N8458R, equipped with dual nav/coms, glide slope, ADF, Mode C transponder, and DME.

INITIAL PLANNING

To begin, consider what limiting factors may prevent you from completing the flight successfully. Are aircraft equipment or performance limitations significant? What about your proficiency level? Will the route you select be approved by ATC? How is the weather? Let's take a closer look at some of these factors.

Start with an overview of the expected weather. If you need a long-range forecast, contact the National Weather Service (NWS). Forecasters there can provide reasonably accurate predictions of general weather patterns up to five days in advance. They also can provide a 6- to 10-day outlook. Keep in mind that the longer the forecast period, the less accurate the information. A more common way to begin checking weather is to watch the local television weather, paying particular attention to forecast frontal movement and precipitation. The day before the flight, call your local FSS and obtain an outlook briefing. This should give you a good idea of expected weather as you begin your actual flight planning.

ROUTE SELECTION

Several factors influence route selection, including the availability of route alternatives, aircraft performance considerations, and fuel economy.

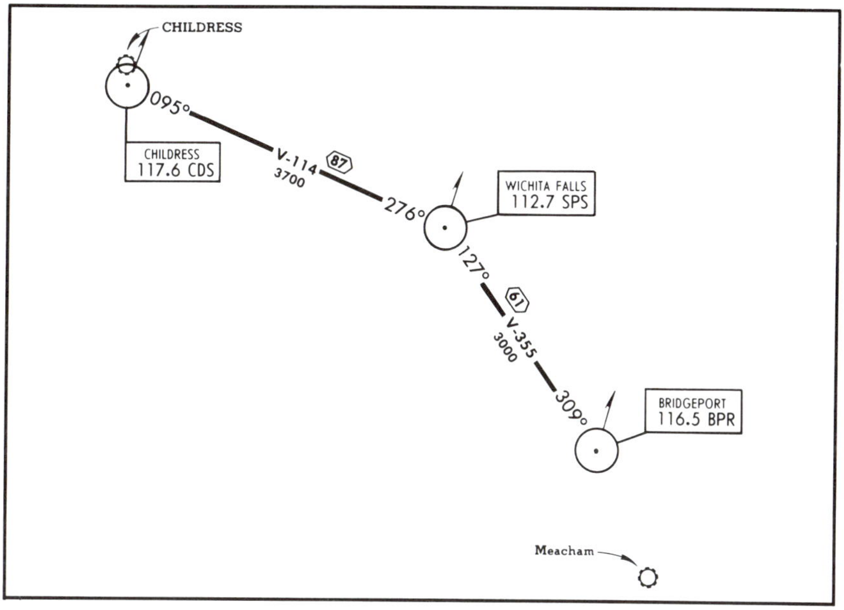

Figure 7-1. A preferred IFR route does not exist between Childress and Meacham, so you must select your own route. This simplified illustration shows that the most practical route is from Childress via V-114 to Wichita Falls, then V-355 to Bridgeport. The enroute chart does not show a route from Bridgeport to Meacham; however, the approach chart will probably show a feeder route. The highest MEA for the flight is 3,700 feet.

If there is a preferred IFR route available, plan to fly it unless weather conditions or aircraft performance warrant a different route. Preferred routes are listed in the Enroute section of the *Jeppesen Airway Manual* and in the *Airport/Facility Directory*. Preferred IFR routes beginning or ending with a fix indicate that aircraft may be routed to or from these fixes via a SID, radar vector, or STAR. If one is not listed, consult the enroute chart to find the most practical route for the flight. In all cases, remember to check applicable MEAs for your flight. In some parts of the country, the minimum enroute altitudes may be beyond your aircraft's climb capabilities or they may require you to use oxygen during the flight. [Figure 7-1]

Preferred routes may begin or end with a fix. This means you will be routed to or from these fixes via a SID, radar vector, or STAR.

Your next step is to review the charts for both Childress Airport and Meacham Field. Be sure to check if the one for Childress indicates an IFR departure procedure. Also, you should review the approaches into Childress, just in case you must return there shortly after takeoff. A check of the charts for Meacham allows you to become familiar with the available approaches and the runways they serve. These factors may be important as you consider minimums during your standard preflight weather briefing. Next, make a check for applicable SIDs and STARs. Although there are no SIDs for Childress, there is a STAR published for Meacham. [Figure 7-2]

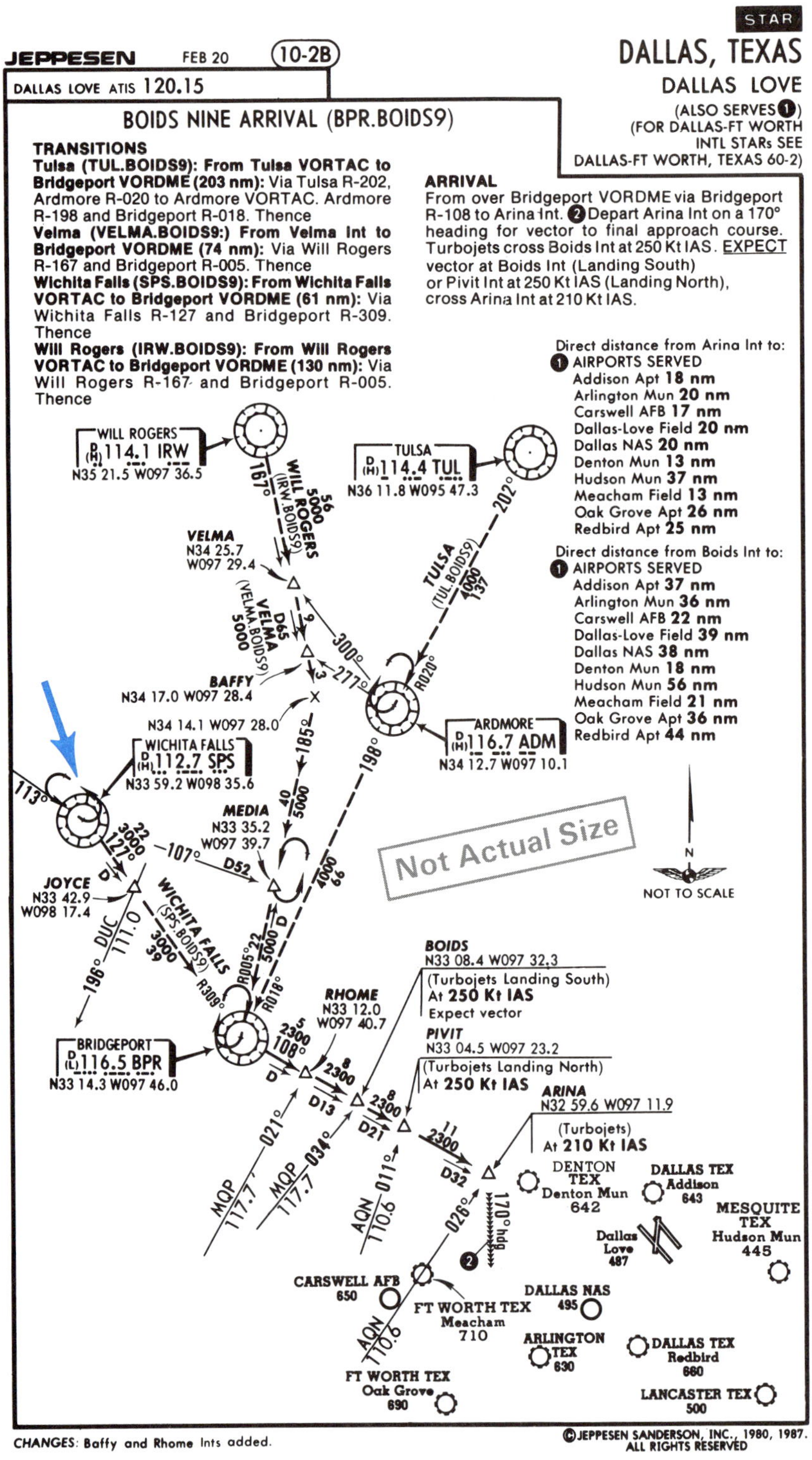

Figure 7-2. The Boids Nine Arrival serves many airports in the Dallas area. Notice that this STAR has a transition at Wichita Falls. Since your route of flight crosses this VORTAC, you can include the Boids Nine Arrival in your route.

You should also consider possible alternate airports. Each alternate should be far enough away to avoid adverse weather that may blanket the area near your destination. Generally, each should have adequate communications, weather reporting, and at least one instrument approach. Other factors being equal, the more approaches an alternate has, the better. Examine the approach charts for alternate minimums and compare them to the forecast weather at your ETA.

FLIGHT INFORMATION PUBLICATIONS

Be sure to check NOTAMs for such information as navaid and lighting outages, runway closures, and limitations on instrument approach procedures. NOTAMs are published in the Chart NOTAMs section of the *Jeppesen Airway Manual* and in the *Notices to Airmen* publication.

Next, check the A/FD for specific information about your departure, destination, and possible alternates. You can find data about such things as runway lengths, fuel availability, pilot-controlled lighting, hours for part-time facilities, and restrictions to navaids. [Figure 7-3]

CHILDRESS MUNI (CDS) 4 W UTC–6(–5DT) N34°26.02′ W100°17.28′ **DALLAS-FT. WORTH**
1953 B **FUEL** 100LL, JET A **H-2E, L-13B**
RWY 17-35: H5949X75 (ASPH-PFC) S-21 RWY LGTS (NSTD) **IAP**
RWY 17: P-line.
RWY 04-22: H4375X60 (ASPH) S-21
RWY 22: Tree.
1 → **AIRPORT REMARKS:** Attended 1400–2300Z‡. Rwy 04–22 has grass encroaching along edges of pavement, loose aggregate on rwy and twy. Rwy 22 has 10–12 inch turf buildup across thld. Rwy 17–35 NSTD MIRL. S 4500′ lgtd, last 2000′ clear lenses. For fuel after hours call 817-937-2259/6015/6451.
FORT WORTH FSS (FTW) TF 1-800-WX-BRIEF. NOTAM FILE CDS.
RCO 122.45 (FORT WORTH FSS)
FORT WORTH CENTER APP/DEP CON 133.5
AIRSPACE: CLASS E svc effective continous.
RADIO AIDS TO NAVIGATION: NOTAM FILE CDS. VHF/DF ctc FSS. Out of svc indefinitely.
(L) VORTACW 117.6 CDS Chan 123 N34°22.14′ W100°17.34′ 351° 3.9 NM to fld. 1920/10E. **HIWAS.**

- -

FORT WORTH MEACHAM (FTW) 5 N UTC–6(–5DT) N32°49.16′ W97°21.71′ **DALLAS-FT. WORTH**
710 B S4 **FUEL** 100LL, JET A OX 1, 3 LRA ARFF Index B **H-2E, 4F, 5B, L-13C, A**
RWY 16L-34R: H7501X150 (CONC-GRVD) S-80, D-100, DT-190 HIRL 0.5% up N **IAP**
RWY 16L: MALSR. Pole. **RWY 34R:** VASI(V4L)—GA 3.0° TCH 54′. Rgt tfc.
RWY 09-27: H4051X100 (ASPH) S-30, D-45, DT-60 MIRL
RWY 09: VASI(V2L)—GA 4.0° TCH 50′. **RWY 27:** VASI(V2L)—GA 4.0° TCH 41′. Pole.
RWY 16R-34L: H4000X75 (ASPH) S-15 MIRL
RWY 16R: VASI(V2L)—GA 4.0° TCH 46′. Dam. Rgt tfc. **RWY 34L:** VASI(V2L)—GA 4.0° TCH 40′.
AIRPORT REMARKS: Attended continuously. Twy E lgts out of svc indefinitely. Twy U lgts out of svc indefinitely. ARFF avbl all users all hours. Flight Notification Service (ADCUS) available. NOTE: See SPECIAL NOTICE—Simultaneous Operations on Intersecting Runways.
COMMUNICATIONS: CTAF 118.3 **ATIS** 120.7 **UNICOM** 122.95
FORT WORTH FSS (FTW) on arpt. TF 1-800-WX-BRIEF. LC 429-6434. NOTAM FILE FTW.
2 → ® **REGIONAL APP/DEP CON** 118.1 (North) 120.5 (South)
3 → **MEACHAM TOWER** 118.3 **GND CON** 121.9 **CLNC DEL** 124.65
RADIO AIDS TO NAVIGATION: NOTAM FILE DFW.
DALLAS–FORT WORTH (H) VORTACW 117.0 DFW Chan 117 N32°51.96′ W97°01.68′ 253° 17.1 NM to fld. 560/8E.
MUFIN NDB (H-SAB/LOM) 365 FT N32°53.58′ W97°22.41′ 164° 4.5 NM to fld. NOTAM FILE FTW. Route forecast only on TWEB 0400–1100Z‡.
ILS/DME 109.9 I-FTW Chan 36 Rwy 16L LOM MUFIN NDB.

Figure 7-3. Notice that runway 04-22 has grass encroaching along edges of pavement, and there is loose aggregate on the runway and taxiway (item 1). Meacham has radar services available (item 2). The tower frequency (item 3) is on 118.3, and the frequencies for ground control and clearance delivery are also given.

Another important item in the A/FD is the location of VOR receiver checkpoints. If you plan to use VOR equipment for navigation under IFR, regulations require that it has been operationally checked within the preceding 30 days and found to be within acceptable limits. Consult the aircraft records to see if a check is required for this flight. If one is needed, you can either use a designated checkpoint or you can check the aircraft's dual VORs against one another when they are tuned to the same facility. In either case, the maximum permissible error is 4°. Remember to make the required logbook entry — date and location of the check, bearing error, and your signature — after you complete the check.

The national airspace system is in a continual state of change. To stay abreast of these changes and to review current procedures, you should periodically consult the AIM. It provides information on such things as navigation aids, lighting and airport markings, airspace, ATC, emergency procedures, safety of flight, medical factors, and charts.

Once you have settled on a route and have consulted the appropriate publications, it is time to begin your navigation log. List the selected checkpoints, routes, courses, distances, appropriate communications and navigation frequencies, and any other information you feel is important.

STANDARD WEATHER BRIEFING

It is apparent that automated systems and computers have revolutionized the way we gather weather information. While many new systems have advantages, their effectiveness depends, to a large extent, on your knowledge of the system and weather data in general. For a standard briefing, there are several points to consider: Will the weather affect your proposed route? What is the weather at your destination? Are you required to designate an alternate airport? Where are the freezing levels? Can you expect any other hazardous conditions such as turbulence? Is the aircraft equipped properly for the expected weather? How does your own experience and currency level compare to the anticipated conditions?

Begin gathering information by obtaining a general overview of the weather patterns. Check the weather section of the local newspaper, and watch the weather portion of the TV news. The weather channel on cable TV also is useful for general weather information.

You can then obtain more detailed information from any of several automated weather briefing sources, such as the telephone access to Transcribed Weather Broadcasts (TEL-TWEB), Telephone Information Briefing Service (TIBS), and Direct User Access Terminal System (DUATS). For more information on these systems, refer to the AIM. When using these services, make sure you are aware of the information they offer and, more importantly, what they do not offer.

Keep in mind that many automated sources do not provide all of the necessary information for an instrument or a cross-country flight. As pilot in command, you are responsible for obtaining a complete briefing. With a good idea of the general weather picture, you can now call or visit an FSS for a detailed briefing. You also can receive a standard briefing through DUATS. In addition, the NWS can provide weather briefings in some areas on a limited basis. However, the NWS does not provide NOTAMs, flow control advisories, or information on MOAs and MTRs, nor do they accept flight plans. As a general rule, an FSS or DUATS are your primary sources for a standard weather briefing.

An FSS is still one of the best sources of current enroute and destination weather information.

During preflight weather briefings, you can request any reports or forecasts which are not routinely available at your service outlet. This is called "request/reply" service and it is available at all FSS and NWS outlets. You may wish to use a weather log to record the briefing. As the briefing proceeds, analyze your information. Does it make sense? For example, if a recent forecast for your destination called for VFR conditions to exist at the present time and the current weather is IFR, how reliable is the forecast going to be at your ETA? If you have the opportunity to examine graphic weather products, compare them to the terminal and area forecasts. Do they agree? The first chart you might examine is the weather depiction chart. This is a simplified version of the surface analysis chart. It is marked to depict areas of IFR and marginal VFR conditions which help you quickly determine the reported weather along your route of flight. [Figure 7-4]

Use request/reply service to obtain reports or forecasts which are not available locally.

Next, you may wish to review the low-level significant weather prognostic chart. The upper panel shows the forecast weather from the surface through 24,000 feet, and the lower panel depicts surface conditions. You can use this chart to quickly identify areas of forecast IFR and marginal

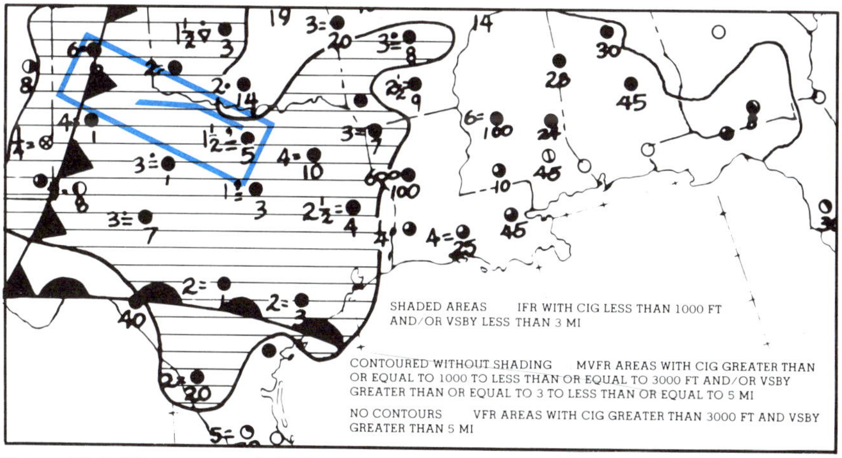

Figure 7-4. The route of flight is shown by the line within the colored box. According to the chart, IFR conditions exist over the entire route. This means the ceiling is less than 1,000 feet and/or the visibility is less than three miles. The station model near the destination indicates a 500-foot overcast with one and one-half miles visibility in intermittent drizzle and fog. A cold front is presently located over western Texas.

VFR weather, frontal movement, turbulence, precipitation, and the location of the freezing level. [Figure 7-5]

The upper panel of the prognostic chart indicates that the weather at 1800Z should be marginal VFR for the entire route. Since the reported weather is IFR, this signifies an improving weather trend. The upper panel also indicates the freezing level for the route to be at about 12,000 feet. If you stay below this altitude, structural icing should not be a factor.

The lower panel shows the cold front should have moved through Childress by 1800Z. This should alert you to check the terminal forecast for Childress and determine the forecast time of frontal passage. This will indicate if the front will be a factor at your planned departure time. Finally, the area enclosed in the dot/dash line, which covers most of your route, is forecast to have showery precipitation and thunderstorms over less than half the area.

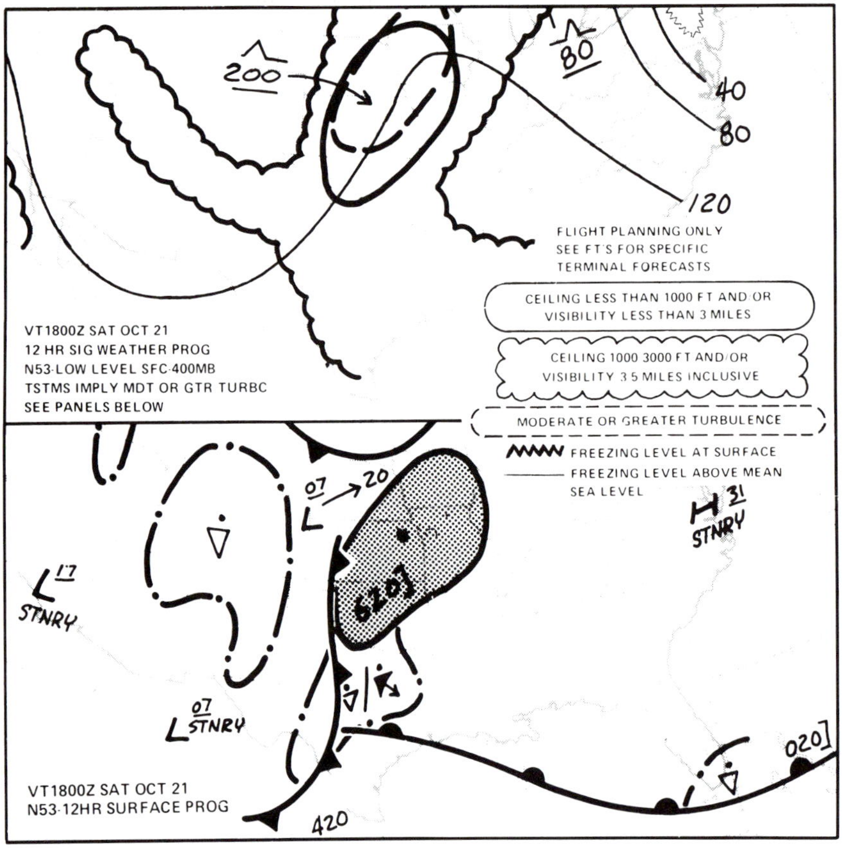

Figure 7-5. The prognostic chart shown here corresponds to your estimated arrival time of approximately 1800Z at Meacham. Compare this chart to the weather depiction chart to evaluate the weather trend.

Since thunderstorms have been forecast, you should also check the radar summary chart for information about thunderstorm activity. If the charts just discussed are not available, you can use the area forecast to obtain much of this same information. You can also use it to confirm and amplify information displayed on the graphic weather charts. The area forecast for the route of the flight indicates: "*. . . Over northern Texas, ceilings frequently below 1,000 feet and visibilities below three miles in rain and fog. Conditions improving to ceilings 2,500 feet to 3,000 feet broken variable to overcast with visibilities three to six miles in rain showers and fog by 1700Z. Ceilings occasionally 1,000 feet to 2,000 feet.*"

With a good understanding of the current and forecast weather along your route, you are ready to get specific information for your departure and destination airports. You should also check the weather at airports you might use as alternates. [Figure 7-6]

The weather at your point of departure (KCDS), along your route (KSPS), and at your destination (KFTW) is reported as IFR. This is in agreement with the weather depiction chart. At your estimated time of departure (1600Z), the weather at Childress is forecast to be a 500-foot overcast with two miles visibility in rain. This weather is an important consideration, because you need to know if it will be possible to return to Childress if mechanical difficulties develop during departure. The forecast weather is above the straight-in and circling minimums for the VOR approach to Childress; therefore, you should be able to return to Childress, if necessary. You will also need to determine whether your destination will be at

METAR KCDS 1454Z 35008KT 2SM RA OVC005 21/17 A2993
METAR KSPS 1455Z 00000KT ½SM -RA FG OVC002 20/19 A2998
METAR KFTW 1453Z 35010KT 1SM DZBR OVC004 19/19 A3001
METAR KDFW 1455Z 35005KT 1½SM DZBR OVC005 19/18 A3002
METAR KDAL 1454Z 35005KT 3SM DZ OVC008 19/18 A3002

TAF KCDS 210940 1010 35012KT 2SM RA OVC005 FM17
35020KT 1SM RA OVC010 BECMG 1820 33010KT 2SM DZ
BKN030 FM22 00000KT 2SM BR SCT040 FM04 VRB04KT P6SM NSW SKC
TAF KSPS 210950 1010 00000KT 1SM RABR VV002 FM16 VRB05KT 2SM DZBR OVC005
BECMG 1921 35010KT 4SM DZ OVC035 FM04 32005KT P6SM NSW SKC
TAF KFTW 210995 1010 VRB03KT 1SM DZBR OVC003 BECMG 1315 00000KT 2SM
DZBR OVC080 PROB40 1417 1SM RABR OVC030 FM17 33005KT 4SM DZ BKN020
PROB40 1722 00000KT 3SM RA OVC010 FM22 33008KT 4SM BR BKN020
TAF KDFW 210950 1010 00000KT 1SM DZ OVC003 FM15 35010KT 3SM RA OVC010
BECMG 1820 33010KT 4SM BR OVC020
TAF KDAL 210940 1010 00000KT 1½SM DZ OVC004 BECMG 1517 34010KT 3SM RA
OVC010 FM22 33010KT 3SM BR BKN020

Figure 7-6. Here are the METARs and TAFs for the route. "KCDS" is Childress, "KSPS" is Wichita Falls, "KFTW" is Meacham Field at Fort Worth, "KDFW" is Dallas/Fort Worth International which is located less than 20 miles east of Meacham, and "KDAL" is Dallas Love Field which is located about 30 miles east of Meacham.

or above landing minimums at your ETA. The terminal aerodrome forecast for Meacham at 17Z shows a visibility of four miles in drizzle, with a 40% chance of three miles and rain. Since the visibility is above that required for the ILS and NDB approaches at Meachan, you may file for this destination.

ALTERNATE AIRPORT REQUIREMENTS

An alternate is required unless the weather at your destination is forecast to be 2,000 and 3 at your ETA plus or minus one hour.

Your next step is to determine if an alternate airport is required for the flight. Regulations require an alternate unless the destination is forecast to have a ceiling of at least 2,000 feet and a visibility of at least three miles at your ETA plus or minus one hour. The prognostic chart and area forecast both indicate that the weather at Meacham should be marginal VFR.

Although this information is useful, you should use all available reports and forecasts to determine the weather at your destination. You will again need to refer to the TAF for Meacham. Between 17Z and 22Z, the forecast is for a ceiling of 2,000 feet and four miles, with a 40% probability of a 1,000-foot ceiling and visibility of three miles. Based on this forecast, you cannot conclude that the ceiling at your destination will be at or above 2,000 feet and the visibility at least three miles. Therefore, you are required to list an alternate on your flight plan.

Alternate minimums are for flight planning purposes only. If you proceed to your alternate, use the normal landing minimums for the approach you use.

Two nearby airports which might be used as alternates are Dallas/Fort Worth International (DFW) and Dallas Love Field (DAL). The standard alternate minimums for flight planning purposes are a ceiling of 600 feet and two miles visibility for a precision approach, and a ceiling of 800 feet and two miles visibility for a nonprecision approach. Since precision approaches are available at both airports, the lower alternate minimums apply. If it becomes necessary to proceed to your alternate, you can use the normal published landing minimums for the approach you plan to use. [Figure 7-7]

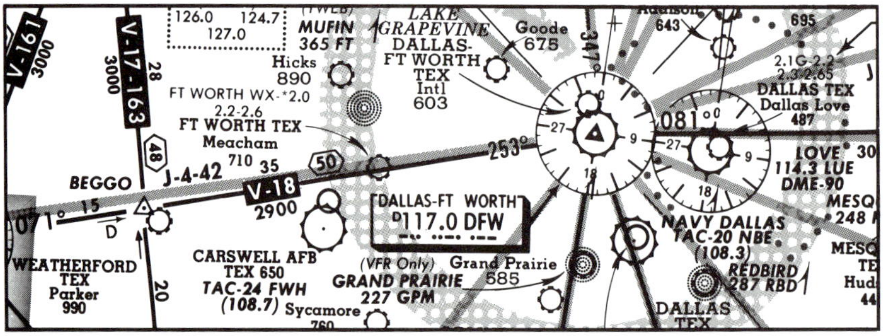

Figure 7-7. The Dallas/Fort Worth Area offers several options for alternatives. Two of the best alternates are Dallas/Fort Worth International (DFW) and Dallas Love Field (DAL). Both airports have a variety of approaches, issue terminal forecasts, and have standard alternate minimums, both precision and nonprecision.

With a 1600Z departure from CDS, you should reach one of these alternates between 1700Z and 1800Z. For this time period, both airports qualify with forecast ceilings of 1,000 feet and three miles, so let's take a closer look at them. DFW is a major international airport and you can expect delays into this field, especially during IFR conditions. On the other hand, DAL is a general aviation airport, and it is better suited to your needs. Therefore, DAL is a reasonable alternate.

With current and forecast weather in hand, your next step is to obtain the forecast winds aloft for the route of flight. In this case, the winds for the route are 190° at 25 knots through 12,000 feet. Pilot reports indicate occasional light chop along the route with tops varying from 6,500 feet to 8,000 feet.

The last item to check is NOTAMs. An inoperative navaid could be critical. Be sure to check NOTAMs for all airports you plan to use. [Figure 7-8]

JEPPESEN WEATHER LOG

		Departure KCDS	Enroute KSPS	Destination KFTW	Alternate KDAL
Ceiling, Visibility and Precipitation	Reported	35008 2RA OVC005 21/17 2993	CALM 1/2-RA FG OVC002 20/19 2998	35010 1DZBR OVC004 19/19 3001	35005 3DZ OVC008 19/18 3002
	Forecast	16Z 35012 2RA OVC005	16Z VRB12 2DZ BR OVC005	17Z 4DZ BKN020 PROB40 1722 3RA OVC010	17Z 3 RA OVC010
Winds Aloft			190°@ 25 Through 12,000		
Icing and Freezing Level			Freezing level 12,000		
Turbulence and Cloud Tops			Light chop Tops 6,500 to 8,000		
NOTAMs			NONE		

Figure 7-8. As you obtain the weather for the flight, be sure to write down the essential information on your weather log. Notice, you can use the weather log as a checklist during your briefing.

ALTITUDE SELECTION

Before you complete your planning, you must select a cruising altitude. You have already determined the highest MEA for the flight is 3,700 feet. Because ATC generally assigns altitudes which correspond with the hemispheric rule, the lowest altitude you should request is 5,000 feet. Since you are not carrying oxygen for this trip, the highest altitude you should request is 11,000 feet.

Next, consider the effects of a headwind or tailwind and select an altitude to optimize groundspeed. For this flight, the winds are constant through 12,000 feet, so winds are not a factor. Other considerations are forecast turbulence and icing. Again, neither of these is a factor for the flight. The cloud tops are reported as 6,500 feet to 8,000 feet, so a cruising altitude of 9,000 feet will put you on top. The prognostic chart indicates thunderstorm activity along your route so you could select 9,000 feet to allow you to observe thunderstorm buildup and avoid flying into embedded storms. A check of current forecasts, however, indicates convective activity is no longer predicted. This is confirmed by pilot reports and current radar observations. Therefore, for this flight, you decide to fly at 5,000 feet to keep your IFR skills sharp. Should convective activity develop, you can request a different altitude while enroute.

COMPLETING THE NAVIGATION LOG

With the briefing complete, you can now make your go/no-go decision. Base this decision on the weather reports and forecasts, and weigh them against your aircraft's capabilities and equipment. You should also consider your personal limitations and level of proficiency. In analyzing this flight, you can see that it can be conducted on top during the enroute phase, if desired. Icing is not a factor nor is turbulence. If you have to proceed to your alternate, the weather there is forecast to be VFR at your ETA. Based on this analysis, you decide the flight is within your capabilities and that of the aircraft. Your next step is to complete the navigation log. [Figure 7-9]

The navigation log offers you a convenient way to complete your preflight planning and to monitor the progress of the flight while enroute. For example, if you start with 48 gallons of fuel, the fuel remaining should agree with the figures listed on the navigation log as you pass each checkpoint. By recording the "time off" and completing the estimated time of arrival (ETA), actual time enroute (ATE), and actual time of arrival (ATA) boxes, you will be quickly alerted to any changes in winds aloft or in your aircraft's performance.

FILING THE FLIGHT PLAN

After you have completed the navigation log, transfer the appropriate information to your IFR flight plan and file it with the local FSS.

JEPPESEN NAVIGATION LOG

Aircraft Number: N8458R	Dep: CDS	Dest: FTW	Date:

Clearance:

Estimated Time Enroute = 1:55

Check Points (Fixes)	Ident / Freq.	Course (Route)	Altitude	Mag. Crs.	FUEL Leg / Rem.	Dist. Leg / Rem.	GS Est. / Act.	Time Off ETE / ATE	ETA / ATA
CHILDRESS A					48	196			
CHILDRESS	CDS	D→	↑5	172		3	68	3	
	117.6					193			
LEVEL OFF		V114	↑5	095	3.0	7	84	5	
					45	186			
WICHITA FALLS	SPS	V114	5	095	5.0	80	120	40	
	112.7				40	106			
BRIDGEPORT	BPR	BOIDS	9	127	4.2	61	107	34	
	116.5				35.8	45			
ARINA Δ		↓		108	2.1	32	114	17	
					33.7	13			
FORT WORTH A		RV			1.2	13		10	
					32.5	—			
Approach					1.0			6	
					31.5				
(Alternate) DALLAS LOVE A	LUE	RV			2.5	27		20	
	114.3				29.0				

FUEL									
Climb	3.0	Cruise	12.5	Apch.	1.0	Alt.	2.5	Res.	5.6
Cruise Burn/Hr	7.5	Block In		Block Out		Log Time			

Figure 7-9. Here is the completed navigation log for the flight. Notice that abbreviations commonly used in IFR shorthand are also used here. Magnetic course is also used since this is what is shown on instrument charts.

Remember to file your IFR flight plan at least 30 minutes prior to the listed departure time. This allows sufficient time for the FSS to transmit your flight plan to the appropriate ARTCC facility and process it. [Figure 7-10]

Be sure to include the aircraft equipment code in block 3 of the flight plan.

An important item to include on IFR flight plans is the aircraft equipment code. This is done with a code letter suffix attached to the aircraft type in block 3. The codes and their meanings are shown at the bottom of this flight plan form. In this case, the "/A" means the aircraft is equipped with DME and a transponder with altitude encoding capability.

Notice that airport identifiers are used instead of the airport names in blocks 5, 9, and 13. Use of these codes expedites the processing of your flight plan. If you do not know the correct code, list the name of the airport. For the route in block 8, use identifer codes for VORs, airways, waypoints, SIDs and STARs.

List the initial cruising altitude in block 7 of the flight plan.

The altitude listed in block 7 should be your initial cruising altitude. If you want to change the altitude, direct your request to the controller during flight. Use the latest forecast winds to compute the estimated time enroute, listed in block 10, to the point of first intended landing. For fuel on board, recorded in block 12, you should include the total usable fuel on board expressed in hours and minutes.

FAA FLIGHT PLAN

1 Type	2 AIRCRAFT IDENTIFICATION	3 AIRCRAFT TYPE/ SPECIAL EQUIPMENT	4 TRUE AIRSPEED	5 DEPARTURE POINT	6 DEPARTURE TIME PROPOSED (Z)	6 DEPARTURE TIME ACTUAL (Z)	7 CRUISING ALTITUDE
VFR / ✓ IFR / DVFR	8458R	PA 28/A	110 KTS	CDS	1600		5,000

8 ROUTE OF FLIGHT: CDS V114 SPS. BOIDS 9

9 DESTINATION (Name of airport and city)	10 EST TIME ENROUTE HOURS	10 EST TIME ENROUTE MINUTES	11 REMARKS
FTW	1	55	

12 FUEL ON BOARD HOURS	12 FUEL ON BOARD MINUTES	13 ALTERNATE AIRPORT(S)	14 PILOTS NAME, ADDRESS & TELEPHONE NUMBER & AIRCRAFT HOME BASE	15 NUMBER ABOARD
6	00	DAL	I. KAHN FTW 555-6834	2
			17 DESTINATION CONTACT/TELEPHONE (OPTIONAL)	

16 COLOR OF AIRCRAFT: GREEN ON WHITE

CLOSE VFR FLIGHT PLAN WITH ______________ FSS ON ARRIVAL

Special Equipment Suffix
X - No Transponder
T - Transponder With No Altitude Encoding Capability
U - Transponder With Altitude Encoding Capability
D - DME, No Transponder
B - DME, Transponder With No Altitude Encoding Capability
A - DME, Transponder With Altitude Encoding Capability
C - RNAV, Transponder With No Altitude Encoding Capability
R - RNAV, Transponder With Altitude Encoding Capability
W - RNAV, No Transponder
G - Global Positioning System (GPS)/Global Navigation Satellite System (GNSS) equipped aircraft with oceanic, en route, terminal, and GPS approach capability.

Figure 7-10. Each block of the IFR flight plan contains specific information. Some of the blocks are discussed in more detail in the accompanying text.

COMPOSITE FLIGHT PLAN

When a flight is to be conducted under both VFR and IFR, check both boxes in block 1 of the flight plan.

Ordinarily, a flight is conducted under either VFR or IFR from departure to arrival. Sometimes, though, you may wish to fly a segment of a route under VFR and another under IFR. When this is the case, be sure to check both the VFR and IFR boxes in block 1 of the flight plan.

If you fly the first portion of the flight under VFR, report your departure time to the local FSS. Be sure to close the VFR portion of the flight plan and request an ATC clearance from the FSS that is nearest the point at which you propose to change from VFR to IFR. Remember, too, you must remain in VFR conditions until you receive an IFR clearance from ATC.

When a flight plan indicates IFR for the first portion of the flight, you are normally cleared to the point at which the change is proposed. Upon arrival at the clearance limit, you should cancel the IFR portion of the flight and contact the nearest FSS to activate the VFR portion. If you desire to continue the flight under IFR past the clearance limit, you should advise ATC at least five minutes prior to reaching the clearance limit and request a further IFR clearance. If the requested clearance is not received prior to reaching the clearance limit, you are expected to establish yourself in a standard holding pattern on the radial or course, as appropriate. If a holding pattern is depicted at that location, you should hold as published.

CHECKLIST

After studying this section, you should have a basic understanding of:

- ✓ **Preliminary weather checks** — What the reasons for preliminary weather checks are and what type of information you need.
- ✓ **Route selection** — What factors you should consider in making the initial route selection, including the use of appropriate SIDs and STARs.
- ✓ **Flight information publications** — Which publications to use and what information each contains.
- ✓ **Weather information** — What types of reports and forecasts to use, how to interpret the trends they indicate, and how to determine if you need to file for an alternate airport.

SECTION B

IFR EMERGENCY PROCEDURES

No one likes to think that they might be involved in an emergency situation, especially during IFR flight. However, emergencies happen, and being prepared is your best defense. In this section, we define various types of emergency conditions, look at procedures for dealing with the loss of two-way radio communications, and discuss a method for handling a directional gyro malfunction. We also discuss applicable regulations, the information ATC requests from you during an emergency, and some of the ways ATC can provide assistance.

Naturally, you should be thoroughly familiar with the aircraft's POH. The POH outlines specific procedures that apply to your aircraft in various emergencies. In spite of your preparation, however, it is not possible to prepare for every situation that can occur. When confronted by an emergency not covered in the regulations or in the aircraft's POH, you are expected to exercise good judgment in whatever action you take.

An emergency can be defined as either a distress or an urgency situation. The AIM defines **distress** as a condition of being threatened by serious and/or imminent danger and of requiring immediate assistance. An **urgency** situation is a condition of being concerned about safety and of requiring timely, but not immediate, assistance — a potential distress condition. If you become apprehensive about your safety for any reason, you should request assistance immediately. An emergency may be declared by either the pilot or a controller.

After an emergency in which you are given priority handling by ATC, you may be requested to submit a detailed report.

Situations associated with weather, such as inadvertent entry into a thunderstorm, hail, severe turbulence, or icing, are potential emergencies. Fuel starvation or inability to maintain the MEA are undeniable emergencies. If, after considering the particular circumstances of the flight, you feel a potentially dangerous or unsafe condition exists, you should declare an emergency. In an emergency, you may deviate from any rule in FAR Part 91, Subparts A and B, to the extent necessary to meet the emergency. If, during the course of the emergency, you are given priority handling by ATC, you may be requested to submit a detailed report to the chief of that ATC facility, even though you violated no rule.

DECLARING AN EMERGENCY

To declare an emergency when operating under an IFR clearance, you should contact ATC on the currently assigned frequency. If you receive no response, try calling the same facility on a different frequency. If they still do not respond, use the emergency frequency of 121.5 MHz. Lastly, you should attempt to contact the nearest tower or FSS. In a distress situation, begin your initial call with the word "MAYDAY" repeated three times. Use "PAN-PAN" repeated three times in an urgency situation. [Figure 7-11]

Another way to declare an emergency is to squawk code 7700 on your transponder. This code triggers an alarm or a special indicator in radar facilities. However, you should not change your transponder code from its current setting when in radio and radar contact with ATC, unless you are instructed to do so.

A special emergency is a condition of air piracy, or other hostile act by a person or persons aboard an aircraft, which threatens the safety of the

Information	Example
Distress or Urgency	*"MAYDAY, MAYDAY, MAYDAY (or PAN-PAN, PAN-PAN, PAN-PAN),*
Name of station addressed	*Seattle Center,*
Identification and type of aircraft	*1114V Piper Comanche,*
Nature of distress or urgency	*severe icing,*
Weather	*IFR,*
Pilot's intentions and request	*request immediate course reversal and lower altitude,*
Present position and heading	*Jimmy Intersection, heading 253°,*
Altitude or flight level	*9,000.*
Fuel remaining in hours and minutes	*Estimate two hours fuel remaining,*
Number of people aboard	*five aboard,*
Any other useful information	*squawking 1146."*

Figure 7-11. This is an example of a complete distress or urgency message. You should provide ATC with only as much of this information as is necessary under prevailing conditions. For example, if you are in radar contact with an ATC facility, you may not need to include your aircraft's type, position, heading, or altitude, since the controller is already aware of these facts.

aircraft or its passengers. Although these incidents rarely involve the average pilot, you should be aware of the recommended ATC procedures.

If possible, you should use normal distress or urgency procedures. When circumstances do not permit you to use these, transmit a message containing as much of the following information as possible on the frequency in use.

1. Name of the station addressed
2. Aircraft identification and present position
3. Nature of the special emergency condition and your intentions

If you are unable to provide the above information, you should alert ATC by transmitting the phrase, "*transponder seven five zero zero*" and/or squawking transponder code 7500. Each of these means, "*I am being hijacked/forced to a new destination.*"

MINIMUM FUEL

Declaring "minimum fuel" is merely an advisory that further delays may cause an emergency situation to develop. It is not the same as declaring an emergency.

If your remaining fuel is such that you can accept little or no delay, you should advise ATC that you have "minimum fuel." This is not an emergency, but only an advisory that any undue delays may cause an emergency situation. If it appears that your fuel supply may be critical, do not hesitate to declare an emergency.

COMMUNICATIONS FAILURE

Two-way radio communications failure procedures for IFR flights are outlined in FAR 91.185 and expanded in the *Aeronautical Information*

DEPARTURE ROUTE DESCRIPTION

TAKE-OFF RUNWAYS 16L/R: Climb via I-RNO Localizer south course to WAGGE INT then via radar vectors to assigned route.

LOST COMMUNICATIONS: If not in contact with departure control within one minute after takeoff, or if communications are lost before reaching 9000', continue climb via I-RNO localizer south course to WAGGE INT, turn left, proceed direct FMG VORTAC. Cross FMG VORTAC at or above MCA, thence via assigned route or climb in holding pattern northeast on FMG R-041, left turns to cross FMG VORTAC at or above MCA for assigned route.

WASHOE TWO DEPARTURE
(VECTOR) (WASH2.WAGGE)

RENO, NEVADA
RENO CANNON INTL

Figure 7-12. Special lost communications procedures are sometimes used with SIDs. If you are flying this SID and lose radio communications capability after departure, you must comply with the lost communications procedures specified here instead of those in the FARs.

Manual. Unless otherwise authorized by ATC, pilots operating under IFR are expected to comply with this regulation. In some cases, special lost communications procedures are established for certain IFR procedures. [Figure 7-12]

ALERTING ATC

To alert ATC to a radio communications failure, you should squawk code 7600. If only your transmitter is inoperative, listen for ATC instructions on any operational receiver. These include your navigation receivers, since ATC may try to contact you over a VOR, VORTAC, NDB, or localizer frequency.

Squawk code 7600 to alert ATC to a communications failure.

The primary objective of the regulations for communications failure is to preclude extended operations of an aircraft experiencing communications failure in the ATC system. If your radio fails in VFR conditions, or if you encounter VFR conditions at any time after the failure, you should continue the flight under VFR, if possible, and land as soon as practical. If IFR conditions prevail, you must comply with procedures designated in the FARs to ensure aircraft separation.

If you lose communications and encounter VFR conditions, you are expected to continue the flight under VFR, if possible, and land as soon as practical.

ROUTE

When you continue your flight under IFR after experiencing two-way radio communications failure, you should fly one of the following routes in the order shown.

1. The route assigned by ATC in your last clearance
2. If being radar vectored, the direct route from the point of radio failure to the fix, route, or airway specified in the radar vector clearance
3. In the absence of an assigned route, the route ATC has advised you to expect in a further clearance
4. In the absence of an assigned or expected route, the route filed in your flight plan

ALTITUDE

The altitude you fly after a communications failure must be the highest of the following altitudes for each route segment flown.

1. The altitude assigned in your last ATC clearance
2. The minimum altitude (or flight level) for IFR operations
3. The altitude ATC has advised you to expect in a further clearance

In some cases, your assigned or expected altitude may not be as high as the MEA on the next route segment. In this situation, you normally begin a climb to the higher MEA when you reach the fix where the MEA rises. If the fix also has a published minimum crossing altitude (MCA), start your climb so you will be at or above the MCA when you reach the fix. If the next succeeding route segment has a lower MEA, descend to the applicable altitude — either the last assigned altitude or the altitude

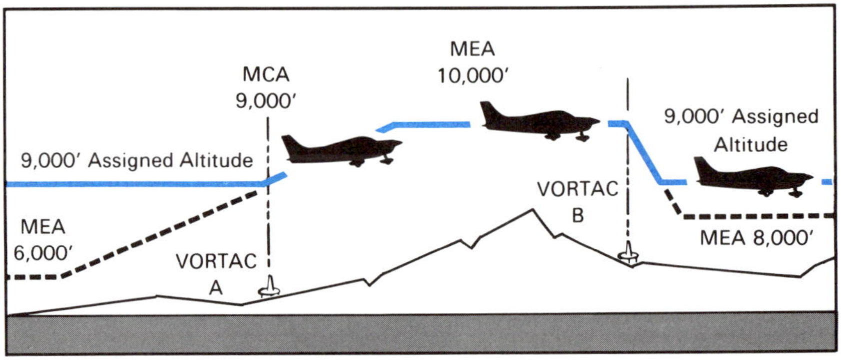

Figure 7-13. Assume your last assigned altitude is 9,000 feet. At VORTAC "A," the MEA increases to 10,000 feet, and an MCA of 9,000 feet is also specified. Since your assigned altitude is 9,000 feet, you need not begin a climb before you reach the VORTAC. Upon reaching VORTAC "A," climb to the higher MEA of 10,000 feet. As you arrive over VORTAC "B," descend back to your assigned altitude of 9,000 feet, since the MEA is now lower than your assigned altitude.

expected in a further clearance — when you reach the fix where the MEA decreases. [Figure 7-13]

LEAVE CLEARANCE LIMIT

Upon arrival at a destination with more than one instrument approach procedure, you may fly the approach of your choice. Similarly, if more than one initial approach fix is available for the approach you choose, you may select whichever fix is appropriate. ATC provides separation for your flight, regardless of the approach selected and the initial approach fix used. You may not, however, begin a descent and approach when you arrive at a point from which the approach begins unless certain conditions have been met.

When the clearance limit specified in the last ATC clearance you received prior to radio failure is also a point from which an approach begins, commence your descent and approach as close as possible to the expect further clearance (EFC) time. If an EFC time was not provided, begin your descent and approach as close as possible to the ETA calculated from your filed or amended (with ATC) time enroute.

When you are holding at a point which is not an approach fix, depart the fix at the EFC time.

When the clearance limit specified in the last clearance is not a fix from which an approach begins, such as a holding pattern, leave the fix at the EFC time specified. After departing the fix and upon arrival at a point from which an approach begins, commence your descent and complete the approach. If you have not received an EFC time, continue past the clearance limit to a point at which an approach begins. You should then commence your descent and approach as close as possible to the ETA calculated from your filed or amended time enroute.

SURVEILLANCE APPROACH

An instrument procedure where a controller provides only azimuth navigational guidance is referred to as an **airport surveillance radar** (ASR) approach. The ASR approach is also referred to as a **surveillance approach**. On this type of procedure, the controller furnishes you with headings to fly to align your aircraft with the extended centerline of the landing runway. You are advised when to start the descent to the MDA or, if appropriate, to an intermediate stepdown fix and then to the prescribed MDA. The ASR approach is available only at airports for which civil radar instrument approach minimums have been published. It is typically available as a backup procedure. Use it when you have lost navigation equipment which allows you to fly the other instrument approaches available at the airport.

You may use a surveillance approach at airports for which civil radar instrument approach minimums have been published.

The published MDA for straight-in ASR approaches is issued to you before you are instructed to begin your descent. When a surveillance approach terminates in a circle-to-land maneuver, you must furnish the aircraft approach category to the controller, who then provides you with the appropriate MDA.

During the approach, you are advised of the location of the missed approach point and, while on final, of your position each mile from the runway or MAP. If you request, you are advised of the recommended altitude each mile. Normally, ATC provides navigation guidance until your aircraft reaches the MAP. At the MAP, ATC terminates guidance and instructs you to execute a missed approach unless you have reported the runway environment in sight. Also, if at any time during the approach the controller considers that safe guidance for the remainder of the approach cannot be provided, guidance is terminated and you are instructed to execute a missed approach. Radar service is automatically terminated at the completion of the approach.

On a surveillance approach, ATC automatically tells you what headings to use, when to begin the descent to MDA, your aircraft's position each mile on final from the runway or MAP, and when you have arrived at the MAP.

NO-GYRO APPROACH

If your directional gyro instrument becomes inoperative or inaccurate, you may request a no-gyro vector/approach. During the vector or approach, ATC instructs you to make turns by saying, "*turn right*," "*stop turn*," and "*turn left*." ATC expects these turns to be made at standard rate and to be executed as soon as you receive the instructions. During a no-gyro approach, you are advised when you have been turned onto final and told to make all turns at one-half standard rate.

Before turning final, make all turns at standard rate. After turning final, all turns must be one-half standard rate.

MALFUNCTION REPORTS

The loss of certain equipment during a particular IFR flight may have little significance to you. For example, you are flying to a destination not served by an NDB, the loss of the ADF receiver may be hardly noticeable. However, if you are operating in controlled airspace under IFR, you are required by regulations to report immediately to ATC the loss of certain equipment. These include any malfunctions of navigational, approach, or communication equipment.

When a report is required, you are expected to include your aircraft identification, equipment affected, and the degree to which your IFR operational capability is impaired. You must also state the nature and extent of assistance you need from ATC. For example, if your ADF receiver fails on the way to an airport served only by an NDB approach, you may have to request an amended clearance to another airport.

CHECKLIST

After studying this section, you should have a basic understanding of:

✓ **Declaring an emergency** — The different types of emergencies, who to contact, and when to use the terms "MAYDAY" or "PAN-PAN."

✓ **Transponder codes** — The codes used to indicate an emergency, radio failure, and air piracy.

✓ **Communications failure** — The procedures to be followed if VFR conditions are encountered, the route and altitudes to fly in IFR conditions, when to depart a hold, and when to begin your descent for an approach.

✓ **Surveillance approach** — The services ATC will provide, when to descend and execute a missed approach, and when radar guidance will terminate.

✓ **No-gyro approach** — What it is, when to ask for it, and the instructions that are issued by ATC.

✓ **Malfunction reports** — What types of equipment malfunctions you must report to ATC and what you must include in the report.

SECTION C

IFR DECISION MAKING AND FLIGHT CONSIDERATIONS

In recent years, the FAA, General Aviation Manufacturer's Association, and Transport Canada have conducted studies concerning decision making. Various other aviation groups also participated in or contributed to these studies. The following section is a summary of the results of this research. The first portion covers basic decision-making information, while the second part addresses information of particular interest to you as an instrument pilot.

THE DECISION-MAKING PROCESS

Flying is a combination of events which requires you to make a continuous stream of decisions. The events in this process are interrelations among people, the aircraft, and the environment. They can be placed in five subject areas: pilot, aircraft, environment, operation, and situation.

Flying requires a continuous stream of decisions about yourself, your airplane, and the environment.

As a pilot, you are continually making decisions about your own competency, state of health, level of fatigue, and many other variables. Your decisions are frequently based on evaluations of the aircraft, such as its power, equipment, or airworthiness. The environment encompasses many of the items not covered in the two previous categories. It can include such things as weather, air traffic control, and runway length or surface. The interactions of you, your aircraft, and the environment are influenced by the purpose of each flight operation. You must evaluate the three previous areas to decide on the desirability of undertaking or continuing the flight operation as planned. Is the trip worth the risks? The situation refers to your situational awareness, which is the accurate perception of the conditions affecting you and your aircraft during a specific period of time. More simply, it is knowing what is going on around you. [Figure 7-14]

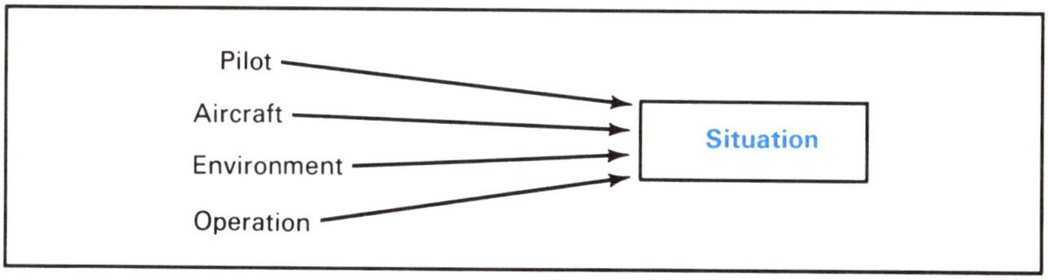

Figure 7-14. Any given situation is affected by the pilot, aircraft, environment, and operation subject areas. There is a direct relationship between your awareness of the situation and safety. The higher your situational awareness, the safer you are as a pilot.

It is important to recognize a change, then take action to control it.

In conventional decision making, the need for a decision is normally triggered by the recognition that something has changed or that an expected change did not occur. Once you have recognized the need for a decision, your selection of the proper response depends on several elements. Your level of skill, knowledge, experience, and training all influence your ability to make the best choice. If you fail to search for and recognize a change, you reduce your chance of controlling it. As time passes, your available alternatives can decrease. For example, if you have already entered an area of hazardous weather, your alternative to circumnavigate the weather is automatically lost.

Aeronautical decision making (ADM) builds on the foundation of the conventional process, but enhances it to decrease the probability of pilot error. It can be defined as:

* The ability to search for and establish the relevance of all available information regarding a flight situation, to specify alternative courses of action, and to determine expected outcomes from each alternative
* The motivation to choose and authoritatively execute a suitable course of action within the time frame permitted by the situation

The first portion of this definition refers to intellectual capabilities. It relies on your abilities to sense, store, retrieve, and integrate information. This part of decision making is purely rational. If used alone, it allows problem solving in much the same manner as a computer. The second portion implies that your decisions are based, in part, on tendencies to use other than safety-related information when choosing courses of action. For example, you might consider such items as job demands, convenience, monetary gain, and self-esteem before taking action. Ideally, this nonsafety information can be eliminated and your decisions can be directed by a more rational process.

The steps in the DECIDE process are detect, estimate, choose, identify, do, and evaluate.

The decision-making process is quite complex; however, it can be condensed into six elements using the acronym DECIDE. The steps in the DECIDE process are as follows:

* **D**etect the fact that a change has occurred.
* **E**stimate the need to counter or react to the change.
* **C**hoose a desirable outcome for the success of the flight.
* **I**dentify actions which could successfully control the change.
* **D**o the necessary action to adapt to the change.
* **E**valuate the effect of the action.

There are many examples of poor aeronautical decision making as the result of the pilot's failure to take charge in a situation where action was needed. A survey of Ohio pilots by students and faculty at Ohio State University helps to illustrate this. In this survey, pilots were asked to give brief descriptions of errors that they had made or that they had seen other pilots make during their flying careers. The descriptions were grouped into eight different categories, roughly corresponding to the most important 1,500 errors revealed by 77 pilots. Of these, 300 were found to be unique. The following abbreviated list of observed pilot errors in the area of "failure to take charge" indicates that this is a serious problem in aviation.

1. Flying into clouds while waiting for an IFR clearance
2. Descent below ATC-assigned altitude while looking for airport on a visual approach
3. Failure to recognize that basic aircraft control is first priority
4. Failure to declare an emergency due to fear of FAA sanction
5. Failure to contact ATIS prior to entry into busy terminal area
6. Failure to ask for information when it is needed (for example, being uncomfortable about talking to FSS about a problem)
7. Unclear division of cockpit responsibilities
8. Not thinking before doing (for example, talking on the radio)
9. Getting "behind" the airplane
10. Not having alternate plans of action formulated in case of weather problems
11. Failure to provide margin of safety
12. Proper decision made too late
13. Failure to recognize that an emergency had occurred
14. Out of order priorities (for example, "get-home-itis")
15. Failure to make decisions concerning weather minimums; expecting ATC to tell them what to do
16. Continued VFR flight into IFR weather
17. False sense of security from having an airplane with two engines, leading to increased risk taking

Several factors are involved in good decision making. These include: identifying personal attitudes which are hazardous to safe flight, developing risk assessment skills, learning to recognize and cope with stress, learning behavior-modification techniques, and evaluating the effectiveness of your ADM skills. First, let's look at the role your attitudes play in decision making.

HAZARDOUS ATTITUDES

There are five hazardous attitudes which affect pilot decision making. It is important to understand them as they apply to your flying. Your *Exercise Book* contains a self-assessment inventory to give you personal insight into these attitudes.

The five hazardous attitudes are anti-authority, impulsivity, invulnerability, macho, and resignation.

1. **Anti-authority:** "Don't tell me!" — People with this attitude may resent having someone tell them what to do, or they may just regard rules, regulations, and procedures as silly or unnecessary. (Remember,

ber, though, it is always your prerogative to question authority if you feel it is in error.)

2. **Impulsivity:** "Do something — quickly!" — This is the thought pattern of people who frequently feel the need to do something — anything — immediately. They do not stop to consider what they are about to do so they can select the best alternative; they do the first thing that comes to mind.
3. **Invulnerability:** "It won't happen to me!" — Many people feel that accidents happen to others but never to them. Pilots who think this way are more likely to take chances and unnecessary risks.
4. **Macho:** "I can do it." — These people are always trying to prove that they are better than anyone else by taking risks and by trying to impress others. While this pattern is thought to be a male characteristic, women are equally susceptible.
5. **Resignation:** "What's the use?" — People with this attitude do not see themselves as making a great deal of difference in what happens to them. When things go well, they think, "That's good luck." When things go badly, they attribute it to bad luck or feel that someone is "out to get them." They leave the action to others — for better or worse. Sometimes, such an individual will even go along with unreasonable requests just to be a "nice guy."

ANTIDOTES FOR HAZARDOUS ATTITUDES

There are ways to overcome the five major hazardous attitudes which contribute to poor pilot decision making. One way is to become thoroughly aware of them by studying the preceding paragraphs and completing the self-assessment inventory in the *Exercise Book*. Another is to use **antidotes**. By telling yourself something which counteracts the hazardous attitude, you're "taking an antidote." Learn to recognize a hazardous attitude, correctly label the thought, and then say its antidote to yourself. [Figure 7-15]

Hazardous Attitude	Antidote
Anti-authority: "Don't tell me!"	"Follow the rules. They are usually right."
Impulsivity: "Do something—quickly!"	"Not so fast. Think first."
Invulnerability: "It won't happen to me!"	"It could happen to me."
Macho: "I can do it."	"Taking chances is foolish."
Resignation: "What's the use?"	"I'm not helpless. I can make a difference."

Figure 7-15. To overcome hazardous attitudes, memorize the antidotes for each of them. Know them so well that they will automatically come to mind if you need them.

RISK ASSESSMENT

Every aspect of life involves some element of risk, regardless of whether you drive a car, ride a motorcycle, or fly an airplane. You must learn to cope with the risks associated with flying to ensure years of safe flying. The five subject areas discussed earlier — pilot, aircraft, environment, operation, situation — are also the five elements of risk. Let's look at each of these separately.

To enjoy safe flying, you must learn to identify and deal with the elements of risk.

As a pilot, your performance can be affected in many ways during a flight. Environmental considerations, your physical condition, and your emotional state can all influence your performance.

The risk element of the aircraft focuses on its equipment, condition, and suitability for the intended purpose of the flight. The best time to make this assessment is on the ground during preflight planning. However, it also needs to be done continuously during flight, since conditions can change at any time. For example, winds aloft may increase your anticipated flight time, making available fuel a factor to be analyzed.

The environment is a far-reaching risk element which includes situations outside the aircraft that might limit, modify, or affect the aircraft, pilot, and operational elements. Weather is a common environmental risk factor. Density altitude, runway length, obstacles, and related factors can also create environmental concerns.

In terms of the purpose of the flight operation, you must evaluate the interaction of you, your aircraft, and the environment. When determining the risks involved with beginning or continuing the flight as planned, be sure to consider all available information.

The combination of these first four risk elements leads into the fifth — the overall situation which you must evaluate continuously. Remember, to become a safer pilot, you should increase your situational awareness.

Good aeronautical decision making requires a continuous assessment of whether to start a particular flight or to continue a flight as planned. You can use the five risk elements to help you make a "go/no go" decision. Unless all five indicate "go," you should reevaluate your decision to make a flight. A good decision maker does not act hastily on "gut" feelings. With an accurate assessment of the risks associated with each of the five elements, you are best able to arrive at decisions that ensure a safe conclusion to a flight, even if it means not departing.

Unless all five risk elements indicate "go," you should reconsider your decision to make a flight.

To assess risks effectively, you must be aware of risk raisers and the possibilities for risk accumulation so you can determine the need to neutralize, or balance, these factors. One way to become aware of this is to look at statistics to see what types of flight activities are most likely to result in accidents. The National Transportation Safety Board (NTSB) has conducted studies of accident rates for the various types of flying in general aviation (1982 data). They reveal that aerial application operations have the highest accident rate, followed by personal and business flying.

Poor skills in decision making are often a factor in aircraft accidents.

You should also be aware that the accident rate for single-engine airplanes is the highest for all general aviation airplane operations. More importantly, studies of the most common cause/factors of accidents in fixed-wing aircraft show that most accidents were the result of an unsafe outcome of the pilot's decision-making process. In fact, 85% of all general aviation accidents can be attributed, at least in part, to pilot error. Some common causes were: failure to maintain directional control, failure to maintain airspeed, misjudged distance, fuel exhaustion, inadequate preflight preparation and/or planning, selection of unsuitable terrain for landing, and inadequate visual scanning.

The NTSB also evaluated the phase of operation in which accidents occurred. The results indicate that the largest number (27.1%) occurred during landing, while 21.5% occurred during takeoff. Other statistics concern accident rates for pilots without instrument ratings flying single-engine airplanes at night in VFR conditions. [Figure 7-16]

IDENTIFYING AND REDUCING STRESS

As a healthy pilot, you should perform at your optimum level and make decisions to the best of your ability. Unfortunately, numerous physical

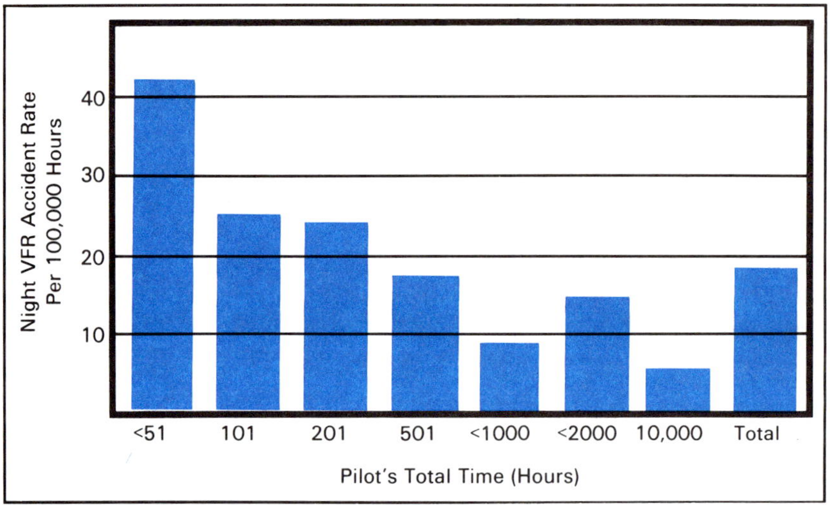

Figure 7-16. The night VFR accident rate is high for pilots with less than 51 hours. It decreases by nearly 50% once a pilot obtains 100 hours, and continues to decrease until the 1,000-hour level. The data suggest that, for the first 500 hours, pilots flying VFR at night should establish higher personal limits than are required by the regulations and apply instrument flying skills in this environment.

and physiological conditions in your life and the nature of flight itself can hamper this ability.

Piloting an aircraft is an individual process. Even though you hold a medical certificate that implies you meet the health requirements for your type of flying, the decision whether you are fit to fly is strictly in your hands. You need to evaluate your well-being from a physical and emotional standpoint. Factors affecting your total health are: physical, physiological, psychological, and sociological. These factors are known as **stressors.** Let's look at each of them individually.

1. **Physical stressors** include conditions associated with the environment, such as temperature and humidity extremes, noise, vibration, and lack of oxygen. The effects of these stressors can affect your physiological being as well as your psychological health. Recall the effects of the last flight you made without a headset or earplugs. After landing, were you a little irritable? Was your hearing somewhat impaired for a few hours?
2. **Physiological stressors** include fatigue (chronic and acute), lack of physical fitness, sleep loss, and missed meals. Have you ever felt light-headed or shaky during a flight when you hadn't eaten properly beforehand?
3. **Psychological stressors** are related to emotional factors such as self-imposed demands and perfectionism. The need to achieve affects your ability to perform and make decisions. The actual decision-making process is a stressor itself. Mental workload such as analyzing a problem, navigating, or handling an emergency are psychological stressors.
4. **Sociological stressors** include such things as a death in the family, divorce, sick child, demotion, or pressure from your boss. Sociological stressors are also part of adhering to the rules of society and the regulations of the government.

Stress, as used here, is a response to a stressor or set of circumstances that induces a change in your ongoing physiological and/or psychological patterns of functioning. Any internal or external stimulus that is perceived as a threat to the body's equilibrium causes a reaction (stress) with which your body must cope.

Acute physiological reactions to stressors include the release of chemical hormones (such as adrenalin) into your blood and the speeding up of your metabolism to provide energy to the muscles. Blood is taken away from your stomach and digestive tract to supply the muscles in your arms and legs with more oxygen. Blood sugar is increased, along with the heart rate, respiration, blood pressure, and perspiration. In other words, your body is preparing to "fight or flee," an ageless physiological response to threat.

Some stress is normal and beneficial, but too much stress can reduce your level of performance.

Stress forces you to adapt to it. It is an inevitable and necessary part of living. It motivates life and heightens your response to meet any challenge. Some individuals actually seek out stressful situations to keep life interesting. Stress can help prevent accidents in some cases. Stress is normal and beneficial; however, it is also cumulative. Performance of a task generally improves with the onset of stress, but peaks and then begins to fall off rapidly as stress levels exceed your ability to adapt to and to handle the situation. [Figure 7-17]

WHO ME? STRESSED?

You didn't sleep well last night; you had the meeting on your mind. You arrive at the airport hoping your aircraft has already been preheated as you had asked. It is extremely cold and damp, though there has been no snowfall, yet.

You had to stop at the bank for some money, so you had no time for breakfast. You grab a candy bar and coffee as you head for the hangar. You wash a couple of aspirin down with the last of your coffee and climb into your plane. You must get going to be at your meeting on time, and you mentally note what time you must leave the meeting to be home for your daughter's birthday party . . .

A generalized stress reaction can develop as a result of these accumulated effects. In this scenario, it is clear that many factors will influence your behavior and decision making throughout the trip.

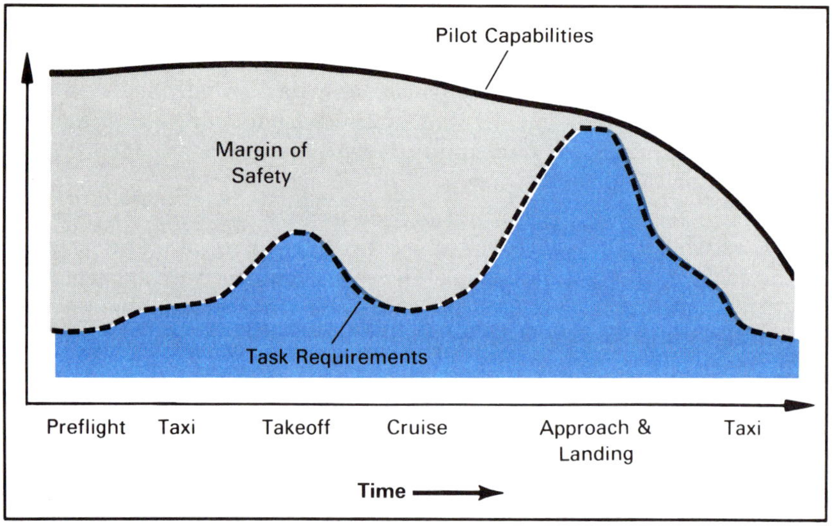

Figure 7-17. Accidents often occur when flying task requirements exceed pilot capabilities. The difference between these two factors is called the margin of safety. Note that in this idealized example, the margin of safety is minimal during the approach and landing. At this point, an emergency or distraction could overtax the pilot's capabilities and cause an accident.

. . . You arrive at the meeting without any problems and are preoccupied all day with whether you will get the money for the project. You are too nervous to even eat lunch, but you go ahead and have another cup of coffee. You are hesitant about leaving early, but you have promised your daughter you will be home for her party. As you make your apologies and leave the meeting, you find the snow just beginning to fall. You visit the flight service station briefly and find the weather to be deteriorating. You note the freezing level and the braking action reported at your home airport. You depart for home . . .

There is a limit to your adaptive nature and this limit is your **stress tolerance level**. It is based on your ability to cope with the situation. If the number or intensity of the stressors becomes too great, you are susceptible to "environmental overload." This is the point where your performance begins to decline and your decision-making skills deteriorate.

. . . Enroute you encounter some light rime ice and, though darkness is setting in, you can see well enough to monitor the buildup. You're glad you are on your way home and know the approaches at the airport well; you would not want to wrestle with an unfamiliar instrument approach right now. As you near your destination airport, you're informed the field has gone below minimums. You opt to make at least one pass and, indeed, you cannot see the field when you reach the MDA. Another airport, 10 miles away, is reporting better conditions. Your head is starting to pound as you throw open the approach chart book to the correct chart for the other airport . . .

Personality and behavior patterns are stress related. That aggressive, dominating personality you are so proud of can go a little too far, emerging as hostility. Rigid and unbending individuals will trigger the stress response when they find themselves between the proverbial "rock and a hard place." Another typical stress-provoking personality trait you may exhibit is your obsessive-compulsiveness and your attention to detail. For the most part, that perfectionism keeps you alive and allows you to be a good pilot. Just realize that it can be physiologically triggering the stress response. To a point it is adaptive, but past that point it becomes maladaptive.

. . . You quickly scan the instrument approach chart while being vectored for an ILS approach. You cannot find the localizer frequency anywhere on the chart and notice that you are using the VOR approach for the same runway. You think to yourself, "how dumb," and that thought just keeps repeating over and over. You turn the page and prepare for the approach . . .

Stress is insidious. It creeps up on you. It is cumulative. You may think you are handling everything quite well when, in fact there are subtle signs you are beyond your ability to respond appropriately. The inadequate strategies employed by most people in trying to cope with stress often impose more stress. When this occurs, people can develop anxieties and become frustrated. Frustration often leads to anger and aggression. Anger may be directed at other people or turned inward, showing up as self-destructive tendencies such as overeating, smoking, alcoholism, and drug abuse. It may take the form of excessive risk-taking and accidents. When an individual directs anger toward others, the resulting problems with interpersonal relationships lead to loss of friends, trouble with the boss, marital problems, divorce, child abuse, assault, and even homicide. Aggression directed at inanimate objects can lead to damage and injury that show up as accidents.

SIGNS OF INADEQUATE COPING

Emotional, physical, or behavioral symptoms may indicate the presence of excessive stress.

The indicators of excessive stress often show as three types of symptoms: emotional, physical, and behavioral. These symptoms differ, depending upon whether aggression is focused inward or outward. Those individuals who typically turn their aggressive feelings inward often demonstrate the emotional symptoms of depression, preoccupation, sadness, and withdrawal. Physical symptoms may show up as headaches, insomnia, appetite changes, weight gain or loss, indigestion, and so on. Behavioral symptoms include hypochondria, self-medication, a reluctance to accept responsibility, tardiness, absenteeism, and poor personal appearance and hygiene.

The individual who typically takes out frustration on other people or objects exhibits few physical symptoms. Emotional symptoms may show up as overcompensation, denial, suspicion, paranoia, agitation, restlessness, defensiveness, excess sensitivity to criticism, argumentativeness, arrogance, and hostility. Behavioral symptoms can include episodes of temper tantrums.

LIFE STRESS MANAGEMENT

It's important to recognize and reduce stress.

There are many techniques available that can help reduce the stress in your life or help you cope with it better. Not all of the following ideas may be the solution, but some of them should be effective.

1. Become knowledgeable about stress. Understand the process and effects of stress. Identify your major sources of stress, then anticipate stressful periods and plan your counterattack. Learn to identify the opportunities for personal growth inherent in periods of stress. Find the level of stress that is best for you, remembering that both insufficient and excessive stress are potentially harmful.
2. Take a realistic assessment of yourself. Evaluate your capabilities and limitations, then work (and fly) within them. Establish realistic goals, and identify the objectives that will lead toward those goals.

3. Take a systematic approach to problem solving. Learn to recognize and avoid the heavy pressures imposed by getting behind schedule and not meeting deadlines. Define your problem specifically, delving beyond symptoms. Divide it into manageable components that can be dealt with easily. Separate your tasks into three categories of action: things that must be done, those that can be delayed, and those that you can forget about. Gather sufficient information about the problem to put it in perspective. Discover why the problem exists for you, then review your experience with the present problem or similar ones. Develop and evaluate a set of alternative courses of action, then select a course of action, and proceed with it.
4. Develop a life style that acts as a buffer against the effects of stress. Whenever possible, avoid stressful situations and encounters. Don't volunteer for stressful jobs when you are already overburdened. Plan your use of time on both a daily and a long-term basis, and use your free time productively. Manage conflicts openly and directly. Deal with problems as soon as possible after they appear; if you procrastinate, they may intensify. Learn to let go of stressful situations and take breaks.
5. Practice behavioral management techniques. Establish a program of physical fitness. Exercise provides the body with an outlet for the energy used by the muscles for "fight or flee," reducing stress effects such as high blood pressure, accelerated pulse, and excess weight. Engage in recreational sports. Beware of increasing your stress levels through excessive competition. If you find this happening to you, try noncompetitive sports such as jogging, swimming, or biking. Eat a balanced diet and maintain your recommended weight. Obtain sufficient rest on a regular basis. Finally, be aware of other, more specialized techniques that can help you cope with stress, such as biofeedback and meditation. Most are designed to evoke what psychologists call the relaxation response. These techniques have been used successfully by athletes, businessmen, and others in high tension professions to maximize performance and minimize the effects of stress.
6. Establish and maintain a strong support network. Ask for direct help, and be receptive to it when it is offered. Consider professional help. Use resources and information that are available through such sources as the library or your physician.

COCKPIT STRESS MANAGEMENT

Good cockpit stress management begins with good life stress management. Many of the stress coping techniques practiced for life stress management are not usually practical in flight. Rather, you must condition yourself to relax and think rationally when stress appears. The following checklist outlines some thoughts on cockpit stress management.

1. Avoid situations that distract you from flying the aircraft. When you are carrying passengers, make sure that they are calm, informed, and prepared. If you encounter an emergency, keep them informed if you can. Avoid family squabbles in flight.
2. Reduce your workload to reduce stress levels and to provide the proper environment in which to make good decisions. If you feel tension mounting, loosen your collar, stretch your arms and legs, or open the aircraft's vents. Don't hesitate to ask controllers to help you by speaking more slowly or telling you your position.
3. If an emergency does occur, remain calm and continue to fly the airplane. Think for a moment, weigh the alternatives, then act. Fear and panic are your greatest enemies during an in-flight emergency. Don't hesitate to declare an emergency when necessary, or to let other people (including passengers) know your situation. Don't delay until it is too late.
4. Maintain proficiency in your aircraft, because proficiency builds confidence. Familiarize yourself thoroughly with your aircraft, its systems, and emergency procedures.
5. Know and respect your own personal limits. Give yourself plenty of leeway for an "out" when needed. Always have a "plan" and an "alternate plan." Make sure your stops allow adequate time for rest, for meals, and to stretch your legs. A good rule of thumb is to stop at least every three hours, especially if you are in a small aircraft where you don't have room to stretch.
6. Don't let little mistakes bother you until they build into a big thing. Wait until after you land, then analyze your past actions.
7. If flying is adding to your stress, either stop flying or seek professional help to manage your stress within acceptable limits.

FLIGHT FITNESS

Use the "I'm Safe" checklist before each flight.

The "go/no-go" decision is made before every flight. Most pilots give their aircraft a thorough preflight, yet many forget to preflight themselves. Ask yourself, "Could I pass my medical examination right now?" If the answer is not an absolute "yes," don't fly. [Figure 7-18]

IFR FLIGHT CONSIDERATIONS

Approach practice is one of the most important aspects of instrument training. However, because it is normally conducted under the hood in VFR conditions, such practice fails to provide some essential conditions for the instrument approach under actual IFR. After you reach decision height or MDA, your instructor tells you to take off the hood and land the airplane. You take off your hood, see the runway in full view ahead, and land VFR as you have done many times before.

This type of training fails to expose you to the situations that you must face in actual conditions, such as seeing the ground but not the runway,

The "I'm Safe" Checklist	
Illness?	Do I have any symptoms?
Medication?	Have I been taking prescription or over-the-counter drugs?
Stress?	Am I under psychological pressure from the job? Worried about financial matters, health problems, or family discord?
Alcohol?	Have I been drinking within eight hours? Within 24 hours?
Fatigue?	Am I tired and not adequately rested?
Eating?	Am I adequately nourished?

Figure 7-18. Check yourself as carefully as you check your airplane before flight. If in doubt, don't go.

or faintly seeing approach lights oriented in a very strange way but no runway. These conditions can lead to very serious problems and have led to many accidents. The final VFR segment of the instrument flight is the most hazardous, and many students are not exposed to these hazards during training.

The airlines recognize the dangers of the instrument approach to published minimums. They do not permit pilots to execute an ILS approach to Category II minimums unless an autopilot or dual-display flight director is used. In addition, they normally require captains who are checking out in new airplanes to use "high minimums" — 100 feet above published minimums — for the first 100 hours in that airplane.

You need to place similar personal minimums on your own instrument flying. If you fly only once a week, you probably don't have the skill to do an approach to published minimums — and it is not good judgment to try to prove that you do. If you fly on instruments very rarely, you should raise your personal minimums farther. If you fly with equipment that is malfunctioning or of questionable reliability, you should raise your minimums even farther.

Preparing to fly an instrument approach is at least as important as actually flying it. This should include studying the approach and the missed approach procedures. Be sure to include the lighting aids for the runway you expect to use, so you will know what to expect when you break out of the clouds in low visibility. If you know the wind direction and speed, it should give you an idea of the angle and the position of the runway in the windscreen — it will not necessarily be straight ahead. Realize that as sparse visual cues become available, they can cause disorientation until they are sufficiently clear to be used for landing.

Also know that there is a natural tendency for you to want to make a landing following an approach. This temptation can lead you to descend below minimums or "duck under" when even a few visual cues become available. However, when you do this, you are changing from the good quality cues that are presented on your flight instruments to uncertain outside visual cues that can cause confusion. When you first see the runway, use it only to confirm the instruments.

When decision height is reached, look outside and determine whether there is enough visibility to make a landing. If not, initiate a missed approach. If the approach can be continued, cross check the instruments and the outside environment to confirm agreement. If there is conflicting information between these two sources, execute the missed approach.

DESIGN FACTORS

You should also be aware of design factors that may increase the risk of mistakes. One example of a potential problem for the IFR pilot is entry of information into an RNAV computer. Some RNAVs do not provide adequate feedback to show you at a glance the point to which you are navigating. You are responsible for the correct input of the RNAV fix locations and for cross-checking your flight's progress using other navigation methods.

Different instrument designs can also cause confusion in the cockpit, particularly if you are flying an unfamiliar aircraft. Many VOR indicators are oriented so you read the course at the top of the instrument. Others, however, present the readout at the bottom. What makes this even more confusing is that most indicators show the reciprocal course at the opposite side from the correct OBS selection. If you set the instrument without looking at it carefully, you may have the desired course and the reciprocal course reversed.

COCKPIT ORGANIZATION

Effective cockpit organization can help you handle routine matters easily and reduce your in-flight workload.

Cockpit resource management (CRM) is an important aspect of air carrier training. It refers to the ability to manage all resources available in the cockpit in an efficient and effective manner for both routine and emergency procedures. You can use some of these same techniques to improve the organization of the cockpit. A few ideas to improve organization in the single-pilot IFR environment are planning ahead for busy times such as terminal area flying, arranging your charts in the sequence to be used, cross-checking navigation aids, updating weather, obtaining ATIS information well ahead of time, and using the autopilot.

Effective use of your nonflying passengers can also help to reduce your workload. For example, you can have them help hold your charts, refold your enroute charts as the flight progresses, or turn to the proper charts as you near the airport. If there are children aboard, you can ask an adult passenger to keep them occupied and quiet, particularly during the approach and landing.

If there is another pilot on board, you can share certain flight duties. For example, you may ask the other pilot to take care of the radios, read you the checklists, or monitor your altitude. In this situation, it is critical that you both have a clear understanding of what is expected. Discuss your expectations carefully before the flight so there is no confusion about who will handle the various duties. You should emphasize, though, that the other pilot should speak freely if there is a question about any aspect of the flight. This "second set of eyes" can help you detect and correct a potential problem before it becomes critical.

AUTOPILOTS

The use of the autopilot has already been mentioned as an effective device to ease pilot workload and improve safety during instrument approaches. There are two precautions concerning autopilot use which should be considered. Some pilots become overly dependent on autopilots to the extent that they have no confidence in conducting an approach without them. This is the result of never flying a "hands-on" approach. It is also possible for autopilots to fail in ways that cannot easily be detected. Too much dependency on a working autopilot may prevent you from detecting a malfunction.

HEADSETS

Communications can be a problem during single-pilot IFR operations, especially in busy terminal areas. This can add to an already high workload. Using a headset with an auxiliary microphone is helpful because it not only eliminates the problem of finding the microphone, but it also reduces the time required to initiate a communication when the frequency is congested. Headsets also reduce pilot fatigue by reducing engine noise levels and making it easier to hear the radio. This can be an important factor toward the end of a long flight.

AVIATION WEATHER

The information provided in weather forecasts is essential to safe operation in instrument flight conditions, but it is not always perfect. In fact, pilots sometimes think that the forecasts are never right, and may not give them proper credit. However, a few statistics regarding the probability of a correct forecast are worth mentioning to improve your assessment of the risks involved in a given flight.

How good are our forecasts? You should understand the capabilities as well as the limitations of present day meteorology. Otherwise, you may request the impossible while overlooking the attainable. Specialists understand many atmospheric behaviors and have watched them long enough to understand that their knowledge of the atmosphere certainly is not complete. Weather is not a precise science, despite rapid progress in recent years. It is much less exact than anyone in aviation would like.

To have complete faith in weather forecasts would be as bad as to have no faith at all. Pilots who understand the limitations of observations and forecasts usually are the ones who make most effective use of the weather forecast service. The safe pilot continually views aviation forecasts with an open mind. He or she knows that weather always is changing and that, consequently, the older the forecast, the greater chance that some part of it will be wrong. The weather-wise pilot looks upon a forecast as professional advice rather than absolute certainty.

LIMITATIONS OF AVIATION FORECASTS

Recent studies of the aviation forecasts indicate the following information.

1. Up to 12 hours and even beyond, a forecast of good weather (ceiling 3,000 feet or more and visibility three miles or greater) is much more likely to be correct than a forecast of conditions below 1,000 feet or below one mile.
2. However, for three to four hours in advance, the probability that below-VFR conditions will occur is more than 80% if below-VFR conditions are forecast.
3. Forecasts of single reportable values of ceiling or visibility instead of a range of values imply an accuracy that the present forecasting system does not possess beyond the first two or three hours of the forecast period.
4. Forecasts of poor flying conditions during the first few hours of the forecast period are most reliable when there is distinctive weather, such as a front, trough, or precipitation, which can be tracked and forecast, although there is a general tendency to forecast too little bad weather in such circumstances.
5. The weather associated with fast-moving cold fronts and squall lines is the most difficult to forecast accurately.
6. Errors in forecasting the time of occurrence of bad weather are more prevalent than errors in forecasting whether it will occur or will not occur within a span of time.
7. Surface visibility is more difficult to forecast than ceiling height. Visibility in snow is the most difficult of all visibility forecasts.

Available evidence shows that forecasters can predict the following at least 75% of the time.

1. Passage of fast-moving cold fronts or squall lines within 2 hours as much as 10 hours in advance
2. Passage of warm fronts or slow-moving cold fronts within 5 hours up to 12 hours in advance
3. Rapid lowering of ceiling below 1,000 feet in pre-warm front conditions within 200 feet and within 4 hours
4. Onset of a thunderstorm 1 to 2 hours in advance if radar is available
5. Time rain or snow will begin within 5 hours

Currently, forecasters cannot predict the following with an accuracy which satisfies aviation operational requirements.

1. Time freezing rain will begin
2. Location and occurrence of severe or extreme turbulence
3. Location of a tornado
4. Ceilings of 100 feet or zero, before they exist
5. Onset of a thunderstorm which has not yet formed
6. Position of a hurricane center to nearer than 80 miles for more than 24 hours in advance
7. Occurrence of ice and fog

Studies at the Severe Local Storm Center (SELS) reveal the following verification statistics concerning severe local storms.

1. One out of two tornado forecasts is verified.
2. The SELS can forecast with reasonable accuracy about 40% of all storms that are expected to occur within the selected areas during the time specified. Another 30% are close to the selected areas and times.

As an added note, can you fly safely through an area for which severe storms have been forecast? Sometimes, with the aid of a briefer. Severe weather rarely occurs simultaneously throughout the forecast area. The briefer can help you select a route and time through the area that should avoid the hazardous weather. But always be ready to fly to safety if the advice proves wrong. There probably is no better investment in personal safety than the effort spent increasing your knowledge of basic weather principles and learning to interpret and use the products provided by flight service.

CHECKLIST

After studying this section, you should have a basic understanding of:

✓ **The decision-making process —** What it is and what factors influence it.

- ✓ **Hazardous attitudes** — What they are and how you can combat them.
- ✓ **Risk** — How to assess and reduce it.
- ✓ **Stress** — What its symptoms are and how you can relieve them.
- ✓ **IFR flight considerations** — How to determine your personal limitations during IFR flight.
- ✓ **Cockpit organization** — How to organize the cockpit to help reduce your workload.
- ✓ **Weather** — How to analyze available weather information, keeping in mind the limitations of weather observations and reports.

CHAPTER 8

AIRPLANE PERFORMANCE REVIEW

INTRODUCTION

As you gain flight experience in larger and more complex airplanes, you will appreciate their improved performance capabilities. You also will see how important it is to have a sound knowledge of factors that affect airplane performance. Section A is intended to expand your knowledge of aerodynamic principles as they relate to performance. In the next section, you will review methods of predicting performance under a variety of flight conditions. Then, you will become familiar with manufacturer's performance data for complex, high-performance aircraft. The last section contains a description of methods used for controlling the weight and balance condition of your aircraft. Finally, you will learn how the balance condition of an aircraft directly affects performance.

SECTION A

AERODYNAMICS

In straight-and-level, unaccelerated flight, the sum of opposing forces equals zero.

The purpose of this section is to provide a review of the basic aerodynamic forces acting on an airplane in flight. These basic forces will be related to the four fundamental flight maneuvers: straight and level, turns, climbs, and descents. It is also intended to expand your knowledge of advanced aerodynamic principles to the level required of a commercial pilot. You already know that opposing aerodynamic forces are balanced in straight-and-level, unaccelerated flight. Lift opposes weight and thrust opposes drag. The sum of opposing forces is zero during steady state flight. Any inequality in opposing forces disturbs this basic equilibrium. As fuel is consumed, for example, the weight of the airplane decreases. If you maintain a constant power setting in level flight, you know from experience that true airspeed will gradually increase as fuel is burned off. This occurs because you are slowly reducing angle of attack to maintain your desired altitude. Because of decreasing lift requirements, induced drag also decreases. The decrease in drag is the real reason the airplane accelerates to a higher speed.

During stabilized climbs and descents, forward-acting forces are balanced by rearward-acting forces.

The interaction of basic forces in climbs and descents is somewhat more complex. In a stabilized, constant-airspeed climb, a force called the "lift of thrust" is involved. In a stabilized descent, the "thrust of weight" becomes a factor. However, in both cases, the sum of the upward and downward forces is zero and a condition of equilibrium, similar to straight-and-level, unaccelerated flight, exists. Forward- and rearward-acting forces are also balanced. For example, while transitioning from straight-and-level flight to a descent, your aircraft will accelerate if you do not reduce power or increase drag. This occurs because the forward component of weight increases as the descent angle increases. In a steady state climb or descent, the sum of all forward forces equals the sum of all rearward forces.

LIFT

A knowledge of how lift is produced is essential for understanding aerodynamics. Production of lift requires relatively negative air pressure on the upper surface of the wing and positive air pressure on the lower surface. This pressure variation experienced by an aircraft in flight results from a difference in airflow velocity above and below the wing. The difference in airflow stems from the wing's camber, or curvature. The upper camber is more pronounced, while the lower camber is comparatively flat. Let's take a moment to review airfoil terminology. [Figure 8-1]

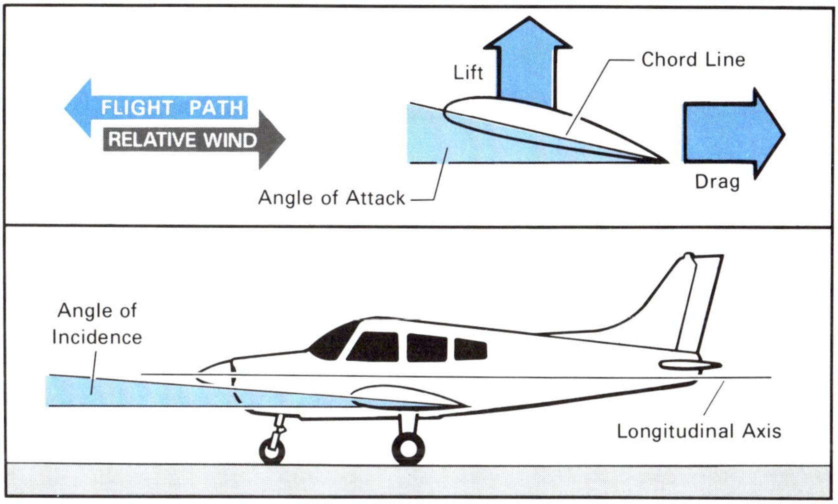

Figure 8-1. Relative wind is parallel and opposite to the flight path of the airplane. The angle between the chord line of the wing and the relative wind is the angle of attack. Lift always acts perpendicular to the flight path (and to the relative wind), regardless of the wing's angle of attack. Drag acts opposite to the flight path. The angle of incidence is formed by the wing's chord line and a line parallel to the longitudinal axis. A small angle of incidence is desirable, since it provides a positive angle of attack in level, cruising flight and improves over-the-nose visibility from the cockpit.

When an airfoil is producing lift, the air pressure over a large portion of the top surface is lower than atmospheric pressure. Some of the lift is provided by positive pressure beneath the forward section of the wing. This is caused by airflow striking the bottom surface of the wing and is referred to as impact, or **dynamic pressure**. At high angles of attack, this may account for as much as 25 to 30% of the total lift. [Figure 8-2]

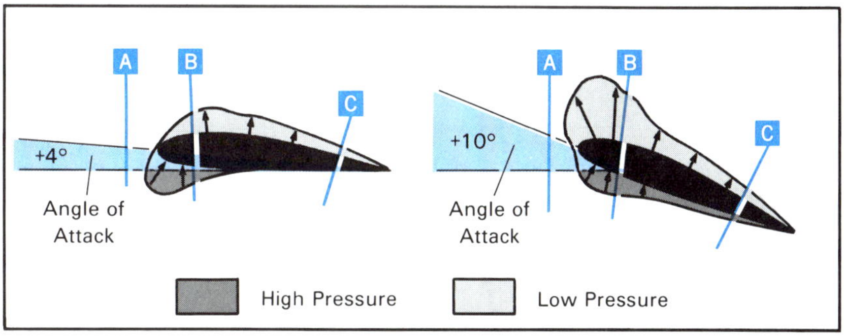

Figure 8-2. This illustration shows the positive and negative pressure patterns affecting an airfoil in flight. In normal cruising flight, where the angle of attack is relatively low, most lift is created by the center section of the wing from B to C, as shown in the left view. When you increase the angle of attack, as shown in the right view, the forward portion of the wing from A to B produces most of the lift. An increase in angle of attack causes a corresponding increase in dynamic pressure and drag.

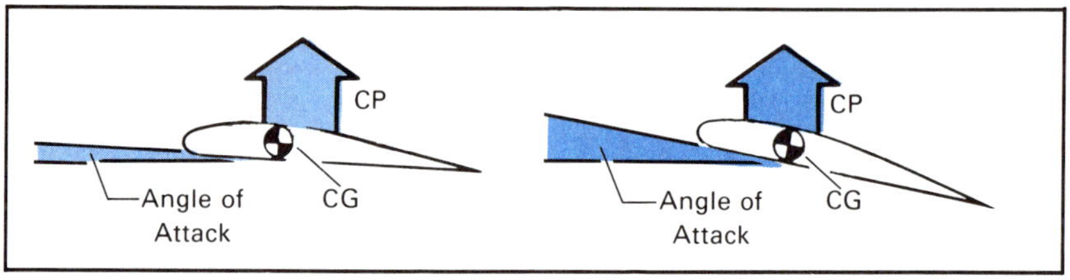

Figure 8-3. The arrow labeled "CP" represents the center of pressure, which is the vector sum of the pressure differences around the airfoil. You can see from the left view that the center of pressure is located behind the center of gravity at low angles of attack. On a typical cambered airfoil, the center of pressure moves ahead of the center of gravity at high angles of attack, as shown in the right view.

You can control lift, airspeed, and drag by changing angle of attack.

The fact that the pressure distribution of the wing changes with angle of attack is important. When you change the angle of attack of the wing, you also change lift, airspeed, and drag forces. [Figure 8-3]

With a cambered airfoil, the CP shifts as the angle of attack changes, but the CP is unaffected by angle of attack changes on airplanes with a symmetrical airfoil design.

When the center of pressure (CP) lifting force is concentrated behind the CG, it tends to cause the aircraft to pitch down and must be counteracted by a tail-down force created by the elevator. As the angle of attack is increased, the center of pressure moves forward. If it moves ahead of the CG, an upward force from the tail is required to maintain attitude. The movement of the center of pressure, therefore, influences the aerodynamic balance and controllability of typical training aircraft that have a cambered wing. However, this CP movement does not occur on a symmetrical airfoil design. With symmetrical airfoils the CP is unaffected by changes in angle of attack.

LIFT EQUATION

The lift equation may be helpful in explaining the relationship between the various factors that affect it.

$$L = C_L S \frac{\rho}{2} V^2$$

L = Lift
C_L = Coefficient of lift (This changes with airfoil design and angle of attack.)
S = Wing surface area (Square feet)
ρ = Air density (The Greek letter rho represents the density of the air in slugs per cubic foot.)
V^2 = Velocity squared (Velocity is in feet per second)

If the angle of attack and other factors remain constant while you double your airspeed, you will quadruple lift.

Since lift varies in proportion to the square of the velocity, doubling the airspeed results in the production of four times as much lift, assuming you use a constant angle of attack. At higher airspeeds, drag also increases in proportion to the velocity squared. Eventually, you will reach a point where the total drag equals the maximum thrust available, and the airplane can't be accelerated further in straight-and-level flight. Since air density decreases with altitude, an aircraft must have either a higher angle of attack or a higher velocity to generate the same amount of lift at higher altitudes. Velocity, as mentioned earlier, is limited by thrust available. In addition, the angle of attack can only be increased to the point of airflow separation that results in a stall. The coefficient of lift, C_L, is determined by both angle of attack and airfoil design. In this simplified equation, C_L is the ratio between the lift pressure and dynamic pressure. The graph of C_L versus angle of attack clarifies this relationship. [Figure 8-4)

Flight at critically slow airspeeds provides you with an application of the lift formula. As your airspeed decreases, the angle of attack has to increase to maintain the same amount of lift necessary to sustain level flight. The angle of attack, however, can be increased only until C_{Lmax} is reached. At this point, you can no longer increase lift by increasing the angle of attack.

The indicated stall speed of an airplane is directly affected by weight, CG location, load factor, and power.

In straight-and-level flight, C_{Lmax} normally is reached as the indicated airspeed approaches the published stall speed. You should note, however, that turbulence can cause abrupt changes in the relative wind which result in an increase in the stall speed. Stall speed also increases rapidly as G-loading increases during maneuvers, such as steep turns. Violent or abrupt maneuvering can cause a stall at speeds twice as high as the level flight, unaccelerated stall speed. Remember, the indicated

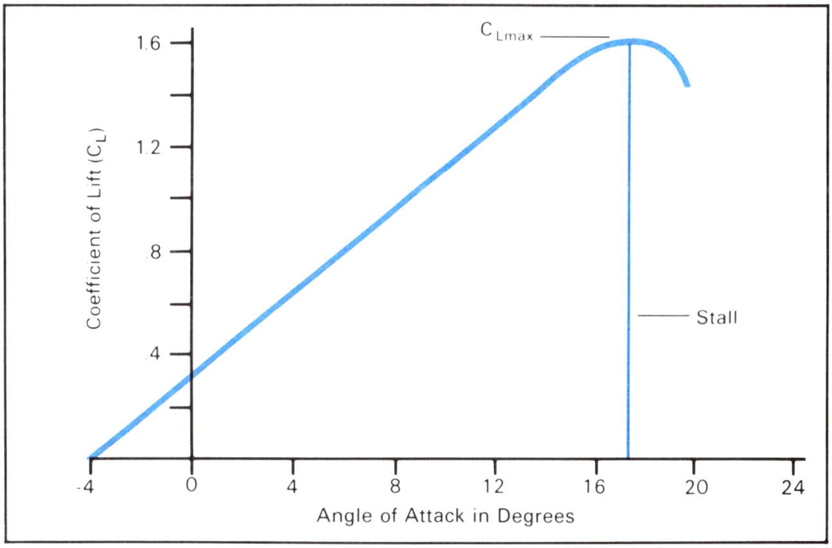

Figure 8-4. Notice that the coefficient of lift increases steadily until C_{Lmax} and then drops suddenly as the wing stalls. It is apparent from this illustration that, for a given airfoil, a stall will always occur at the same angle of attack regardless of weight, dynamic pressure, bank angle, or pitch attitude. However, the indicated stall speed at the critical angle of attack will vary, since stall speed is not a fixed value.

stall speed can be affected by a number of things such as weight, weight distribution, load factor, and power. Finally, you should never intentionally stall an aircraft at a speed above the design maneuvering speed, because this can easily result in excessive G-loading.

HIGH-LIFT DEVICES

Two of the variables in the lift formula that you can control are the angle of attack and velocity. In addition, wing design features may allow you to change the overall camber of the wing through use of trailing-edge flaps and leading-edge high-lift devices.

TRAILING-EDGE FLAPS

When extended, flaps decrease the stall speed and provide an increase in your rate of descent without increasing airspeed.

The most common high-lift device is the trailing-edge flap, which produces a large increase in airfoil camber and moves the effective lift force toward the trailing edge of the wing. Flaps are useful because they increase your rate of descent without increasing your airspeed. Since flaps increase the coefficient of lift, they also decrease the stall speed. Four basic types of flaps are found on general aviation airplanes. [Figure 8-5]

The plain flap is the simplest of the four types. It increases the airfoil camber, resulting in a significant increase in the coefficient of lift at a given angle of attack. At the same time, it greatly increases drag and moves the center of pressure aft on the airfoil, resulting in a nose-down pitching moment.

The split flap is deflected from the lower surface of the airfoil and produces a slightly greater increase in lift than does the plain flap. However, more drag is created because of the turbulent air pattern produced behind the airfoil.

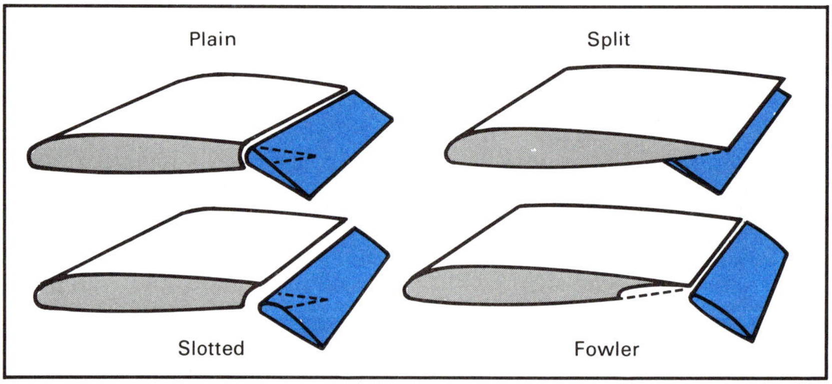

Figure 8-5. The types of trailing-edge flaps that are typically found on general aviation airplanes are the plain, split, slotted, and Fowler flaps. Flaps increase lift (and drag) by increasing the wing's effective camber and changing the chord line, which increases the angle of attack. In some cases, flaps also increase the area of the wing. When fully extended, most flaps form an angle of 35° to 40° relative to the wing's chord.

The slotted flap provides an even greater increase in lift. This flap configuration allows air from below the airfoil to flow over the upper flap surface delaying separation at high lift coefficients. This flap also provides an increase in wing camber to aid in increasing lift.

Of the four common flap types, the Fowler flap is the most efficient. This flap not only moves down to increase wing camber, but also moves aft to increase the wing area. These two factors combine to provide large increases in lift with comparatively small increases in drag.

LEADING-EDGE HIGH-LIFT DEVICES

High-lift devices also can be applied to the leading edge of the airfoil. The most common types are fixed slots, movable slats, and leading-edge flaps. [Figure 8-6]

Fixed slots direct airflow to the upper wing surface and delay airflow separation at higher angles of attack. The slot does not increase the wing camber, but does allow a higher maximum coefficient of lift, because stalls are delayed until the wing reaches a higher angle of attack.

A **movable slat** consists of a leading-edge segment which is free to move on tracks. At low angles of attack, the slat is held flush against the wing's leading edge. At high angles of attack, either a low pressure area at the wing's leading edge or pilot operated controls force the slat to move forward. This opens a slot and allows the air below the wing to flow over the wing's upper surface, delaying the airflow separation.

Leading-edge flaps, like trailing-edge flaps, are used to increase the wing camber to effect an increase in C_{Lmax}. This type of leading-edge device is used frequently in conjunction with trailing-edge flaps and can reduce the nose-down pitching moment.

The use of high-lift devices is limited by the airplane's performance characteristics. Because they increase lift and drag, they place additional stresses on the wing structure. For this reason, the maximum operating speed with flaps extended is typically lower than the maximum cruising

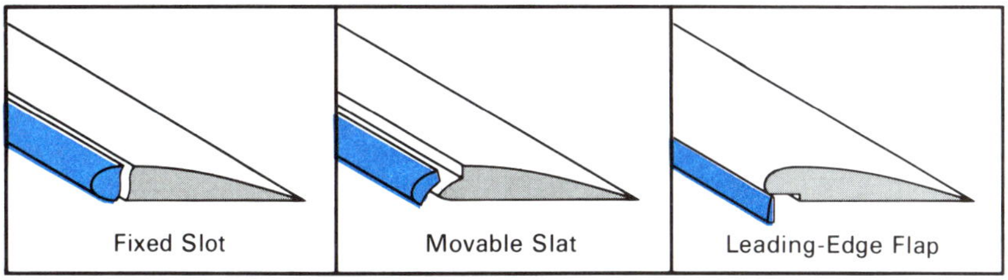

Figure 8-6. Leading-edge devices are less common than flaps on general aviation aircraft. However, when they are used, they give you greater control over the coefficient of lift.

speed for a particular aircraft. Additionally, any device which increases the maximum lift coefficient also increases induced drag, which is of prime importance to the airplane's overall performance. In fact, gliders and some airplanes employ high-drag devices, such as spoilers, which may be added to the upper wing surface. When deployed, spoilers decrease lift and increase drag.

WING PLANFORM

The shape of a wing is typically a design compromise based on the intended operational use of the aircraft. The wing planform is important, because it determines the three-dimensional flow of air around the wings and, therefore, the stall, handling, and cruise performance of the aircraft. Of particular importance is the aspect ratio, which is the numeric relationship of the wing span to the chord line. [Figure 8-7]

DRAG

From previous training, you probably are familiar with induced and parasite drag. The fact that total drag has two components adds to the complexity of the effects it has on an aircraft in flight. Induced drag decreases with speed, while parasite drag increases. The following discussion provides an in-depth look at these two aerodynamic components.

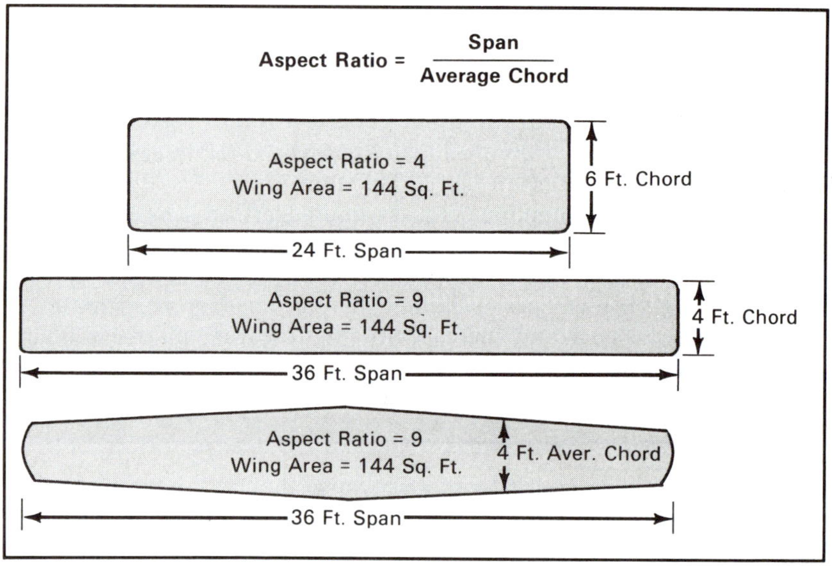

Figure 8-7. A high aspect ratio produces comparatively less drag, especially at high angles of attack. A wing with a low aspect ratio typically has higher wing loading (the amount of weight supported by each square foot of wing area) and, consequently, a higher stall speed. The rectangular wing planform is popular in general aviation aircraft for many reasons. For example, one desirable feature is its tendency to stall at the wing roots first, so it provides adequate stall warning while still maintaining aileron effectiveness.

INDUCED DRAG

The portion of the total drag force that is created by the production of lift is called **induced drag**. At positive angles of attack, low pressure air from above the wing joins high pressure air from beneath the wing at the wingtip and wing trailing edges, creating vortices and eddies. The overall effect is a downward deflection of air which creates a downward force on the relative wind. [Figure 8-8]

Since most of the mixing of low and high pressure air occurs at the wingtips, high aspect ratio wings characteristically produce less induced drag than low aspect ratio wings. Induced drag also varies with airspeed and angle of attack. [Figure 8-9]

PARASITE DRAG

Any drag not associated with the production of lift is **parasite drag**. It is created by airplane surfaces that disrupt the smooth flow of air and create turbulence. The elements of parasite drag include form drag, skin friction drag, and interference drag.

Structures that protrude into the relative wind or impede the smooth flow of air, such as landing gear and radio antennas, create a type of parasite drag called **form drag**. Aircraft designers can minimize this type of drag by streamlining objects which protrude into the airstream. Objects, such as tires, that can't be streamlined are often partially covered with teardrop-shaped shells that are aerodynamically superior in form. The propeller spinner performs a similar function by helping smooth the flow of air over the front of the engine.

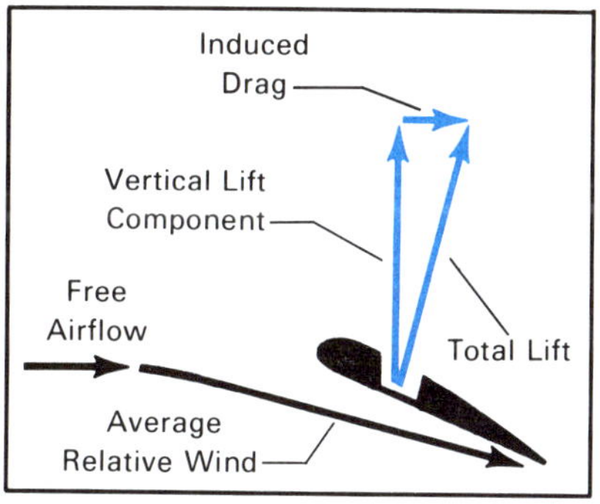

Figure 8-8. With the average relative wind tilted downward, total lift is no longer perpendicular to the free airstream. Instead, it's perpendicular to the average relative wind, while the vertical component of lift is perpendicular to the free airstream. Induced drag, which is parallel to the relative wind, increases as the average relative wind tilts further downward with increases in angle of attack.

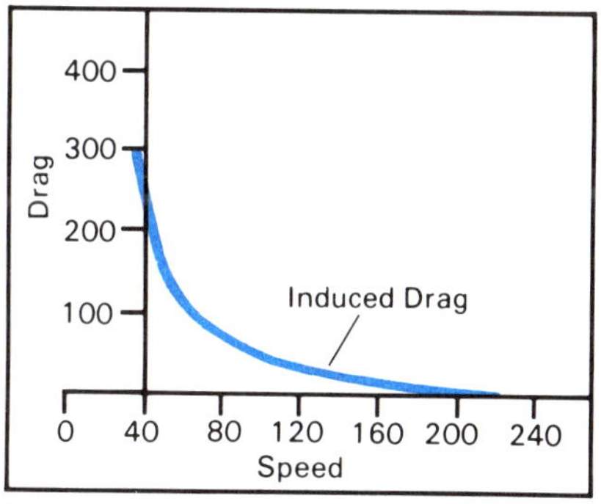

Figure 8-9. The induced drag curve shows that induced drag is highest at slow speeds and decreases rapidly as speed increases. If all other factors remain constant, induced drag varies inversely with the square of the airspeed. For example, if airspeed doubles, induced drag decreases by a factor of four.

At the microscopic level, a wing's surface appears quite rough. A boundary layer of air clings to this surface, creating small whirling eddies called **skin friction drag**. National Aeronautics and Space Administration (NASA) research has pioneered techniques of boundary layer flow control, laminar airflow wing design, and microscopic surface refinements that are finding their way into general aviation aircraft. However, skin friction drag can never be totally eliminated.

Another contribution to parasite drag is **interference drag**, which occurs when the various airflows over an airplane meet and interact. For instance, air flowing along the wing root is deflected by both the wing and the fuselage. In multi-engine aircraft, the engine nacelles deflect air that also interacts with the wing and fuselage. The designer must take interference drag into account when selecting the shape of components and positioning them on the airplane. [Figure 8-10]

TOTAL DRAG

The total drag of an airplane is the sum of induced and parasite drag. Induced drag is the predominant factor at low airspeeds, and parasite drag is the predominant factor at high airspeeds. The intersection of the induced drag and parasite drag lines corresponds to a point on the total drag line where drag is at a minimum. This is also the point where the aircraft is operating at the best ratio of lift to drag, or L/D_{max}. [Figure 8-11]

MAXIMUM RANGE

To achieve maximum range, airspeed must be reduced as fuel is consumed.

The maximum range of an airplane is obtained at L/D_{max}. You should note that L/D_{max} is applicable only to a particular airplane in steady, unaccelerated flight at a specified weight and configuration. As fuel is burned, the weight decreases, which alters the values of airspeed and

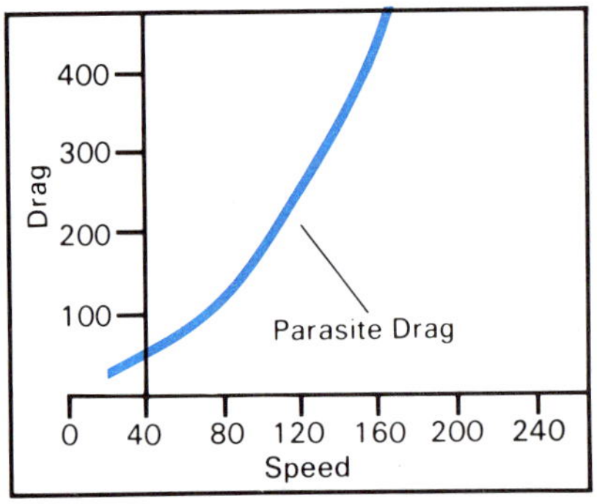

Figure 8-10. The curve illustrates that parasite drag varies proportionately to the square of the airspeed. For example, the same airplane at a constant altitude has four times as much parasite drag at 160 knots as it does at 80 knots.

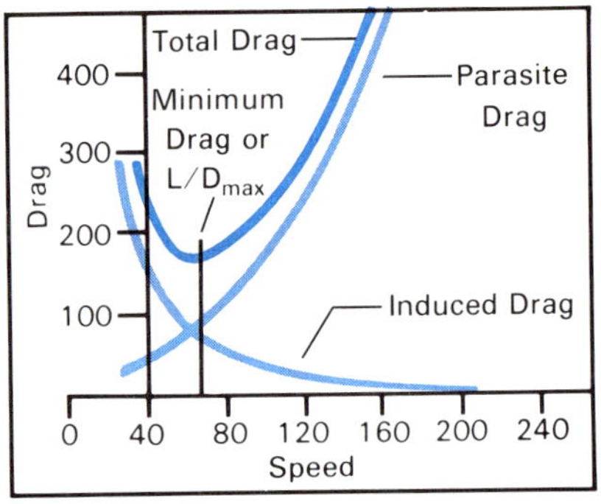

Figure 8-11. Flying your aircraft at L/D_{max} provides both maximum range and the best power-off glide speed. Any decrease or increase in airspeed from that specified for L/D_{max} will result in reduced range or gliding distance.

power required to achieve maximum range. As weight decreases, airspeed (and power) must be reduced or range will decrease from the maximum possible. For propeller-driven airplanes, altitude has little effect on maximum range, and the angle of attack required to maintain L/D_{max} (and maximum range) remains constant. However, variations in weight do not affect the maximum glide angle, provided you use the correct airspeed for each weight.

You can also see that if you are in level flight and you increase your airspeed above L/D_{max}, the total drag of the airplane increases, because you have increased parasite drag. Likewise, if you slow to a speed less than L/D_{max}, your total drag also increases, but this time it is because you have increased induced drag.

Total drag of the airplane increases if you fly at a speed other than L/D_{max}.

THRUST

Thrust is the force that must be produced to overcome total drag. In a steady flight condition, thrust must equal total drag. If thrust exceeds drag, your airplane accelerates until the thrust and drag forces reach equilibrium.

Jet engines produce thrust by increasing the velocity of the air between the inlet and the exhaust of the engine. The thrust leaving the engine results in forward motion of the aircraft through an equal and opposite reaction. On the other hand, the propeller produces thrust by deflecting air rearward. Each propeller blade is cambered like the airfoil shape of a wing. This shape, plus the angle of attack of the blades, causes the dynamic pressure on the engine side of the propeller to be greater than the atmospheric pressure forward of the propeller. This pressure differential produces thrust. For the purpose of this discussion, only propeller thrust and the factors which affect it will be considered.

PROPELLER EFFICIENCY

The propeller converts the shaft horsepower of the engine into thrust horsepower. How efficiently this power is converted to thrust is important to airplane performance. A spinning propeller is subject to all the factors which affect airfoil efficiency. Its maximum efficiency is greatly affected by the angle of attack and rotational speed.

A fixed-pitch propeller with low blade angles (climb propeller) obtains maximum efficiency at low airspeeds; one with high blade angles (cruise propeller) obtains maximum efficiency at high airspeeds. The climb propeller provides greater efficiency during takeoff and climb, while the cruise propeller is most efficient at cruise airspeeds.

Variable-pitch or constant-speed propellers normally are used with high-performance engines. With a constant-speed propeller, you can select the appropriate blade angle and r.p.m. for each phase of flight.

MAXIMUM LEVEL FLIGHT SPEED

In level flight, the maximum speed of your airplane is limited by the amount of power produced by the engine and the total drag generated by the airplane. If thrust exceeds total drag when you apply full power, the airplane accelerates. This continues until the force of total drag equals the force of thrust. At this point, the airplane is flying at its maximum level flight speed.

Remember, thrust and power are not synonymous terms, because thrust is the force created by the propeller to overcome drag, while power is the measure of work performed by the engine to turn the propeller. Although thrust and power are not the same, the curve representing power closely resembles that of the drag or thrust curve. [Figure 8-12]

Some other conclusions can be drawn from the power-required curve. For instance, in selecting your cruise airspeed, you usually want to cover the distance to be traveled in the shortest period of time. However, just as an airplane can fly only so fast, it can go only so far and stay up only so long before it runs out of fuel. When endurance or range are important

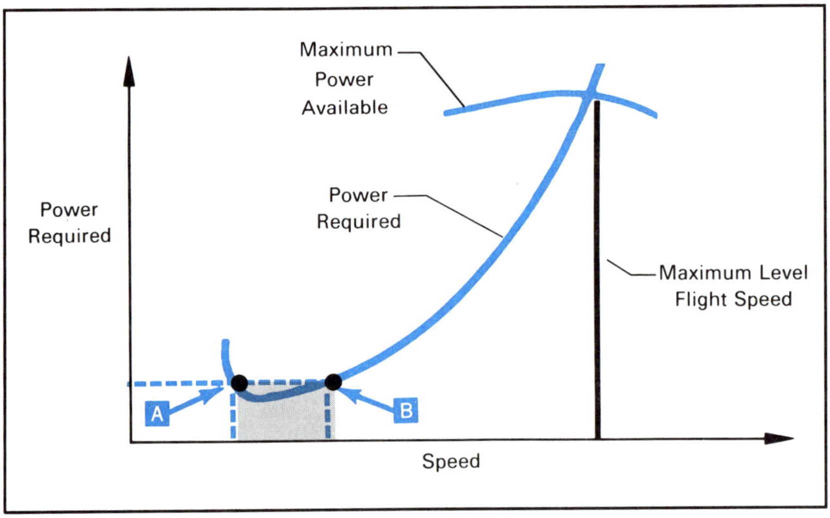

Figure 8-12. The power-required curve shows the amount of power necessary to maintain level flight at various speeds. The power-available line also is curved, since power available in a typical single-engine airplane is a function of airspeed. The speed at which these two curves cross is where the forces of thrust and drag are in balance and where maximum level flight speed occurs. Note that the power required to operate the airplane at point A is the same as that required at point B, even though the speeds are different. If you were flying at point A, you would be operating on the back side of the power curve, where any decrease in airspeed requires an increase in power. At point A, you are very close to minimum flying speed.

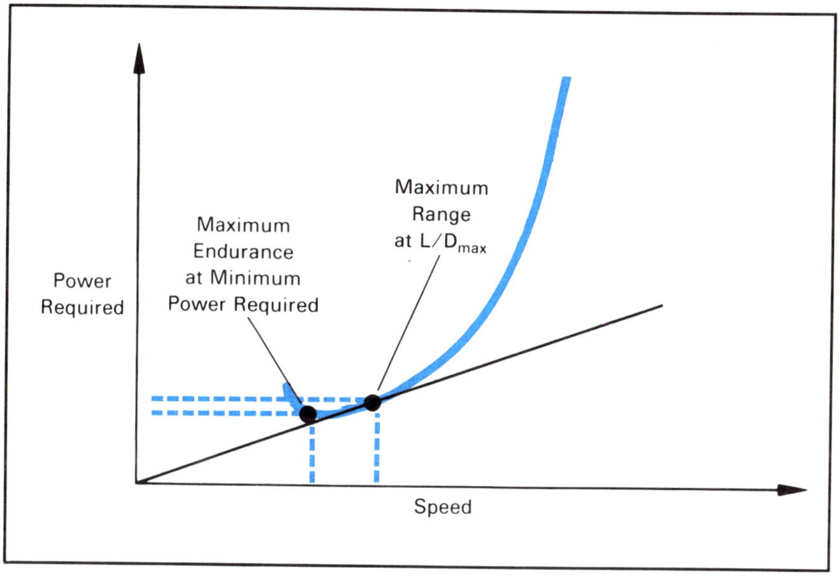

Figure 8-13. Maximum endurance speed occurs at the lowest point on the power-required curve. This is the speed where the lowest power setting will sustain an airplane in steady, level flight. Maximum range speed is a higher indicated airspeed that minimizes total drag.

factors, you must select the proper speed for the flight. Maximum endurance and maximum range require different speeds.

Maximum endurance speed is the speed which allows the airplane to remain aloft for the longest period of time. This speed requires the least amount of power necessary to sustain level flight. The minimum power setting provides the lowest rate of fuel consumption.

In contrast, the **maximum range speed** provides the greatest distance for a given amount of fuel. You can think of it as getting the most miles per gallon out of the airplane. To determine this speed, you must consider the speed and rate of fuel consumption for any given power setting. The setting which yields the greatest distance traveled per gallon of fuel burned is the power setting which provides maximum range speed. As mentioned previously, this speed produces the minimum total drag and occurs at L/D_{max}. [Figure 8-13]

WEIGHT

Weight, or gravity, is the simplest of the four forces. This is the actual weight of the airplane, termed "one G," and it always acts downward, toward the center of the earth. Load factor can be increased above the normal one G by flight maneuvers or turbulence.

Load factor is the ratio between the total airload imposed on the wings and the actual weight of the aircraft.

Remember that the ratio between the load imposed on the wings and the airplane's actual weight is the effective weight, or the load factor. If a load factor of three is imposed on a 3,000-pound airplane, the load on the airframe is 9,000 pounds. For this reason, airplane structures are designed to withstand load factors which are greater than the one-G weight of the airplane. Load factor and operating limitations are discussed in greater detail later in this section, as well as in the next section on performance.

AIRCRAFT STABILITY

An airplane that is said to be inherently stable will require less effort to control.

Although no airplane is completely stable, all airplanes must have desirable stability and handling characteristics. An inherently stable airplane is easy to fly, and it reduces pilot fatigue. You will see that an airplane's handling characteristics are directly related to its stability. In fact, stability, controllability, and maneuverability are all interrelated design characteristics.

Maneuverability is the characteristic of an airplane that permits you to maneuver easily and it suggests the capability to withstand the stress resulting from the maneuvers. An airplane's size, weight, flight control system, structural strength, and thrust determine its maneuverability. **Controllability** is the capability of an airplane to respond to your control inputs, especially with regard to attitude and flight path. Both maneuverability and controllability contribute to effective ease of control.

Stability is a characteristic of an airplane in flight that causes it to return to a condition of equilibrium, or steady flight, after it has been disturbed. For example, if you are flying a stable airplane that is disrupted while in straight-and-level flight, it has a tendency to return to the same attitude. The types of stability include static, dynamic, longitudinal, lateral, and directional.

STATIC STABILITY

An airplane displays negative static stability if its initial tendency is to move farther from its original position following a displacement.

Static stability is the initial tendency of an airplane to return to a state of equilibrium following a displacement from that condition. If positive static stability is present, an airplane has a tendency to return to the original point of equilibrium. If an airplane has negative static stability, it will have a tendency to move farther away from the point of equilibrium when it is displaced. If an airplane tends to remain in a displaced attitude, it has neutral static stability. [Figure 8-14)

DYNAMIC STABILITY

Dynamic stability describes the time required for an airplane to respond to its static stability following a displacement from a condition of equilibrium. It is determined by the airplane's tendency to oscillate and damp

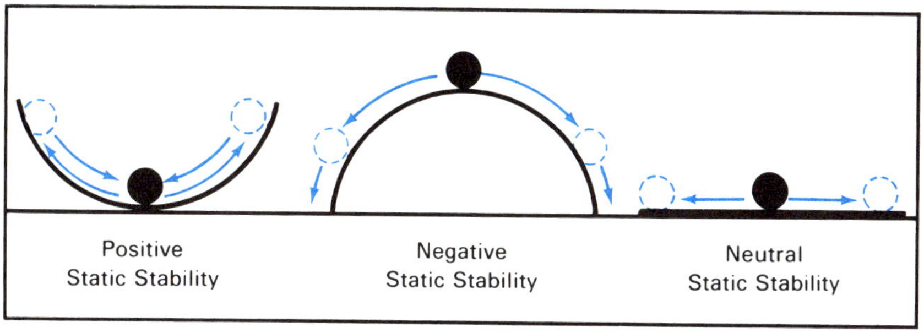

Figure 8-14. Positive static stability is the most desirable characteristic, because the airplane initially attempts to return to its original trimmed attitude. Almost any properly loaded airplane you fly exhibits this characteristic. Both negative and neutral static stability are undesirable characteristics.

out successive oscillations after its initial displacement. Although an airplane may be designed with positive static stability, it could have positive, negative, or neutral dynamic stability.

Assume the airplane you're flying is displaced from an established attitude. If its tendency is to return to the original attitude directly, or through a series of decreasing oscillations, it exhibits **positive dynamic stability**. If you find the oscillations increasing in magnitude as time progresses, **negative dynamic stability** is exhibited. **Neutral dynamic stability** is indicated if the airplane attempts to return to its original state of equilibrium but the oscillations neither increase nor decrease in magnitude as time passes. [Figure 8-15]

LONGITUDINAL STABILITY

Longitudinal stability involves the pitching motion, or tendency, of the airplane about its lateral axis. This means your aircraft will resist any force which might cause it to pitch, and it will return to level flight when the force is removed. This is the most important of the three types of stability of an aircraft in flight.

Longitudinal stability is the quality that gives an airplane stability about the lateral axis.

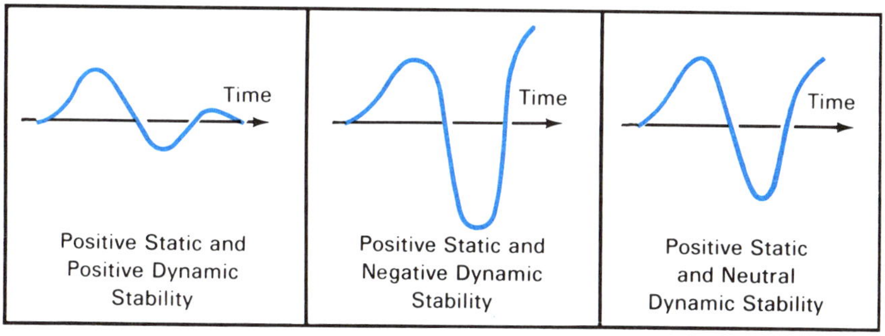

Figure 8-15. The most desirable condition is a combination of positive static and positive dynamic stability. In this situation, you need less effort to return the airplane to its original attitude because it "wants" to return there.

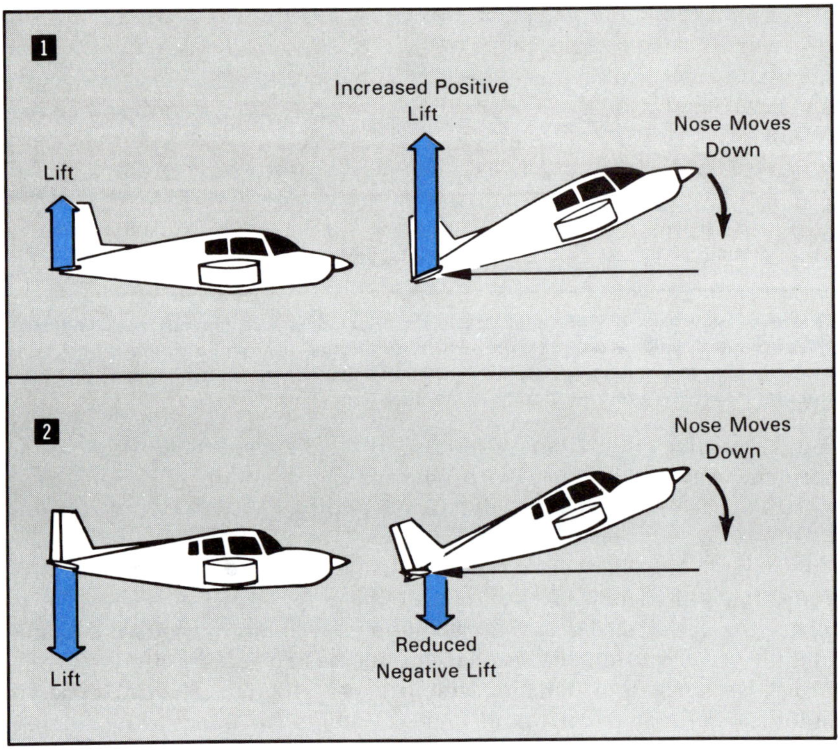

Figure 8-16. If an airplane is tail heavy during straight-and-level flight, the horizontal stabilizer produces positive lift to maintain longitudinal stability (airplane 1). If an airplane is nose heavy, the stabilizer produces negative lift (airplane 2). If either turbulence or a control input pitches the nose up, the horizontal stabilizer experiences an increased angle of attack. This results in an increase in positive lift for airplane 1 and a decrease in negative lift for airplane 2. In both cases, the tail moves up and the nose moves down.

To obtain longitudinal stability, an airplane may be designed to be either slightly tail heavy or nose heavy while trimmed in straight-and-level flight. Additionally, the CG location can affect the nose-heavy or tail-heavy condition. If the airplane is loaded near the forward CG limit, a nose-heavy condition can occur. Consequently, the horizontal stabilizer is designed with the capability of producing either the positive or negative lift necessary to stabilize the airplane longitudinally. [Figure 8-16]

The opposite reaction takes place if the nose of your airplane pitches downward. A horizontal stabilizer producing positive lift experiences a decrease in angle of attack; therefore, less tail lift is developed and the tail lowers as the nose raises. If the stabilizer is producing negative lift to maintain stability, the pitch change results in an increase in the angle of attack, causing the stabilizer to produce greater negative lift. This situation also forces the tail down, returning the airplane to the trimmed condition.

An airplane which is longitudinally unstable cannot be type certificated, because it has dangerous flight characteristics. However, you can create longitudinal instability by loading an airplane so its CG is beyond acceptable limits. For example, assume that after control pressures are released following a pitch displacement, an airplane remains in the new pitch attitude. This is an example of neutral longitudinal static stability. If the airplane corrects itself but enters a series of progressively steeper pitch oscillations, it exhibits longitudinal dynamic instability.

A longitudinally unstable airplane has dangerous flight characteristics.

LATERAL STABILITY

Lateral stability refers to an airplane's tendency to resist lateral, or roll, movement about the longitudinal axis and to return to wings-level flight following a displacement. Generally, it is considered undesirable to design an airplane with too much lateral stability. Strong lateral stability has a detrimental effect on rolling performance as well as handling characteristics during crosswind takeoffs and landings.

The tendency of an airplane to right itself after being displaced from wings-level flight is actually the result of a side slip induced by corrective control movements. If a wing is displaced, another force, such as a control input, must be introduced before the airplane's lateral stability becomes evident. A wing develops stable rolling moments when a side slip is introduced. [Figure 8-17]

Sweepback, which frequently is used on corporate jet airplanes, has a lateral stabilizing effect. When an aircraft with sweptback wings begins

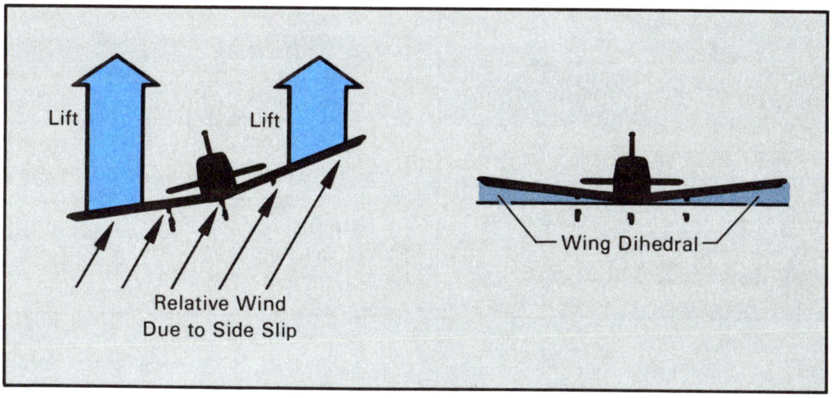

Figure 8-17. When the relative wind strikes the airplane from the side, the upwind wing experiences an increased angle of attack and increased lift. The downwind wing has a reduced angle of attack and decreased lift. The differential in lift results in a rolling moment, which tends to raise the low wing. Dihedral also is used to obtain lateral stability. It increases the stabilizing effect of side slips by increasing the lift differential between the high and the low wing during a slip.

to slip, the leading edge of the low wing meets the relative wind more nearly perpendicular than the higher wing. [Figure 8-18]

DIRECTIONAL STABILITY

The tendency of your airplane to remain stationary about the vertical or yaw axis is known as **directional stability**. When the relative wind is parallel to the longitudinal axis, the airplane is in equilibrium. If some force yaws your airplane, it produces a side slip. A positive yawing moment also is developed which returns the airplane to equilibrium.

To obtain directional stability, the side area of fuselage ahead of the center of gravity must be less than that behind the center of gravity. This gives the airplane a tendency to "weathervane" with the relative wind. However, this difference in fuselage area usually is insufficient for a high degree of stability, so a vertical stabilizer is added. The vertical stabilizer is a symmetrical airfoil, capable of producing lift in either direction. [Figure 8-19]

EFFECTS OF LATERAL AND DIRECTIONAL STABILITY

Although airplanes are designed with stabilizing characteristics which lighten your workload while you are flying, there are normally some undesirable side effects. Two of the most common ones are Dutch roll and spiral instability.

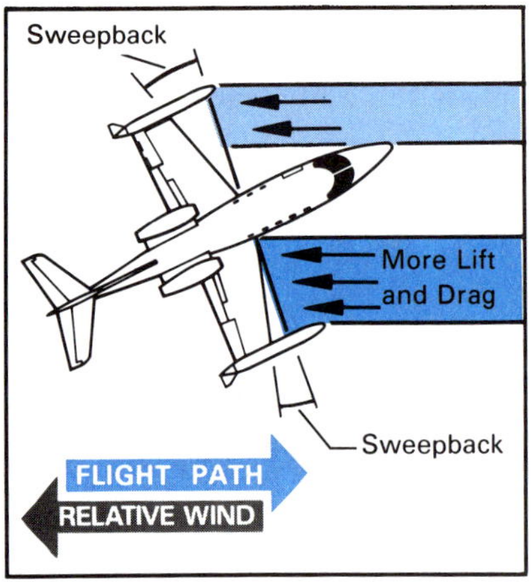

Figure 8-18. In a slip, the difference in the placement of the wings results in more lift and, therefore, more drag on the low wing. This causes a restoring force to return the wing to a level position.

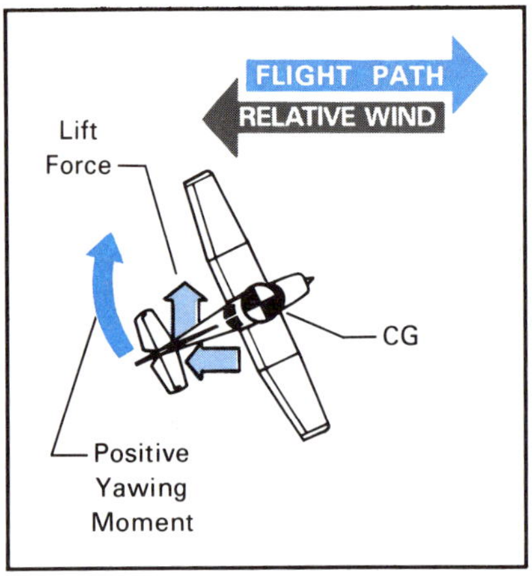

Figure 8-19. A positive side slip angle changes the relative wind on the vertical stabilizer, creating a higher coefficient of lift in one direction. In this example, the airplane is in a side slip to the right, resulting in positive lift to the left of the stabilizer. Additional lift moves the tail to the left, causing the longitudinal axis to align with the relative wind.

Dutch roll is a combination of rolling/yawing oscillations caused either by your control input or by wind gusts. In a typical case, when equilibrium is disturbed, the rolling reaction precedes the yaw, and the roll motion is more noticeable than the yaw motion. When the airplane rolls back toward level flight in response to the dihedral effect, it continues to roll too far and side slips the other way. Each oscillation overshoots the wings-level position because of the strong dihedral effect. Dutch roll is actually the back-and-forth, rolling/yawing motion. If the Dutch roll tendency is not effectively dampened, it is considered objectionable. The alternative to an airplane that exhibits Dutch roll tendencies is a design that has better directional stability than lateral stability. If directional stability is increased and lateral stability is decreased, the Dutch roll motion is adequately suppressed; however, this design arrangement tends to cause spiral instability.

Spiral instability is associated with airplanes that have strong directional stability in comparison with lateral stability. When an airplane with spiral instability is disturbed from a condition of equilibrium, a side slip is introduced. In this case, the strong directional stability tends to yaw the airplane back into alignment with the relative wind. At the same time, the comparatively weak dihedral effect lags in restoring lateral stability. Due to the yaw back into the relative wind, the outside wing travels faster than the inside wing, resulting in more lift being generated by the outside wing. The yaw forces the nose of the airplane down as it swings into alignment with the relative wind. The net result is an overbanking and nose-down tendency, which generally is considered less objectionable than a Dutch roll. As you can see, even a well-designed airplane may have some undesirable characteristics. Generally, designers attempt to minimize the Dutch roll tendency, since it is less tolerable than spiral instability. Because of this, some degree of spiral instability is considered acceptable.

AERODYNAMICS AND FLIGHT MANEUVERS

The performance of any aircraft is directly related to its aerodynamic design. The difference in appearance and flight performance between individual airplanes stems from aerodynamic considerations and the purpose for which the airplane is designed. While individual airplanes vary widely in design features, all are subject to the same principles of aerodynamics.

STRAIGHT-AND-LEVEL FLIGHT

For every angle of attack, there is a given indicated airspeed required to maintain level flight.

When your airplane is in straight-and-level, unaccelerated flight, a condition of equilibrium exists. Lift equals weight, and thrust equals total drag. To accelerate or decelerate in level flight, you must change the power setting to produce an unbalanced condition. But, as the power is changed and the airspeed changes, you must simultaneously adjust the angle of attack or you will not maintain level flight. The indicated airspeed required to maintain level flight is unique to each angle of attack (all other factors remaining constant).

Generally, an increase in airspeed requires an increase in engine power. This relates back to the basic lift equation; a change in velocity results in a change in total lift, unless one of the other variable factors is changed. [Figure 8-20]

CLIMBS

When transitioning from straight-and-level flight to a climb, the aircraft's angle of attack is increased and lift momentarily increases.

Normally, you initiate a climb by applying back pressure to the control wheel, which increases the angle of attack and momentarily increases lift. However, you sustain the climb with excess thrust horsepower rather than by excess lift. The angle of climb is a function of thrust; rate of climb is a function of thrust horsepower, a product of shaft horsepower and propeller efficiency. [Figure 8-21]

For any given airplane weight, the maximum angle of climb (the greatest altitude gain in the least horizontal or forward distance) depends upon the amount of excess thrust available. This speed occurs where there is the greatest difference between thrust and drag. [Figure 8-22]

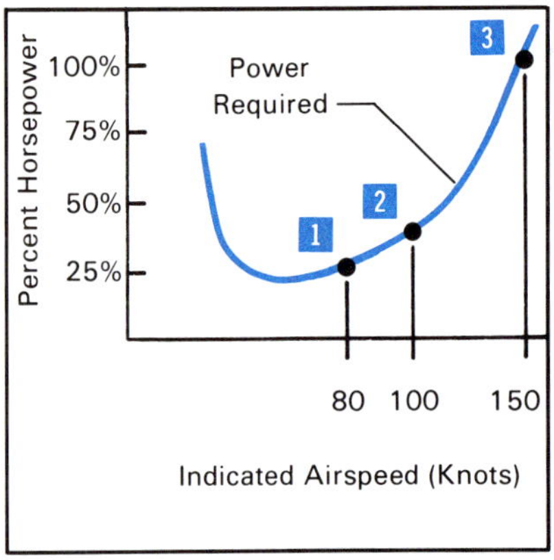

Figure 8-20. This graph shows the power required to achieve equilibrium at various airspeeds. For a given airspeed (item 1), the power-required curve shows the engine power that must be available to maintain equilibrium. If you desire a higher speed (item 2), the power required for equilibrium is considerably greater. The maximum level flight speed (item 3) is obtained when the power required equals the power available from the powerplant.

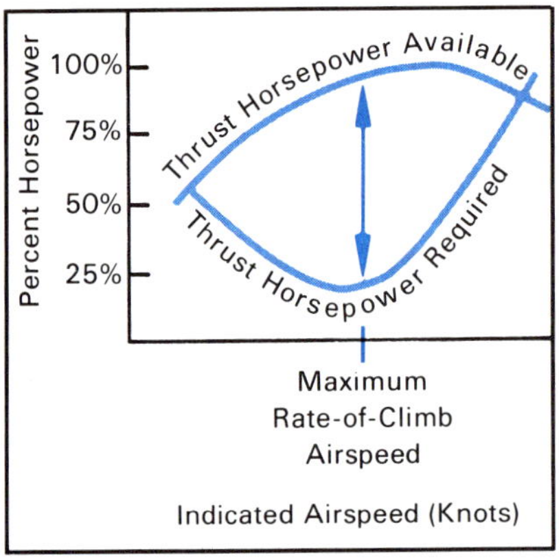

Figure 8-21. The maximum rate of climb occurs at the airspeed with the greatest difference between thrust horsepower available and thrust horsepower required. If excess thrust horsepower is not available, the airplane is in steady, level flight. If excess thrust horsepower is available, the rate of climb produced is proportional to the amount of excess power. The thrust horsepower available line is curved, because propeller efficiency changes with changes in airspeed and pitch attitude.

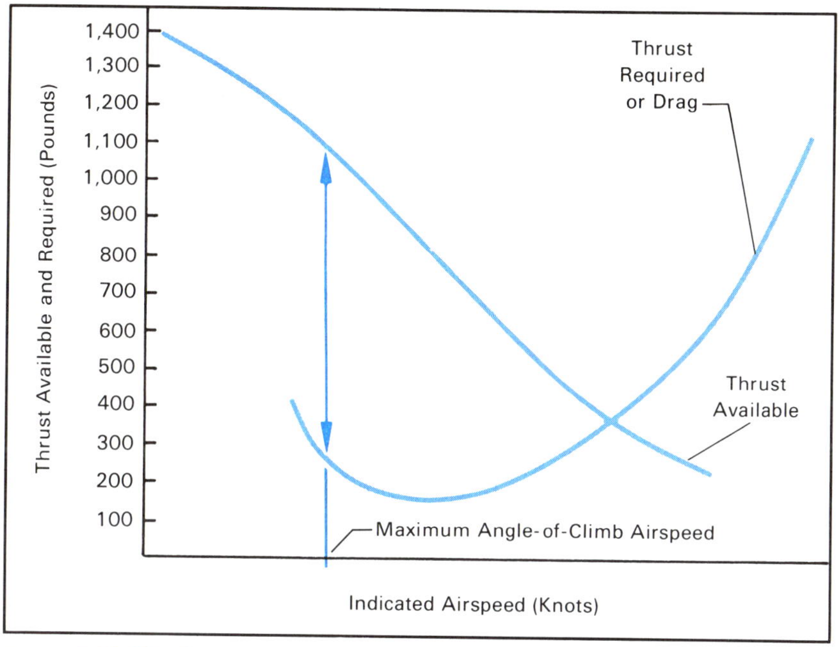

Figure 8-22. The thrust-available and thrust-required curves are shown here in relation to airspeed. Note that thrust available decreases as the airspeed increases, but the thrust-required curve increases with airspeed. Therefore, the maximum angle-of-climb airspeed occurs at the point with the greatest difference between thrust available and thrust required.

The thrust-available curve shows that propeller thrust is the highest at low airspeeds and decreases as airspeed increases. From this, you can see that maximum excess thrust is available at an airspeed very close to stall speed.

FACTORS AFFECTING CLIMB PERFORMANCE

Airspeed affects both the angle and rate of climb. At a given airplane weight, a specific airspeed is required for maximum performance. If you vary your airspeed from climb speed, you will have a decrease in performance. Generally, the larger the speed variation, the larger the performance variation. Weight also affects climb performance. If you change the weight of the airplane, you also change the drag and power required. As you increase weight, you reduce the maximum rate of climb and maximum angle of climb.

Increases in altitude generally have the greatest effect on climb performance. This occurs because a higher altitude increases the power and the thrust required, but decreases the power and the thrust available. Therefore, performance for both angle of climb and rate of climb decreases as altitude increases. At the same time, the airspeed necessary to obtain the maximum angle of climb increases with altitude, but the airspeed for the best rate of climb decreases. The point where these two airspeeds converge is referred to as the airplane's **absolute ceiling**.

GLIDES

In the event of an engine failure, you will probably be interested in flying the airplane at the minimum glide angle so you can travel the maximum distance for the altitude lost. When wind is not a significant factor, the minimum glide angle is obtained under the aerodynamic conditions which produce the least drag. Since lift is basically equal to weight, you will obtain the minimum drag at the maximum lift-to-drag ratio (L/D_{max}). This occurs at a specific value of the lift coefficient and, therefore, at a specific angle of attack. [Figure 8-23]

An angle of attack different from L/D_{max} will result in a decreased glide range. You can also see that more than one angle of attack can produce the same glide range. In this example, both four degrees and nine degrees will provide an L/D ratio of approximately 11.6.

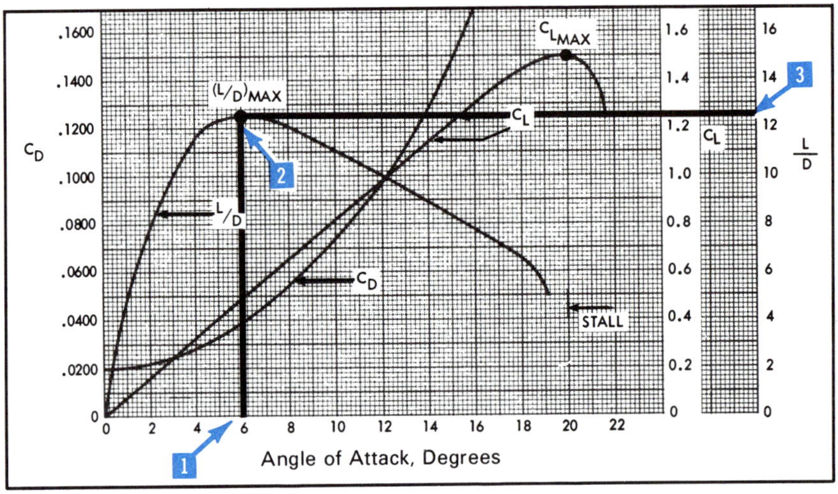

Figure 8-23. This graph is constructed by plotting the coefficients of lift and drag for each angle of attack. The L/D curve can be used to show the relationship between angle of attack and glide range. For example, an angle of attack of six degrees yields the best lift-to-drag ratio. To find the lift-to-drag ratio, locate six degrees on the angle of attack axis (item 1), move up to the intersection of the six-degree line with the L/D line (item 2). Then move horizontally to the far right edge to find the answer of 12.5 (item 3). This means the airplane will glide 12.5 n.m. horizontally for each mile of altitude lost. To find the altitude lost per nautical mile, divide 6,076 by 12.5 for an answer of approximately 486 feet. Statute miles may also be used, since the glide ratio is unaffected by units. However, you must divide by 5,280 when using statute miles.

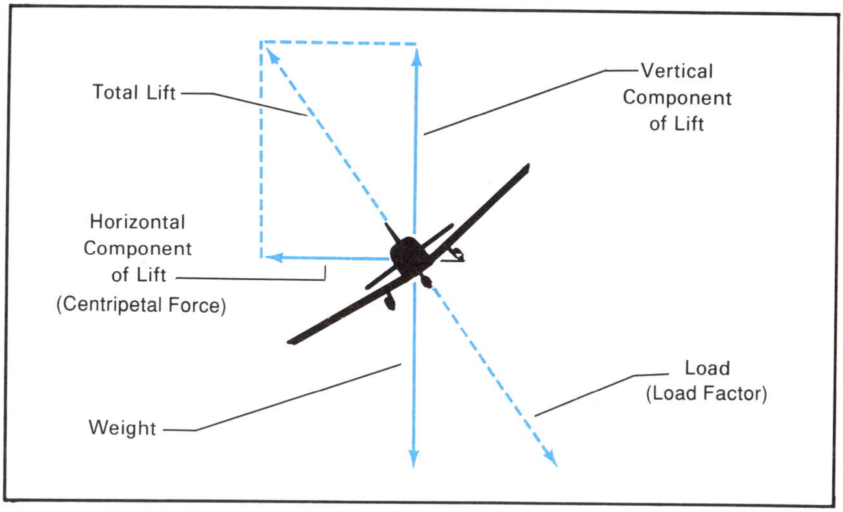

Figure 8-24. The horizontal component of lift provides the turning force which accelerates your aircraft toward the center of the turn, while the vertical component of lift overcomes weight. Total lift required to maintain altitude in a turn is the resultant of the horizontal and vertical components of lift. The horizontal component of lift is also known as centripetal force. This is the center-seeking force which causes an airplane to turn. In turns, centripetal force must continuously overcome the tendency of the airplane to continue in a straight line. Thus, centripetal force overcomes inertia. When centripetal force is opposed by inertia, the opposing force is often referred to as centrifugal force. Actually, centrifugal force is not a true aerodynamic force; it is an apparent force that results from the effect of overcoming inertia. Finally, total lift must be equal to the resulting load in order to maintain altitude during the turn. This is usually expressed as load factor.

TURNS

The basic relationships between aerodynamic forces determine the turn performance of an airplane. Any object traveling in a curved path is being constantly accelerated by unbalanced forces. If this were not so, the object would travel in a straight line at a constant speed, in accordance with Newton's first law of motion, which is also known as the law of inertia.

You make turns by canting the vertical component of lift to one side, which produces the turning force that changes the aircraft's flight path. The lift force can be subdivided and represented as two forces — one acting vertically and one acting horizontally. [Figure 8-24]

LOAD FACTOR

The constant load factor generated in a coordinated, level turn does not depend on the rate or radius of turn but, rather, the bank angle. Load factor remains constant if there is no change in bank angle. At a bank angle of 60°, the load factor is two G's, which means that the wings are supporting twice the weight of the aircraft. To maintain altitude as you increase bank, you must increase the angle of attack to make up for the loss in the vertical lift component. The relationships

To maintain altitude in a turn while steepening the bank, you must increase the angle of attack to compensate for the decrease in the vertical component of lift.

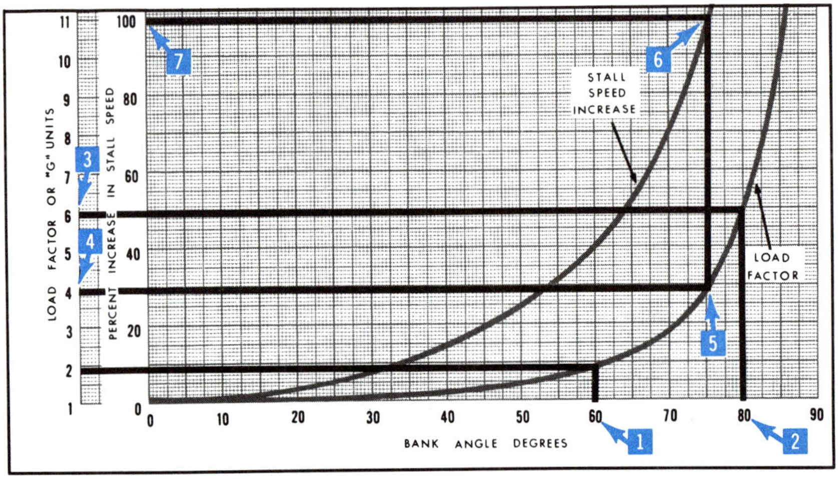

Figure 8-25. As bank angle increases, you can see that load factor increases rapidly. An increase in bank angle from 60° (item 1) to 80° (item 2) increases the load factor by four G's (item 3). High bank angles also increase stall speed. Stall speed increases in proportion to the square root of the load factor. An airplane which normally stalls at 50 knots in level flight stalls at 100 knots under a four-G load factor. Enter the chart on the left at four G's (item 4) and proceed right to the load factor line (item 5). Then, move up to the stall speed increase line (item 6) and left to the percent increase in stall speed (item 7), which is 100%. This means the stall speed doubles in a four-G turn.

between the load factor, angle of bank, and stall speed are the same for all airplanes. [Figure 8-25]

COORDINATION IN TURNS

During a turn, the relationship between bank angle and turning rate is reflected by the ball in the inclinometer. The ball gives an indication of the quality of the turn. If the ball is centered, the turn is coordinated. When the airplane is in a slip, it is yawed to the outside of the turn and the ball moves to the inside of the turn. This means the airplane is not turning at a rate fast enough for the angle of bank. In a skidding turn, the airplane is yawing toward the inside of the turn, and the ball of the inclinometer is to the outside of the turn. In this case, the rate of turn is too great for the bank angle being used. [Figure 8-26]

RADIUS AND RATE OF TURN

The radius of turn at any given bank angle varies directly with the square of the airspeed. Therefore, if the airspeed is doubled at any given bank angle, the radius of turn is quadrupled. The rate of turn for a given bank angle also varies with airspeed. If speed is increased at a constant bank angle, the rate of turn is reduced. Slower airplanes require less time and area to complete a turn.

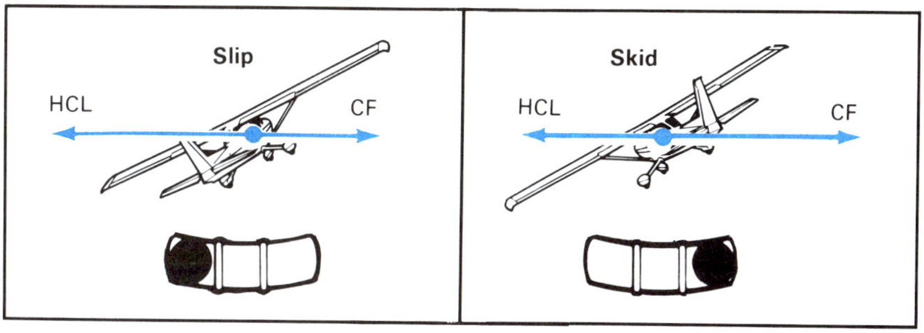

Figure 8-26. During a slip, the rate of turn is too slow for the angle of bank used, and the horizontal component of lift (HCL) exceeds centrifugal force (CF). You can reestablish equilibrium by decreasing the angle of bank, increasing the rate of turn by applying rudder to center the ball, or a combination of the two. During a skid, CF exceeds the HCL. An increase in centrifugal force also increases the load factor. To reestablish equilibrium, increase the bank angle or reduce the rate of turn by reducing the rudder force and allowing the ball to move to the center. You may also use a combination of these two corrections.

Since turn performance depends on the horizontal component of lift, the two variables that affect it are velocity and angle of bank. There are two equations used for computing radius and rate for a steady, coordinated, level turn.

$$\text{Turn Radius} = \frac{(\text{Velocity})^2}{11.26 \times \text{Tangent (Bank Angle)}}$$

$$\text{Rate of Turn} = \frac{1{,}091 \times \text{Tangent (Bank Angle)}}{(\text{Velocity})}$$

A specific angle of bank and true airspeed will produce the same rate and radius of turn regardless of weight, CG location, or airplane type. You can also see from these equations that increasing the velocity increases the turn radius and decreases the turn rate. The load factor on the airplane remains the same because you have not changed the angle of bank. To increase the rate and decrease the radius of a turn, you should steepen the bank and decrease your airspeed.

ADVERSE YAW

As you begin a turn, the aileron on the inside wing is raised and the aileron on the outside wing is lowered relative to the desired direction of turn. The lowered aileron causes a slight increase in wing camber, momentarily producing more lift on the outside wing. The increase in lift causes a corresponding increase in induced drag; this causes the airplane to yaw to the outside of the turn. At the same time, induced drag decreases on the inside wing. Rudder force in the direction of the turn is required to overcome adverse yaw so you can maintain coordination during turn entry.

GROUND EFFECT

An airplane is usually in ground effect when it is less than the height of the airplane's wingspan above the surface. Ground effect reduces induced angle of attack and induced drag.

Another significant aerodynamic consideration is the phenomenon of **ground effect.** During takeoffs or landings, when you are flying very close to the surface, the ground alters the three-dimensional airflow pattern around the airplane. This causes a reduction in wingtip vortices and a decrease in upwash and downwash. **Wingtip vortices** are caused by the air beneath the wing rolling up and around the wingtip. This, in turn, causes a spiral or vortex that trails behind each wingtip whenever lift is produced. Wingtip vortices also contribute to induced drag. Upwash and downwash refer to the effect an airfoil exerts on the free airstream. **Upwash** is the deflection of the oncoming airstream upward and over the wing. **Downwash** is the downward deflection of the airstream as it passes over the wing and past the trailing edge. During flight, the downwash of the airstream causes the relative wind to be inclined downward in the vicinity of the wing. This is referred to as the **average relative wind**. The angle between the free airstream relative wind and the average relative wind is the **induced angle of attack**. In effect, the greater the downward deflection of the airstream, the higher the induced angle of attack and the higher the induced drag. Since ground effect restricts the downward deflection of the airstream, both the induced angle of attack and induced drag decrease. When the wing is at a height equal to its span, the decline in induced drag is only about 1.4%; when the wing is at a height equal to one-tenth its span, the loss of induced drag is about 48%. [Figure 8-27]

When an airplane climbs out of ground effect, it suddenly experiences greater induced drag and requires more thrust.

With the reduction of induced drag in ground effect, the amount of thrust required to produce lift is reduced. What this means is that your airplane is capable of lifting off at a lower than normal speed. Although you might initially think this is desirable, consider what happens as you climb out of ground effect. The power (thrust) required to sustain flight increases significantly as the normal airflow around the wing returns, and induced drag is suddenly increased. If you attempt to climb out of ground effect before reaching the speed for normal climb, you might sink back to the runway.

A lower than normal angle of attack produces the same amount of lift when you are flying in ground effect.

Ground effect probably is most noticeable in the landing phase of flight. Within one wingspan above the ground, the decrease in induced drag makes your airplane seem to float on the cushion of air beneath it. This is why a power reduction usually is required during the flare to help the airplane land. In ground effect, you also need a lower angle of attack to produce the same amount of lift. Although all airplanes may experience ground effect, it is more noticeable in low-wing airplanes, simply because the wings are closer to the ground.

STALL AND SPIN AWARENESS

National Transportation Safety Board (NTSB) statistics indicate that most stall/spin accidents are associated with incidents where a pilot is momentarily distracted from the primary task of flying the aircraft. Because of

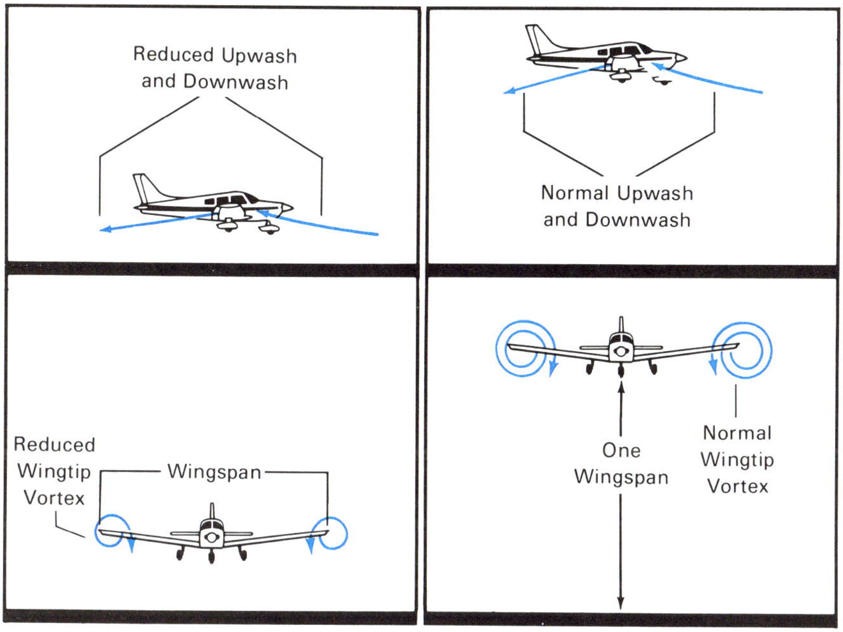

Figure 8-27. When you are flying in ground effect, upwash, downwash, and wingtip vortices decrease. This results in a reduction of induced drag. Ground effect is most noticeable near the surface; it decreases rapidly until, at a height approximately equal to the wingspan of the aircraft, it becomes negligible.

this, stall and spin awareness is included in the training requirements for commercial pilots. According to the FAA, the training is intended to emphasize recognition of situations that could lead to an unintentional stall and/or spin. This is accomplished by use of instructional exercises that include realistic distractions during flight at slow airspeeds. The practice of intentional stalls and recovery techniques is also included in this training. However, only flight instructor candidates are required to demonstrate proficiency in actual spin entry, spins, and spin recovery techniques as a requirement for certification.

Stall awareness is stressed during flight training. The use of realistic distractions during flight at slow airspeeds is part of this required training.

STALLS

In addition to the training requirements, common sense dictates that you be knowledgeable about the causes and effects of stalling. This is particularly important during flight at slow airspeeds where the margin above the stall speed is small. Keep in mind, slow airspeeds are necessary for every takeoff and landing.

CAUSES OF STALLS

As indicated earlier, a stall occurs when the maximum lift angle of attack (C_{Lmax}) is exceeded. This can occur at any airspeed, in any attitude, and at any power setting when the air can no longer flow smoothly over the top of the wing. The airflow separation, which causes the stall, is really

The cause of a stall is exceeding the maximum lift angle of attack.

a progressive event. It begins when the airfoil angle of attack approaches C_{Lmax}; and then, as the angle of attack increases to C_{Lmax} and beyond, the airflow separation increases, lift progressively decreases, and drag increases. If you do not recover promptly by reducing the angle of attack, a secondary stall and/or a spin may result.

While it is true that a stall can occur at any airspeed, you should not conclude that airspeed has nothing to do with stalls or stall avoidance. Airspeed, when analyzed in conjunction with other factors, can be an indicator of an impending stall. The speed at which C_{Lmax} is exceeded is the stall speed for the existing flight conditions. Stall speeds for specific conditions are published in the airplane flight manual or pilot's operating handbook (POH). [Figure 8-28]

A fully loaded airplane stalls at a higher speed than a lightly loaded airplane; and a forward CG causes a higher stalling speed than a more rearward CG. Besides load factor, flap setting (configuration), weight, and CG position, other factors can affect the actual stalling speed. Turbulence, coordination, and snow, ice, or frost on the wings are examples. As you might expect, turbulence, lack of control coordination, and the presence of snow, ice, or frost on the wings can increase the stall speed. Although not often considered, the condition of an aircraft can also be a factor. A poorly maintained an/or an improperly rigged airplane adversely affects stalling speeds.

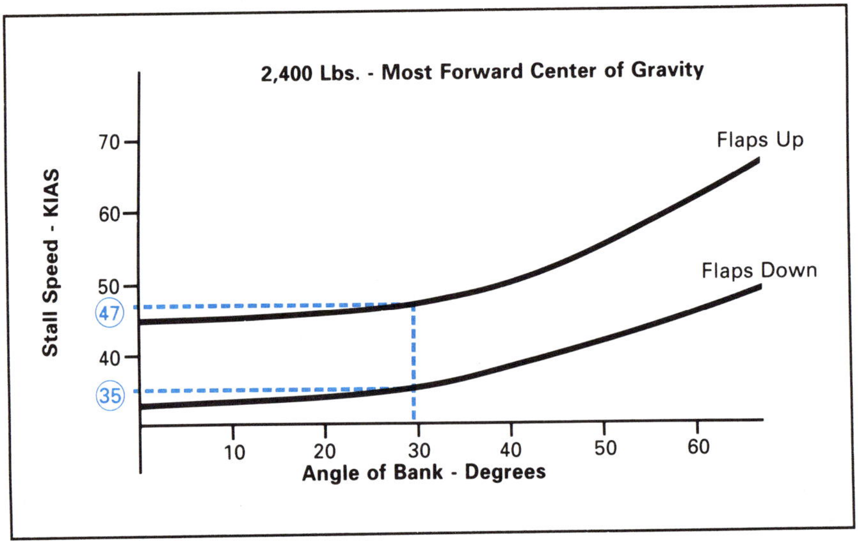

Figure 8-28. The stall speed for a level, 30° angle of bank is 47 knots indicated airspeed (KIAS) with the flaps up and 35 KIAS with the flaps extended. It is important to remember that the speeds published in a POH apply only to the stated conditions. In this example, the conditions specify a weight of 2,400 pounds and the most forward center of gravity position.

TYPES OF STALLS

General classifications include power-on/power-off and straight ahead or turning stalls. Commercial pilot applicants are required to be proficient in recognizing, performing, and recovering from these stall maneuvers. Normally, **power-on stalls** are associated with a takeoff and departure (or climb) configuration and **power-off stalls** are considered most likely in an approach configuration. The takeoff or departure configuration is usually gear up (if retractable) and flaps at the takeoff setting. The approach configuration suggests gear and flaps down. Straight ahead and turning are simply descriptive terms which refer to the flight attitude during the onset of a stall.

Additional types or classifications include accelerated, secondary, cross-controlled, and elevator trim stalls. These terms further define a stall. Thus, you may be able to identify a stall by both general and specific terms. For example, a turning, power-on, cross-controlled stall might describe a particular maneuver related to an aircraft accident or incident.

Accelerated stalls are caused by abrupt or excessive control movement. In fact, they are sometimes called excessive back-pressure stalls. They commonly occur during maneuvers such as steep turns or pullups involving an abrupt change in the flight path. Accelerated stalls are usually more severe than unaccelerated stalls and they are often unexpected because of the comparatively high airspeed.

Secondary stalls may occur during an improper recovery from any stall. The usual cause is an attempt to speed up a stall recovery before you decrease the angle of attack and regain sufficient flying speed.

As the name implies, **cross-controlled stalls** occur when the flight controls are crossed. This means aileron pressure in one direction and rudder pressure in the opposite direction. Along with the uncoordinated use of the ailerons and rudder, excessive back pressure is the other ingredient for a cross-controlled stall. The most likely place for this is during a poorly executed turn to the final approach for a landing. In a typical cross-controlled stall, excessive inside rudder is used to help turn, along with opposite aileron pressure to keep the bank angle from becoming too steep. When elevator back pressure is added to avoid flying too low, the aircraft is set up for the stall. If the aircraft does stall, the nose may pitch down and the wing on the inside of the turn may suddenly drop. Often, these stalls occur with little warning.

An **elevator trim stall** is most likely to occur during a go-around from a landing approach. In this case, the aircraft's trim is adjusted to a slow speed for the approach with considerable nose-up trim. As power is applied for the go-around, the normal tendency is for the nose to pitch up. If positive pressure is not used to counteract the strong trim forces, the nose of the aircraft will continue to pitch up. In addition, since full power is required during a go-around, the aircraft is affected by left-turning forces which can easily result in an uncoordinated flight condition. Now you have the conditions that can cause a stall — a high angle of attack, uncoordinated flight, and a decreasing airspeed.

STALL RECOGNITION AND RECOVERY

One part of stall recognition is knowing the flight characteristics of the airplane you are flying. Then, you must maintain an awareness of the flight conditions where various types of stalls are most likely to occur.

In training airplanes, a stall warning horn usually provides the most obvious indication of an impending stall.

The most obvious stall warning indicator in light, general aviation airplanes is the warning horn. This warning system is a simplified angle of attack indicator installed on the leading edge of the wing. Normally, it consists of a vane or pressure-sensing device that is activated by airflow as the wing approaches C_{Lmax}. Most systems will trigger the stall warning horn and/or a warning light about 5 to 10 knots above the stall speed. A disadvantage of this type of system is that it is attached to only one wing. An uncoordinated or asymmetric stall on the other wing may not actuate the warning signal.

Other indications of a stall include a mushy feeling in the flight controls, as speed decreases and the controls become less effective. This is due in part to reduced airflow over the control surfaces. In airplanes with a fixed-pitch propeller, a loss of r.p.m. may be noticeable as you approach a stall in power-on conditions. Also, a reduction in the sound of air flowing along the fuselage is usually evident. Just before the stall occurs, buffeting, uncontrollable pitching, or vibrations may begin. Finally, your kinesthetic sense (ability to recognize changes in direction or speed) may also provide you with a warning of decreased speed or the beginning of a sinking feeling. These preliminary indications serve as a warning for you to increase airspeed by decreasing the angle of attack and by adding power. Ideally, you should be able to detect the signs of an impending stall and make appropriate corrections before it actually occurs. Recovery at the first indication is quite simple; but, if you allow the stalled condition to progress, recovery becomes more difficult.

The first step in stall recovery is to reduce the angle of attack.

The key action in recovery from a stall is to promptly reduce the angle of attack to allow the wings to regain lift. You decrease the angle of attack by forward movement of the elevator (stabilator) control. The amount of movement may consist only of releasing back elevator control pressure; or you may have to firmly move the control column forward, depending on the aircraft design, severity of the stall, and other factors. Excessive forward movement of the control column, however, may impose a negative load on the wings and delay the stall recovery. The next step, which should be accomplished almost simultaneously, is to smoothly apply maximum power to increase the airspeed and to minimize the loss of altitude. As airspeed increases, adjust power to return the airplane to the desired flight condition. In most training situations, this is straight-and-level flight or a climb. During the recovery, full coordinated use of the controls is especially important. Uncoordinated flight control inputs can aggravate a stall and result in a spin. You also must avoid exceeding the airspeed and r.p.m. limits of the airplane.

SPINS

Since commercial pilot applicants are not required to demonstrate flight proficiency in spins, this discussion is primarily concerned with fundamental concepts and relationships associated with inadvertent spins. As indicated earlier, the regulations require you to be aware of what causes these spins, how to avoid them, and how to recover, if necessary.

The spin is the most complex of all flight maneuvers. There are actually hundreds of factors that contribute to the spinning of an airplane. In a light, training airplane, a **spin** may be defined as an aggravated stall which results in autorotation. During the spin, the airplane descends in a helical, or corkscrew, path while the angle of attack is greater than the critical angle of attack.

PRIMARY CAUSES

A stall must occur before a spin can develop. However, a stall is essentially a coordinated maneuver where both wings are equally or almost equally stalled. In contrast, a spin is an uncoordinated maneuver with the wings unequally stalled. In this case, the wing that is more completely stalled will often drop before the other, and the nose of the aircraft will yaw in the direction of the low wing.

To enter a spin, an airplane must first be stalled. In a spin, both wings are in a stalled condition.

Typically, the cause of an inadvertent spin is exceeding the critical angle of attack while performing an uncoordinated maneuver. The lack of control is normally caused by either too much or not enough rudder control for yawing forces induced by aileron deflection or P-factor. The result is a cross-controlled condition. If you do not initiate the stall recovery promptly, the airplane is more likely to enter a full stall that may develop into a spin. The spin that occurs from cross controlling usually results in rotation in the direction of the rudder being applied, regardless of which wing is raised.

When an airplane is in stalled or nearly stalled condition, uncoordinated use of the flight controls (aileron and rudder) can easily cause a spin. In most cases, the spin rotation will be in the direction of rudder deflection.

During a cross-controlled, uncoordinated maneuver, the pitot/static instruments, especially the altimeter and airspeed indicator, may be unreliable due to the uneven distribution of air pressure over the fuselage. This error may lead you to believe that your airspeed is sufficient to maintain smooth airflow over the wing when, in fact, you are very close to stalling. Your only warnings of the stall may be aerodynamic buffeting and/or the stall warning horn.

Although most pilots are able to maintain coordination of the flight controls during routine maneuvers, this ability often deteriorates when distractions occur and their attention is divided between important tasks. Distractions that have caused problems include preoccupation with situations inside or outside the cockpit, maneuvering to avoid other aircraft, and maneuvering to clear obstacles during takeoffs, climbs, approaches, or landings. Because of this, you will be required to learn how to recognize and cope with these distractions by practicing "flight at slow airspeeds with realistic distractions" during flight lessons under the guidance of your instructor.

PHASES OF A SPIN

In light, training airplanes, a complete spin maneuver consists of three phases — incipient, developed, and recovery. The **incipient spin** is that portion of a spin from the time the airplane stalls and rotation starts until the spin is fully developed. A **fully developed** spin means the angular rotation rates, airspeed, and vertical speed are stabilized from turn to turn and the flight path is close to vertical. The **spin recovery** begins when anti-spin forces overcome the pro-spin forces. [Figure 8-29]

Normally, the recovery from an incipient spin requires less time (and altitude) than the recovery from a fully developed spin. Higher altitudes tend to prolong a recovery. Anti-spin controls are not as effective in the less dense air as they are at lower altitudes. You also should keep in mind that, although some characteristics of a spin are predictable, every airplane spins differently. In addition, the same airplane's spin behavior changes with variations in configuration, loading, and other factors.

WEIGHT AND BALANCE

Airplanes with an aft CG tend to be less stall/spin resistant than airplanes with a forward CG.

Even minor weight and balance changes can affect an aircraft's spin characteristics. Heavier weights generally result in slower spin rates initially; but, as the spin progresses, heavier weights tend to cause an increasing spin rate and longer recovery time. Distribution of weight is even more significant. Forward center of gravity positions usually inhibit the high angles of attack necessary for a stall. Thus, an airplane with a forward CG tends to be more stall and spin resistant than an aircraft with an aft CG. In addition, spins with aft CG positions are flatter than ordinary spins. A **flat spin** is characterized by a near level pitch and roll attitude with the spin axis near the CG of the airplane. Recovery from a flat spin may be extremely difficult or even impossible.

In a training airplane, the addition of a back seat passenger or a single suitcase to an aft baggage compartment can affect the CG enough to change the characteristics of a spin. Instead of an ordinary spin, a flat spin could easily occur. There also are other unfavorable loading possibilities. A concentration of weight, or unbalanced weight distribution, a long way

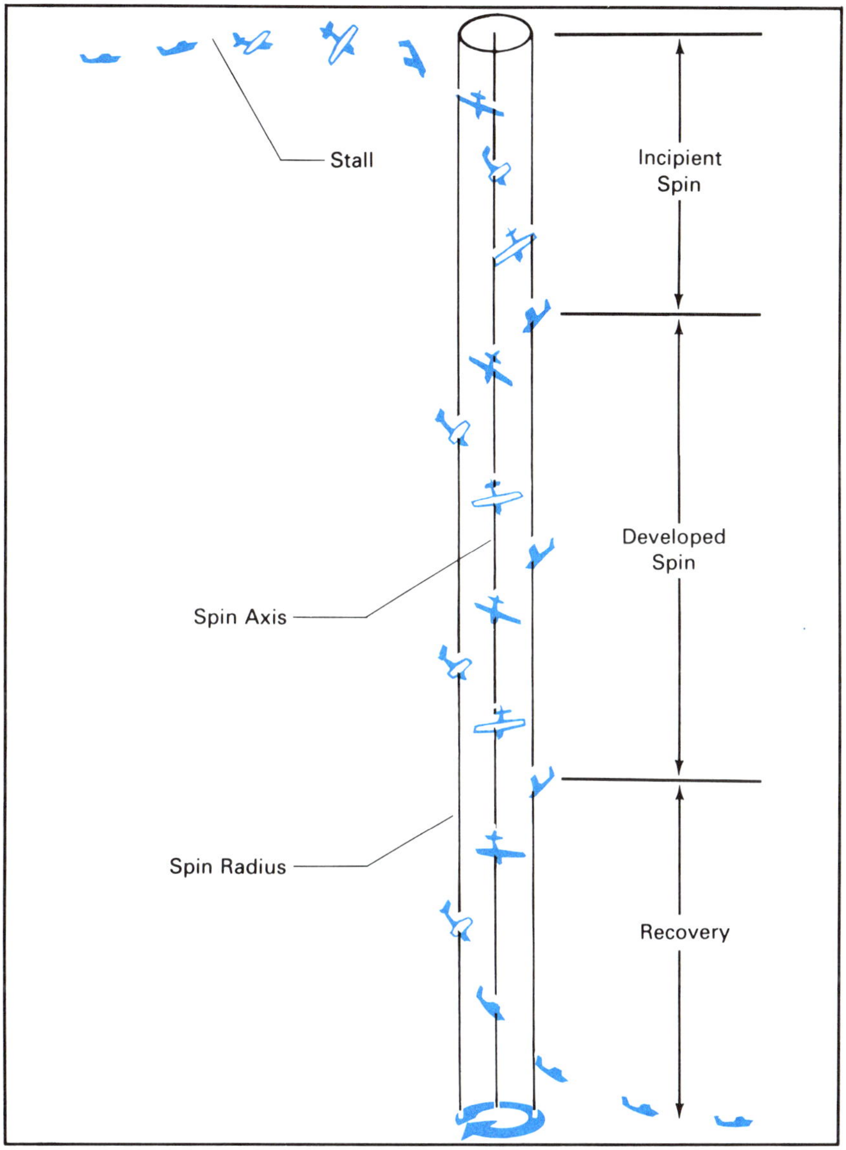

Figure 8-29. The incipient spin usually occurs rapidly in light airplanes (about 4 to 6 seconds) and consists of approximately the first two turns. At about the half-turn point, the airplane is pointed almost straight down but the angle of attack is above that of the stall because of the inclined flight path. As the one-turn point approaches, the nose may come back up and the angle of attack continues to increase. As the airplane continues to rotate into the second turn, the flight path becomes more nearly vertical, and the pitching, rolling, and yawing motions become more repeatable. This is the beginning of the fully developed spin. The last phase, recovery, occurs when anti-spin forces overcome pro-spin forces. During recovery, the angle of attack on both wings decreases below the critical angle and the rotation rate slows. This phase can range from one quarter of a turn to several turns.

from the CG is undesirable. This type of loading may occur with tip tanks or outboard wing tanks. If the fuel in these tanks becomes unbalanced, an asymmetrical condition exists. The worst asymmetric condition is full fuel in the wing on the outside of the spin and no fuel in the tanks on the inside of the turn. Once the spin is developed, the momentum (inertial force) makes recovery unlikely under these conditions.

There is no guarantee that a normal category airplane will recover from a fully developed spin.

Single-engine, normal category airplanes are prohibited from intentional spins. This is indicated by a placard with words such as "No acrobatic maneuvers, including spins, approved." However, during aircraft certification tests, normal category airplanes must demonstrate recovery from a one-turn spin or a three-second spin, whichever takes longer. The recovery must take place within one additional turn with normal control inputs. Since airplanes in the normal category have not been tested for more than one-turn/three-second spins, their performance characteristics beyond these limits are unknown. The one-turn margin of safety only provides a check of the airplane's controllability in a delayed or aggravated stall which is early in the incipient phase of a spin. This does not provide any assurance that recovery is possible from a fully developed spin.

Acrobatic category airplanes must recover from fully developed spins within one and one-half additional turns. Certification in this category also requires six turns or three seconds, whichever takes longer, before the recovery controls are applied.

Utility category airplanes may be tested under the one-turn (normal) criteria or they may satisfy the six-turn (acrobatic) spin requirements. However, spins in utility category airplanes may be approved only with specific loading, such as at a reduced weight and with a forward CG position. It is extremely important for you to understand all of the operating limitations for your airplane. Applicable limitations are placarded and/or included in the POH.

SPIN RECOVERY

Spin recovery techniques vary for different aircraft; therefore, you must follow the recovery procedures outlined in the POH for your airplane. The following is a general recovery procedure, but it should not be applied arbitrarily without regard for the manufacturer's recommendations.

Since an airplane must be in a stalled condition before it will spin, the first thing you should do is try to recover from the stall before the spin develops. If your reaction is too slow and a spin develops, move the throttle to idle, neutralize the ailerons, and raise the flaps. Next, apply full rudder deflection opposite to the direction of the turn. When the rotation rate slows, briskly position the elevator forward to approximately the neutral position to decrease the angle of attack. As the rotation stops, neutralize the rudder and gradually apply enough back pressure to recover from the steep nose-down attitude. During the recovery, you should avoid excessive airspeed and G-forces. Too much or too abrupt use of back elevator pressure and/or application of rudder and ailerons during the pull-up can result in a secondary stall and possibly another spin. In a fully developed spin, the engine will sometimes stop due to centrifugal force acting on the fuel system. Therefore, power may not be immediately available during the recovery.

CHECKLIST

After studying this section, you should have a basic understanding of:

✓ **Lift** — How angle of attack affects the center of pressure on an airfoil.

✓ **Center of pressure** — How its location and movement in relation to the center of gravity influence the aerodynamic balance of an airplane.

✓ **High-lift devices** — How the various types of flaps and leading-edge devices can be used to change the wing's camber and increase the coefficient of lift.

✓ **Wing planform** — How an airfoil's shape and aspect ratio affect its stall and performance characteristics.

✓ **Drag** — How the proportions of induced and parasite drag vary with speed.

✓ **Thrust** — How propeller efficiency and engine power in relation to total drag determine performance speeds, including the maximum level flight speed.

✓ **Weight** — What the relationship is between load factor and weight.

✓ **Stability** — What the various types are and why an airplane must be designed to achieve stability about the three axes of flight.

✓ **Turns** — What balanced forces are involved in turns and why load factor, angle of bank, and stall speed relationships are the same for all airplanes.

✓ **Ground effect** — How an airplane is affected when it is less than the height of its wingspan above the surface.

✓ **Stall and Spin Awareness** — Causes, recovery techniques, and how to prevent them.

SECTION

B

PREDICTING AIRPLANE PERFORMANCE

Your ability to accurately predict the performance of an airplane is based on the assumption that you will operate it within the limits specified by the manufacturer. As you know, the airplane's operating limitations establish the boundaries within which it must be flown. They are often referred to as the flight, or performance, envelope. Operating within the envelope is safe, while operating outside the envelope may cause structural damage or even failure. If you fully understand an airplane's operating limitations, you are unlikely to fly outside its performance envelope.

V-G DIAGRAM

A useful method for showing an airplane's operating limitations is the **V-g diagram**. It relates velocity (V) to load factor (g). Although it is seldom included in the POH, a V-g diagram often is used during the initial testing and certification of an airplane. Each diagram applies to one airplane type and the information is valid only for a specific weight, configuration, and altitude. V-g diagrams show the maximum amount of positive or negative lift the airplane is capable of generating at a given speed. They also show the safe load factor limits and the load factor, or number of G's, the airplane can sustain at various speeds. Airplane designers use a diagram such as this to establish the important speed and load factor limits that define the flight or performance envelope. [Figure 8-30]

Major points of the V-g diagram include the curved lines representing positive and negative maximum lift capability. These lines portray the maximum amount of lift the airplane can generate at the specified speed. The intersection of these lines with the vertical speed lines indicates the maximum G-load capability at that speed. If you attempt to exceed this G-load limit at that speed, the airplane will stall. For example, at a speed of about 92, this airplane will stall with a load factor of two G's under the conditions represented. The horizontal positive and negative load factor limits represent the structural limitations of the airplane. Exceeding these values may cause structural damage.

As altitude increases, the indicated airspeed at which a given airplane stalls in a specific configuration remains the same.

Another important point on the V-g diagram is the normal stall speed, V_{S1}. Note that at this speed, the airplane stalls at a load of one positive G. This is the same speed shown by the lower limit of the green arc on the airspeed indicator. Although V_{S1} changes with weight and configuration, it is not affected by altitude. However, above the one-G load factor, stalling speed

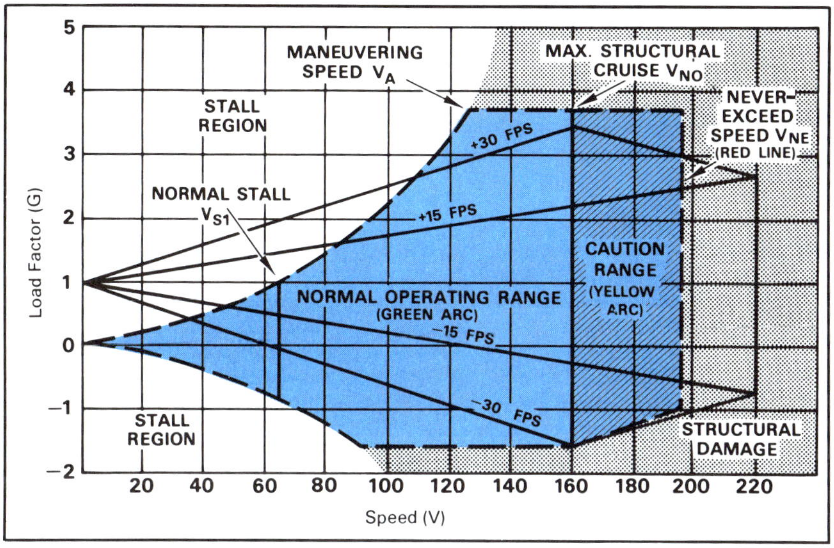

Figure 8-30. The horizontal scale indicates speed (V), and the vertical scale is load factor (g). Speed is plotted against the load factor or G-loading. The curved lines extending upward and downward from a load factor of zero are the positive and negative lift capability lines. The horizontal line at 3.8 G's indicates the positive load factor limit, and the line at -1.52 G's indicates the negative load factor limit. These are the load factor limits for a normal category airplane. Remember, load factor limits for utility category aircraft are different and allow you to perform limited aerobatics, including spins in some airplane types. You may want to refer back to this figure throughout the remaining V-g diagram discussion.

increases. Any stall which occurs above the straight-and-level load of one G is an accelerated stall.

V_A is the maximum speed at which full or abrupt control movements may be used without overstressing the airplane.

Maneuvering speed (V_A) is another critical speed. It is defined as the maximum speed at which you can safely use full and abrupt control movement without causing structural damage. The term also suggests a speed you can use for maximum maneuverability without overstressing the airplane. You can see that V_A occurs at the point where the curved line representing maximum positive lift capability intersects the maximum positive load factor limit at the top of the envelope. At this speed, a load in excess of 3.8 G's will result in a stall. In fact, the airplane always stalls before structural damage occurs provided the airspeed does not exceed V_A. If, however, you allow the speed to exceed V_A and you exceed 3.8 G's, structural damage may occur.

V_{NO} is the maximum structural cruising speed during normal operations while V_{NE} is the never-exceed speed. Above this speed, load factors may exceed design limits, causing structural damage or failure.

The vertical line at a speed of 160 is the maximum structural cruising speed, V_{NO}. It should not be exceeded in rough air. The speed range from V_{S1} to V_{NO} is the normal operating range and corresponds to the green arc on the airspeed indicator. The vertical line at a speed of about 195 represents the never-exceed speed, V_{NE}. This speed corresponds to the red line. If you fly faster than V_{NE}, there is a possibility of control surface flutter, airframe structural damage, or failure, because the design load limit factors may be easily exceeded during maneuvers. The range from V_{NO} to V_{NE} is the caution range represented by the yellow arc on an airspeed indicator. It should be used only in smooth air.

V-g diagrams often include diagonal **gust lines**, which are important design considerations. In the sample V-g diagram, gust lines are plotted for plus and minus 15 and 30 feet per second. By using these lines, you can determine the load factor imposed on the airplane when it encounters a sudden vertical gust. For example, an upward gust of 15 feet per second at a speed of 140 results in a total positive load factor of approximately 2.1 G's. Since the total load that may be imposed on an aircraft includes the maneuvering load factor plus the gust factor, you could easily exceed the 3.8-G limit if you encounter a gust during a 2-G steep turn. You can see why it is so important to reduce your speed to V_A or less in turbulence.

FACTORS AFFECTING PERFORMANCE

Anything that affects an engine's operating efficiency affects performance. Full power for a normally aspirated engine is based on conditions at sea level in a standard atmosphere. In fact, standard atmospheric conditions are the basis for estimating airplane performance. However, airplanes are rarely flown at sea level and are almost never operated in weather conditions that exactly duplicate a model of the standard atmosphere. Temperature, pressure, humidity, and wind conditions vary constantly. In addition, the airport elevation, runway slope, and the condition of the runway surface also affect takeoff performance.

DENSITY ALTITUDE

When temperature increases, density altitude increases, causing a decrease in aircraft performance.

By now, you know that density altitude is pressure altitude corrected for nonstandard temperature. In other words, the density altitude at a given level is equal to the pressure altitude only when standard atmospheric conditions exist at that level. You should be concerned about density altitude, because the density of the air directly affects the performance capability of your airplane. In fact, the primary reason for computing density altitude is to determine aircraft performance. Low density altitude (dense air) is favorable, and high density altitude (thin air) is unfavorable. As density altitude increases, engine power output, propeller efficiency, and aerodynamic lift decrease. The performance degradation occurs throughout the airplane's operating envelope, but it is especially noticeable in takeoff and climb performance.

One of the ways you can determine density altitude is by using a flight computer. Another way is through the use of charts designed for that purpose. For example, assume you are planning to depart an airport where the field elevation is 1,165 feet MSL, the altimeter setting is 30.10, and the temperature is 70°F. What is the density altitude? [Figure 8-31]

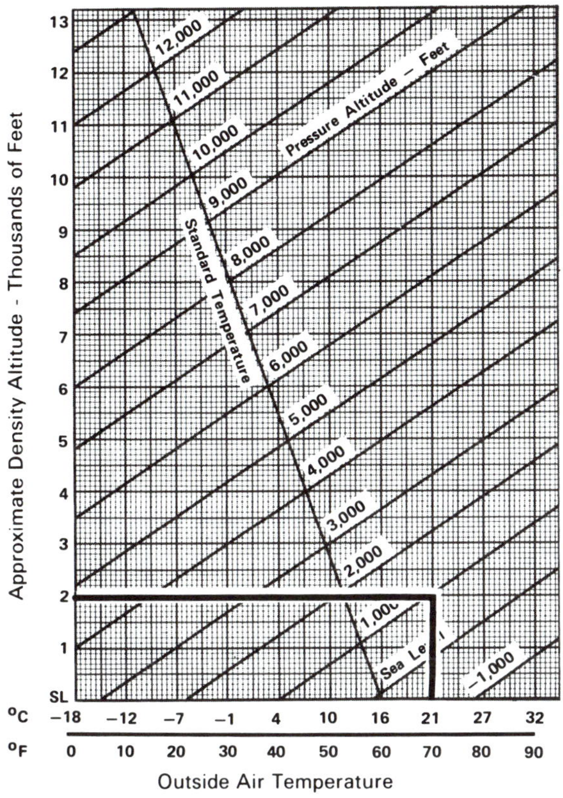

Altimeter Setting	Pressure Altitude Conversion Factor
28.0	1,824
28.1	1,727
28.2	1,630
28.3	1,533
28.4	1,436
28.5	1,340
28.6	1,244
28.7	1,148
28.8	1,053
28.9	957
29.0	863
29.1	768
29.2	673
29.3	579
29.4	485
29.5	392
29.6	298
29.7	205
29.8	112
29.9	20
29.92	0
30.0	-73
30.1	-165
30.2	-257
30.3	-348
30.4	-440
30.5	-531
30.6	-622
30.7	-712
30.8	-803
30.9	-893
31.0	-983

Figure 8-31. First, correct for nonstandard pressure by subtracting 165 feet from the field elevation to obtain the pressure altitude of 1,000 feet. Then, enter the chart at the bottom, just above the temperature of 70°F (21°C). Proceed up the chart until you intercept the diagonal 1,000-foot pressure altitude line, then move horizontally to the left and read the density altitude of approximately 2,000 feet. This means your airplane will perform as if it were at 2,000 feet MSL on a standard day.

Most modern performance charts do not require you to compute density altitude. Instead, the computation is built into the performance chart itself. All you have to do is enter the chart with the correct pressure altitude and the temperature. Older charts, however, may require you to compute density altitude before entering them.

HUMIDITY

Humidity refers to the amount of water vapor contained in the atmosphere and is expressed as a percentage of the maximum amount of vapor the air can hold. This amount varies with air temperature. As air temperature increases, the maximum amount of water vapor the air can hold increases. At a given temperature, perfectly dry air has a humidity of 0%. When the same air is saturated, it cannot hold any more water vapor at the same temperature, and its humidity is 100%.

You should consider the effects of humidity, since very high humidity can increase total takeoff distance and reduce climb performance by as much as 10%.

Although the effects of humidity are not shown on performance charts, it does reduce airplane performance. For one thing, it takes up airspace that is normally available for vaporized fuel. As humidity increases, less air enters the engine. This has the effect of causing a small increase in density altitude. The moist air also tends to retard even burning of fuel in the cylinder. When the relative humidity is very high, the engine power loss may be as high as seven percent, and the airplane's total takeoff and climb performance may be reduced by as much as 10%.

The most adverse effects on airplane performance occur at high altitudes when it is hot and humid.

Under conditions of high humidity, take extra time in setting the mixture control to the manufacturer's specifications in order to obtain the maximum available power. Anytime the humidity and other atmospheric conditions are conducive to increased density altitude, you should anticipate a decrease in your acceleration rate because of reduced engine power and propeller efficiency. Always provide yourself with an extra margin of safety by using longer runways and by expecting reduced takeoff and climb performance. Remember, conditions that cause high density altitude are "high, hot, and humid."

SURFACE WINDS

Winds can have a significant impact on airplane operations. When they are used to your advantage, they reduce takeoff and landing distances. Since surface winds may not always be aligned with the runway in use, you need a method for determining what portion of the wind is acting along the runway and what portion is acting across it.

If a headwind equals 10% of your takeoff airspeed, it will reduce the takeoff distance by approximately 19%.

Headwind or headwind component refers to that portion of the wind which acts straight down the runway toward the airplane. Tailwind or tailwind component describes that portion of the wind which acts directly on the tail of the airplane. For example, a headwind decreases takeoff distance and runway length requirements. A headwind that is equal to 10% of your takeoff airspeed will reduce your takeoff distance by approximately 19%. However, a tailwind which is 10% of your takeoff speed will increase takeoff distance by about 21%.

An airplane's maximum demonstrated crosswind component must equal 0.2 V_{SO}.

Crosswind component refers to that portion of the wind which acts perpendicular to the runway. Most airplanes have a maximum demonstrated crosswind component listed in the POH, although this is not an operating limitation. FARs require that all airplanes type-certificated since 1962 have safe ground handling characteristics in 90° crosswinds equal to 0.2 V_{SO}. If V_{SO} for a particular airplane is 65 knots, the manufacturer's demonstrated crosswind component is 13 knots ($0.2 \times 65 = 13$ knots). Your personal crosswind limit is actually based on your skill level in taking off and landing safely in a particular airplane type.

You can easily compute headwind and crosswind components by use of a wind component chart. When you use the chart, remember that for takeoff or landing, surface wind is reported in magnetic direction, so it corresponds with the runway number. To use a sample wind component chart, you first need to know the wind direction and the velocity. Next, you find the angle between the runway in use and the wind. For example, assume you are departing on runway 3 and the wind is from 060° at 20 knots. The angle between the runway and the wind is 30°. In this case, you have a headwind 30° off the runway heading. Since the headwind is from the right, you also have a crosswind component. [Figure 8-32]

WEIGHT

The amount of lift required to maintain altitude during flight is directly related to the total weight of the airplane and its contents. With a constant angle of attack, the airspeed necessary to provide sufficient lift to

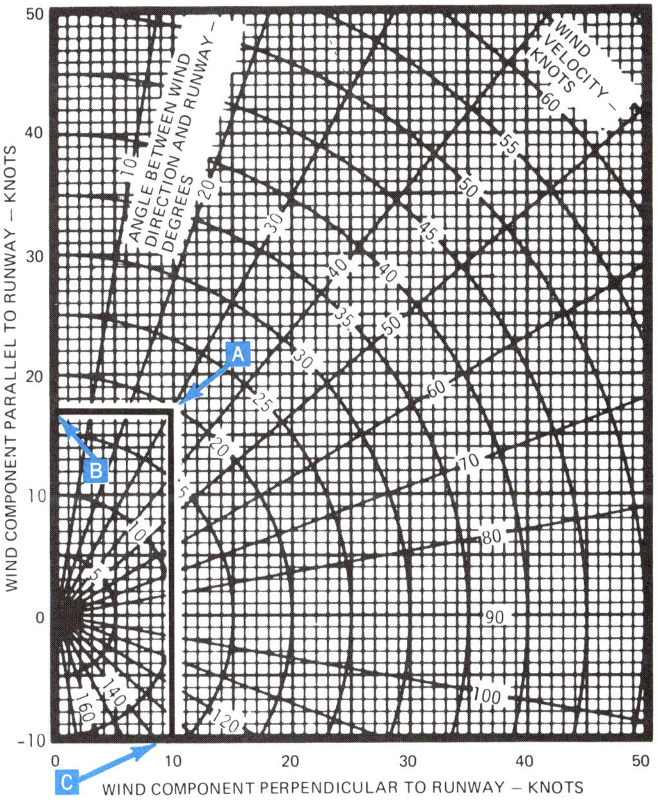

Figure 8-32. To determine the headwind and crosswind components, enter the sample chart at the point where the wind angle and the wind velocity arc meet (point A). Proceed horizontally to the left to find the headwind component of 17 knots (point B). Find the crosswind component by following the vertical lines down from point A to the bottom of the chart. The crosswind component is 10 knots (point C).

maintain level flight increases as weight increases. An overloaded airplane must accelerate to a higher than normal speed to generate sufficient lift for flight. This means the airplane requires more runway for takeoff, since it must attain a higher speed. The additional weight also reduces acceleration during the takeoff roll, and this adds to the total takeoff distance. Problems do not end once the airplane is airborne. An overloaded airplane suffers a reduction in climb performance, and its service ceiling is decreased. This is because excess power and thrust are limited. In addition, exceeding the maximum allowable weight may cause the airplane to become unstable and difficult to fly. For an airplane to be safe, it must be operated within weight limitations. In addition, the weight must be distributed, or balanced, properly. Procedures for computing weight and balance are covered in the next section.

RUNWAY CONDITIONS

When airplane manufacturers develop performance data for their airplanes, they normally specify distances based on a paved, level runway with a smooth, dry surface. If any one of these conditions does not exist for the runway you use, the takeoff and landing distances will not agree with the values shown in the performance charts.

A positive runway gradient adversely affects takeoff performance, but it is beneficial during landing. A negative gradient shortens the takeoff roll, but lengthens the landing distance.

Completely level runways are rare. The gradient, or slope, of the runway refers to the amount of change in runway height over its length. Gradient is usually expressed as a percentage. For example, a gradient of two percent means the runway height changes two feet for each 100 feet of runway length (100 × 2% = 2). A positive gradient indicates the height of the runway increases as you move away from the approach end, while a negative value means it decreases. A positive gradient is unfavorable for takeoff, because the airplane must take off uphill. It is desirable for landings, however, since landing uphill reduces the landing roll. A negative gradient has the opposite effects. Runway gradient is listed in the A/FD when it is three-tenths of one percent or more.

The runway condition affects the landing roll. A major concern is braking effectiveness, which refers to how much braking power you can apply to the runway. It depends, to a large extent, on the amount of friction between the tires and the runway. A great deal of friction (and braking power) is normal on a dry surface. However, if the runway is wet, less friction is available and your roll-out increases. In some cases, you may lose braking effectiveness due to hydroplaning, which is caused by a thin layer of water that separates the tires from the runway.

Dynamic hydroplaning occurs when there is standing water or slush on the runway about one-tenth of an inch or more in depth. A wedge of water builds up, finally lifting the tire away from contact with the runway surface. The greater the speed of the aircraft, the depth of the water, and the pressure in the tires, the more likely this is to take place. **Viscous hydroplaning** can occur on just a thin film of water, not more than one-thousandth of an inch deep, if it covers a smooth or smooth-acting surface. **Reverted rubber hydroplaning** is the result of a prolonged locked-wheel skid, in which reverted rubber acts as a seal between the tire and the runway. Entrapped water is heated to form steam, which supports the tire off the pavement. Nil braking action could result in an aircraft running off the end of the runway. Braking effectiveness also may be completely lost on ice-covered runways. If you must operate in conditions where braking effectiveness is poor or nil, be sure the runway length is adequate and surface wind is favorable.

THE PILOT'S OPERATING HANDBOOK

According to FARs, you can find your aircraft's operating limitations in the approved airplane flight manual, approved manual materials, markings and placards, or any combination of these. For most light airplanes manufactured in the U.S. after March 1, 1979, the pilot's operating handbook is the approved flight manual, and a page in the front of the handbook will contain a statement to that effect.

PERFORMANCE CHARTS

In developing performance charts, manufacturers make assumptions about the condition of the airplane and ability of the pilot. They assume the airplane is in good condition, and that the engine has been properly tuned and is developing its rated power. Also, the brakes are considered to be in good condition. The pilot is expected to follow normal checklist procedures and to have average abilities. "Average" means the person is not a highly trained test pilot with exceptional reaction time. At the same time, a student pilot or a person who has not flown for a long time is not considered average. In this context, average means a pilot who can do each of the required tasks correctly and at the right times.

Performance charts assume a properly tuned airplane in good condition with good brakes; the values are based on average piloting abilities.

Based on these assumptions, the manufacturer develops performance criteria for the airplane using data from actual flight tests. Manufacturers, however, do not test airplanes under each and every condition shown on a performance chart. Instead, they evaluate specific flight data and, based on those values, mathematically derive the remaining data. As a practical consideration, you should remember that the original flight tests were conducted by test pilots flying new airplanes. With this in mind, you should look at performance data with a skeptical eye to ensure yourself a margin of safety. For example, departing on a 2,000-foot runway when performance charts indicate 1,900 feet are required for takeoff is very dangerous and reflects poor judgment.

Performance charts are developed by deriving the data from actual flight tests.

Keep in mind that all performance charts in this presentation are samples only. They must never be used with any specific airplane you may be flying. When determining performance data for an airplane, refer only to the POH for that airplane. Performance data can vary significantly, even between similar models. The POH presents numerous charts which allow you to predict the airplane's performance accurately.

Performance charts generally present their information in either table or graph format. The table format usually contains several notes which require you to make adjustments for various conditions not accounted for in the body of the chart. Graph presentations, though somewhat harder to use, usually incorporate more variables, and adjustments may not be required. Almost every chart, whether a table or a graph, will specify a set of conditions under which the chart is valid. For example, a takeoff chart may only be valid for a particular weight and flap setting. Therefore, you should always check the chart conditions before using it.

TAKEOFF CHARTS

First, let's look at some samples of takeoff charts. For this first sample, assume you are planning to depart an airport under the following conditions and want to determine the takeoff ground roll. [Figure 8-33]

Pressure altitude . . . . 1,400 ft.
Temperature . . . . 15°C
Takeoff weight . . . . .3,480 lbs.
Headwind . . . . 10 kts.
Propeller . . . . .3 blades

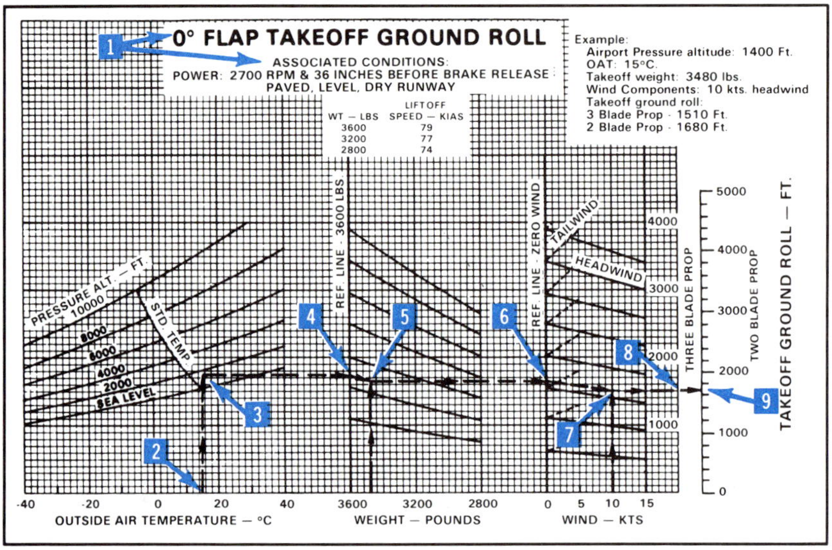

Figure 8-33. Always check the associated conditions (item 1) to see that you are using the correct chart. Enter the graph at a temperature of 15°C (item 2) and proceed vertically to the pressure altitude of 1,400 feet (item 3). Next, proceed horizontally to the first reference line (item 4), then diagonally down until you intercept the takeoff weight of 3,480 pounds (item 5). Continue across to the second reference line (item 6), and correct for the headwind by paralleling the diagonal line downward until you intercept the 10-knot mark (item 7). You find the takeoff ground roll of 1,510 feet on the right side of the graph (item 8). Since the conditions stipulate that the aircraft has a three-bladed propeller, you can stop here. If the aircraft had a two-bladed propeller, you would need to continue over to the next line and read 1,680 feet (item 9).

If you are departing from a runway with an obstacle, don't forget to use the appropriate chart for the associated conditions. For our next sample, assume you are planning to depart an airport under the following conditions, and need to determine the weight reduction necessary to take off over a 50-foot obstacle in 1,000 feet of runway. [Figure 8-34]

Temperature .... 75°F
Pressure altitude .... 6,000 ft.
Takeoff weight ....2,900 lbs.
Headwind .... 20 kts.

OBSTACLE TAKEOFF

ASSOCIATED CONDITIONS:
POWER — TAKEOFF POWER SET BEFORE BRAKE RELEASE
FLAPS — 20°
RUNWAY — PAVED, LEVEL, DRY
TAKEOFF SPEED — IAS AS TABULATED

NOTE: GROUND ROLL IS APPROX. 73% OF TOTAL TAKEOFF DISTANCE OVER A 50 FT. OBSTACLE

EXAMPLE:
OAT — 75°F
PRESSURE ALTITUDE — 4000 FT
TAKEOFF WEIGHT — 3100 LBS
HEADWIND — 20 KNOTS

TOTAL TAKEOFF DISTANCE OVER A 50 FT OBSTACLE — 1350 FT
GROUND ROLL (73% OF 1350) — 986 FT
IAS TAKEOFF SPEED
LIFTOFF — 74 MPH
AT 50 FT — 74 MPH

WARNING
Obstacle takeoff is not a recommended procedure, as it utilizes speeds at or below power-off stall speed. In the event of engine failure or wind speed fluctuations, a stall may occur which may cause uncontrolled contact with the ground.

WEIGHT POUNDS	IAS TAKEOFF SPEED (ASSUMES ZERO INSTR. ERROR)			
	LIFTOFF		50 FEET	
	MPH	KNOTS	MPH	KNOTS
3400	77	67	77	67
3200	75	65	75	65
3000	72	63	72	63
2800	69	60	69	60

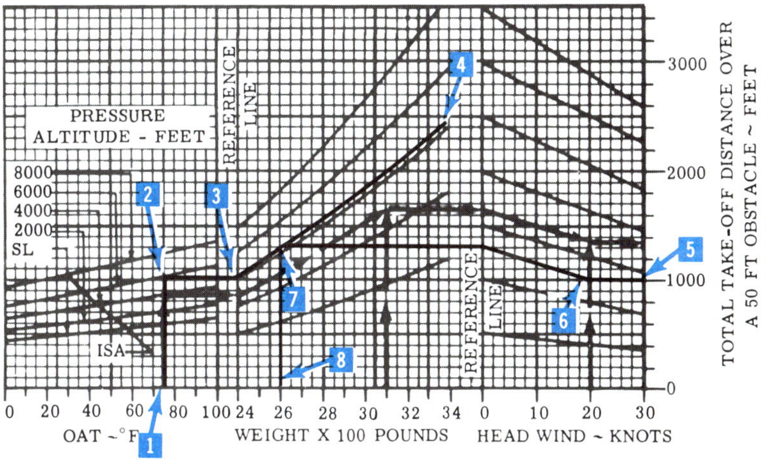

Figure 8-34. Enter the chart at 75°F (item 1) and move up until you intersect the 6,000-foot pressure altitude line (item 2). From this point, go horizontally to the right until you reach the weight reference line (item 3). From here, draw a line parallel to the diagonal lines up and to the right (item 4). This will be used as a reference line for coming in from the takeoff distance side of the chart. Now, enter the chart from the right side at the 1,000-foot takeoff distance mark (item 5). Proceed horizontally to the left and intercept the 20-knot headwind line (item 6). Proceed up and to the left to the reference line. Move horizontally to the line you drew (item 7), and from this point, proceed vertically downward and read the weight limit of 2,600 pounds for a takeoff distance of 1,000 feet with a 20-knot headwind (item 8).

The difference between the weight of the aircraft and the weight required to meet the conditions is 300 pounds (2,900 - 2,600 = 300). This is the amount that must be taken from the airplane to meet the takeoff distance requirement. This is one method for determining the maximum weight for departing on a particular runway.

Since a graphic presentation often includes many variables, you must exercise care when determining performance values. There may be a tendency to estimate where two lines meet. This means that instead of drawing out your lines carefully, you simply select line positions and intersections visually. This can lead to substantial errors. A straightedge will help you derive more precise values.

TIME, FUEL, AND DISTANCE TO CLIMB CHART

With this chart you first determine the time, fuel, and distance to climb to your cruise altitude, then you apply a credit for departing any airport above sea level. For this example, assume the following conditions, and determine the time, fuel, and distance required to climb from the airport elevation of 4,000 feet MSL to 12,000 feet MSL. [Figure 8-35]

Weight .... 4,100 lbs.
Airport pressure altitude .... 4,000 ft.
Temperature .... 7°C (Standard)

TIME, FUEL, AND DISTANCE TO CLIMB

MAXIMUM RATE OF CLIMB

[1] CONDITIONS:
Flaps Up
Gear Up
2700 RPM

37.0 Inches Hg (or Full Throttle)
Fuel Flow Set For Climb
Cowl Flaps Open
Standard Temperature

[4] NOTES:
1. Add 18 pounds of fuel for engine start, taxi, and takeoff allowance.
2. Increase time, fuel and distance by 10% for each 10° above standard temp.
3. Distances shown are based on zero wind.

WEIGHT LBS	PRESS ALT FT	CLIMB SPEED KIAS	RATE OF CLIMB FPM	FROM SEA LEVEL		
				TIME MIN	FUEL USED POUNDS	DISTANCE NM
4100	S.L.	110	1150	0	0	0
	4000	110	1120	4	13	7
	8000	110	1070	7	26	14
	12,000	110	1000	11	41	22
	16,000	110	910	15	56	32
	20,000	110	810	20	73	43
	24,000	107	580	26	93	57
3700	S.L.	108	1335	0	0	0
	4000	108	1305	3	11	6

Figure 8-35. First, check the conditions (item 1). Then, read the time, fuel, and distance to climb to 12,000 feet. In this case, they are 11 minutes, 41 pounds, and 22 miles, respectively (item 2). Next, determine the time, fuel, and distance credits to be applied for departing an airport at 4,000 feet. These values are 4 minutes, 13 pounds, and 7 miles, respectively (item 3). After you subtract the credits, the net values are 7 minutes, 28 pounds, and 15 miles. However, a check of the notes indicates you must add an additional 18 pounds of fuel for engine start, taxi, and takeoff allowances (item 4). This changes the fuel burned to 46 pounds.

Although similar to the other graphs you have used, the next sample chart is a little different from the others. Notice that there is a stipulation for using maximum continuous power in the associated conditions. The maximum r.p.m. depends on whether you are flying an airplane with a two- or three-blade propeller. [Figure 8-36]

Airport pressure altitude	3,000 ft.
Airport temperature	30°C
Cruise pressure altitude	16,000 ft.
Cruise temperature	2°C

RATE-OF-CLIMB CHART

This type of chart is somewhat unique to higher performance aircraft. You may find it useful for determining such things as what rate of climb is required to clear an obstacle, or what the service ceiling is for a particular aircraft weight. Remember, service ceiling refers to the altitude where a single-engine airplane is able to maintain a maximum climb rate of 100

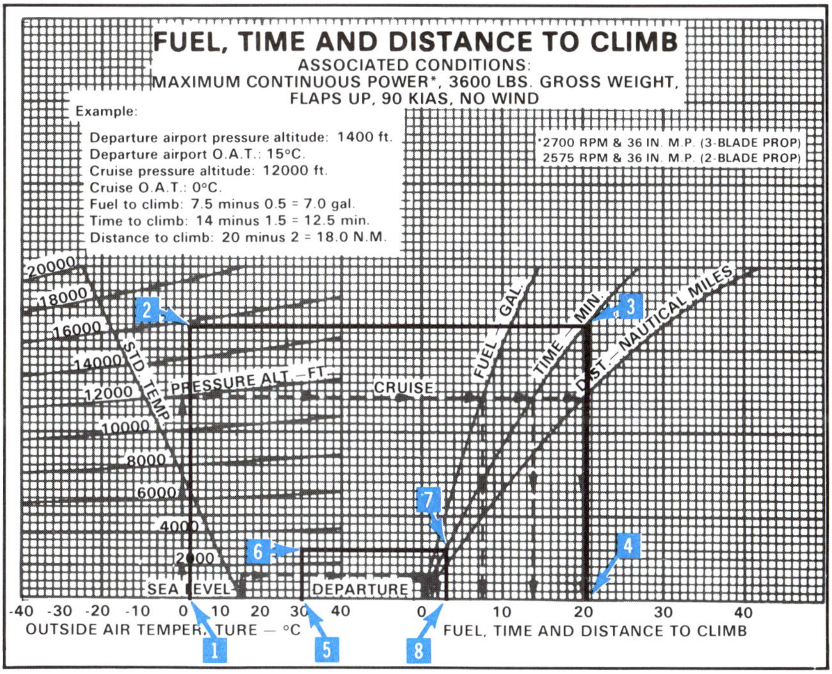

Figure 8-36. Enter the chart at your cruise temperature of +2°C (item 1) and proceed upward to intersect the 16,000-foot pressure altitude line (item 2). From here, move to the right until you intersect the curved time line (item 3). Move down vertically and read the time of 20.5 minutes (item 4). Enter the chart again, but this time at 30°C (item 5), and proceed up to the 3,000-foot pressure altitude line (item 6). Then, move horizontally to the time line (item 7), and follow it vertically to the bottom of the chart (item 8). This gives you a value of 3 minutes. The difference between these two (17.5 minutes) is the time it will take you to climb to cruise. For fuel and distance, use the same steps with the appropriate reference line.

feet per minute. This chart also permits you to estimate the time to climb to altitude by calculating your average rate of climb. [Figure 8-37]

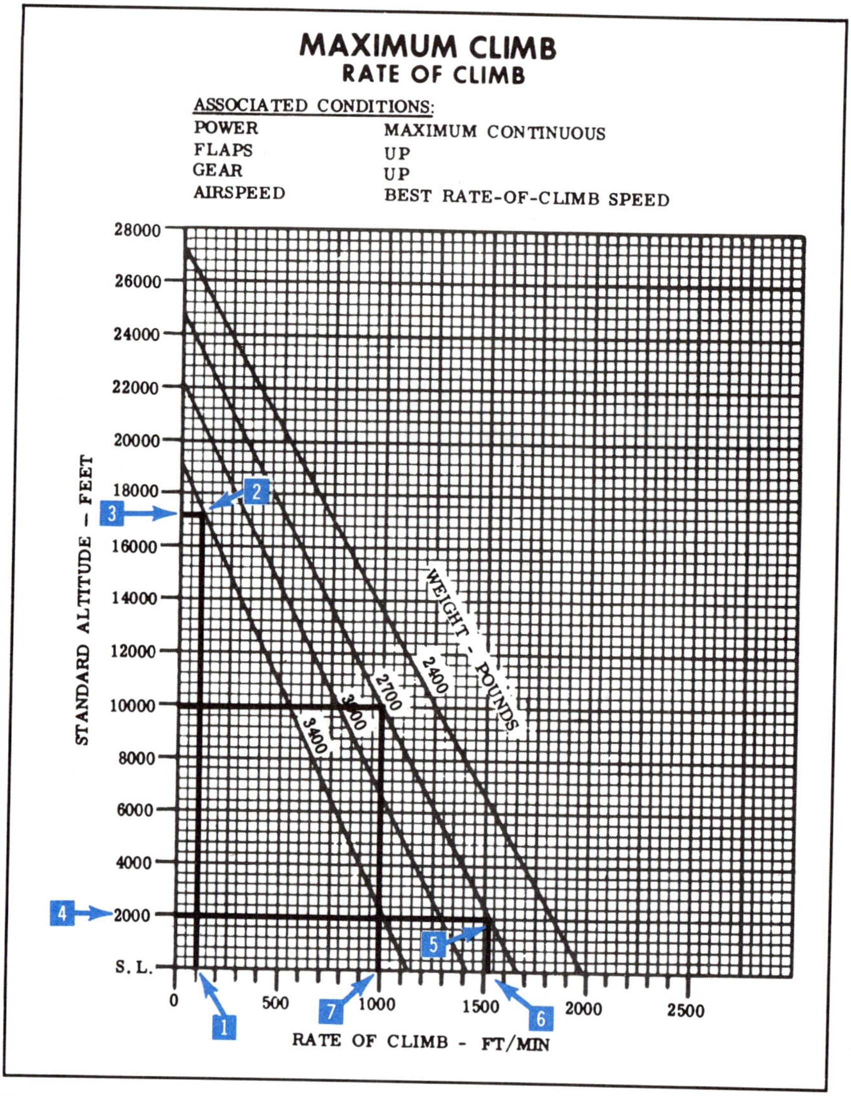

Figure 8-37. To determine the service ceiling of an aircraft weighing 3,400 pounds, locate the 100 f.p.m. climb rate line (item 1) and follow it vertically until it intersects the 3,400-pound diagonal line (item 2). Read the altitude of 17,200 feet on the left (item 3). As another example, assume you want to know how long it will take you to climb from 2,000 feet to 10,000 feet at an aircraft weight of 2,700 pounds. Enter the chart at the left at 2,000 feet (item 4) and proceed horizontally to intersect the 2,700-pound diagonal line (item 5). Move vertically to the bottom of the graph and read a climb rate of about 1,510 f.p.m. (item 6). Repeat this for an altitude of 10,000 feet and 2,700 pounds, and read a climb rate of 1,000 f.p.m. (item 7). Add these values together and divide by 2 for an average climb rate of approximately 1,255 f.p.m.

Another way to solve for average climb rate with this chart is to select an altitude which is midway between the starting and ending altitudes. Since you will be climbing 8,000 feet, the midpoint of the climb is 6,000 feet (8,000 ÷ 2 + 2,000). According to the chart, 1,250 f.p.m. is the climb rate at 6,000 feet and the approximate average for the climb. From this chart you can also determine the amount of time it will take to climb to altitude. First, divide the total number of feet you will be climbing by your average rate of climb (8,000 ÷ 1,250 = 6.4). Your time to climb is 6 minutes and 24 seconds.

The next sample climb performance chart allows you to correct for non-standard temperature. Using the following conditions, solve for rate of climb. Also note the 60 f.p.m. decrease in performance if the aircraft is not equipped with wheel fairings. [Figure 8-38]

Pressure altitude	6,000 ft.
Temperature	8°C
Weight	3,400 lbs.
Wheel fairings	Not installed

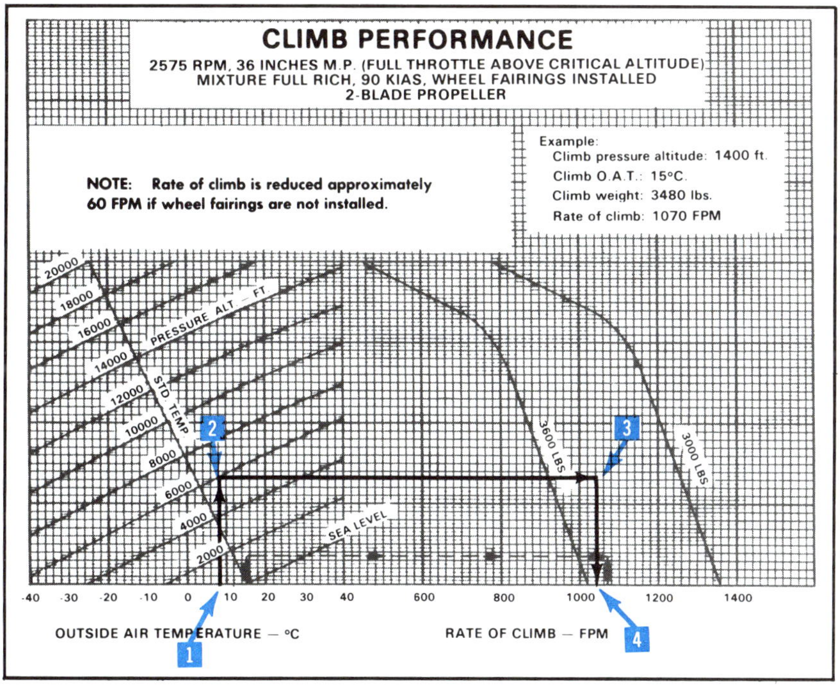

Figure 8-38. With the given set of conditions, enter the chart at 8°C (item 1), proceed vertically to intersect the 6,000-foot pressure altitude line (item 2). Then, proceed over to the right and interpolate to locate the 3,400-pound line (item 3), and draw the line down from this position to the bottom of the chart to read a rate of climb of 1,040 feet (item 4). Since the airplane does not have wheel fairings, subtract 60 f.p.m. from this value for an answer of 980 f.p.m.

CRUISE PERFORMANCE CHART

Cruise performance data vary significantly between manufacturers. Before you use a particular graph or chart, make sure you understand what information is portrayed. Temperatures for cruise may include International Standard Atmosphere (ISA) values, as well as colder and warmer temperatures. For instance, categories may specify ISA −20°C, standard day ISA, and ISA +20°C. Both Celsius and Fahrenheit values may be included. In addition, temperature figures for some manufacturers may be adjusted for frictional heating of the temperature probe.

As an example, assume you are planning a flight under the following conditions and wish to determine the approximate flight time available (endurance). Since this will be a day VFR trip, you will need to consider appropriate fuel reserves. [Figure 8-39]

Pressure altitude .... 18,000 ft.
Temperature .... −1°C
Power .... 2,200 r.p.m., 20″ MAP
Fuel economy .... Best
Usable fuel .... 344 lbs.
Fuel reserve .... 30 min.

CRUISE PERFORMANCE
PRESSURE ALTITUDE 18,000 FEET 4

CONDITIONS:
4000 Pounds
Recommended Lean Mixture
Cowl Flaps Closed

NOTE
For best fuel economy at 70% power or less, operate at 6 PPH leaner than shown in this chart or at peak EGT.

		20°C BELOW STANDARD TEMP -41°C			STANDARD TEMPERATURE -21°C			20°C ABOVE STANDARD TEMP -1°C (2)		
RPM	MP	% BHP	KTAS	PPH	% BHP	KTAS	PPH	% BHP	KTAS	PPH
2500	30	- - -	- - -	- - -	81	188	106	76	185	100
	28	80	184	105	76	182	99	71	178	93
2200 (1)	26	66	166	87	62	163	82	58	159	77
	24	61	158	80	57	154	76	54	150	72
	22	55	148	73	51	144	69	48	138	66
	20	49	136	66	46	131	63	43	124	59 (3)

Figure 8-39. First, locate 2,200 r.p.m. and 20″ MAP in the column on the far left (item 1). Then, select the column that corresponds with the given temperature of 20°C above standard (item 2). Move down this column to locate the values that apply to 2,200 r.p.m. and 20″ MAP (item 3). The corresponding fuel flow is 59 pounds per hour at a true airspeed of 124 knots and a BHP of 43%. Since the BHP is below 70%, the note at the top of the chart (item 4) applies and you may subtract six pounds per hour from the fuel flow (59 − 6 = 53). Take the usable fuel of 344 pounds and divide it by the fuel flow of 53 pounds per hour, and this gives you an endurance of 6 hours and 29 minutes. After you subtract 30 minutes for day VFR fuel reserve, you have 5 hours and 59 minutes of flight time available.

This sample graph shows a different format for finding TAS and brake horsepower. The box at the top of the chart correlates percent power, engine r.p.m., and BHP to the performance lines numbered at the bottom of the chart. Use the following conditions to find TAS and BHP. [Figure 8-40]

Weight .... 3,400 lbs.
Standard altitude .... 5,000 ft.
Power 75%

Since this graph doesn't give you a fuel consumption rate, you need to use a fuel flow graph to complete your cruise performance calculations. Let's work two samples using this graph. Based on the following conditions, find the fuel that will be consumed during takeoff and climb, then

CRUISE OPERATION

STANDARD DAY

WEIGHT 3400 LBS

NO.	% POWER	ENG SPEED RPM	BHP
1	45	2100	128
2	50	2100	142
3	55	2100	157
4	60	2200	171
5	65	2300	185
6	70	2400	200
7	75	2500	214

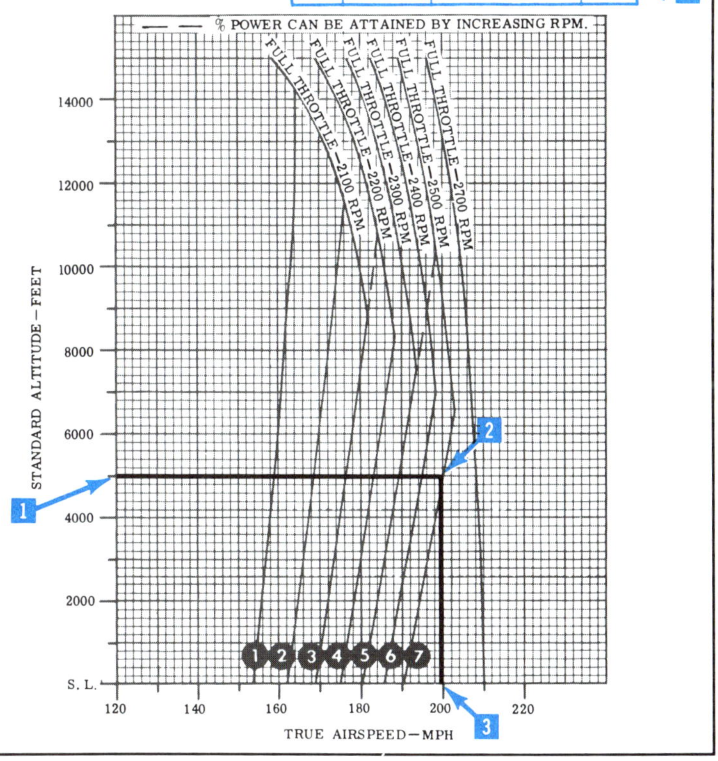

Figure 8-40. Begin at the left side of the graph at your cruising, or standard, altitude of 5,000 feet (item 1). Move horizontally to the line numbered seven which is the 75% power line (item 2). Now, go straight down to the bottom of the chart to find your TAS of 200 m.p.h. (item 3). Your BHP at 75% is 214 (item 4).

find how much flight time is available based on fuel quantity and use of cruise of power. [Figure 8-41]

Climb power 70%
Time to climb . . . . 10 min.
Fuel quantity . . . . 47 gal.
Power — cruise (lean) 55%

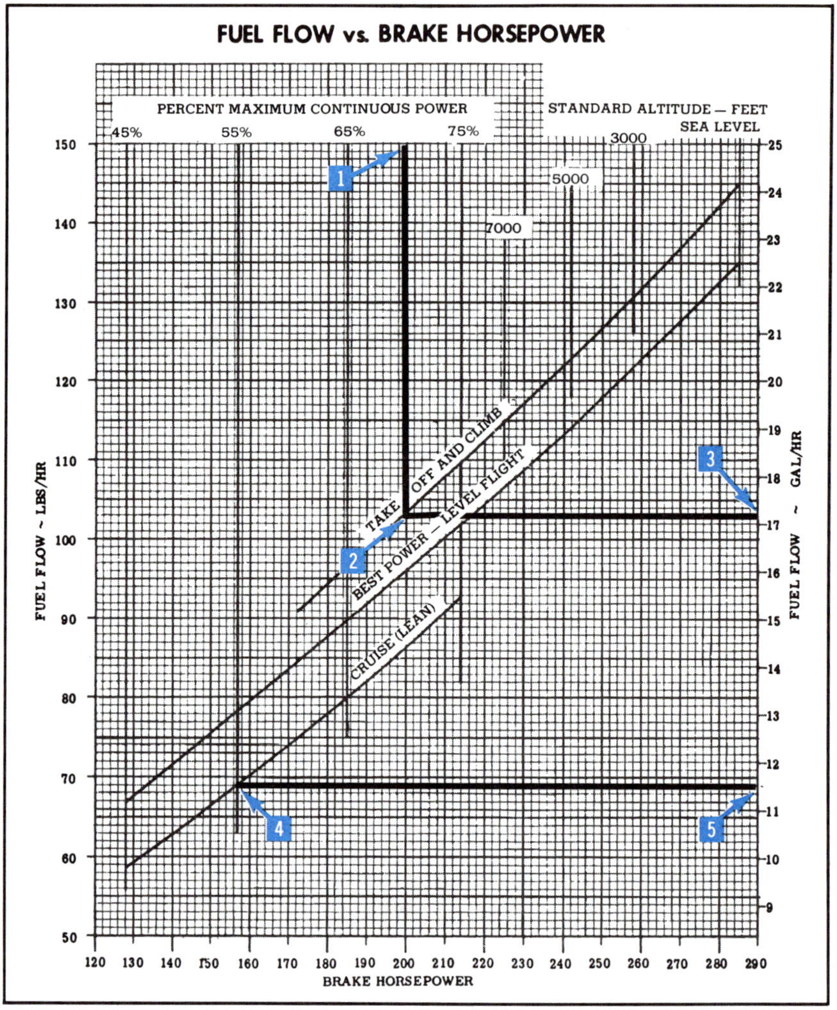

Figure 8-41. Locate the 70% maximum continuous power line (item 1). Enter the chart and move down to the point where the power line intersects the slightly curved takeoff and climb line (item 2). Proceed to the right and read the fuel flow of 17.2 g.p.h. (item 3). Climbing for 10 minutes at 17.2 g.p.h. consumes 2.9 gallons of fuel. To find the flight time available during cruise, enter the chart where the 55% maximum continuous power line intersects the cruise line (item 4). Proceed to the right and read the fuel flow of 11.5 g.p.h. (item 5). With 47 gallons of fuel and a consumption rate of 11.5 g.p.h., you have 4 hours and 5 minutes endurance. Be sure to consider fuel needed for takeoff and climb, as well as for fuel reserve.

Notice the curve on the graph labeled "best power — level flight." This performance specification is also known as the **best power mixture**. It is the fuel-air ratio at which the most power can be obtained for any given throttle setting.

The best power mixture is the ratio that provides the most power for a given throttle setting.

RANGE PROFILE

This type of chart provides a method of calculating your aircraft's range at different cruise power settings. Once again, be sure to read the notes that accompany the chart. Use the given conditions to determine the range for this flight. [Figure 8-42]

Altitude . . . . 7,000 ft.
Power 50%
Tailwind . . . . 18 kts.

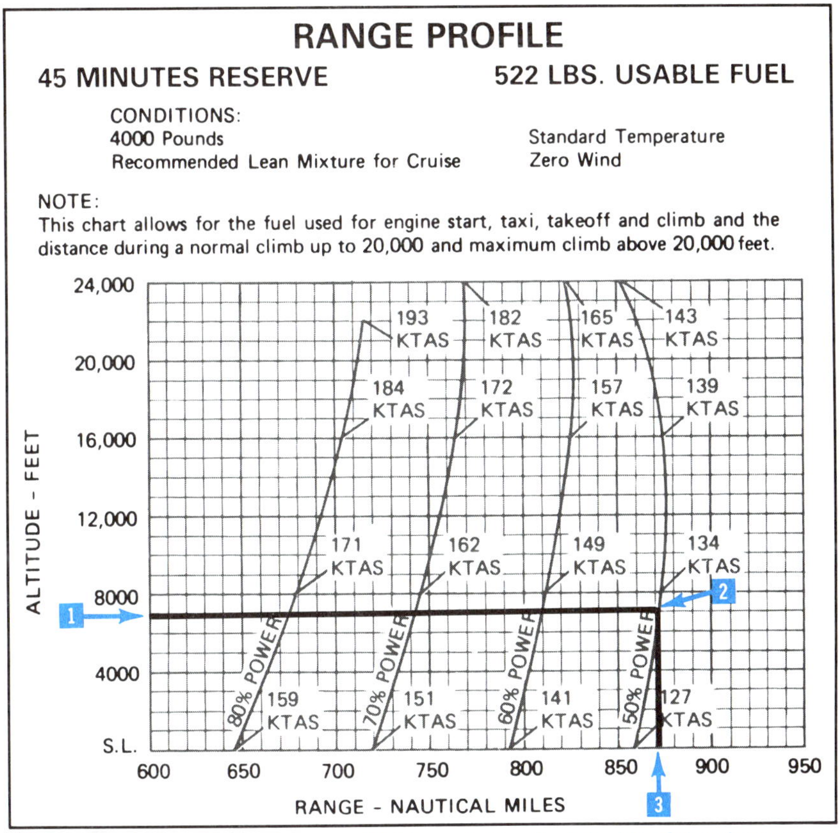

Figure 8-42. First, determine your TAS by entering the left side of the chart at your pressure altitude of 7,000 feet (item 1). Then, move to the right to the 50% power line and interpolate to calculate the TAS of 133 knots (item 2). Move straight down and read the no-wind range of 872 n.m. at the bottom of the chart (item 3). An 18-knot tailwind added to your TAS results in a groundspeed of 151 knots. Since this increases the speed by 13.5%, you can expect a proportional effect on range. Multiply the no-wind range by the same percentage increase. This will result in a total range of about 990 n.m.

LANDING DISTANCE CHART

Landing distance charts are similar to takeoff distance charts. For this sample chart, use the following conditions to determine the distance to land and bring the airplane to a complete stop. [Figure 8-43]

Airport pressure altitude .... 4,100 ft.
Airport temperature................................ -10°C
Landing weight ....3,000 lbs.
Headwind .... 6 kts.

GLIDE DISTANCE

This type of chart allows you to estimate the distance that your aircraft can glide. Remember, you must fly the aircraft at the manufacturer's recommended speed to obtain maximum glide performance. [Figure 8-44]

Height above terrain .... 12,000 ft.
Headwind .... 20 kts.

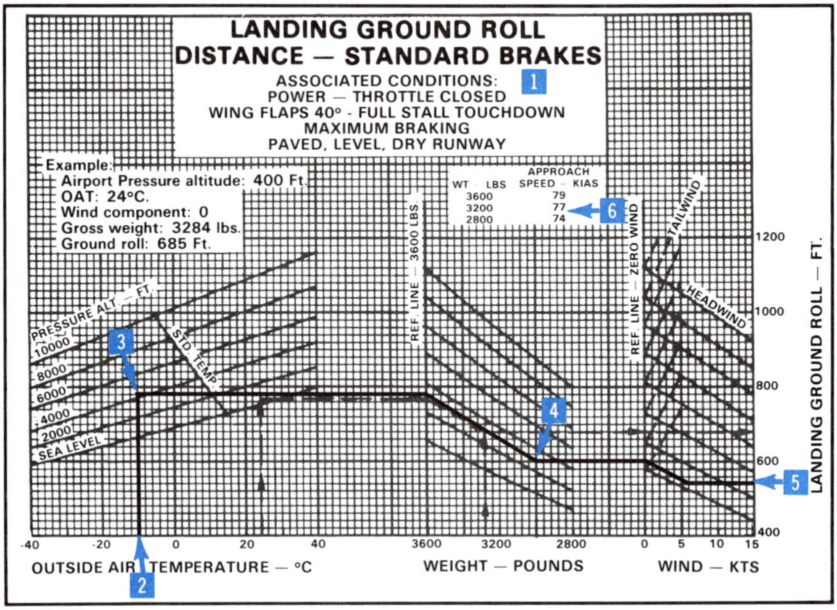

Figure 8-43. First, check the associated conditions listed on the sample chart (item 1). Next, enter the chart at the given temperature of -10°C (item 2) and proceed vertically to the pressure altitude of 4,100 feet (item 3). Proceed horizontally to the first reference line, then diagonally down to a weight of 3,000 pounds (item 4). Continue horizontally to the right to the second reference line. Follow the headwind line to 6 knots, and then read the distance required of 540 feet on the right side of the chart (item 5). The approach speeds are indicated near the top of the chart below the associated conditions. The approximate approach speed for this aircraft at 3,000 pounds is 76 knots (item 6).

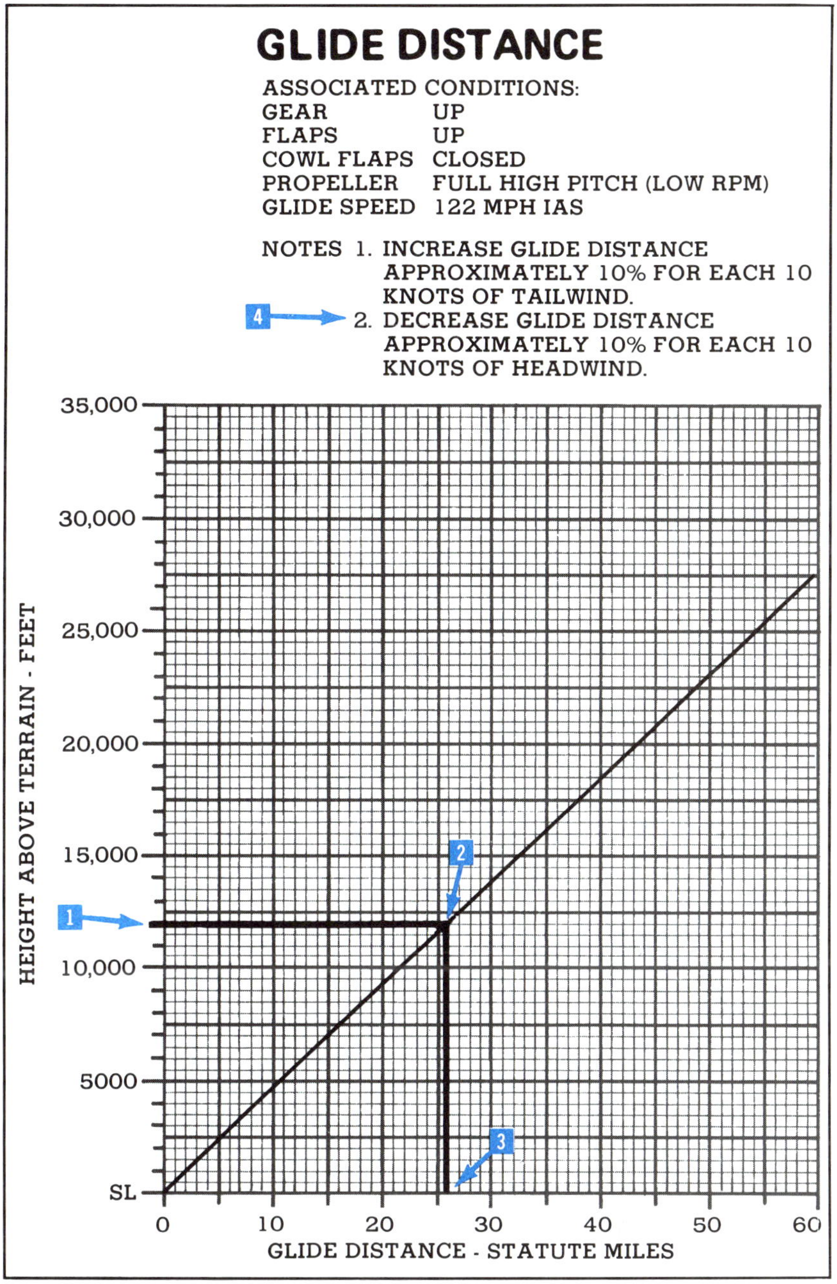

Figure 8-44. Start on the left hand side of the graph at 12,000 feet (item 1) and go across to the reference line (item 2). Now, go down to the bottom of the chart and find the glide distance of 26 s.m. (item 3). Note that the distance must be decreased 10% for each 10 knots of headwind (item 4). Therefore, the 20-knot wind requires a 20% decrease in the gliding distance. The total glide distance is 20.8 s.m. (80% × 26 = 20.8).

STALL SPEEDS					
GROSS WEIGHT 2750 LBS		ANGLE OF BANK			
		LEVEL	30°	45°	60°
POWER		GEAR AND FLAPS UP			
ON	MPH	62	67	74	88
	KTS	54	58	64	76
OFF	MPH	75	81	89	106
	KTS	65	70	77	92
		GEAR AND FLAPS DOWN			
ON	MPH	54	58	64	76
	KTS	47	50	56	66
OFF	MPH	66	71	78	93
	KTS	57	62	68	81

Figure 8-45. This chart shows the increase in stall speed that accompanies an increase in angle of bank. With gear and flaps up, the power-on stall speed in a 30° banked turn is seven miles per hour less than it is in a 45° banked turn (item 1). The power-on stall speed with the gear and flaps down causes a significant decrease in stall speeds for both 30° and 45° banked turns (item 2).

STALL SPEEDS

Although an airplane always stalls at the same angle of attack, the speed at which it stalls varies with a number of factors, such as weight, CG location, configuration, angle of bank, and whether power is on or off. Manufacturers provide charts that tell you the stall speed for various conditions. [Figure 8-45]

INTERPOLATION

Sample problems often use convenient altitudes and temperatures which permit you to extract your answers directly. To see how to solve a problem when this is not the case, let's work another takeoff performance problem that requires interpolation. In this example, assume that you need to know the ground roll distance under the following conditions. [Figure 8-46]

Pressure altitude .... 1,500 ft.
Temperature .... 30°C
Weight .... 1,670 lbs.

WEIGHT LBS	TAKEOFF SPEED KIAS		PRESS ALT FT	0°C		10°C		20°C		30°C		40°C	
	LIFT OFF	AT 50 FT		GRND ROLL FT	TOTAL TO CLEAR 50 FT OBS	GRND ROLL FT	TOTAL TO CLEAR 50 FT OBS	GRND ROLL FT	TOTAL TO CLEAR 50 FT OBS	GRND ROLL FT	TOTAL TO CLEAR 50 FT OBS	GRND ROLL FT	TOTAL TO CLEAR 50 FT OBS
1670	50	54	S.L.	640	1190	695	1290	755	1390	810	1495	875	1605
			1000	705	1310	765	1420	825	1530	890	1645	960	1770
			2000	775	1445	840	1565	910	1690	980	1820	1055	1960
			3000	855	1600	925	1730	1000	1870	1080	2020	1165	2185

Figure 8-46. Looking at the sample takeoff chart excerpt, you can see that the given pressure altitude of 1,500 feet falls between the 1,000- and 2,000-foot pressure altitude values. This means your ground roll distance will fall between 890 and 980 feet. Interpolation is required.

Interpolation refers to the process of finding an unknown value between two known values. Often this is necessary when you use charts with tabular information. In this case, you know the ground roll figures for 1,000 and 2,000 feet and want to determine the ground roll for 1,500 feet. The first step in interpolation is to compute the differences between known values.

	Pressure Altitude	**Ground Roll**
	2,000 ft.	980 ft.
	1,000 ft.	890 ft.
Difference	1,000 ft.	90 ft.

The 1,500-foot airport pressure altitude is 50% of the way between 1,000 and 2,000 feet. Therefore, the ground roll also is 50% of the way between 890 and 980 feet. The answer then, is 935 feet (90-foot difference × 0.5 + 890 feet = 935 feet). In practice, pilots often round off values derived from table charts to the more conservative figure. This practice gives you a good estimate of performance data, and it also provides a margin of safety.

CHECKLIST

After studying this section, you should have a basic understanding of:

✓ **V-g diagram** — How it relates velocity to load factor, and what important airspeeds and operating limitations it portrays.

✓ **Performance factors** — How density altitude, humidity, wind, and runway conditions affect airplane performance.

✓ **Performance charts** — How to use various chart formats including multiple value graphs and table presentations to determine performance values.

✓ **Takeoff and landing distance charts** — How they are used and what types of conditions are common to these performance charts.

- ✓ **Climb charts —** What chart presentations and methods are used to determine average climb rates and how to determine the time, fuel, and distance to climb.

- ✓ **Cruise and range charts —** How to extract various types of performance data from these charts including power settings, TAS, fuel consumption, endurance, and range.

SECTION C

CONTROLLING WEIGHT AND BALANCE

This last section of the chapter is a review of weight and balance theory and the practical applications that are important to commercial pilots. As part of your preflight preparations, you should always verify the weight and balance condition for your aircraft. Flying an overloaded or out-of-balance airplane is extremely hazardous.

Keep in mind that the weight and balance charts, graphs, and tables in this presentation are samples only. They must never be used with any specific airplane you may be flying. When determining the weight and balance condition for a given airplane, refer only to the pilot's operating handbook for that airplane.

AIRPLANE WEIGHT

Let's begin with a review of terminology and definitions. When you refer to airplane weight, keep in mind that it includes the weight of the basic airplane, installed equipment, passengers, cargo, and fuel. Of course, the effective weight (load factor) varies when you maneuver the airplane in flight. The starting point for weight computation is the weight of the airplane before passengers, cargo, and fuel are added. Most manufacturers use the term **basic empty weight,** which includes the weight of the standard airplane, optional equipment, unusable fuel, and full operating fluids including full engine oil.

Empty weight is the airplane's weight before passengers, cargo, or fuel are added, and it usually includes full operating fluids such as engine oil and hydraulic fluids.

Older airplanes use the term **licensed empty weight,** which is similar to basic empty weight except that it does not include full engine oil. Instead, it includes only the weight of undrainable (residual) oil. If you fly an airplane that lists a licensed empty weight, be sure to add the weight of oil to your computations. Remember, avgas weighs 6 pounds per gallon, and oil weighs 7.5 pounds per gallon. The weight of turbine fuel varies with grade and temperature, so use the weight(s) specified in the pilot's operating handbook for the airplane you are operating.

Another term, **payload,** refers to the weight of the flight crew, passengers, and any cargo or baggage which you will carry; **useful load** is the difference between the maximum takeoff weight and the basic empty weight. Useful load includes payload and usable fuel. **Usable fuel** is the fuel available for the flight. It does not include **unusable fuel,** which is the quantity of fuel that cannot be safely used during flight. The following summary should help you remember how various weights are calculated:

Basic Empty Weight
\+ Payload

= Zero Fuel Weight
\+ Usable Fuel

= Ramp Weight
\- Fuel Used for Start, Taxi, and Engine Runup

= Takeoff Weight
\- Fuel Used During Flight

= Landing Weight

WEIGHT LIMITATIONS

Some weight and balance terms may be preceded by the word maximum. When this word is used, it indicates the manufacturer has established a limit which must not be exceeded. For example, the maximum ramp weight is the maximum weight approved for ground operations, while maximum takeoff weight is the maximum weight approved for the start of the takeoff roll.

Operating an aircraft at higher weights requires more time to accelerate, higher airspeeds for takeoff, and different power settings for specific airspeeds.

At higher weights, the airplane accelerates more slowly with the same power output, and a higher airspeed is required to generate the lift necessary for takeoff. Higher weights also change the power setting required to produce a specific airspeed. Weight limitations are necessary to guarantee the structural integrity of the airplane and enable you to accurately predict performance. These structural limits are based on an airplane operated at or below the maximum weight, as specified by the airplane manufacturer. Operating the airplane above this weight could result in structural deformation or failure during flight if you encounter excessive load factors, strong wind gusts, or turbulence. Operating in excess of the manufacturer's maximum weight also results in a longer takeoff roll, a reduced climb rate, reduced cruising speed, increased fuel consumption, an increase in stall speed, and an increase in the landing roll.

From time to time, the owner or operator of an airplane may have equipment removed, replaced, or additional equipment installed. These changes must be included in the weight and balance records. In addition, major repairs or alterations to the airplane itself must be recorded by a certificated mechanic. When the revised weight and moment are recorded on a new form, the old record is marked with the word "superseded" and dated with the effective date of the new record.

AIRPLANE BALANCE

The location of the CG is critical to an airplane's stability and elevator effectiveness. Your airplane must be loaded within acceptable CG limits before flight, and you must maintain proper distribution of the load during flight. Improper balance of the airplane's load can result in serious control problems. You can avoid these problems by taking the time to determine the CG location prior to flight and then by comparing the computed location to the acceptable limits prescribed by the manufacturer.

WEIGHT AND BALANCE CONTROL

Whenever practical, ask your passengers for their actual weights and use a scale to weigh baggage.

Whether you use the computational, graph, or table method to determine weight and balance, it is important to note that each method is only as accurate as the information you provide. Be sure to ask passengers what they weigh and, during the colder months, add a few pounds to cover the additional weight of winter clothing. If you must estimate weights, you can use the standard weights designated by the FAA. A person of average stature is considered to weigh 170 pounds, and a child between 2 and 12 years of age is considered to weigh 80 pounds. However, if the aircraft is loaded near its weight and balance limit, or a passenger's weight is obviously different from these standard weights, you should use actual weights. Use a scale to check the weight of baggage, if practical. If a scale is not available, be conservative and overestimate the weight.

Items to the rear of the datum have positive values, while those forward of it have negative values.

Remember that the reference datum is an imaginary vertical plane, arbitrarily fixed somewhere along the longitudinal axis of the airplane, from which all horizontal distances are measured for weight and balance purposes. The horizontal distance from the datum to any component of the airplane or to any object located within the airplane is called the **arm,** or **station**. If the component or object is located to the rear of the datum, it is measured as a positive number and usually is referred to as inches aft of the datum. Conversely, if the component or object is located forward of the datum, it is indicated as a negative number and is usually referred to as inches forward of the datum. Some manufacturers may locate the reference datum ahead of the airplane. Since there can be no components ahead of the datum plane, all items used to compute the weight and balance of such an airplane will have positive values.

Moment = Weight × Arm

Once you have found the arm of each object, you multiply the object by its weight to determine its **moment**. By totaling the weights and moments of all components and objects carried, you can determine the point where a loaded airplane would balance on an imaginary fulcrum. This is the airplane's CG.

CENTER OF GRAVITY POSITION

As you might expect, for an airplane to be controllable during flight, the CG must be located within a reasonable distance forward or aft of an optimum position. All airplanes have forward and aft limits for the position of the CG. The CG will remain within the approved CG range if you load the aircraft so the distribution of weight is acceptable. If the CG is located near the forward or aft limit of the approved CG range, a slight loss of longitudinal stability may be noticeable. However, elevator effectiveness will still be adequate to control the airplane during all maneuvers.

CG TOO FAR FORWARD

As you know, an airplane that is loaded forward of the forward CG limit will be too nose heavy. This condition becomes progressively worse as the CG moves to an extreme forward position where the airplane actually becomes too stable. Eventually, elevator effectiveness will not be sufficient to allow you to lift the nose. Elevator effectiveness also may be insufficient to exert the required tail-down force needed for a nose-high landing attitude. In this event, the nosewheel will touch down on the runway before the main gear. In cruise flight, additional tail-down force is required to counteract the nose-heavy condition. Since the wings must support the weight of the airplane plus the tail-down force, an extra load is placed on the airplane. As a result, more lift is necessary to sustain level flight. Greater lift causes greater induced drag which can be overcome only with increased thrust.

CG TOO FAR AFT

If an airplane is loaded to the rear of its CG range, it will tend to be unstable about its lateral axis.

A CG located aft of the approved CG range is even more dangerous than a CG that is too far forward. With an aft CG, the airplane becomes tail heavy and very unstable in pitch, regardless of its speed. If the CG is too far aft, you will not have enough elevator effectiveness to raise the tail and lower the nose of the airplane. During takeoff, the nose may pitch up unexpectedly before the airplane has reached safe flying speed, resulting in a stall. On landings, there may not be enough elevator force to lower the nose to a normal landing attitude. In cruise flight, the longitudinal stability of the airplane will be reduced, possibly requiring you to continually adjust the pitch downward in an attempt to maintain level flight.

In any airplane, stall recovery becomes progressively more difficult as the CG moves farther aft.

CG limits are established during initial testing and airworthiness certification. One of the criteria for determining the CG range in light airplanes is its spin recovery capability. If the CG is within limits, a normal category airplane must demonstrate that it can be recovered from a one-turn spin. A utility category airplane that is approved for spins must be recoverable from a fully developed spin. The aft CG limit is the most critical factor. As the CG moves aft, more elevator effectiveness is needed to maintain longitudinal stability. When the CG is at the aft limit, elevator effectiveness is still adequate; when the CG is beyond the aft limit, the elevator cannot exert enough force for stall or spin recovery.

When you are loading an airplane, make sure the heaviest passengers and baggage, or cargo, are located as far forward as practical. The main thing you must do is follow the airplane manufacturer's loading recommendations in the POH. If you do this, your airplane will be loaded so

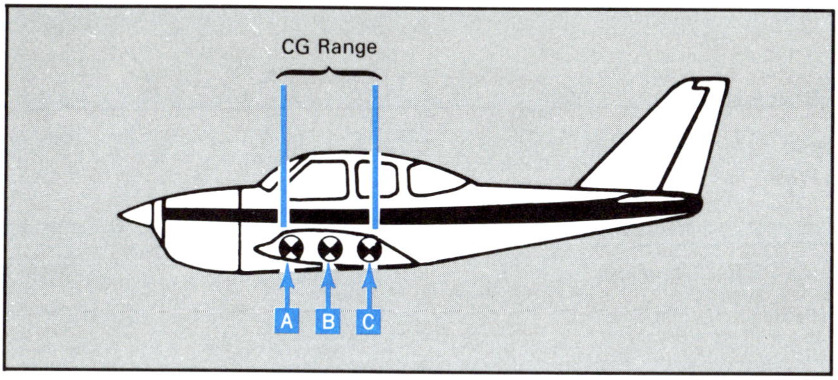

Figure 8-47. An airplane loaded with the CG at point A requires increased lift and thrust to maintain the same airspeed as an airplane with the CG located farther aft. At point A, the airplane will stall at a higher airspeed because of increased wing loading. The optimum location for the CG is at point B. If the CG is moved from point A to point C, the airplane will become less controllable at slow speeds and less stable at any speed. At point C, less total lift is required to maintain altitude and less thrust is required to maintain the same airspeed as an airplane with the CG forward. The airplane also will cruise faster with the same power setting, because a reduced angle of attack produces less lift which results in less induced drag.

the CG is within the approved range where longitudinal stability is adequate and where you can control the airplane during all approved maneuvers. Some important points to remember are that a CG beyond acceptable limits adversely affects longitudinal stability, and that the most hazardous condition is an extreme aft CG position. Performance is affected even when the CG is within the approved range but near the forward or aft limits. [Figure 8-47]

COMPUTATIONAL METHOD

When you determine weight and balance using the computational method, the first step is to look up the basic empty weight and total moment for the particular airplane you plan to fly. If the center of gravity is given, it should be noted. The empty weight CG can be considered the "arm" of the empty airplane.

Next, record the weights of the fuel, pilot and passengers, and baggage. Multiply the gallons of usable fuel by six to get the total fuel weight. Use care in recording the weight of each item, since many complex and high performance airplanes have numerous stations and several baggage compartments. Once you have recorded all of the weights, add them together to determine the total weight of the loaded airplane.

$$CG = \frac{\text{Total Moment}}{\text{Total Weight}}$$

Now, check to see that the total weight does not exceed the maximum allowable weight under existing conditions. If it does, one solution is to reduce the amount of fuel you are carrying and plan an extra fuel stop. Once you are satisfied that the total weight is within prescribed limits,

	Weight (Lbs)	Arm (Inches)	Moment (In-Lbs)
Basic Empty Weight	2,364	82.5	195,030
Pilot and Front Passenger	340.0	85.5	29,070
Passengers (Center Seats) (Forward Facing)		118.1	
Passengers (Center Seats) (Aft Facing) (Optional)		119.1	
Passengers (Rear Seats)	340.0	157.6	53,584
Passenger (Jump Seat) (Opt.)		118.1	
Fuel (90 gal.)	540	94.0	50,760
Baggage (Forward) (100 Lb. Limit)		42.0	
Baggage (Aft) (100 Lb. Limit)	33	178.7	5,897
Ramp Weight (3617 Lbs. Max.)	3,617	92.4	334,341
Fuel Allowance for Engine Start, Taxi & Runup	-17.0	94.0	-1598
Takeoff Weight (3600 Lbs. Max.)	3,600		332,743
CG = 92.4 inches			

Figure 8-48. Here is a sample of a completed weight and balance form. The appropriate weight is recorded next to each item. Notice that the amount of fuel in gallons also has been recorded on the fuel line. The CG, filled in at the bottom of the form, is determined by dividing the total moment of the airplane by its total weight.

multiply each individual weight by its associated arm to determine its moment. Then, add the moments together to arrive at the total moment for the airplane. Your final computation is to find the center of gravity of the loaded airplane by dividing the total moment by the total weight. [Figure 8-48]

After you determine the airplane's weight and center of gravity location, consult the airplane's CG envelope to see if the CG is within limits. If the CG falls outside of the envelope, you will have to adjust the loading of the airplane. When using a CG envelope, plot the point at which the airplane's total weight and CG meet. [Figure 8-49]

GRAPH METHOD

The graph method is a simplification of the computational method. Instead of multiplying weight times arm to determine moment, you simply use a graph supplied by the manufacturer. In this method, however, the CG envelope usually reflects weight and moment instead of weight and CG.

The graph method uses the same first step as the computational method. Record the basic empty weight of the airplane, along with its total

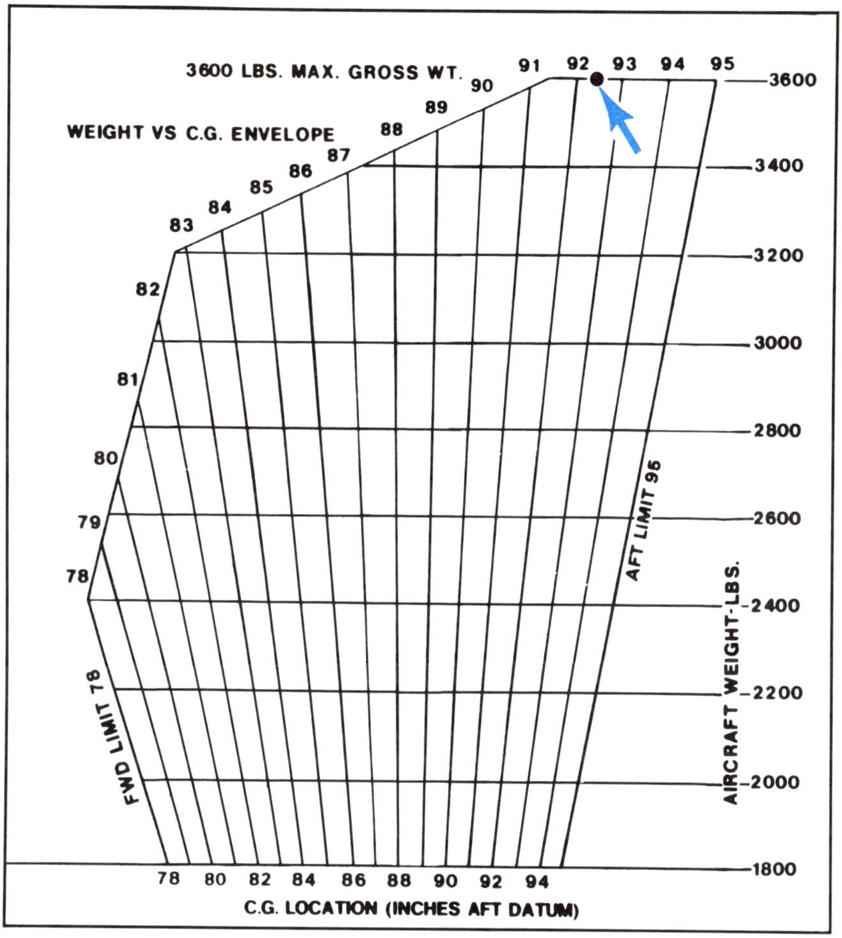

Figure 8-49. In this example, notice that a total weight of 3,600 pounds and a CG location of 92.4 inches are within the acceptable limits for flight operations in this aircraft.

moment; the empty weight CG location is not required. Remember to use the actual weight and moment of the airplane you will be flying. Next, record the weights of the fuel, pilot, passengers, and baggage on the weight and balance worksheet. Then, determine the total weight of the airplane.

If this weight exceeds the maximum allowable weight, you will have to make some adjustments to fuel, baggage, or passengers. For example, assume the maximum allowable weight for the airplane is 4,100 pounds and the total weight of the loaded airplane comes to 4,250 pounds. To make this flight, you will need to carry 25 gallons less fuel (25 gal. × 6 lb./gal. = 150 lb.) or leave 150 pounds of baggage behind. Other options may be available depending on what items you have in the payload. Once

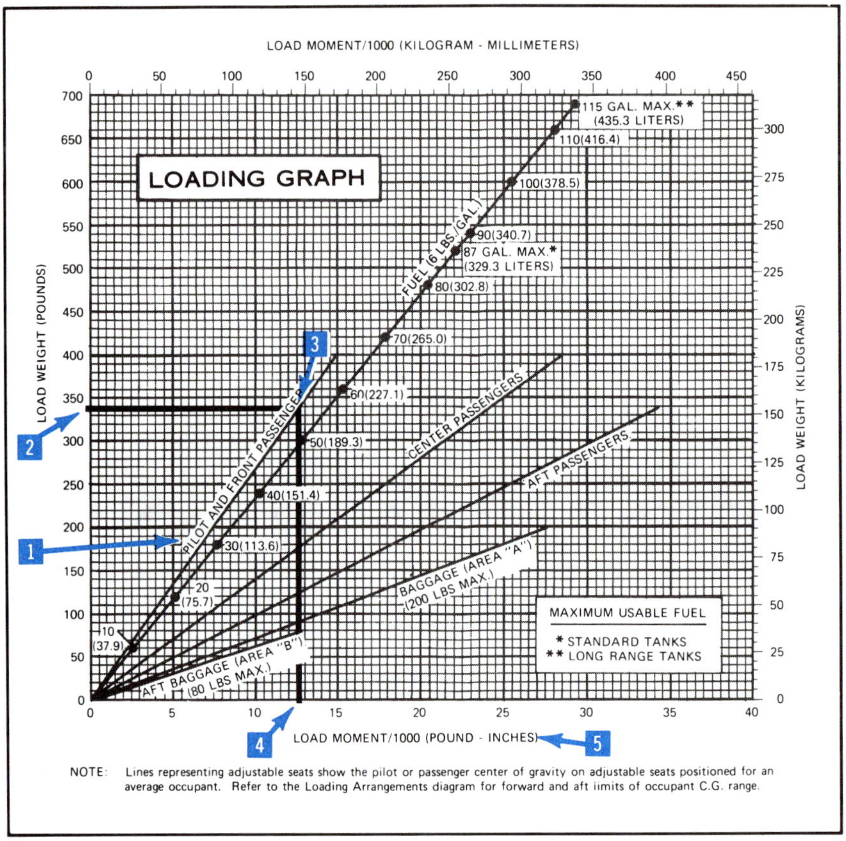

Figure 8-50. Let's assume the pilot and front seat passenger have a combined weight of 340 pounds. First, find the line that is labeled "pilot and front seat passenger" (item 1). Now, enter the graph from the left side at a weight of 340 pounds (item 2) and proceed horizontally to the appropriate line (item 3). Next, drop straight down and read the moment of 12.6 (item 4). Notice that numbers on this scale represent the load moment/1,000 (item 5). This is referred to as a reduction factor. In this case, the moment of 12.6 actually represents a moment of approximately 12,600 pound-inches.

you have determined the weight to be within prescribed limits, compute the moment for each weight and for the loaded airplane. [Figure 8-50]

Reduction factors are commonly used to reduce the size of large numbers to manageable levels. In most cases, you need not be concerned with a reduction factor because the CG moment envelope chart normally uses the same reduction factor. [Figure 8-51]

TABLE METHOD

The table method is a further refinement of the computational and graph methods. Using this method, you simply locate a given weight on the

SAMPLE LOADING PROBLEM	SAMPLE AIRPLANE		YOUR AIRPLANE	
	Weight (lbs.)	Moment (lb.-ins. /1000)	Weight (lbs.)	Moment (lb.-ins. /1000)
1. Basic Empty Weight (Use the data pertaining to your airplane as it is presently equipped. Includes unusable fuel and full oil)	2525	104.5	2,525	104.5
8. RAMP WEIGHT AND MOMENT	4118	210.4	3,972	194.0
9. Fuel allowance for engine start, taxi and runup	-18	-.8	-18	-.8
10. TAKEOFF WEIGHT AND MOMENT (Subtract step 9 from step 8)	4100	209.6	3,954	193.2

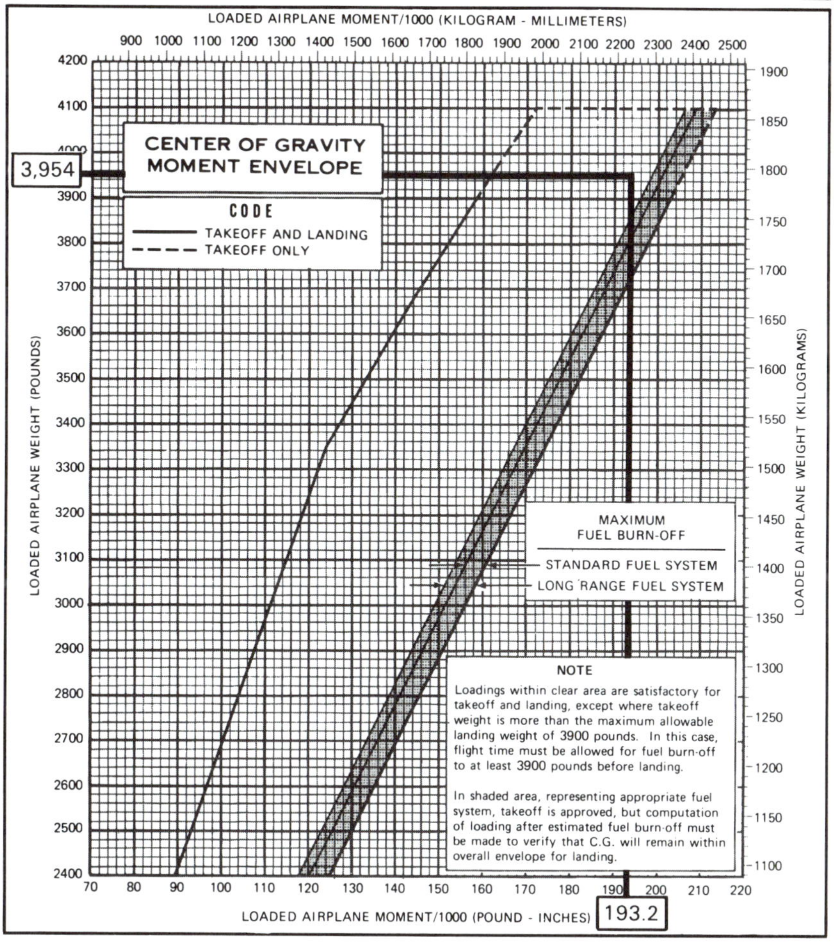

Figure 8-51. After recording and totaling weights and moments of the airplane and its contents, apply the credit for fuel used during engine start, taxi, and runup. Next, plot the listed takeoff weight and moment on the sample moment envelope graph. Based on a weight of 3,954 pounds and a moment/1,000 of 193.2 pound-inches, the airplane is properly loaded.

	FRONT SEATS			3RD AND 4TH SEATS	
	FWD POS.		AFT POS.	BENCH SEAT	SPLIT SEAT
	ARM 104	ARM 105	ARM 112	ARM 142	ARM 144
WEIGHT	MOM/100	MOM/100	MOM/100	MOM/100	MOM/100
120	125	126	134	170	173
130	135	137	146	185	187
140	146	147	157	199	202
150	156	158	168	213	216
160	166	168	179	227	230
170	177	179	190	241	245
180	187	189	202	256	259
190	198	200	213	270	274
200	208	210	224	284	288

Figure 8-52. Assume you need to determine the moment for the third and fourth seat occupants whose combined weight is 340 pounds. Also, assume this airplane has a bench seat. Note that this table uses a reduction factor of 100 and the highest weight given is 200 pounds. To determine the moment/100 for 340 pounds, look up the moments for 200 pounds and 140 pounds (200 lb. + 140 lb. = 340 lb.). In this case, the moment/100 is 483 pound-inches (284 lb.-in. + 199 lb.-in.).

appropriate table, then look up its corresponding moment. Often, however, the exact weight you are looking for is not found on the table. In these situations, look up two smaller weights which add up to the total weight you want, and add their corresponding moments. [Figure 8-52]

After the weight and balance worksheet is complete, you must check to see if the airplane is loaded within acceptable limits for flight. You have already seen one method of doing this — consulting a center of gravity moment envelope. Another method is to check a "moment limits versus weight table" provided by some manufacturers. To use this table, first find the loaded weight of the airplane. Then, check to see if the loaded moment falls within the minimum and maximum values for that weight. [Figure 8-53]

WEIGHT SHIFT

During weight and balance computations, you may find that either the weight of the airplane or its CG location is beyond acceptable limits. Usually, decreasing weight is a fairly simple matter, especially if you accomplish it before the airplane's CG is calculated. However, if you change the load to adjust the center of gravity, the computations are

ITEM	WEIGHT	MOM/100
1. BASIC EMPTY CONDITION	1,723	1,912
2. FRONT SEAT OCCUPANTS	340	357
3. 3rd & 4th SEAT OCCUPANTS	340	483
4. 5th & 6th SEAT OCCUPANTS		
5. BAGGAGE	100	167
6. CARGO		
7. **SUB TOTAL**	2,503	2,919
8. FUEL LOADING (40 gal.)	240	281
9. **SUB TOTAL** RAMP CONDITION	2,748	3,200
10. *LESS FUEL FOR START, TAXI, AND TAKE-OFF	-8	-9
11. **SUB TOTAL** TAKE-OFF CONDITION	2,740	3,191
12. LESS FUEL TO DESTINATION	162	189
13. **LANDING CONDITION**	2,578	3,002

*Fuel for start, taxi and take-off is normally 8 lbs at an average mom/100 of 9.

MOMENT LIMITS vs WEIGHT

Weight	Minimum Moment 100	Maximum Moment 100
2500	2775	2958
2510	2788	2969
2520	2801	2981
2530	2814	2993
2540	2828	3005
2550	2841	3017
2560	2854	3028
2570	2867	3040
2580	2880	3052
2590	2894	3064
2600	2907	3076
2610	2920	3088
2620	2933	3099
2630	2947	3111
2640	2960	3123
2650	2973	3135
2660	2987	3147
2670	3000	3159
2680	3013	3170
2690	3027	3182
2700	3040	3194
2710	3054	3206
2720	3067	3218
2730	3081	3230
2740	3094	3241
2750	3108	3253

Figure 8-53. On the left is a completed weight and balance worksheet using given values for a sample flight. Note the takeoff weight of 2,740 pounds and moment/100 of 3,191 pound-inches. Using the table on the right for a weight of 2,740 pounds, the minimum acceptable moment/100 is 3,094 pound-inches and the maximum is 3,241 pound-inches. Since our moment is well within these limits, the airplane is safely loaded for flight.

somewhat more complex. You can use the following formula to solve weight shift problems:

$$\frac{\text{Weight of Cargo Moved}}{\text{Weight of Airplane}} = \frac{\text{Distance CG Moves}}{\text{Distance Between Arm Locations}}$$

Perhaps the easiest way to see how to apply the weight shift formula is to work two different weight shift problems. In the first, assume you have a loaded airplane weight of 3,200 pounds and a center of gravity of 45.6 inches. A box weighing 120 pounds is moved from cargo area "B," which has an arm of 99 inches, to cargo area "A," whose arm is 70

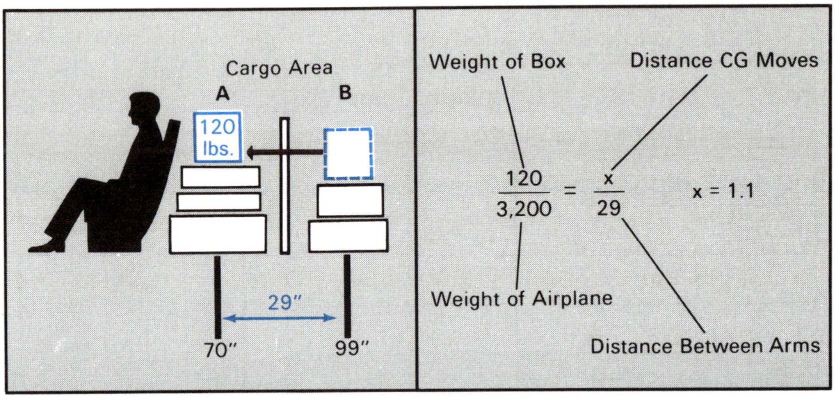

Figure 8-54. As shown in this illustration, the first step is to determine the known values. For example, the distance between cargo areas "A" and "B" is 29 inches. Next, insert the other known values into the formula and solve for the unknown. In this case, the unknown value is the distance the center of gravity moves forward when the weight is shifted. Solving this equation results in the answer of 1.1 inches.

inches. How many inches will the CG move when this weight is shifted? [Figure 8-54]

For the next example, assume that the center of gravity is located aft of acceptable limits and you need to determine the amount of weight to be shifted from the aft cargo area to the forward one. Again, the same weight shift formula is used; the only difference is the unknown value. [Figure 8-55]

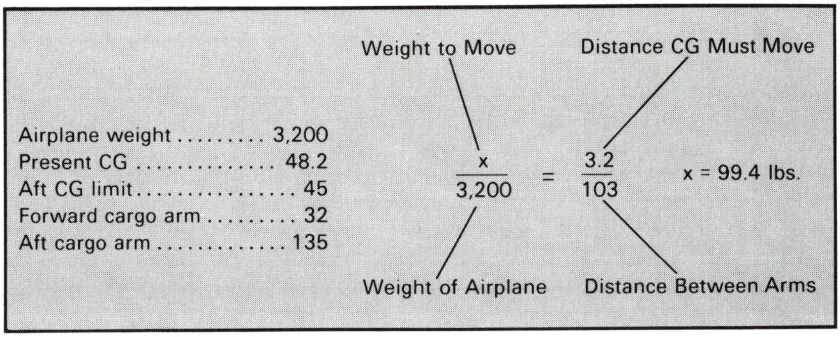

Figure 8-55. As with the previous problem, insert the known values into the formula. In this case, you are solving for the weight of the cargo to be moved. Solving this equation results in an answer of 99.4 pounds.

CHECKLIST

After studying this section, you should have a basic understanding of:

✓ **Airplane weight —** What terms are used to describe various weights and what the hazards of overloading an airplane are.

- ✓ **Empty weight** — What the difference is between basic empty weight and licensed empty weight.

- ✓ **Airplane balance** — What effects on airplane performance can be expected when the CG location is near forward or aft limits.

- ✓ **Determining weight and balance** — How the weight and balance condition of an airplane can be determined and what three methods are used.

- ✓ **Weight shift** — How to use the weight shift formula to quickly solve loading problems during preflight planning.

CHAPTER 9

ADVANCED AIRPLANE SYSTEMS

INTRODUCTION

The advanced systems presented in this chapter are designed to expand your knowledge of complex, high-performance airplanes. Fuel injection improves the power output of your engine by providing a better distribution of fuel to the cylinders than a carbureted engine. A constant-speed propeller allows you to select the manifold pressure and r.p.m. that best suits the type of operation you're conducting. Turbocharging enables you to maintain sea level manifold pressure at much higher altitudes than is possible with a normally aspirated engine. Environmental systems, such as oxygen and pressurization, provide increased utility for high-altitude operations. Ice control systems give you that additional margin of safety in IFR operations. A retractable landing gear reduces parasite drag, which improves the aerodynamics of your airplane.

The objective of this chapter is to provide you with a basic understanding of the theory behind each system. It is not our intention to teach you how to operate a particular system. Your official sources for operating characteristics, performance limitations, and procedures are the airplane's flight manual, pilot's operating handbook, and engine manufacturer's manual. Keep in mind that the systems differ from one another. A procedure that works well for one fuel-injected engine, for example, may cause problems in another. In addition, when you combine some systems, such as fuel injection and turbocharging, the operating procedures may be different than for the one system alone. It is your responsibility, as the pilot in command, to be thoroughly familiar with the operation of the airplane you're flying. This is extremely important when you are flying airplanes that have the advanced systems presented in this chapter.

SECTION A

FUEL INJECTION

One of the advantages of fuel injection is that it provides an even distribution of fuel to the cylinders.

Fuel injection has several advantages over float carburetion. The most significant benefit is the improvement in the fuel distribution to each cylinder. Some of the other benefits you can expect from this system include lower fuel consumption per unit of horsepower, increased horsepower per unit of weight, lower operating temperatures, longer engine life, and increased reliability. On the other hand, you may experience some problems, such as difficulty in starting a hot engine, vapor lock during ground operations on hot days, and difficulty restarting an engine that has failed because of fuel starvation.

To better understand the benefits of fuel injection, let's review some problems that may arise with carburetors and similar fuel distribution systems. In an ideal engine, each cylinder would receive the same amount of fuel. However, when the fuel is mixed with the inflow of air at the carburetor, it is difficult to achieve this even distribution. Some of the fuel sprayed from the discharge nozzle inside the carburetor is not entirely vaporized and remains as minute, liquid droplets of different sizes. These droplets flow unevenly around the bends and restrictions of the intake manifold. As a result, they are not evenly distributed to each cylinder, and they may or may not be vaporized before entering the combustion chamber. These small differences can reduce your engine's power output at peak power settings.

Since engine power depends on gas expansion, the temperature of the vaporized fuel entering the combustion chamber is extremely important. You'll obtain the greatest engine power when the vaporized fuel temperature is low. With float carburetion, intake air is cooled when the fuel is vaporized inside the carburetor. However, the mixture is reheated as it passes through the induction system. This problem is reduced with a fuel injection system, because vaporization occurs at each discharge nozzle before the mixture enters the combustion chamber. Let's take a closer look at how this system works.

FUEL INJECTION SYSTEM COMPONENTS

Fuel injection systems vary with engine manufacturers, as well as with airplane models and types. However, the operating principles are similar. You should review the fuel system coverage in the pilot's operating handbook before you fly an unfamiliar aircraft. [Figure 9-1]

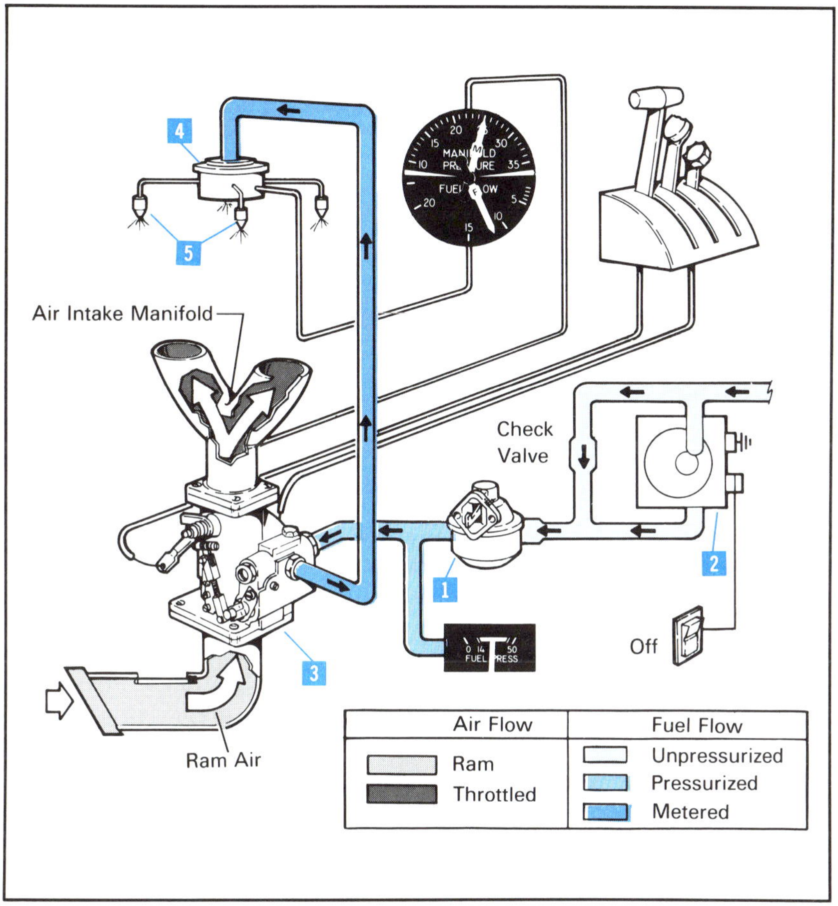

Figure 9-1. This is an example of one type of a fuel injection system. It has several basic components: an engine-driven fuel pump (item 1), electric auxiliary "boost" pump (item 2), fuel injector unit (item 3), fuel flow divider (item 4), and individual fuel discharge nozzles (item 5).

In a typical system, fuel is delivered from the aircraft's tanks to the engine-driven pump, where it is put under pressure. The pressurized fuel is then routed through the injector unit. A set of valves meters the fuel flow according to the mixture setting and flow of air to the engine. A throttle valve controls the amount of air intake. The metered fuel is then sent to the fuel flow divider, where it is distributed equally among individual lines that lead to each cylinder. A discharge nozzle is located at the intake port of each cylinder, where the fuel is readily atomized, then vaporized.

The **engine-driven fuel pump** is the only constantly moving part in the fuel injection system. This pump pressurizes the system and sends the fuel to the injector unit. In some systems, the pump supplies more fuel than the engine can use. In this case, the excess fuel is routed back to the tank through a return line. In other systems, the fuel is strictly metered through the system and does not require a return line.

The electrically driven **auxiliary "boost" pump** serves several functions. It pressurizes the system for engine starting, removes unwanted fuel vapor, and serves as a backup if the engine-driven pump fails.

The **fuel flow divider**, sometimes called the fuel manifold, is designed to divide the fuel evenly between each of the cylinders. Valves are built into the flow divider to provide an automatic fuel shutoff when you move the mixture control to idle-cutoff.

The **fuel injection nozzles** "inject" the metered fuel into the air induction manifold. This occurs in the cylinder head just outside the intake valves and ports which lead to the combustion chamber. Thus, the fuel is mixed with the intake air just before it enters the chamber.

A **fuel flow indicator** enables you to monitor the operation of the system and to set the mixture correctly. It senses the pressure at which fuel is delivered to the injection nozzles. [Figure 9-2]

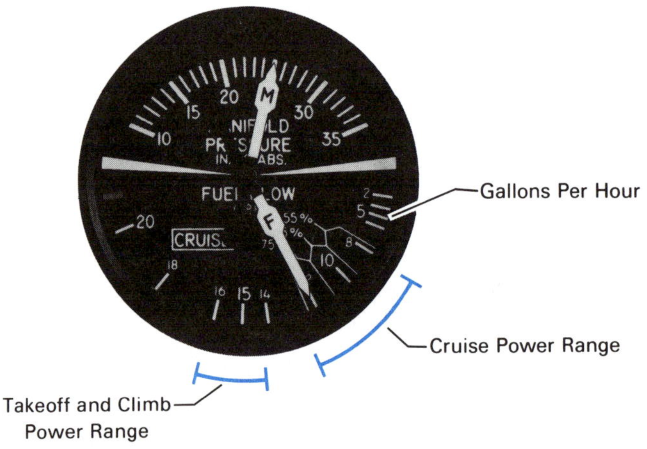

Figure 9-2. The fuel flow indicator is sometimes combined with the manifold pressure gauge. A manifold pressure gauge is found on airplanes with constant-speed propellers. It shows engine power and is controlled by the throttle. The fuel flow indicator displays the rate of fuel consumption in either gallons per hour or pounds per hour. Some gauges have indications for pounds per square inch and ranges for various flight operations.

STARTING PROCEDURES

When the proper amounts of fuel and air are delivered to the combustion chamber, a spark from the ignition system will almost always start the engine. In this respect, a fuel injected engine is no different from an engine with a float carburetor. But, there are differences between the two systems that you must understand to help you during engine starting. Fuel injection systems also vary among engine manufacturers, with some engines being easier to start than others. The following are general guidelines for starting typical fuel injected engines. You should always be familiar with your particular engine design and starting procedures as outlined in the pilot's operating handbook.

NORMAL STARTING

Although the actual manipulation of the throttle, propeller, and mixture controls varies with engine design, all fuel injected engines have one thing in common. An engine that is cool to the touch or one that is being started for the first flight of the day must be primed. The auxiliary fuel pump is used to prime and pressurize the injection system. With the throttle and mixture control set according to the POH, turn on the electric fuel pump and monitor the fuel flow indicator. When a sufficient fuel flow rate is noted, continue with the normal starting procedures.

The auxiliary fuel pump primes and pressurizes the injection system.

HOT STARTING

A common area of difficulty with fuel injection is starting a hot engine. Fuel vaporization causes most of the problems. When the engine is shut down, the air temperature inside the cowling increases rapidly. Within approximately 10 to 15 minutes, the temperature inside the fuel flow divider, distribution lines, and injector nozzles reaches a point where the fuel vaporizes. This creates a vapor lock, which effectively blocks the flow of fuel to the nozzles. An attempt to start the engine under these conditions would be unsuccessful, since there is not enough fuel for combustion.

A hot fuel injected engine is more difficult to start than a hot carbureted engine.

The auxiliary fuel pump is normally used to remove the fuel vapor and cool the distribution lines. Again, the exact procedure varies with the engine design.

Once the engine is started, you may still have vaporization problems, especially on very hot days. During ground operations, the propeller slipstream is providing almost all of the airflow for cooling. Since most fuel injected engines are closely cowled, airflow is decreased and the air temperature inside the engine compartment may increase. This may increase the temperature of the injection system to a point where a vapor lock forms in the distribution line. This interrupts the normal fuel flow and causes fuel starvation. If you notice a fluctuation in the fuel flow and/or the engine r.p.m. starts to decrease, activate the auxiliary fuel pump to purge the vapor lock.

On hot days, vapor lock can be a problem during ground operations.

ABNORMAL COMBUSTION

Always monitor your engine instruments to ensure the temperatures and pressures remain within specified limits. You can minimize engine problems by using the recommended fuel grade and by properly maintaining and servicing your airplane. Improper operation and maintenance could result in abnormal combustion, such as detonation or preignition.

Detonation is the result of fuel exploding within the cylinder, rather than burning normally.

Detonation is the uncontrolled, explosive combustion of fuel. During normal combustion, the fuel burns evenly and progressively inside the combustion chamber. Detonation causes the fuel to explode, producing excessive pressures and temperatures within the engine. It can cause overheating, loss of power, and roughness, which can shorten the life of your engine. In extreme cases, it can destroy a piston, a valve, or part of the cylinder, causing your engine to fail suddenly.

Conditions that produce excessive engine temperatures can cause detonation.

Detonation is most likely to occur with an overheated engine and when operating at high power settings. Some common causes of overheating include using a grade of fuel lower than that recommended, operating with extremely high manifold pressure and extremely low r.p.m., and operating at over 75% power with a lean mixture setting that produces high exhaust gas temperature. An engine that is overheated on the ground may exceed its temperature limits during takeoff and initial climb because of the high power setting and slow airspeed.

If you suspect detonation is occurring during climbout, a slight reduction in power may correct the problem.

If you suspect the engine is detonating during climbout, reduce the power slightly and climb at a slower rate to help cool the engine. In most cases, it is difficult to recognize a detonating engine through sound or roughness. However, be alert for high temperature and a decrease in engine performance.

Preignition is the result of an uncontrolled firing of the fuel/air mixture before normal spark ignition.

Preignition can occur when the fuel/air mixture ignites prior to its normal spark point. This premature burning of fuel is usually caused by a residual hot spot in the cylinder. The hot spot can be a small carbon deposit on a spark plug, a cracked ceramic spark plug insulator, or almost any damage around the combustion chamber.

Preignition and detonation often occur simultaneously, and one may cause the other. Both can cause a decrease in engine performance and high temperatures, so you may not be able to tell which one is occurring. You can reduce the chance of combustion problems by maintaining engine temperatures within the recommended limits, operating with a sufficiently rich mixture, and using only the proper grade of fuel.

EXHAUST GAS TEMPERATURE GAUGE

The exhaust gas temperature (EGT) gauge measures the temperature of the exhaust at a point just outside the combustion chamber. Some systems measure the temperature of only one cylinder, while others have probes located at each cylinder. Generally speaking, the temperature of the exhaust gas increases as you lean the mixture and decreases as you enrich it. By measuring the change, the EGT allows you to adjust the fuel/air ratio with improved accuracy and efficiency. Remember, at altitude you adjust the fuel/air mixture to decrease the fuel flow to compensate for decreased air density. When you use the proper fuel/air ratio for a particular operating condition, you can significantly increase your engine's performance. If you lean the engine too much, it can cause detonation and/or preignition. On the other hand, if you operate the engine with an excessively rich mixture, your spark plugs can foul and cause carbon to build up inside the cylinder. [Figure 9-3]

The EGT allows you to adjust the mixture accurately to obtain the proper fuel/air ratio for various power settings and altitudes.

Takeoff power requires a rich mixture to prevent detonation. Normally, you should use a full rich setting when applying takeoff power. At density altitudes above 5,000 feet, however, you may need to lean the mixture to achieve smooth engine operation. A common procedure is to use a setting somewhat richer than that required for peak EGT to obtain best fuel economy, especially if the engine is turbocharged. To obtain best cruise power, you should set your mixture slightly richer than the mixture for best fuel economy. Regardless of the operation, you should always follow the manufacturer's recommendations for your engine.

INDUCTION ICING

Internal icing is more common in engines with a float carburetor than in those with fuel injection. This is mainly the result of the location of the vaporization process. With a carburetor, intake air cools as it passes through the venturi. Further cooling occurs when the vaporized fuel is

A fuel injection system is less susceptible to evaporative icing than is a float carburetor.

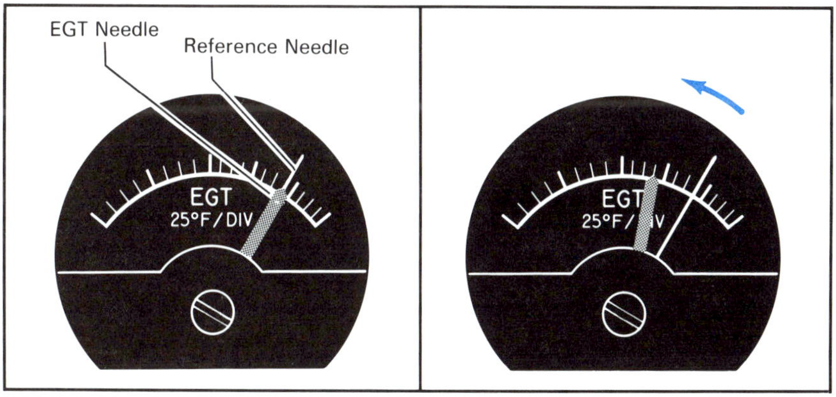

Figure 9-3. To use the EGT, slowly lean the mixture while observing the needle of the gauge. As you lean, the exhaust temperature increases until it reaches its peak. At this point, align the reference needle with the EGT needle, as shown on the left. Then, as shown on the right, adjust the mixture to lower the temperature to the setting recommended by the manufacturer. To prevent engine damage and excessive temperatures, avoid leaning to peak EGT when your engine is developing more than 75% power.

added. This expansion and vaporization causes water vapor to sublimate and form ice on the internal parts of the carburetor. In a fuel injected system, the vaporization and mixture of the fuel and air occurs just outside the combustion chamber where temperatures are high and ice is not likely to develop.

Both fuel injected and carbureted engines are affected by impact icing.

Although fuel injection is less susceptible to internal icing, it is still affected by **impact ice** that can build up on the external air scoop or filter. Impact ice affects both fuel injected and carbureted engines when you fly in visible moisture with the air temperature near freezing. An alternate air source on fuel injected engines eliminates this problem. [Figure 9-4]

Fuel injection has numerous advantages over carburetion; however, to use it efficiently, you must be familiar with its operating characteristics, components, and procedures. Before you fly an airplane with fuel injection, you should obtain a thorough checkout on the system from a qualified flight instructor.

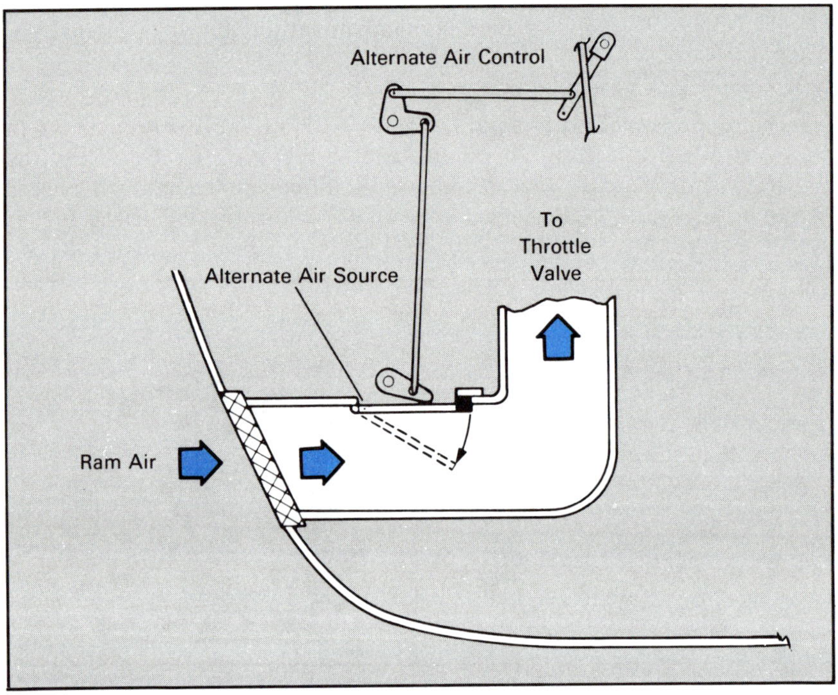

Figure 9-4. In a fuel injection system, ram air enters through an air scoop and is filtered before passing into the intake manifold and throttle valve. If the inlet or filter becomes clogged, an alternate air door opens to provide engine compartment air to the induction system. This door operates either manually or automatically, depending on the system design. Without this alternate air source, induction icing could cause engine failure.

CHECKLIST

After studying this section, you should have a basic understanding of:

✓ **Fuel injection** — Its advantages and disadvantages and how it differs from float carburetion.

✓ **Fuel injection system components** — What they are and how they operate.

✓ **Starting procedures** — What normal starting procedures are and what process to use for hot starts.

✓ **Abnormal combustion** — The causes of detonation and preignition and how to prevent them.

✓ **Exhaust gas temperature gauge** — What it is and how it can be used to adjust the mixture accurately.

✓ **Induction icing** — How it affects carburetors and how it differs with fuel injection.

SECTION B

HIGH-PERFORMANCE POWERPLANTS

Aircraft performance is largely determined by the amount of thrust the powerplant develops. If a propeller could perform with 100% efficiency, all of the engine's power output to the propeller would be converted to thrust; however, this efficiency rate is rarely achieved. Some engine systems, such as constant-speed propellers and turbochargers, are designed to improve the performance of your aircraft.

CONSTANT-SPEED PROPELLER

Propeller efficiency is the ratio of thrust horsepower to brake horsepower.

The principal advantage of a constant-speed propeller is that it converts a very high percentage of engine power, called brake horsepower (BHP), into thrust over a wide range of conditions. The efficiency of a given propeller is the ratio of thrust horsepower to BHP. In addition to being more efficient than a fixed-pitch propeller, a constant-speed propeller enables you to select the optimum combination of engine power (manifold pressure and r.p.m.) for the type of operation you are conducting. This means you can obtain the best overall performance by achieving a high rate of climb, increasing cruise airspeed, or extending endurance by reducing fuel consumption.

POWER CONTROLS

The manifold pressure gauge measures the absolute atmospheric pressure inside the intake manifold.

With a constant-speed propeller, you use separate controls to determine engine power output. The throttle controls the **manifold pressure**, which is indicated on the manifold pressure gauge in inches of mercury. Manifold pressure increases as you move the throttle forward and decreases as you move it back. The gauge measures the absolute pressure of the fuel/air mixture inside the intake manifold and is commonly referred to as manifold absolute pressure (MAP). At a constant r.p.m. and altitude, the amount of power produced is directly related to the fuel/air flow being delivered to the combustion chamber. As you increase the throttle setting, more fuel and air is flowing to the engine; therefore, MAP increases.

In a standard atmosphere, MAP decreases one inch of mercury for each 1,000 feet of altitude gain.

Since the manifold pressure gauge is essentially an aneroid barometer, it indicates ambient air pressure when the engine is shut down. For example, at sea level on a standard day, the gauge indicates 29.92 in. Hg. with the engine stopped. When the engine is running, MAP is also influenced by changes in air density. In standard atmospheric conditions, manifold pressure decreases approximately one inch for every 1,000 feet of altitude gain. In nonstandard conditions, a change in density altitude produces the same effect.

Carburetor ice reduces the power output of the engine. On an airplane equipped with a constant-speed propeller, carburetor icing is indicated by a reduction in manifold pressure. This differs from an airplane with a fixed-pitch propeller, where carburetor ice is indicated by a drop in r.p.m. When you apply carburetor heat on an airplane with a constant-speed propeller and ice is present, you will notice a decrease in MAP followed by a gradual increase.

With a constant-speed propeller, carburetor ice is indicated by a decrease in manifold pressure.

The **propeller control** is used to set the r.p.m. of the engine. Moving the control forward increases r.p.m., while moving it aft decreases r.p.m. Some prop controls have what is called a vernier setting mechanism that allows you to make small, very precise changes in the r.p.m. Rotating the vernier control clockwise causes a small increase in r.p.m. When you set the propeller control, a governor mounted on the engine and a mechanism inside the propeller hub automatically change the pitch of the propeller to maintain the corresponding r.p.m. [Figure 9-5]

Some aircraft have a vernier control to set the r.p.m. precisely.

At a constant power setting, the load imposed on a propeller increases or decreases as you change airspeed. With a constant-speed propeller, the governor senses this change and automatically increases or decreases the blade angle to maintain the proper blade angle of attack and r.p.m. Since

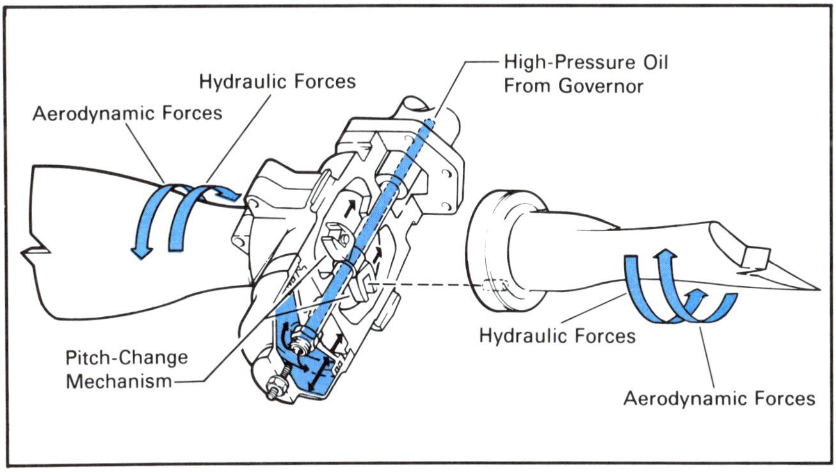

Figure 9-5. The governor senses engine speed and delivers high-pressure engine oil to a piston located inside the propeller hub. Piston movement causes the pitch-change mechanism to alter the angle of the propeller blades to maintain the r.p.m. The aerodynamic force tends to move the blades to low pitch (high r.p.m.), while hydraulic pressure opposes the force and tends to move the blades to high pitch (low r.p.m.).

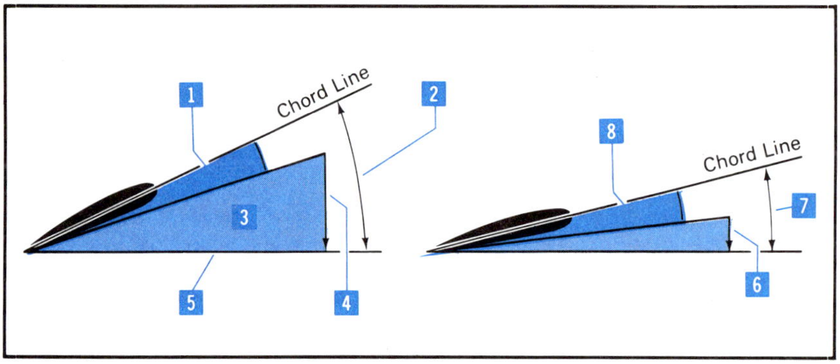

Figure 9-6. As shown on the left, the angle of attack of a propeller (item 1) is determined by the blade angle or pitch (item 2), and the effective pitch angle (area 3). This angle represents the combined effects of airspeed (item 4) and the rotational velocity of the propeller (item 5). As shown on the right, at a constant prop setting, a decrease in airspeed (item 6), produced by either raising the nose or decreasing MAP, causes the blade angle (item 7) to automatically decrease so the same angle of attack (item 8) is maintained. This keeps the load on the propeller constant and the r.p.m. the same. Conversely, an increase in airspeed by either lowering the nose or increasing MAP causes the blade angle to automatically increase and the r.p.m. to remain the same.

a propeller is essentially a rotating airfoil, its angle of attack is the principal factor affecting its efficiency. [Figure 9-6]

A fixed-pitch propeller achieves peak efficiency only at a given combination of airspeed and r.p.m., while a constant-speed propeller can achieve peak efficiency over a wider range of airspeeds by varying propeller pitch.

To produce the effective pitch angle that provides the angle of attack necessary for optimum thrust or peak efficiency, a fixed-pitch propeller must be operated at a specific r.p.m. and airspeed. For example, a climb prop has a small blade angle which produces high r.p.m. and maximum thrust for takeoff and climb. Conversely, a cruise prop has a large blade angle which produces low r.p.m. and higher cruise airspeeds. With a constant-speed propeller, you can change the propeller blade angle to achieve the optimum, effective pitch angle. In this way, you can maintain a high plateau of propeller efficiency over a wide range of airspeed and r.p.m. combinations. [Figure 9-7]

CONSTANT-SPEED PROPELLER OPERATION

For takeoff, you should set the propeller to a blade angle that produces a small angle of attack and high r.p.m.

Obtaining maximum thrust is critically important for takeoff and climb performance; therefore, you should use the maximum permissible or obtainable MAP and r.p.m. In most cases, this requires you to place the propeller in the low pitch setting (full forward) and set the manifold pressure as high as possible without exceeding the manufacturer's limits. You should avoid any power settings that are beyond red line limits. A caution range is usually identified by yellow arcs on either the manifold pressure gauge or tachometer, or both. Always use the recommendations specified in your pilot's operating handbook or engine manual for the aircraft you are flying.

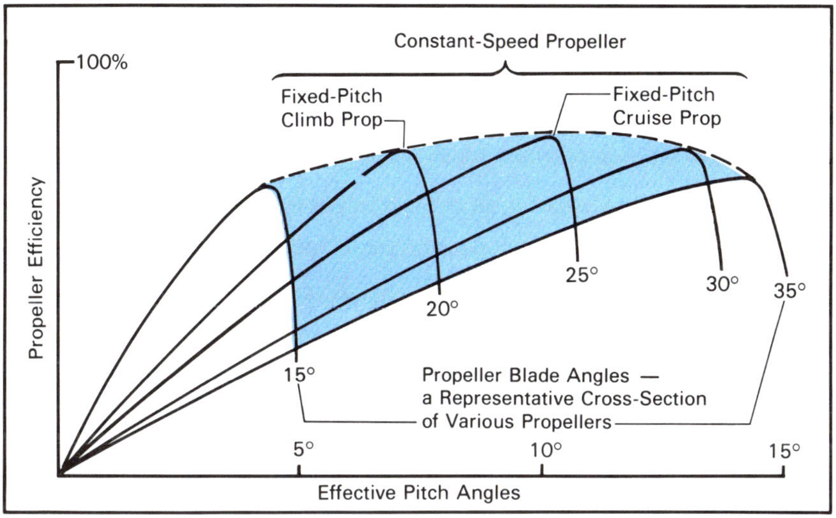

Figure 9-7. The efficiency of a fixed-pitch propeller increases with effective pitch angle until the optimum angle of attack is reached. Then, the efficiency of the fixed-pitch propeller decreases rapidly. The shaded area represents the range for a constant-speed propeller. By varying the blade angle, the constant-speed propeller can maintain higher efficiency over the entire range shown for fixed-pitch propellers. This allows you to use a high pitch setting (low r.p.m.) for cruise and a low pitch (high r.p.m.) for takeoff.

When you have reached an altitude where maximum climb performance is no longer critical, you can decrease the horsepower output slightly. The correct sequence is to reduce power, lower the MAP first, then reduce the r.p.m. by increasing the propeller blade angle. Once the power is set, adjust the mixture control, if necessary. When you increase power, begin by enriching the mixture and increasing the r.p.m. by decreasing the propeller blade angle. Then, advance the throttle to increase the manifold pressure. Many aircraft engines are equipped with balance weights as part of the crankshaft assembly. These counterweights can be detuned by rapidly opening and closing the throttle or excessive speed or power settings. When these counterweights are "detuned," they cannot dampen the crankshaft vibrations that they are designed to absorb and can lead to failure of the crankshaft and engine.

To avoid detuning the crankshaft counterweights and undue stress on the engine, make power changes smoothly and in the correct sequence.

Although engine designs vary, as a general rule, you should avoid excessively high manifold pressures in combination with extremely low r.p.m. settings. This situation produces excessive cylinder pressure and can lead to overheating and detonation that can severely damage the engine. When you use a low MAP with a high r.p.m., the chance of creating excessive cylinder pressure is reduced. However, some engines are designed to operate with very high MAP. These engines are usually equipped with turbochargers.

High MAP and low r.p.m. produces excessive cylinder pressure, while low MAP and high r.p.m. reduces the pressure.

TURBOCHARGING

The maximum amount of horsepower your engine can develop depends on engine r.p.m. and its ability to create the highest manifold pressure at a given altitude. When you climb to higher altitudes, the ambient air pressure decreases (approximately one inch of mercury per 1,000 feet). This means that less air is available for the combustion process than at a lower altitude. With a normally aspirated engine, which is one without turbocharging, you experience a decrease in the maximum MAP with an increase in altitude. When a turbocharger is installed on the engine, the intake air is compressed before it enters the combustion chamber. This enables an engine to obtain much higher manifold pressure readings than with a normally aspirated engine. It also allows you to maintain sea level manifold pressure up to a much higher altitude. The operating principles of the turbocharger are surprisingly simple; however, it is a carefully designed, very sophisticated piece of machinery. [Figure 9-8]

The speed of the turbine and, therefore, the amount of air compressed by the compressor depends on the position of the waste gate. When the waste gate is open, most of the exhaust bypasses the turbine and is routed out the exhaust port. In this situation, the engine operates much

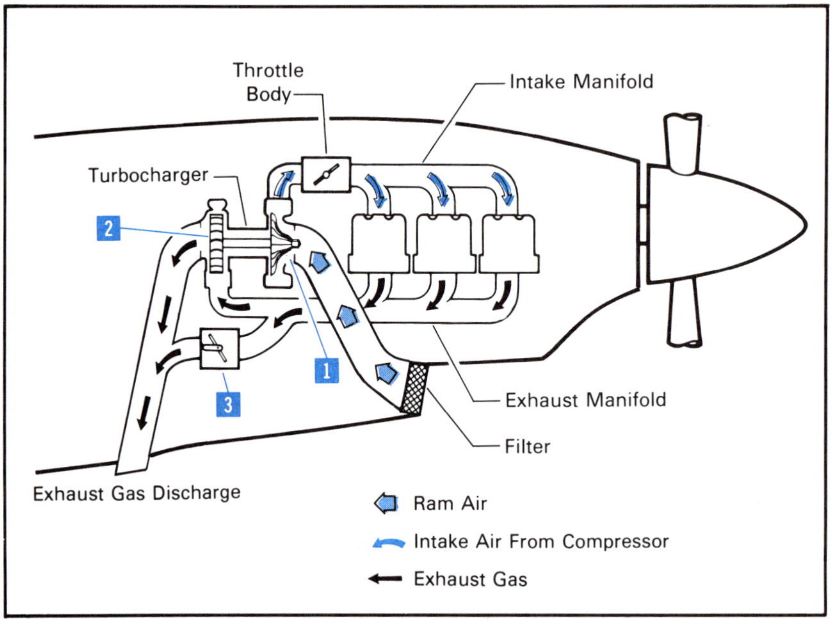

Figure 9-8. The components of a typical turbocharging system are the compressor (item 1), turbine (item 2), and the waste gate (item 3). Ram air is ducted into the compressor, where it is compressed. The pressurized air passes through the throttle body and into the intake manifold. Exhaust gas is ducted through the exhaust manifold and is used to spin the turbine which, in turn, drives the compressor. The waste gate controls the amount of exhaust passing through the turbine.

like a normally aspirated engine, since very little air is being compressed. When the waste gate is closed, all of the exhaust gas is used to spin the turbine. In this case, the maximum amount of compressed air is being delivered at the MAP set on the manifold pressure gauge.

SYSTEM OPERATION

During a climb, the turbocharger maintains a constant manifold pressure up to the critical altitude.

On most turbocharged engines, the position of the waste gate is governed by a pressure-sensing control mechanism and a waste gate actuator. This helps to simplify the system operation to one control — the throttle. Once you set the desired manifold pressure, virtually no throttle adjustment is required with changes in altitude. The mechanism senses manifold pressure requirements for various altitudes and regulates oil pressure to the actuator, which adjusts the waste gate. Therefore, the turbocharger maintains only the manifold pressure called for by the throttle setting. As you climb, the waste gate gradually closes and the turbine speed increases to maintain MAP. When the waste gate is completely closed, any increase in altitude will require an increase in the throttle setting to maintain the desired manifold pressure. The altitude where the waste gate is fully closed and the turbine is operating at its maximum speed is called the **critical altitude**.

Above the critical altitude, there are certain operational characteristics which you must understand to fully use the advantages and capabilities of the turbocharged engine. Any change in r.p.m. results in a change in manifold pressure. When the waste gate is closed, a decrease in r.p.m. normally produces a decrease in manifold pressure. However, with some aircraft, the opposite may be true. Consult the POH for your airplane.

When you operate above the critical altitude, an increase in airspeed results in a corresponding increase in manifold pressure. The increase in ram air pressure is magnified by the compressor, resulting in an increase in manifold pressure. You can use this characteristic to your advantage by allowing the aircraft to accelerate to cruise speed after leveling off and prior to reducing power. Above the critical altitude, with r.p.m. and manifold pressure established for cruise, leaning may cause a slight decrease in manifold pressure. A slight reduction in manifold pressure may be necessary when the mixture reaches the recommended fuel flow. Leaning a turbocharged engine requires special considerations. Since the response to leaning may vary with different aircraft, always follow the manufacturer's recommendations. On most turbocharged engines, the limiting factor is the temperature of the air entering the turbine. Some airplanes are equipped with a turbine inlet temperature (TIT) gauge. Never lean the mixture beyond the maximum temperature limits specified by the manufacturer. Sudden, large power changes at altitude with rich mixtures can cause loss of engine power. Power changes should be made slowly, with necessary mixture adjustments, in a series of two or three steps.

A common problem associated with turbocharging is exceeding the maximum allowable manifold pressure, or overboosting the engine, when power is increased. To avoid exceeding normal manifold pressure limits, particularly in cold weather, apply the throttle slowly and smoothly while observing the manifold pressure. A momentary overboost is normally not considered detrimental to the engine. If the overboost is more than momentary, however, you should have the engine checked by an authorized maintenance facility.

Shutting down a turbocharged engine also requires special considerations. Since the turbine is usually spinning at a very high rate and its temperature is extremely hot, it's important to allow the turbocharger bearings time to cool down. Let the engine idle for the amount of time recommended by the manufacturer before you shut it down.

Turbocharging allows you to maintain the engine's rated horsepower at much higher altitudes than is possible with a normally aspirated engine. It also gives you the added advantages of higher climb rates and faster cruise speeds. On some airplanes, the compressed air from the turbocharger is used to pressurize the cabin, which eliminates the need to use supplemental oxygen. When you fly an airplane with turbocharging, you should be familiar with the operating procedures of the system itself, as well as the regulations pertaining to high altitude flight. These include the use of oxygen, aircraft equipment requirements, and pilot certification limitations.

CHECKLIST

After studying this section, you should have a basic understanding of:

✓ **Constant-speed propeller** — The advantages it has over fixed-pitch propellers, how it operates, and how power is converted to thrust.

✓ **Power controls** — How manifold pressure and r.p.m. are used to achieve optimum power and proper use of each control.

✓ **Propeller blade** — Its principle of operation and the importance of blade angle and angle of attack.

✓ **Propeller efficiency** — How it varies with different operations and how a constant-speed propeller maintains efficiency over a wide range of conditions.

✓ **Turbocharging** — Its components and what its advantages are.

✓ **Turbocharger operation** — How the waste gate functions and how the engine operates above and below the critical altitude.

ENVIRONMENTAL AND ICE CONTROL SYSTEMS

This section contains a general discussion of oxygen, cabin pressurization, and ice control systems. Since these systems vary widely on different aircraft, you should be completely familiar with the equipment on the aircraft you are flying.

OXYGEN SYSTEMS

FAR Part 91 requires you to use supplemental oxygen above certain altitudes and have it available for passengers. The following is a brief discussion of three basic types of oxygen equipment. Of these, the most common in general aviation aircraft is the continuous-flow system. It may be installed as optional equipment in some airplanes and it also is available in smaller, portable units. The second type is the diluter-demand system, which increases the utility of the basic continuous-flow equipment by conserving oxygen at lower altitudes and increasing the oxygen flow at higher altitudes. The third type is the pressure-demand oxygen system, normally installed in higher performance turboprop and jet aircraft.

CONTINUOUS FLOW

Continuous-flow oxygen delivery systems usually consist of four main components. They are the high-pressure storage cylinder, regulator, distribution system, and continuous-flow masks. [Figure 9-9]

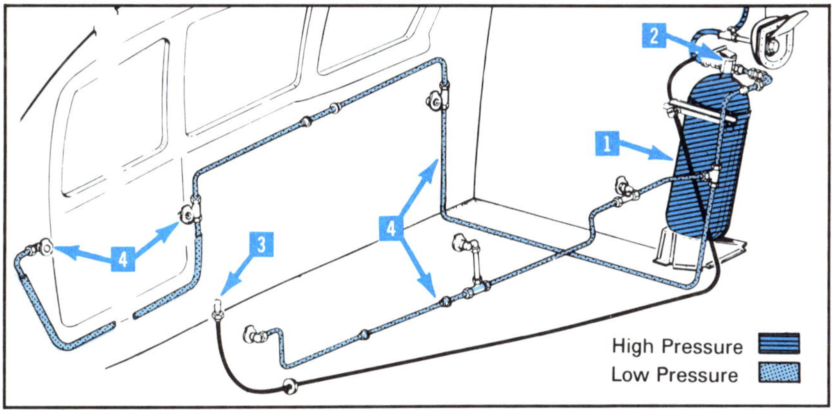

Figure 9-9. The supply of high-pressure oxygen is maintained in a storage cylinder (item 1). A regulator (item 2) reduces the high-pressure oxygen to a lower, usable pressure. It usually incorporates an ON/OFF valve in the cabin (item 3). When the regulator is ON, the distribution system (item 4) routes the low-pressure oxygen to outlets throughout the cabin. You and your passengers may then plug individual masks into outlets when you need supplemental oxygen.

During servicing, it is important to observe general safety precautions and use a checklist.

Most oxygen systems have an external filler valve which allows you to fill the storage cylinder without removing it from the aircraft. Make sure you follow the operating procedures exactly. Although oxygen, by itself, is not flammable, it makes other materials burn rapidly by supporting combustion. Do not handle oxygen equipment with greasy hands or permit an accumulation of oily waste or residue in the vicinity of pure oxygen. In addition, do not smoke or allow passengers to smoke while the oxygen system is being used.

Use only aviator's breathing oxygen to fill aircraft oxygen cylinders.

Make sure the aircraft oxygen system is filled with **aviator's breathing oxygen**. Specifications for this type of oxygen are 99.5% pure oxygen and not more than .005 milligrams of water per liter. Medical oxygen contains too much moisture, which can collect in the valves and lines of the system and freeze. This may stop the flow of oxygen.

Several types of manual and automatic regulators are used to control the flow of oxygen through the distribution system. Manual regulators incorporate a pilot-operated flow control valve, while automatic regulators use atmospheric and cylinder pressure differential to maintain proper flow rate. Oxygen is routed from the regulator to the oxygen outlets. These outlets usually contain spring-loaded shutoff valves which close automatically when the mask is disconnected.

The amount of oxygen that flows into the mask depends on two factors — the oxygen pressure supplied to the outlet and the size of the restricted opening in each outlet. Since the oxygen reaches the outlet at a constant rate, the mask's plug-in fitting and the outlet work together to control the oxygen flow into the face mask by means of an oxygen flow restrictor.

The flow restrictor in the oxygen mask plug-in connector may be provided with a dense fiberglass packing or a small, calibrated orifice. The density of the fiberglass packing or the size of the calibrated orifice determines the flow rate, which is usually measured in liters per hour (l.p.h.). The rate of oxygen flow may be reduced either by increasing the density of the fiberglass packing or by reducing the size of the calibrated orifice. Conversely, a less dense fiberglass packing or a larger calibrated orifice increases the flow rate. [Figure 9-10]

A continuous-flow oxygen system can provide adequate oxygen up to an altitude of 25,000 feet. Above 17,000 feet, however, you must be careful to maintain a tight mask seal. For flights above 25,000 feet, all occupants should use one of the demand oxygen systems. [Figure 9-11]

With automatic regulators, the oxygen flow rate increases as pressure decreases. When you inhale, you draw oxygen out of the bag and into your lungs. As the rebreather bag collapses, some outside air is drawn

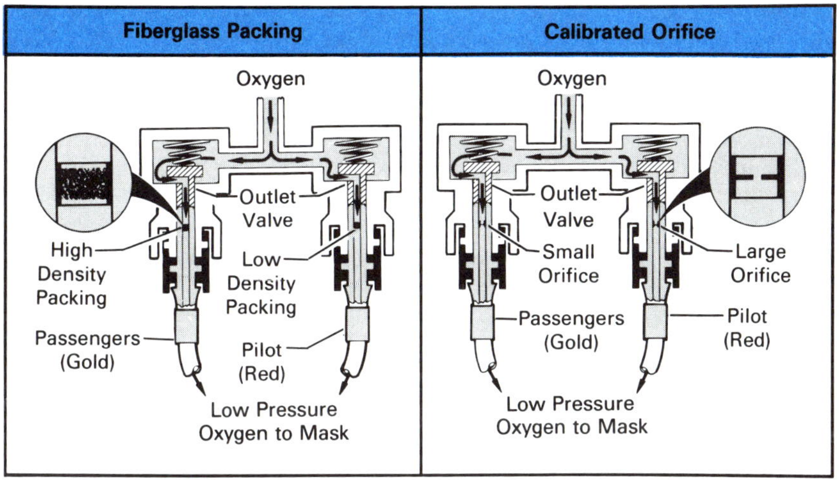

Figure 9-10. Since pilots usually require more oxygen than passengers, the plug-in for the pilot's mask has a lower density fiberglass packing or a larger calibrated orifice. The pilot's mask plug-in usually has a red band to denote its greater flow rate. This distinguishes it from the other oxygen masks, which have higher density packing or a small calibrated orifice and are marked with gold or orange bands.

into the face mask through small holes. When you exhale, air is partially expelled through the mask and the rebreather bag is filled by fresh oxygen and exhaled air.

The amount of outside air that you inhale depends on the rate of oxygen flowing into the mask. The higher the flow rate, the more the bag and mask are pressurized, resulting in less outside air entering the mask during inhalation. In addition, the higher the altitude, the more the rebreather bag expands due to lower surrounding air pressure. Thus, a higher percentage of oxygen is available as air pressure diminishes. The rebreather

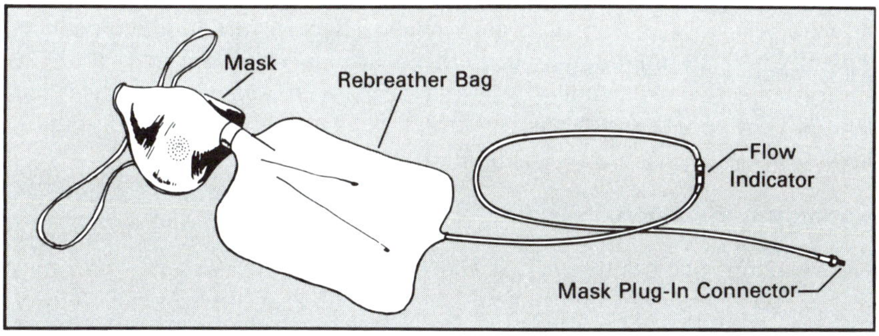

Figure 9-11. The continuous-flow oxygen mask assembly consists of a plug-in connector, flow indicator, rebreather bag, and mask. When the connector is plugged into the outlet, oxygen flows through the plug-in connector, past the flow indicator, into the rebreather bag, and then to the face mask. The flow indicator is a simple check valve which deflects toward the mask to indicate when oxygen is flowing.

bag becomes repressurized, both by incoming oxygen from the distribution system and by the initial exhalation from your lungs. The remaining volume of air you exhale is forced out through vent holes.

In general, the maximum safe altitude for different oxygen systems is based on the actual altitude of the flight environment. Although the cabin altitude in a pressurized aircraft may be considerably less than the actual altitude, the oxygen system should be capable of providing an adequate supply of oxygen to all occupants after a loss of cabin pressurization.

At altitudes above 25,000 feet, the continuous-flow system with the rebreather bag is unsafe for normal operations. Above 28,000 feet, you cannot maintain the oxygen pressure within your lungs without an increase in the pressure of the oxygen you inhale.

DILUTER DEMAND

The diluter-demand oxygen system provides oxygen when you demand it, or when you inhale. This type of system is required above 25,000 feet. It has the same type of oxygen supply cylinder, regulator, and distribution system as the continuous-flow oxygen system. However, instead of a rebreather bag, it incorporates a second regulator in the mask or hose assembly. A small lever adjustment allows you to select either normal or 100% oxygen.

Most of these systems automatically furnish more oxygen as altitude increases. The regulators normally provide 100% oxygen by approximately 30,000 feet, although you can select the 100% setting at any altitude. The mask, when properly adjusted, is designed to provide an airtight seal. You can use a diluter-demand system safely up to 35,000 feet.

Flights above 35,000 feet require you to use a special oxygen system that delivers pressurized oxygen to your lungs. The pressure-demand system typically is used in jet and some high-performance turboprop aircraft. It differs from the continuous-flow and diluter-demand systems because oxygen is supplied under positive pressure. Therefore, you must exhale against the pressure of the incoming oxygen.

OXYGEN SUPPLY

An important preflight consideration is determining how long your aircraft's oxygen supply will last. The duration of an oxygen supply depends on the amount of oxygen available and the total consumption rate. You measure the amount of oxygen available in terms of oxygen pressure within the storage cylinder, then you determine the total consumption rate by adding the consumption rate of each user. Measure the total rate in terms of the amount of pressure in pounds per square inch consumed per hour. Your aircraft flight manual provides oxygen consumption and system duration data in the form of tables or graphs. [Figure 9-12]

The graph method of presenting oxygen duration data has a slight disadvantage, because it is accurate only if you use the red plug-in connector and all the passengers use gold plug-in connectors. You may encounter other methods of presenting oxygen duration, but the ready use of sample calculations or tables makes the computation of oxygen duration relatively simple.

HYPOXIA

Hypoxia is caused by a shortage of oxygen; a typical symptom is euphoria, or a false sense of security.

Hypoxia occurs when the tissues in your body don't receive enough oxygen. **Hypoxic hypoxia** occurs when there is a lack of available oxygen in the atmosphere. It is considered to be the most lethal factor of all physiological causes of accidents. It can occur very suddenly at high altitudes during rapid decompression, or it can occur slowly at lower altitudes when you are exposed to insufficient oxygen over an extended period of time. The symptoms of hypoxia vary with the individual. Some of the common ones include an increase in breathing rate, lightheaded or dizzy

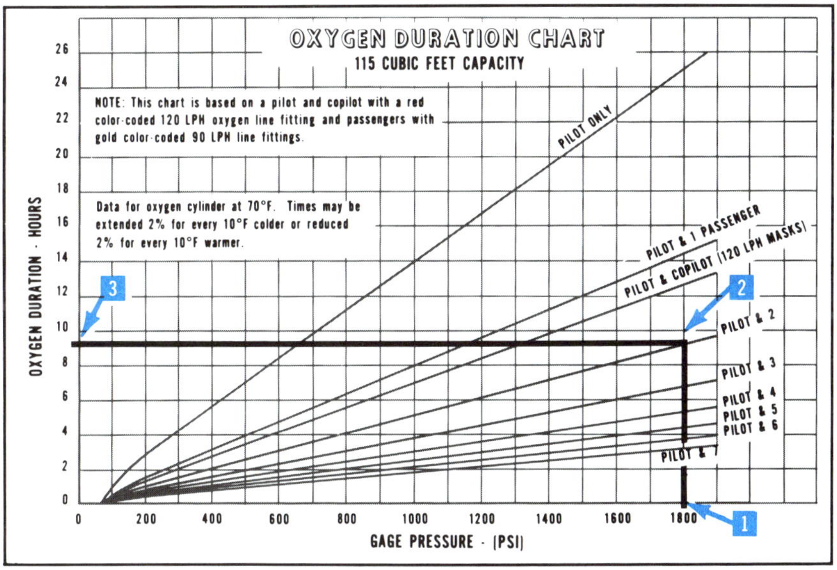

Figure 9-12. This oxygen duration graph eliminates the need for mathematical computations and enables you to read the system duration directly. You need to know only the pressure in the oxygen cylinder and the number of passengers using the system. If, for example, you are planning a trip carrying two passengers and the oxygen cylinder pressure is 1,800 p.s.i., use the following method to determine how long the oxygen supply will last. Locate the intersection of the vertical line on the graph representing 1,800 p.s.i. (item 1) and the diagonal line representing the pilot and two passengers (item 2). Then move to the left to determine the oxygen duration of a little more than nine hours (item 3).

sensation, headache, sweating, tingling or warm sensation, blue fingernails and lips, reduced visual field, sleepiness or frequent yawning, impaired judgment, a slowing of the decision-making processes, a feeling of euphoria (false sense of security), and changes in personality traits.

When your blood is not able to carry a sufficient amount of oxygen to the cells in your body, a condition called **anemic hypoxia** occurs. As the name implies, this type of hypoxia is a result of a deficiency in the blood (anemia), rather than a lack of inhaled oxygen. Anemia can occur as a result of excessive bleeding, a stomach ulcer, or a diet deficiency. It also can occur when oxygen is not able to attach itself to hemoglobin.

Recovery from hypoxia usually occurs rapidly after a person has been given oxygen. However, if you have suffered severe hypoxia, your mental and physical performance may be reduced for several hours.

SUPPLEMENTAL OXYGEN

To avoid the effects of hypoxia, do not fly for prolonged periods above 10,000 feet MSL during the day or 5,000 feet MSL at night without breathing supplemental oxygen.

Since judgment and rationality can be impaired when you are suffering from hypoxia, prevention is the best approach. There are two ways to prevent hypoxia. First, you can fly at low altitudes where hypoxia will not be a factor. This, of course, is not always practical. The most effective way to prevent hypoxia is through the use of supplemental oxygen. If you are planning a flight with a cruise altitude over 12,000 feet, you should consult these regulations. As a general rule, consider using supplemental oxygen when you fly above 10,000 feet during the day or above 5,000 feet at night.

HYPERVENTILATION

A rapid and deep breathing rate, or hyperventilation, depletes the supply of carbon dioxide in your body.

Hyperventilation is the term used to describe a breathing rate that is too rapid and too deep. This process forces too much carbon dioxide from your body and creates a chemical imbalance in the blood. In severe cases it can lead to unconsciousness. It is possible to suffer from hyperventilation even if you are using supplemental oxygen.

Hyperventilation usually is an involuntary response to a stressful situation when you are tense, anxious, apprehensive, fearful, or overworked. Since many symptoms of hyperventilation are similar to those of hypoxia, it is important to correctly diagnose and treat the proper condition. If you are using supplemental oxygen, check the equipment and flow rate to ensure you are not suffering from hypoxia.

Some of the symptoms of hyperventilation include dizziness, tingling of the fingers and toes, muscle spasms, coolness, drowsiness, weakness or numbness, rapid heart rate, apprehension, mental confusion, and a loss of consciousness. The treatment involves restoring the proper carbon dioxide level in the body. Breathing normally is both the best prevention and the best cure. In addition to slowing the breathing rate, you also can breathe into a paper bag or talk aloud to overcome hyperventilation. Recovery is usually rapid when the breathing rate returns to normal.

CABIN PRESSURIZATION

When an aircraft cabin is pressurized, you can fly at much higher altitudes without using supplemental oxygen. You will find that this is much more comfortable for both you and your passengers. Although the individual systems may differ, the basic principles of operation are similar.

These systems pressurize the cabin to some value above the outside air pressure and then maintain that pressure throughout a wide range of altitudes. Two standard types of systems are in general use. A basic system provides pressurization only above a preset altitude, and a controllable system allows you to control the altitude above which pressurization occurs.

A basic system automatically begins pressurizing the cabin as the aircraft climbs through a preset altitude. Although this altitude may vary with different manufacturers, 8,000 feet is commonly used. As you continue to climb, the cabin pressure altitude remains at 8,000 feet until you reach approximately 20,000 feet. If you continue the climb above this altitude, the cabin pressure slowly decreases; however, a cabin pressure regulator maintains a differential pressure up to the aircraft's service ceiling. The maximum differential pressure is normally in the range of 3.35 to 4.5 p.s.i. [Figure 9-13]

STANDARD ATMOSPHERIC PRESSURES			
Altitude (ft.)	**Pressure (p.s.i.)**	**Altitude (ft.)**	**Pressure (p.s.i.)**
Sea Level	14.7	16,000	8.0
2,000	13.7	18,000	7.3
4,000	12.7	20,000	6.8
8,000	10.9	24,000	5.7
10,000	10.1	26,000	5.2
12,000	9.4	28,000	4.8
14,000 ← 2 →	8.6	30,000	4.4 ← 1

Figure 9-13. You can figure your cabin pressure altitude if you know your cruising altitude and the differential pressure and you have a chart similar to this. For example, assume you are cruising at FL300 and the differential pressure is 4.2 p.s.i. At 30,000 feet, the standard pressure is 4.4 p.s.i. (item 1). Adding this to the differential pressure of 4.2 p.s.i. gives a total pressure of 8.6 p.s.i., and a cabin pressure altitude of approximately 14,000 feet (item 2).

The basic system is easy to operate. You simply turn the pressurization controls on before your flight; no other actions are necessary. Use of the basic system is limited, however, because you cannot obtain pressurization below the altitude set by the manufacturer.

A controllable system is quite different. You can set the altitude at which pressurization begins. For example, if you are departing from a sea level airport, you can start pressurization at traffic pattern altitude. The controllable system will maintain this altitude until you climb to the altitude at which the maximum differential pressure occurs. Above this height, the cabin altitude is determined by the relationship between atmospheric pressure and the maximum differential pressure.

With a controllable system, you normally set the cabin altitude selector 1,000 feet above the field elevation of the departure or destination airport, whichever is highest. Then, turn on the pressurization switch and close the dump valve so the system will operate automatically during climb and descent. Additional components of a controllable system include a cabin altitude and differential pressure indicator and a cabin rate-of-climb indicator. [Figure 9-14]

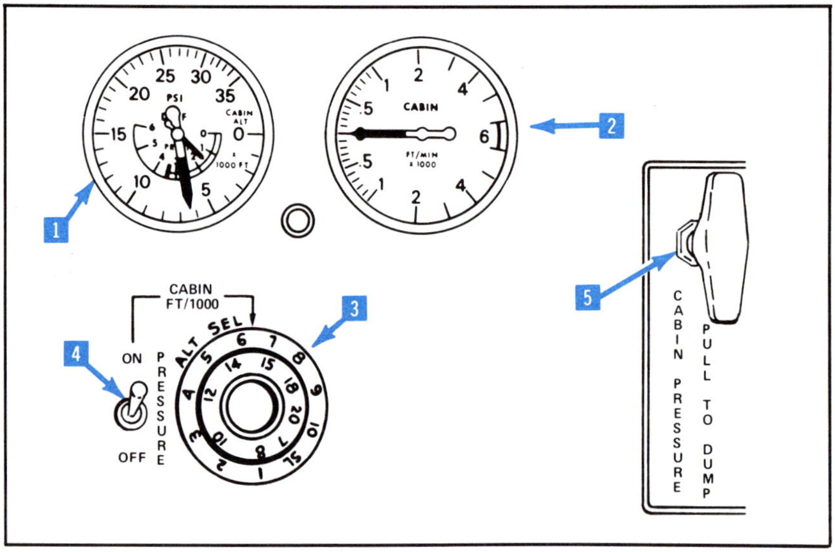

Figure 9-14. In this example of a 3.35 differential pressure system, the outer scale on the cabin altitude and differential pressure indicator (item 1) shows the pressure altitude in the cabin. The inner scale indicates the pressure differential between the cabin and the atmosphere. The cabin rate-of-climb indicator (item 2) shows a climb or descent when you are below the selected altitude or above the maximum differential pressure altitude. At the current setting, the airplane can maintain a cabin pressure of about 6,300 feet while it is at or below 14,500 feet. The cabin altitude selector (item 3) allows you to select the altitude at which pressurization begins. An electrical switch (item 4) controls the safety dump valve. When it is in the ON position, it closes the valve and pressurizes the cabin. When the pressure dump valve (item 5) is pulled, it opens and allows pressurization air to flow overboard.

The pressurization system on most turbocharged aircraft uses compressed air from the compressor side of the turbocharger. Since the compression of air in the turbocharger generates considerable heat, a heat exchanger cools the air to the desired temperature before it actually enters the cabin. The heat exchanger is exposed to the outside air by an airscoop or inlet. After the hot, compressed air is cooled in the heat exchanger, system components route it to the cabin. During cold weather operations, a mixer valve routes hot, compressed air directly from the turbocharger to the blower of the ventilation system rather than the heat exchanger. From the ventilation blower, air is routed through the cabin heat and defroster outlets.

Additional safety features normally include an outflow valve and a safety valve. The outflow valve functions automatically and exhausts "used" cabin air at a controlled rate to maintain the desired cabin altitude. You control the safety valve from the instrument panel and use it for normal decompression. It also prevents the cabin altitude from exceeding the preset differential between the cabin air and outside air in the event the outflow valve fails in the closed position.

With all of the safety devices on pressurization systems, rapid decompression, or even a gradual loss of cabin pressure, seldom occurs. In spite of this, you must consider the consequences of a rapid decompression at high altitudes. **Time of useful consciousness** is the term used to describe the maximum time you have to make a rational, lifesaving decision and carry it out after a loss of oxygen. [Figure 9-15]

TIME OF USEFUL CONSCIOUSNESS		
Altitude	**While Sitting Quietly**	**During Moderate Activity**
40,000 Ft.	30 Sec.	18 Sec.
35,000 Ft.	45 Sec.	30 Sec.
30,000 Ft.	1 Min. and 15 Sec.	45 Sec.
25,000 Ft.	3 Min.	2 Min.
22,000 Ft.	10 Min.	5 Min.
20,000 Ft.	12 Min.	5 Min.

Figure 9-15. Time of useful consciousness varies with altitude and your level of activity. For example, at a cabin pressure altitude of 20,000 feet without supplemental oxygen, it is 5 to 12 minutes.

Above 30,000 feet, your average time of useful consciousness after a rapid decompression is around 30 seconds.

Above 30,000 feet, your time of useful consciousness decreases to a minute or less. In fact, following a rapid decompression, it is approximately 30 seconds. After this, you may not be able to place an oxygen mask over your face, even if you try. The primary danger of rapid decompression is hypoxia. If you experience a rapid decompression at a high altitude, immediately don your oxygen mask, select 100% oxygen, and make a rapid descent to an appropriate lower altitude.

ICE CONTROL SYSTEMS

Ice control systems are used in high-performance aircraft to remove or prevent the buildup of ice. **De-icing equipment** removes ice after it is formed, while **anti-icing equipment** prevents ice formation. Some of the systems that may be available to you are as follows:

1. Propeller (alcohol or electrical)
2. Wing and tail surfaces (pneumatic boots or hot air)
3. Windshield, pitot tube, and fuel vent heaters (electrical)

Shielded antennas (to ensure adequate reception and transmission) and wing ice lights (to illuminate areas of possible ice accumulation) may also be available for use in conjunction with these systems. The following discussion of de-icing and anti-icing pertains to general operation of systems which are typical of most light/medium aircraft used in general aviation. As is the case with all aircraft equipment, you should be thoroughly familiar with the equipment available on the airplane you are flying.

PROPELLER ICE CONTROL

The electrical propeller anti-ice system uses heating elements to reduce the adhesion of ice to the propeller. In icing conditions, the heating elements soften the layer of ice forming on the propeller blades. The heating pads usually contain two elements — an inboard heater (near the propeller hub) and an outboard heater. On some aircraft, a timer provides automatic cycling between the two elements. With a manual system, you use these heating elements individually and cycle them to reduce the amount of electrical current required at any one time. Centrifugal force and airflow over the propeller throw the ice from the blades. [Figure 9-16]

Some aircraft use a fluid (usually isopropyl alcohol) to keep the propeller free of ice. This system is based on the prevention, rather than the removal, of propeller icing. [Figure 9-17]

AIRFRAME ICE CONTROL

Most wing de-icing and anti-icing equipment on aircraft fall within two categories — pneumatic boots and hot air. Pneumatic boots break ice from the wing by expanding, and the hot air system prevents ice formation by channeling hot air through the wing's leading edge. Light and

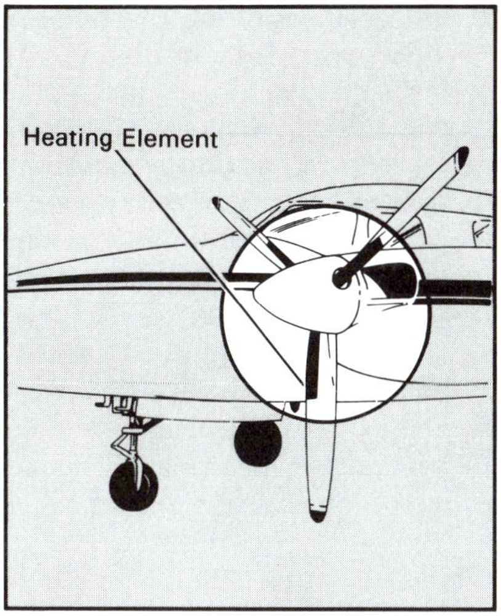

Figure 9-16. The heating elements consist of heater wires enclosed in rubber pads which are attached to the leading edge of the propeller, near the hub. The dull, porous side of the rubber pad is cemented to the leading edge of the propeller blades. The outer portion of the heating pad has a very smooth, glossy surface which reduces ice adherence.

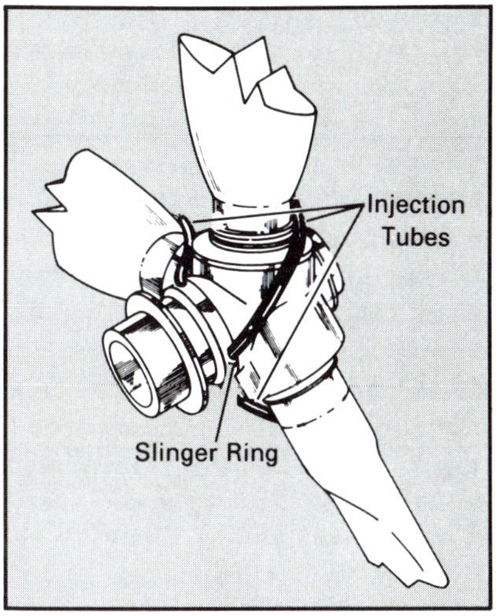

Figure 9-17. Alcohol piped to the propeller slinger ring assembly directs the fluid to the leading edge of the propeller. Alcohol flowing over the propeller's leading edge prevents ice from forming.

medium reciprocating-engine aircraft usually use the pneumatic system, while jet and turbine aircraft normally use the hot-air system.

A typical pneumatic wing de-ice system consists of inflation boots, a timer, pneumatic regulators, and engine-driven pressure and vacuum sources. Inflation boots are fabric-reinforced rubber sheets containing inflation tubes. The inflation boots are cemented to the leading edges of the wings, horizontal stabilizer, and vertical stabilizer. When you turn on the de-ice switch, all of the boots may operate simultaneously or alternately.

During normal operation, a slight vacuum pressure holds the boots in the deflated position. When you energize the system, positive pressure is applied to the inflation tubes. This pressure inflates the boots, separating the ice from the leading edge. The airflow over the airfoil carries the ice away.

Always follow the manufacturer's instructions for use of this type of equipment. As a general procedure, you normally operate the de-ice boots after one-fourth to one-half inch of ice has accumulated on the airfoil's leading edge. If you operate the boots with less ice accumulation, they tend to mold the ice to the new shape rather than breaking it from the airfoil's surface. If this occurs, ice will accumulate on the contour formed by the inflated boot and further operation of the de-ice system will have no effect. [Figure 9-18]

A wing de-ice system may use the same pressure/vacuum source that powers some of the gyro instruments. On twin-engine aircraft, a pneumatic pump is usually located on each engine. In case of a single pump failure, the remaining pump supplies adequate pressure and vacuum to operate the gyro flight instruments and wing de-ice systems.

Hot-air anti-ice systems are commonly installed on turbojet and turboprop aircraft. This is because the jet engine has a ready source of hot air that can be used to heat the wing. In this system, hot air is bled from one of the later stages of the compressor section of the engine and channeled to the aircraft wings. [Figure 9-19]

OTHER ICE CONTROL SYSTEMS

Most aircraft are equipped with a defogger or a defroster consisting of vents which direct hot air across the windshield on the inside of the cabin. Although this system is adequate for some operations, flight in icing conditions may cause ice to adhere to the outside of the windshield, severely restricting your visibility. The electrically heated windshield has been developed for this reason.

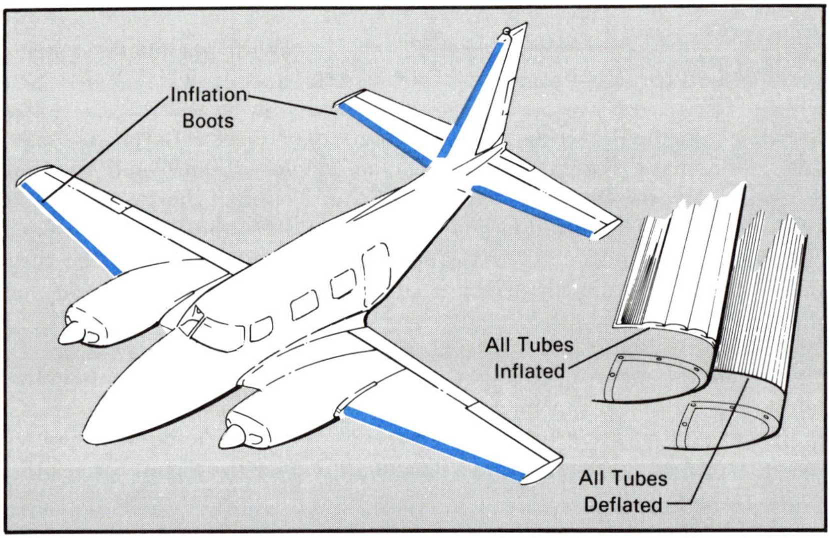

Figure 9-18. When you activate the de-ice switch, the timer energizes the pneumatic pressure control valves for a few seconds, expanding the boots. The timer then automatically de-energizes the control valves, and the pressurized air is vented overboard. A vacuum is reapplied to the boots to hold them in the deflated position, completing the cycle. In some aircraft, if you leave the control switch in the ON position, the timer will repeat the cycle automatically at a preset interval.

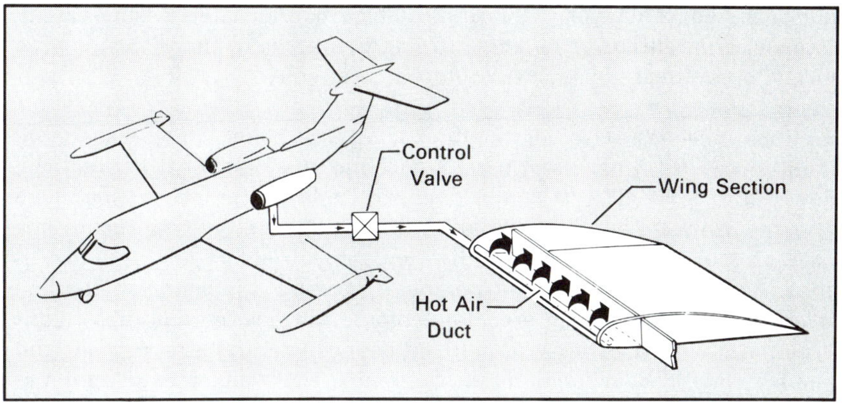

Figure 9-19. The enclosed leading edges become a heated chamber. As the hot air from the engines flows through this chamber, it warms the leading edge to a temperature above freezing, thus preventing ice formation.

In this system, heat wires are either embedded in the windshield or in a piece of transparent material which is bonded to the windshield. In either system, electrical current flowing through the wires heats the windshield and impedes ice formation. To prevent overheating of the windshield, operate the system only during flight. If the windshield does overheat, it may become clouded or distorted and severely limit your visibility.

Certain anti-icing chemicals are also available for general aviation use. These chemicals provide a very smooth, slick surface which ice will not adhere to readily. They are most effective when they are used in conjunction with de-icing systems. For example, wing de-ice boots coated with an anti-icing chemical are more effective than uncoated ones.

Before you fly into known icing conditions, or conditions likely to produce icing, make sure your aircraft is approved for such operation. Even when it is approved for these operations, make sure you are aware of limitations that may apply. In addition, you should also be knowledgeable in the use of the ice control systems in your aircraft and follow procedures recommended by the manufacturer.

Do not fly into known icing conditions unless your aircraft is approved for this type of operation.

Check equipment in flight before you enter possible icing conditions. Make this test in addition to the ground checks that you perform before takeoff. Cycle the wing de-ice boots and visually check for correct operation. During this time, turn on the electric propeller anti-icer and observe the ammeter for proper indications.

Since each piece of icing equipment falls within one of two categories — prevention or removal — you will use slightly different techniques with each system. Turn on the prevention (anti-ice) equipment before you enter icing conditions or at the first indication of icing. This is necessary because most electrical or hot anti-icing equipment must "heat up" to normal operating temperatures before it is effective.

Whether your aircraft is approved for icing conditions or not, you should avoid prolonged flight in an icing environment whenever it is practical. However, if your aircraft is approved for flight in these conditions, you have more time to explore available options. Generally, icing occurs in a zone, so a change in altitude will often get you out of the icing conditions. A descent to a temperature above freezing is recommended when MEAs permit. A climb to a colder temperature also may improve your situation. Freezing rain presents the most severe icing environment and should be avoided. When in doubt, remember that you have the option of turning around and flying to an area where there is no icing. For additional information about weather conditions that may cause icing, refer to Chapter 6, Section B.

CHECKLIST

After studying this section, you should have a basic understanding of:

✓ **Oxygen systems** — What the types and operating principles of each type are, when use of oxygen is necessary, and the hazards of not using it.

✓ **Cabin pressurization** — What the two basic types of systems are, their operating principles, and the significance of time of useful consciousness.

✓ **Ice control systems** — What the difference between de-icing and anti-ice systems is, how propeller, airframe, and other control systems function, and general operational procedures.

✓ **Icing conditions** — What the methods are for avoiding and getting out of icing conditions.

SECTION D

RETRACTABLE LANDING GEAR

Retractable landing gear systems improve airplane performance by reducing the parasite drag created by the wheels and struts extending into the airflow. When the landing gear is retracted, the aircraft is capable of climbing at a greater rate and cruising at a higher speed than a comparable fixed-gear aircraft. You can also use the landing gear as an airbrake during high rates of descent.

During ground operations, nosewheel steering and braking are very similar to those found on a fixed-gear airplane. As with any system, you should be thoroughly familiar with the procedures, limitations, and operating characteristics of the retractable landing gear system on your aircraft. It is very important to follow the recommendations listed in your pilot's operating handbook.

Gear retraction systems are classified according to the power source used for retraction and extension. Electrical and hydraulic actuating mechanisms are used most frequently.

LANDING GEAR SYSTEMS

The **electrical landing gear system** uses a reversible electric motor to drive a series of rods, levers, cables, and bellcranks that extend and retract the gear. In the **hydraulic landing gear system**, the energy that extends and retracts the gear is transmitted by hydraulic fluid through a series of valves, pipes, and actuator cylinders. Some hydraulic systems use an engine-driven pump to supply hydraulic pressure, while others use a separate, electrically controlled pump. The **electrohydraulic system** uses both electrical power and hydraulic pressure to operate the extension and retraction mechanisms. On some electrohydraulic systems, the gear selector operates a reversible electric motor, which controls the direction of hydraulic pressure. Other systems incorporate a hydraulic power pack that is controlled by the gear selector.

A common feature on all retractable landing gear airplanes is an emergency extension system. It allows you to lower the gear if the main electrical or hydraulic system fails. System designs vary among aircraft and can incorporate mechanical handcranks, manually operated hydraulic pumps, or compressed air. On most emergency extension systems, the gear cannot be retracted once it is lowered.

SYSTEM OPERATIONS

On some airplanes, the main gear retracts into the wings, while on others, it retracts into the fuselage. The nose gear usually retracts into the cowling under the engine compartment or forward fuselage [Figure 9-20]

You control the position of the landing gear with a gear handle located in the cabin. When you place the handle in the UP position, the gear retracts. When you place it in the DOWN position, the gear extends. In most systems, you must lift the handle over a safety detent to move it from one position to the other. Position indicator lights are located near the gear handle to help you monitor the system. [Figure 9-21]

Most retractable landing gear systems have a mechanical "squat" switch on the landing gear strut to prevent inadvertent landing gear retraction while the airplane is on the ground. When the strut is compressed by the weight of the airplane, the switch breaks the electrical circuit to the retraction mechanism. This prevents the activation of the hydraulic pump or electric motor until the strut is fully extended.

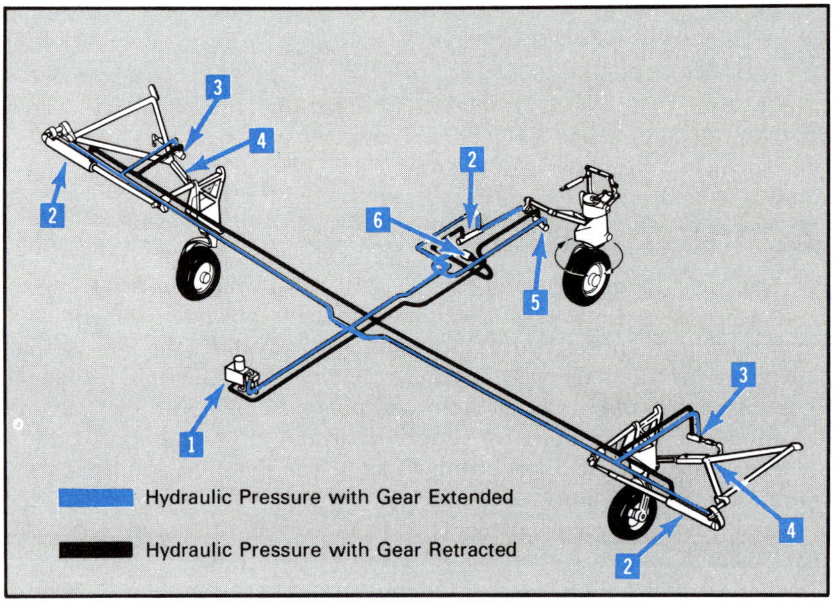

Figure 9-20. This is an example of a typical electrohydraulic retractable landing gear system. Pressure is supplied by a hydraulic pump (item 1) which is driven by a reversible electric motor. Pressure in one direction extends the gear, while pressure in the other direction retracts it. Actuator cylinders (item 2) are used to extend and retract the landing gear. Each gear is held up by hydraulic pressure, but the main gear also has a hydraulically actuated uplock (item 3). Overcenter travel of a spring-held side brace, in conjunction with hydraulic pressure, locks the main gear down (item 4). The nose gear is locked down by overcenter travel of a draglink and a hydraulically actuated downlock (item 5). A pressure dump valve (item 6) is provided for emergency extension of the landing gear.

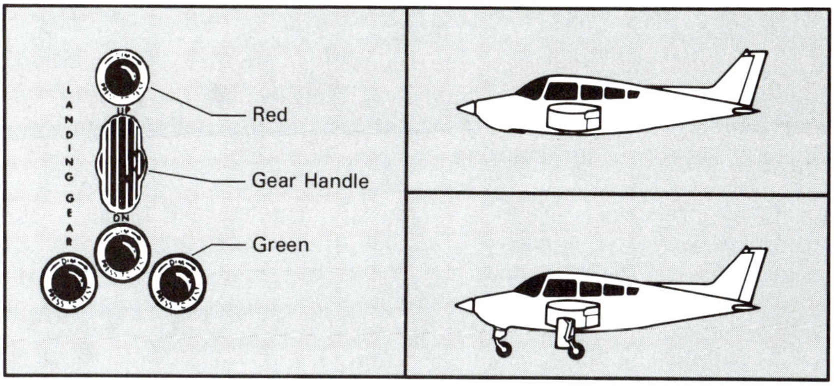

Figure 9-21. A typical system uses three green lights that are on when the respective landing gear is down and locked. A red light is on whenever one or more of the landing gear are in transit or in any intermediate position. When each gear is up and locked, all the lights are off. Each light has a press-to-test feature to verify that it is working.

A warning horn is usually incorporated to alert you when the gear is retracted during an approach. The method used to activate the horn varies with the design of the system. On some airplanes, the horn sounds when the throttle is retarded past a certain point. Other systems use a combination of flap setting, airspeed, and throttle position to activate the horn. Circuit breakers are used to protect the landing gear circuitry. Some airplanes protect the landing gear motor with one circuit breaker, while another protects the remainder of the circuitry, including the gear warning system. Other airplanes may have a single breaker for the entire system. You should not intentionally pull a circuit breaker to silence the warning horn, since you might disable a critical circuit needed for extension, or land gear-up because of a lack of warning.

AIRSPEED LIMITATIONS

Your POH lists the airspeed limitations used during landing gear operations.

During normal operations, you should observe the limiting airspeeds that apply to the landing gear. These speeds are not shown on the airspeed indicator, but are listed in the pilot's operating handbook and on placards in the airplane. The maximum landing gear extended speed, V_{LE}, is the maximum speed at which you can fly an aircraft safely with the landing gear extended. The maximum landing gear operating speed, V_{LO}, is the maximum speed for cycling the landing gear. Some airplanes have a different speed for retracting the gear than they do for extending it. These limitations are due to the additional operating loads placed on the gear mechanism as it opposes gravitational and airstream forces. Since an extended landing gear creates additional drag, it can be used to slow the airplane or begin a descent. For example, when you reach the glide slope intercept point on an instrument approach, lowering the landing gear helps to start a descent without a change in power or airspeed.

OPERATING PROCEDURES

The guidelines presented here are designed to provide general procedures that apply to most retractable gear systems. Since systems vary, you should always follow the procedures specified in your pilot's operating handbook for the airplane.

During the preflight inspection, check the components of the gear for general condition and security. Make sure the wheel wells are free of obstructions and the gear doors are secure. Any accumulation of ice or snow should be removed, since it may prevent normal retraction.

Recycling the landing gear helps prevent ice from adhering to movable parts.

In cold weather, moisture can splash onto gear linkages and freeze. If water, slush, or snow is present on the ramp area, you should taxi the airplane at a slow speed. Avoid isolated areas of water, slush, or snow whenever possible. If moisture is unavoidable, cycle the landing gear several times after takeoff to prevent a solid bond of ice from adhering to movable parts. If gear retraction is prevented by ice formation, return for a landing and have the landing gear deiced.

The retraction point following a normal takeoff should occur when you have a positive rate of climb and you can no longer make a safe landing on the remaining runway. Always follow the manufacturer's recommendations for gear retraction, especially during maximum performance takeoffs. Before you select the gear-up position, gently apply the brakes to stop the wheels from spinning. This prevents tire damage during the retraction into the wheel wells.

A prelanding checklist will help you prevent gear-up landings.

Landing with the gear up when you have a perfectly good landing gear system available is probably the least excusable of all aircraft accidents. The most reliable method to ensure you lower the gear is using the prelanding checklist. This is especially true if you are remaining in the pattern for touch-and-go landings. It also helps to standardize the point during the approach where you normally lower the gear. During VFR operations, this may be located when you are opposite your touchdown point on the downwind leg. On an IFR approach, the glide slope intercept point or final approach fix is usually a good place to lower the gear.

GEAR SYSTEM MALFUNCTIONS

The landing gear position lights are your primary means for determining whether the gear has extended or retracted properly. However, several flight characteristics may provide useful indications of a gear malfunction. When the gear extends, for example, the airspeed decreases and the pitch attitude may change slightly. Conversely, when the gear is retracted, the airspeed increases, assuming the pitch and power are constant. If only one main gear extends, you may notice a change in flight control pressures because of induced yaw. Any one of these characteristics may indicate a gear malfunction.

If a position light does not come on when you move the gear handle down and you notice an airspeed and pitch change, the bulb might be burned out. The press-to-test feature on the position light may reveal the problem. You should also check the landing gear circuit breakers and reset them if necessary. The electric gear motor may not be receiving power due to an overload, or a malfunction in the remaining circuitry may have caused a circuit breaker to trip.

A variety of emergency landing gear systems is available. Some of the most common types are the handcrank, handpump hydraulic, freefall, and carbon dioxide (CO_2) pressurized systems.

The **handcrank system** is commonly found with electrical systems. It uses a crank to turn the gears that operate the actuators. With this system, the crank is turned a specific number of revolutions to fully extend the landing gear. The **handpump hydraulic system** uses a small hydraulic handpump to supply pressure to the actuator cylinders. With some systems, you must place the gear handle in the DOWN position before you operate the pump.

The **freefall system** normally is used on an aircraft employing a hydraulic landing gear system. A control handle is used to open a valve that relieves the system pressure. This permits the hydraulic pressure to equalize on both sides of the actuators and allows the landing gear to fall into position. On some aircraft, a spring assists the gear extension.

In the **carbon dioxide (CO_2) system**, compressed gas supplies the pressure to extend the gear. This system is commonly used in conjunction with hydraulic landing gear. When the CO_2 extension handle is pulled, pressure is released from a cylinder and directed through piping to each landing gear actuator, which extends the landing gear. Some CO_2 cylinders operate only once per flight, so be sure to consult the pilot's operating handbook for specific instructions.

CHECKLIST

After studying this section, you should have a basic understanding of:

- ✓ **Retractable landing gear** — What its advantages are and how they are classified.
- ✓ **Position indicators** — How lights indicate the position of the landing gear.
- ✓ **Warning horn** — How it is activated.

- ✓ **Airspeed limitations —** What they are and how they apply to landing gear operations.
- ✓ **Gear system malfunctions —** What the various indications are and what you should look for.
- ✓ **Emergency gear extension —** What the common types are and how they operate.

CHAPTER 10

COMMERCIAL FLIGHT MANEUVERS

INTRODUCTION

This chapter contains general information on the maximum performance and ground reference maneuvers which are required for commercial pilot certification. Specific procedures and maneuvers that are used to test your competency are outlined in the *Commercial Pilot Practical Test Standards* (PTS). Although some tasks are similar to those in the private pilot standards, a higher level of competency is expected of a commercial pilot. At the completion of your training in each maneuver, you should meet or exceed all PTS requirements.

SECTION A

STEEP TURNS AND CHANDELLES

Steep turns and chandelles are described in this section. Both are required for commercial pilot certification, and are referred to as performance maneuvers. Proficiency in these maneuvers improves your ability to recognize the capabilities, as well as the limitations, of the training airplane you are flying. In addition, they are effective for teaching you the skills and techniques you can use to obtain the best performance from other airplanes.

STEEP TURNS

The steep turn helps you develop the ability to accurately control an aircraft in a high-performance turning maneuver. It also increases your knowledge of the associated performance factors, including load factor, angle-of-bank limitations, effect on stall speed, power required, and the overbanking tendency.

The actual turning performance of an airplane is limited by the amount of power the engine is developing, load limit (structural strength), and aerodynamic design. As you increase the bank angle, you eventually approach maximum performance or the load limit. In most light airplanes, the maximum bank angle you can maintain with full power is 50° to 60°. If you exceed the maximum performance limit while maintaining your airspeed at or below V_A, the airplane will either stall or you will lose altitude. With airspeed above V_A, it is possible to exceed the load limit.

DESCRIPTION

To avoid overstressing the aircraft during steep turns, you must maintain your speed at or below the design maneuvering speed.

During a steep turn, you maintain a bank of 50° through 360° of turn. For most training airplanes, this bank angle is steep enough to cause an overbanking tendency. Turning performance is near the maximum rate, and relatively high load factors are imposed on the aircraft. Because of this, you must be careful to avoid exceeding structural limitations. You should not exceed the aircraft's design maneuvering speed (V_A) during this maneuver.

Steep turns usually are flown with a turn in one direction immediately followed by another turn in the opposite direction. During the early phases of training, you may briefly fly straight and level between the two turns. This allows you to stabilize the airplane before the next turn; however, as you develop proficiency, you will be expected to roll from one turn directly into the other.

PROCEDURE

As is the case with all training maneuvers, you must be aware of other traffic within the area. Before you start the maneuver, make clearing turns to ensure the practice area is free of conflicting traffic. Then, during the maneuver, remain vigilant and avoid other traffic in the area.

You should prepare for steep turns by stabilizing the aircraft altitude, airspeed, and heading. This allows you to do a much better job of precision flying. Using section lines or prominent land features to establish an entry heading will help you roll out on the appropriate heading at the end of the maneuver.

Another important planning factor in any high-performance maneuver is to make sure you have enough altitude to recover in the event a stall occurs. Select an altitude that will permit you to complete the entire maneuver no lower than 1,500 feet AGL. Generally, the best entry results from a smooth roll-in that requires about 25° of heading change. This rule is based on using one-half of the desired bank as the degrees of heading change for roll-in ($50° \div 2 = 25°$). Smoothly apply sufficient power to help maintain level flight in the steep turn. The additional power, in conjunction with a higher pitch attitude, provides lift to balance the increased load factor created by centrifugal force. Remember, an aircraft in a 60° bank must develop sufficient lift to support two times its weight (2 G's). In a 50° bank, an airplane requires about 1.6 G's to maintain altitude. [Figure 10-1]

Enter at an altitude that allows you to complete the maneuver no lower than 1,500 feet AGL.

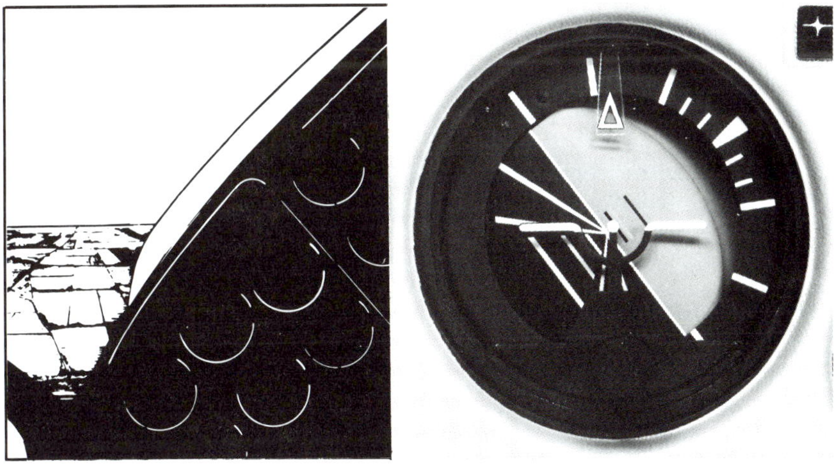

Figure 10-1. The steep turn on the left will result in the instrument indications shown on the right. Notice that approximately a 5° nose-high pitch attitude is required to maintain the level flight attitude.

As you reach the desired bank, briefly check the attitude indicator to confirm the degree of bank, and the altimeter and vertical speed indicators to confirm pitch. You should refer to the aircraft instruments only as a cross-check to confirm the indications from outside visual references.

The overbanking tendency in most light airplanes becomes apparent at higher bank angles. It is caused by the greater lift developed by the wing on the outside of the turn. At a constant bank angle of 50°, you may find that a very slight amount of opposite aileron is necessary to control this tendency.

Some instructors encourage the use of trim to help maintain the correct pitch attitude. This will simplify your task during the turn; however, you must anticipate the need for forward elevator control pressure on recovery and readjust the trim as needed.

Throughout the maneuver, adjust the control pressures, as needed, to maintain a pitch attitude that results in level flight. The easiest way to do this is to select a spot on the windshield directly in front of your eyes that is aligned with the horizon. This spot provides the pitch reference in the same manner as does the dot in the center of the miniature aircraft on the attitude indicator. The pitch change required for the turn is comparatively small. You should not confuse the pitch change with control pressures which change significantly during entry and recovery from the turn.

Since the effects of the bank change are cumulative, make sure you maintain a consistent bank angle during a steep turn. When you vary the bank angle, pitch requirements change and, therefore, the airspeed and load factor also change.

If you allow the bank angle to decrease, the back pressure on the control column that you used for the proper bank angle is too great, and the normal tendency is to gain altitude. Conversely, if your bank angle increases beyond 50°, you probably will lose altitude. Remember to use coordinated control pressures (aileron and rudder) when you adjust the bank angle.

Stalling speed increases significantly during steep turns.

Altitude corrections in steep turns deserve special consideration. If you are losing altitude because of a nose-low attitude, simply pulling back on the control wheel is not a satisfactory correction. Keep in mind that stall speeds increase rapidly with bank angles over 45°. For example, a light airplane that normally stalls at 50 knots in straight-and-level flight will stall at about 70 knots in a level 60° bank turn. Trying to stop a descent by pulling back on the control wheel simply tightens the turn and may cause an accelerated stall. Throughout this maneuver, you should avoid any indication of a stall. In addition, you must stay within the operating limitations of the aircraft.

The proper correction from a nose-low attitude is to decrease the angle of bank first. This normally slows or stops your rate of descent and helps you regain lost altitude. When you have returned to your original altitude, establish a level flight attitude and return to the desired angle of bank. As you increase the bank, apply slightly more back pressure on the control column to maintain your altitude.

You usually can correct for a nose-high attitude by increasing the bank angle slightly; however, you should avoid exceeding 55° of bank. Generally, small changes in your bank angle during steep turns help you maintain the entry altitude. You increase the bank angle to lose altitude and decrease the bank to gain altitude.

Recovery from a steep turn should be smooth and precise, using the approximate rate of roll that you used for entry. As you decrease the bank, you must release enough back pressure to maintain altitude. If you release the control pressure in proportion to the rate of roll, the spot on the windshield will seem to move downward in relation to the horizon, and your altitude will remain constant throughout the recovery. Plan the recovery so you roll the wings level as the aircraft reaches the original heading.

During a roll from one turn to another turn in the opposite direction, deflect the ailerons at a constant rate. As you begin the roll, gradually reduce the back pressure until the aircraft is in wings-level flight. As the aircraft passes through wings-level flight, begin to increase aft control pressure gradually to adjust the pitch attitude and maintain altitude. You will change the pitch attitude slightly throughout the roll from one turn to the other.

Spatial disorientation may occur during initial training because of the effect of the rapid turn rate on the inner ear. After adequate exposure, you will usually develop some degree of immunity to this type of spatial disorientation.

CHANDELLES

The chandelle is another maneuver that requires advance planning, accuracy, coordination, and smoothness. It involves changes in pitch, bank, airspeed, and control surface pressures. When practiced correctly, the chandelle helps you develop good coordination habits and finesse in the use of aircraft controls.

DESCRIPTION

A chandelle may be described as a high-performance climbing turn of 180° with full power. During the maneuver, the aircraft's speed gradually decreases from the entry speed to a few knots above the stall speed. You control the decrease in speed by adjusting the pitch attitude. Even though

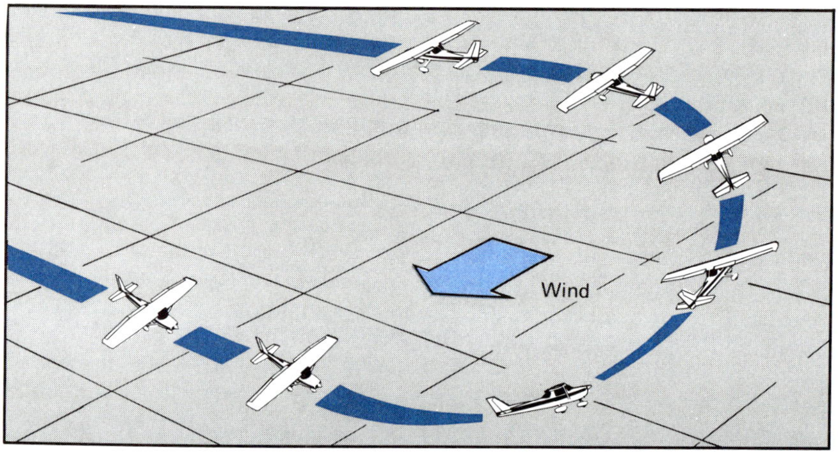

Figure 10-2. Although the prevailing winds have little or no effect on the chandelle, you will find it best to begin the maneuver by turning into the wind. This minimizes wind drift and helps you remain in the practice area. With all training maneuvers, you should make the necessary clearing turns before you start, then continually check the training area for other traffic.

maximum climb performance is an objective of the maneuver, altitude gain is not a criterion. This is due to wide variations in atmospheric conditions and the performance capabilities of different aircraft. [Figure 10-2]

PROCEDURE

During a chandelle, maintain airspeed at or below the design maneuvering speed of the aircraft.

Enter the chandelle from level flight at cruising speed with the flaps and gear (if retractable) in the UP position. Select a prominent feature on the ground to help you maintain orientation. If the cruising speed of the aircraft is higher than the maneuvering speed, slow the aircraft to maneuvering speed or the recommended entry speed, whichever is less. If you enter a chandelle at or below V_A, you should not exceed the limit load factor. In fact, the load factor during a chandelle with a maximum bank angle of 30° should be less than 1.5 G's. Make sure the entry altitude is 1,500 feet AGL, or above. [Figure 10-3]

FIRST 90° OF TURN

Although you may use more bank in some airplanes, 30° is recommended for most light aircraft. First, establish a 30° bank. Then, apply back pressure to the elevator control to begin the climb, and add power. In airplanes with a fixed-pitch propeller, use full power, but add it gradually to avoid exceeding the maximum allowable r.p.m. If you are flying an airplane with a constant- speed propeller, increase r.p.m. to the takeoff or climb setting before you begin the maneuver, then increase the power as you start the climb. You should increase the pitch attitude at a constant rate once the climb is established. Maintain the 30° bank until the 90° point.

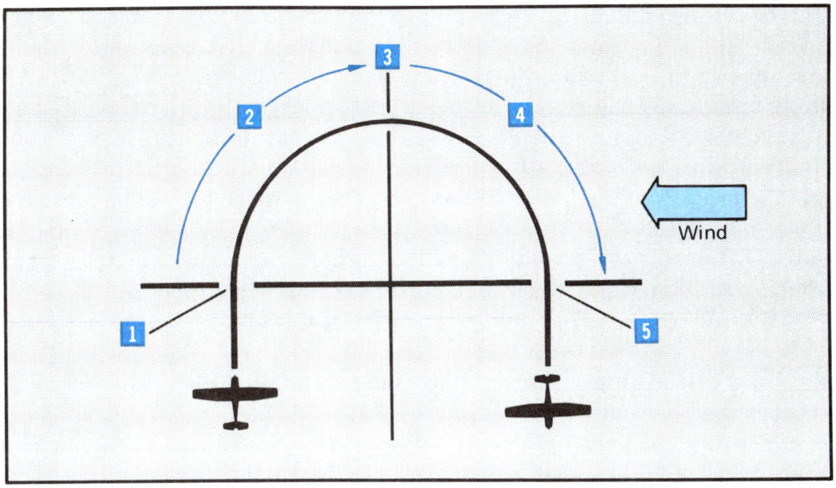

Figure 10-3. You start at position 1 in a level flight attitude and slowly begin to roll into a 30° bank. After you establish the bank, you begin to increase the pitch attitude. At position 2, pitch attitude continues to increase and you maintain 30° of bank throughout the first 90° of turn. As you pass the 90° point (position 3), you maintain the nose-high pitch attitude and slowly begin to roll out of the bank. As you move to position 4, maintain the nose-high attitude and continue the slow decrease in your bank angle. Recovery (position 5) is in a wings-level, nose-high attitude at a speed just above a stall.

You may find that you have a tendency to roll into an initial bank that is too steep, resulting in a loss of performance. With too steep a bank, the aircraft will turn more rapidly and you may arrive at the recovery point before the airspeed has slowed to the desired speed, or you may find that you are rolling all the bank out in the last 10° to 15° of turn. During a turn with an excessive angle of bank, lift that would otherwise result in altitude gain is required to offset the increased bank.

SECOND 90° OF TURN

At the 90° point, you should still be in a 30° bank with a nose-high pitch attitude. Actually, this is where you should reach the steepest nose-high attitude. [Figure 10-4]

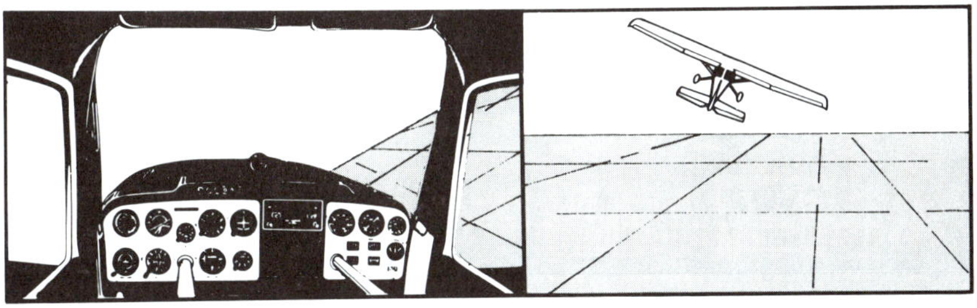

Figure 10-4. This illustration shows the approximate pitch and bank attitude at the 90° point in a chandelle to the right.

The left-turning tendency becomes more pronounced as airspeed decreases and your angle of attack increases.

During the second 90° of turn, time the roll-out rate so you reach wings level at the 180° point. You should maintain a constant pitch attitude throughout the second 90° of turn. To maintain this pitch attitude, slowly increase back pressure. As speed decreases, the elevators become less effective and, as a result, you need more elevator control deflection. You will also notice that the left-turning tendency caused by P-factor and the propeller slipstream is more prevalent, and right rudder pressure is needed to coordinate both right and left turns. This is particularly important during the second 90° of turn, where you are at a high angle of attack with a high power setting.

If your pitch attitude is too steep, you may stall the aircraft before it reaches the recovery point. With a pitch attitude that is too low, you will arrive at the recovery point with an airspeed well above stalling speed. Since you use full power throughout the last part of the chandelle, you must control airspeed by the pitch attitude of the aircraft. Because of this, maintaining the proper pitch attitude is a key element of this maneuver.

At the 180° point, hold the nose-high pitch attitude momentarily, and then gradually lower the nose to a level flight attitude. Allow the aircraft to accelerate while maintaining a constant altitude. [Figure 10-5]

In a chandelle to the right, the aileron on the right wing is lowered slightly during the roll-out. This causes more drag on the right wing and tends to make the airplane yaw slightly to the right. At the same time, the left-turning tendency is pulling the nose to the left. As a result, aileron drag and the left-turning tendency counteract each other and very little left rudder pressure is required. In contrast, when you roll out from a chandelle to the left, two left-turning forces are pulling the nose of the airplane to the left. In this case, you need a significant amount of right rudder pressure.

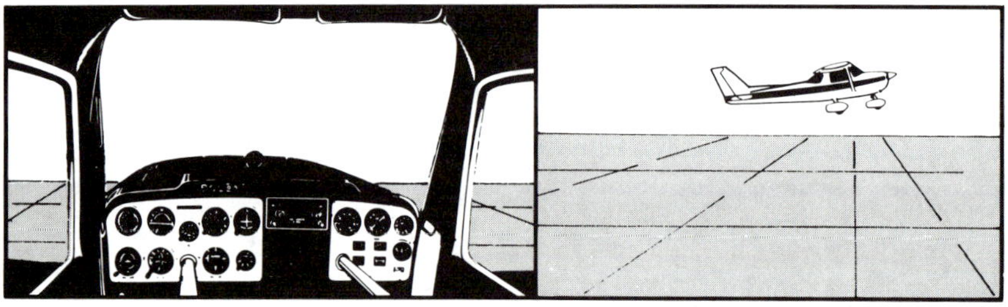

Figure 10-5. At the 180° point, the wings are level and the airspeed is just above the stall speed. You should complete the recovery with a minimum loss of altitude and without stalling the airplane.

CHECKLIST

After studying this section, you should have a basic understanding of:

- ✓ **Steep turns —** How performance factors, such as the recommended airspeed, load limit, power limitations, the effect of bank angle on stall speed, and the overbanking tendency apply.
- ✓ **Entry and recovery procedures —** How to enter and recover from steep turns.
- ✓ **Chandelles —** What the main performance considerations are and how to achieve the desired performance.
- ✓ **Entry and recovery procedures —** How to enter and recover from chandelles.

SECTION

B

STEEP SPIRALS

Like other maximum performance maneuvers, the steep spiral is seldom used during routine flight operations. However, it is a practical training exercise that combines many elements of maximum performance with the requirements of ground reference maneuvers. The steep spiral is valuable for teaching you coordination, planning, orientation, aircraft control, corrections for wind effect, and division of attention. Although not required for commercial pilot certification, you will see that the steep spiral may also be a useful procedure during some situations, such as an emergency descent.

DESCRIPTION

The maximum angle of bank during a steep spiral is 50° to 55°.

The steep spiral is simply a steep bank held for a specified number of gliding turns. To compensate for wind drift, you vary the angle of bank as necessary; however, you should use no more than 50° to 55° of bank at the steepest point in the maneuver. [Figure 10-6]

PROCEDURE

Since you will be descending throughout the steep spiral, it is important to clear the area below as well as around the aircraft. A good way to accomplish this is to perform clearing turns and observe the affected airspace as you approach the entry position. You should continue to watch for other aircraft throughout the maneuver.

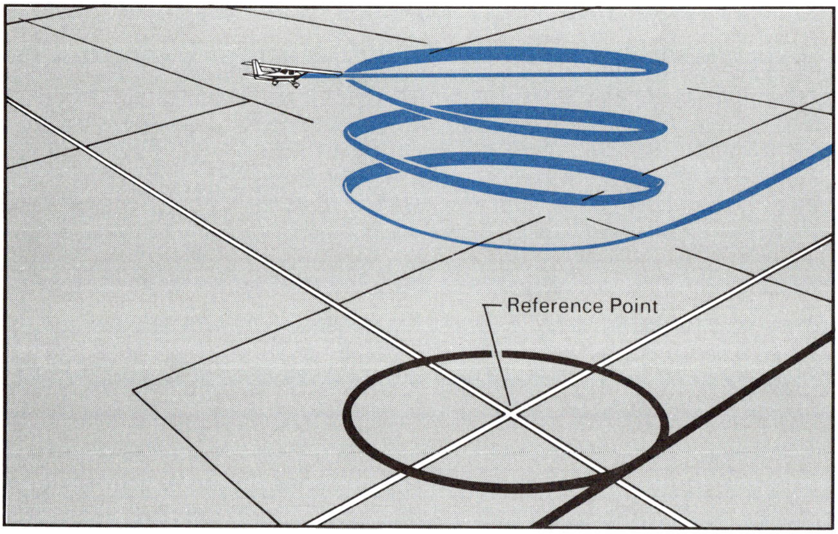

Figure 10-6. During the spiral, you maintain a uniform radius about a reference point on the ground, such as an intersection of roads or fields.

Before attempting this maneuver, you should have sufficient altitude to make at least three 360° steep spirals (1080° of turn) and recover at a safe altitude above the ground. The altitude required to complete the full 1080° of turn will vary with different types of aircraft, as well as with the density of the air. Airspeed at the entry point should be stabilized at the recommended entry speed, which is usually well above normal stall speed of the aircraft. This speed provides you with a solid, responsive feel of the flight controls. Maintain the entry speed throughout the spiral.

Enter at an altitude that will allow you to complete the maneuver and still recover at a safe altitude.

Although you can enter this maneuver from any position, enter downwind and slightly to one side of your ground reference point when practical. If the entry is downwind, the steepest angle of bank will occur at the beginning of the maneuver. This allows you to adjust your bank to control the radius of turn as necessary for a uniform circle around the reference point. Enter the maneuver slightly offset to the right of the reference point for a left spiral and offset to the left for a right spiral. You should practice the maneuver both ways.

If you know the wind direction, enter downwind so the steepest bank angle will be required during the first part of the maneuver.

Anytime you are flying in a steep bank with consecutive turns, you are subject to disorientation. Outside visual references are vital for good orientation. Select a prominent landmark near your entry heading, then count the number of times you pass it. Outside visual cues are also useful for your bank and pitch attitude. During your initial training in this maneuver, you can use shallower bank angles and a larger radius of turn to help with orientation.

Outside visual references help you remain oriented during the steep spiral.

As you approach the entry point, apply carburetor heat (if required) and close the throttle. Prolonged operation of the engine at idle may cause excessive cooling and spark plug fouling. You should periodically clear the engine by briefly advancing the throttle to the cruise power range. When you advance the throttle, make an adjustment to your pitch attitude so your airspeed remains constant. Normally, you should clear the engine while you are headed into the wind. This minimizes groundspeed variations and helps you maintain a uniform radius of turn.

Remember to clear the engine periodically during the descent.

With the power at idle, you control airspeed by adjusting your pitch attitude. As you roll into the bank, you must lower the nose to maintain the correct airspeed. During the spiral, establish a pitch attitude that provides the desired speed, then use a reference spot on the windshield to maintain that attitude. Refer to the airspeed indicator periodically as a cross-check. Airspeed control is important because your radius of turn for a given bank angle changes as airspeed changes. If you concentrate too heavily on wind drift corrections or bank control, you will find yourself constantly changing your pitch attitude. When you raise or lower the nose of the aircraft, the airspeed changes appreciably.

Vary your bank angle, as necessary, to maintain a constant radius of turn around the ground reference point.

Since you must maintain a constant radius about the selected point, wind drift corrections are essential. You maintain the proper radius of turn by varying your angle of bank, as in the turns around a point and S-turns. You have the highest groundspeed on the downwind side and the slowest groundspeed on the upwind side of the turn. [Figure 10-7]

As you change your bank angle during the spiral, you will notice that some variation in control pressure is necessary to maintain the correct airspeed. For example, on the downwind side, you need a steeper bank angle because of a higher groundspeed and more back pressure to maintain your airspeed. On the upwind side, you need less bank because of a lower groundspeed and not quite as much back pressure.

As indicated earlier, a constant altitude turn with 60° of bank requires a load factor of 2 G's. Since the steep spiral is not a constant altitude

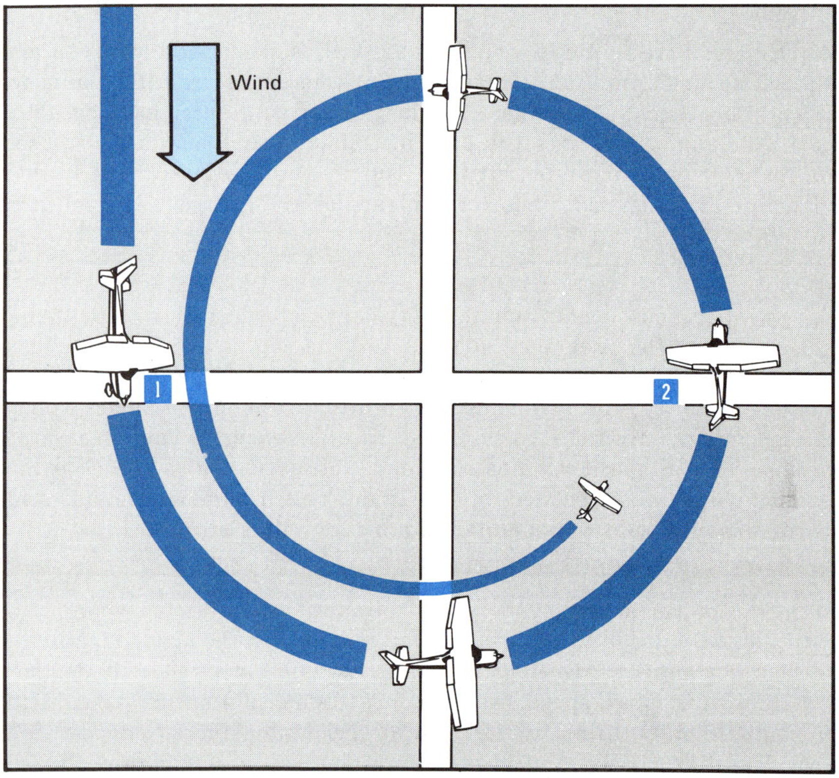

Figure 10-7. The aircraft at position 1 is downwind and, as a result, it requires the steepest bank. From position 1 to position 2, you gradually decrease the bank angle from the steepest to the shallowest point as you fly upwind. As you continue around toward 360° of turn, you slowly increase the bank until it is once again at the steepest point when you are abeam position 1.

maneuver and your maximum bank is 55°, less than 2 G's is exerted upon the aircraft. It is possible, however, for the load factor to increase significantly, depending on how you use the controls. If you let the airspeed build up and then try to slow the aircraft by pulling back on the control wheel, the load factor can increase rapidly. Pulling back on the control column when you are in a steeply banked, descending spiral usually tightens the spiral, while airspeed and rate of descent may continue to increase. The correct procedure is to decrease the angle of bank before you apply back pressure. In an extreme case, with high airspeed and an improper control technique, you can exceed the critical load factor, which is 3.8 G's for a normal category airplane. If your airspeed is below V_A, you will not exceed the critical load factor, but you can inadvertently stall the aircraft. Remember, an increase in load factor is characteristic of maneuvers with high bank angles, and the associated increase in the stall speed can cause an accelerated stall. [Figure 10-8]

To recover from a steep, nose-low spiral where airspeed is increasing, correct the bank attitude first, then raise the nose of the airplane.

LANDING APPROACH

If you lose power and have sufficient altitude, you can use a steep spiral to position your aircraft for an emergency or forced landing. You will find that it is more practical and easier to dissipate altitude as you descend over the point where you intend to land than it is to make S-turns or other maneuvers away from the landing area. In the event of a power failure, fly directly to a point over your planned landing site and then begin the steep spiral. [Figure 10-9]

Load Factor	Increase In Normal Stall Speed
1.0	0%
1.5	22%
2.0	41%
2.5	58%
3.0	73%
3.5	87%
4.0	100%

Figure 10-8. An aircraft with a 2-G load factor has a stall speed 41% greater than normal. With a 3-G load factor, the stall speed increases by 73%. In terms of airspeed, this means that an aircraft with a normal indicated stall speed of 53 knots will stall at 75 knots with a 2-G load factor and at 92 knots with a 3-G load factor.

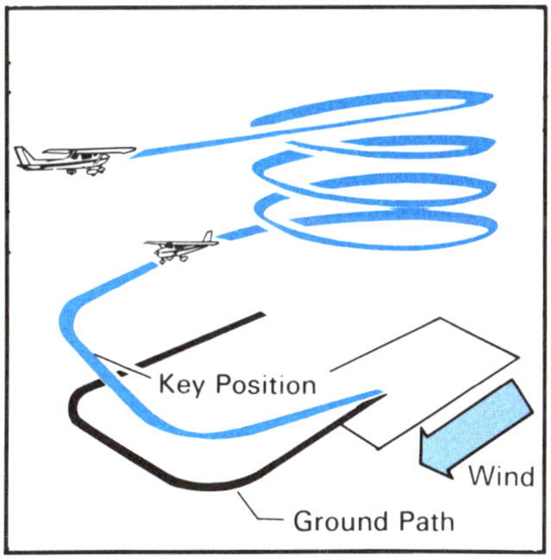

Figure 10-9. If you do not know the surface wind conditions, make an estimate of the wind direction and velocity during the steep spiral. Roll out of the spiral so you are positioned on downwind, like you would be for a normal approach. Plan to arrive at the key position on base leg with sufficient altitude to tly the final approach into the wind.

During a steep spiral descent for an emergency landing, inspect the landing surface.

The spiral approach also gives you a chance to inspect the landing surface for obstructions or other hazards before you are committed to an approach. Once you are on the final approach, it usually is too late to change your mind and try to reach another field.

RECOVERY WITHOUT A LANDING APPROACH

When you are practicing a steep spiral without a landing approach, recover toward a definite object or on a specific heading. For example, you may plan to recover on a heading which leads into a pattern over an area that could be used for a forced landing. When you roll out to resume a straight glide, smoothly coordinate flight control and power changes, as required, to maintain your airspeed and a safe altitude.

In summary, the steep spiral is a high-performance maneuver that also includes the requirement to fly a precise flight path around a reference point on the ground. It is a useful training maneuver, and, in some cases, it may be a practical way to perform an emergency descent.

CHECKLIST

After studying this section, you should have a basic understanding of:

- ✓ **Steep spiral —** How to describe the maneuver and what the basic requirements are.
- ✓ **Steep spiral landing approach —** When and how you might fly this type of approach for an emergency or forced landing.
- ✓ **Steep spiral without a landing approach —** What the primary requirements are to perform this maneuver as a training exercise.

MAXIMUM PERFORMANCE TAKEOFFS AND LANDINGS

As a commercial pilot applicant, you will be required to demonstrate competency in a variety of takeoffs, climbs, approaches, and landings. This is necessary because many airports have runways with limited length, obstructions in the departure path, or soft surfaces. To prepare you for these types of operations, short-field and soft-field procedures are covered in this section. Since elements of maximum performance are involved, we have grouped both short-field and soft-field operations under the title "maximum performance."

SHORT-FIELD TAKEOFF AND CLIMB

The procedures listed here are general. You should always follow your airplane manufacturer's recommendations for use of flaps, mixture/propeller settings, application of power, brake release, and any other procedures that apply to that airplane. When you use performance information from the approved flight manual or POH, keep in mind that the figures are valid only for the listed conditions.

DESCRIPTION

A short-field takeoff and climb requires you to obtain a high level of performance from your airplane. You must lift off in a minimum distance and climb at the best angle. This procedure may be necessary when you have either a short runway or an obstructed departure path. Sometimes both conditions may exist.

Be sure to consider runway surface conditions and all the other variables that affect takeoff performance, such as wind, temperature, pressure altitude, and takeoff weight. You also need to know exactly how much runway is available, as well as the height and location of any obstructions. This information is required before you can properly estimate takeoff performance.

Before you attempt a short-field takeoff and climb, make sure the procedure is safely within the performance capabilities of your airplane.

In addition, you must understand the significance of the rotation speed, best angle-of-climb speed (V_X) or the recommended obstacle clearance speed, and the best rate-of-climb speed (V_Y) for the airplane you are flying. In some circumstances, a variation of five knots from the optimum speed may make the difference between clearing an obstacle or not clearing it. Knowledge of the key airspeeds and your ability to maintain them is very important.

Under adverse conditions, you may have to postpone the departure or, at least, decrease your takeoff weight. A good practice is to be conservative whenever runway limitations are such that you may not be able to take off and clear all obstacles in the departure path.

PROCEDURE

Proper use of the appropriate checklist helps you remember to complete all critical tasks.

When aircraft performance data indicate you can safely take off and climb under the existing conditions, follow the pretakeoff checklist. Select the recommended flap setting and take advantage of all the available runway. Do not take off immediately after a sharp taxi turn, because the fuel level in the tanks may not be stabilized.

For a short-field takeoff and climb, you normally use partial flaps, hold the brakes, apply full power, and then release the brakes. This allows you to check engine operation before you begin the takeoff roll. After brake release, maintain positive directional control on the runway centerline. Use nosewheel steering until the rudder becomes effective. [Figure 10-10]

After you reach V_X or the recommended speed, increase the nose-high pitch attitude slightly to maintain that speed. When you are clear of all obstacles, accelerate to V_Y and retract the flaps as recommended. For aircraft with a retractable landing gear, raise the gear after you have established a safe rate of climb and can no longer make a safe landing on the remaining runway. Maintain a track on the extended runway centerline until a turn is required, and use takeoff power until you reach a safe maneuvering altitude. Finally, complete the after-takeoff checklist.

To achieve maximum performance, you must establish and maintain the correct airspeeds. You control the airspeed by adjustments to your pitch attitude.

As noted earlier, precise airspeed control is necessary before you can achieve maximum performance during a short-field takeoff and climb. Since you use full power throughout the takeoff and the first part of the

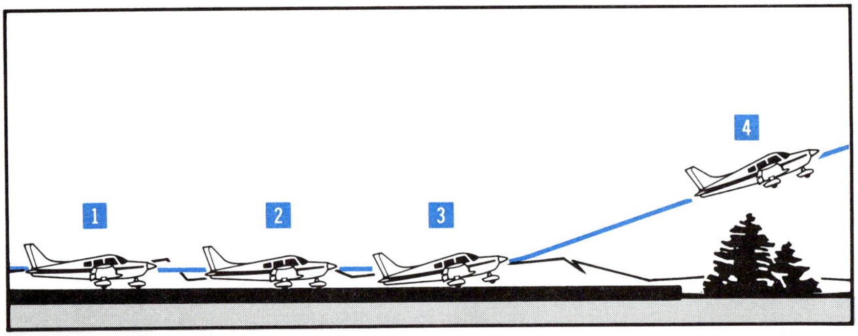

Figure 10-10. Allow the airplane to roll with the full weight on its landing gear in an attitude that results in minimum drag (positions 1 and 2). Smoothly and firmly rotate at the recommended speed to establish the takeoff attitude (position 3). If your pitch attitude is too high, induced drag becomes excessive and your acceleration is significantly decreased. Too low a pitch attitude delays the liftoff and increases the takeoff roll. After liftoff, accelerate to V_X or the recommended speed (position 4), and maintain this speed until you clear all obstacles.

climb, the only way you can change the airspeed after liftoff is to adjust the pitch attitude. Attitude control is critical from liftoff through the climb at V_X.

SHORT-FIELD APPROACHES AND LANDINGS

As the name implies, the short-field approach and landing is for short fields with confined landing areas. The landing area may also be limited by obstacles in the approach path. Under these circumstances, you have to fly the final approach so you can stay within the confined area and minimize the landing roll. To do this, you must touch down very close to the approach end of the runway at a minimum airspeed.

DESCRIPTION

Like the short-field takeoff, this is a maximum performance maneuver. You are required to fly the airplane close to the ground at minimum airspeed. In most cases, you use full flaps for this type of approach, but you should check your POH for specific procedures. [Figure 10-11]

PROCEDURE

A successful short-field approach and landing begins with good planning. General planning factors that you should consider include airspeeds, configuration, obstructions, landing surface, and wind conditions. Based on the existing conditions, you should select a suitable touchdown point.

THE APPROACH

For training, you normally will simulate a short field, as well as a 50-foot obstacle in the approach path. The first part of the approach is normal. Enter downwind, complete the before-landing checklist, and lower the landing gear, if appropriate. You may want to extend the first increment of flaps on the downwind leg, the second on base, and full flaps on final approach, while progressively reducing the airspeed. This provides

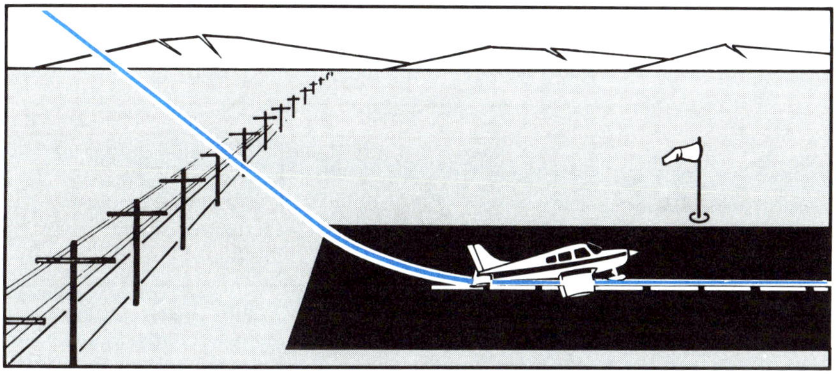

Figure 10-11. You should clear the obstacle with a relatively steep descent angle and touch down near the end of the runway at a slow airspeed.

a smooth transition, in easy steps, to the short-field approach speed. During the transition, use trim to relieve control pressures. When established on final, make sure the aircraft is aligned with the runway centerline. You should also maintain enough altitude to allow a constant angle of descent over the obstacle to your selected touchdown point. Adjust the power and the pitch attitude to establish and maintain the correct airspeed and descent angle.

Since strong, gusty winds may affect aircraft controllability, add approximately 50% of the gust factor to the final approach speed.

Normally, the final approach speed recommended by the manufacturer is not more than 1.3 V_{SO}. If you are flying an airplane that stalls at 50 knots (at the landing weight) with power off, full flaps, and the gear extended, the approach speed should be no more than 65 knots (50 × 1.3 = 65). In gusty wind conditions, you may wish to increase your approach speed to help maintain control. For example, you can add one half of the gust factor to the final approach speed. In other words, if gusts to 20 knots are reported, you would add 10 knots, making the final approach speed 75 knots.

If you maintain adequate airspeed throughout the approach, you can adjust your pitch attitude to establish and maintain the desired rate or angle of descent, and adjust power for the desired airspeed. In most cases, you make a combination of pitch and power adjustments. Once your pitch attitude is established for the correct angle of descent, you may only need small power changes to control your airspeed. Use at least some power until you flare for the landing. [Figure 10-12]

During the approach, be careful to avoid an excessively low airspeed. If it becomes too slow, increasing the pitch attitude and using full power may only result in a higher rate of descent. This occurs when the angle of attack is so great that maximum power is not enough to overcome the associated drag. Under these conditions, you will be flying on the "back side of the power curve," and you cannot decrease the descent rate just by pulling back on the control wheel. Although this is a typical reaction, it will result in a further decrease in airspeed and the requirement for more power. When you are behind the power curve, the only way to regain airspeed is to decrease the angle of attack and descend.

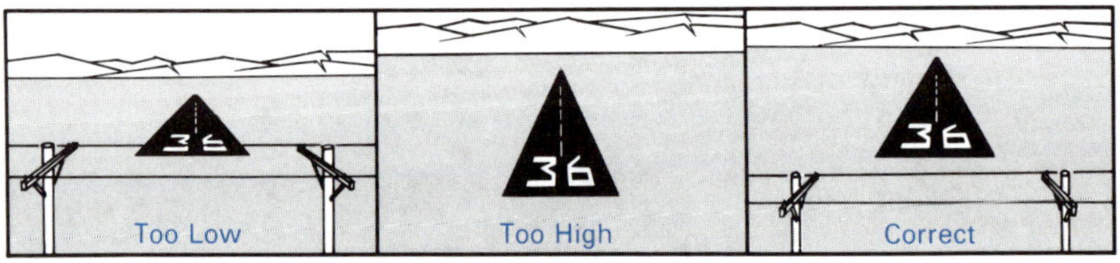

Figure 10-12. If your approach appears like the one on the left, you are too low, and you will need to increase the pitch attitude while simultaneously adding power to slow the rate of descent. The center example indicates you should decrease the pitch attitude and adjust power, as required, because the aircraft is too high. The example on the right shows the correct descent angle.

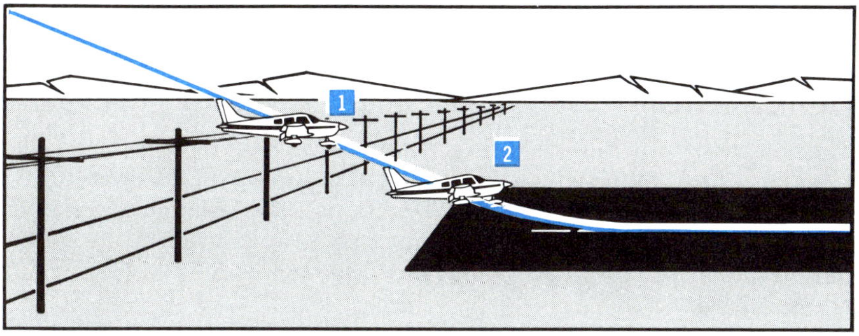

Figure 10-13. After you clear the obstacle, maintain your rate and angle of descent as you approach the runway (position 1). Gradually reduce power and flare at approximately the same height as you do for a normal landing (position 2). If you maintain the correct speed, you will have sufficient elevator effectiveness to control the airplane to the touchdown point, and floating will be minimized.

THE LANDING

As you might expect, the key to flying a precise short-field approach is to maintain a consistent, on-speed descent angle. If your airspeed is too high, you will land well down the runway. If it is too slow, you may stall before your intended touchdown point and land hard. The correct speed and descent angle help you judge obstacle clearance, as well as the touchdown point. [Figure 10-13]

After touchdown, lower the nose as soon as practical and apply maximum effective braking. In most cases, braking is more effective when you apply back pressure to the control wheel. This tends to transfer more of the aircraft's weight to the main wheels. You should avoid locking the brakes or skidding, and you should maintain positive directional control during the landing roll. Follow the manufacturer's recommendations for flap position.

If you touch down at the correct speed and smoothly apply maximum braking, the landing roll will be extremely short. Since you will normally practice on a runway with adequate length, you will probably have to add power to get to the nearest exit taxiway. After you are clear of the runway, remember to complete the after-landing checklist.

Before you make an actual short-field landing, evaluate the feasibility of taking off from that field. Takeoff distances often are greater than landing distances. The time to consider if you can take off after landing on a short field is during preflight planning for the initial flight.

SOFT-FIELD TAKEOFF AND CLIMB

Soft-field takeoff and climb procedures are necessary when the takeoff surface retards acceleration. The surface conditions which normally produce this effect are snow, soft sand, mud, or tall grass. You may also use soft-field procedures to minimize the effects of taking off from a rough surface.

DESCRIPTION

The objective of the soft-field takeoff procedure is to transfer the weight from the wheels to the wings and lift the aircraft clear of the retarding or rough surface as soon as possible. To do this, establish and maintain a fairly high angle of attack or nose-high pitch attitude as soon as possible during the takeoff roll. As a result, the airplane lifts off at a lower than normal speed.

PROCEDURE

Prior to a soft-field takeoff, use enough power to keep the aircraft rolling during all taxi operations.

When the pretakeoff check is complete and the flaps are set as recommended, taxi the aircraft without stopping toward the takeoff position. Use full back elevator pressure and sufficient power to help lift some of the weight from the nosewheel. Before you reach the takeoff position, make sure you check the area for other aircraft and any other hazards in the departure path. If the approach and departure paths are clear, taxi onto the takeoff area as quickly as you can, consistent with normal safety precautions and the surface conditions. Maintain sufficient power to avoid "bogging down." As you align the airplane for takeoff, apply full takeoff power as smoothly and rapidly as the engine will accept it without faltering.

TAKEOFF

Positive directional control is essential during the takeoff roll on a soft field.

Make sure you check the engine instruments as the engine reaches full power. Remember, right rudder is required to counteract the normal left-turning tendency on takeoff. During acceleration, maintain positive directional control and continue to apply full back pressure on the control column to decrease weight on the nosewheel. [Figure 10-14]

As your speed increases and the elevator becomes more effective, decrease back pressure gradually to maintain a constant pitch attitude

Figure 10-14. You begin acceleration in a fairly level attitude (position 1), establish a nose-high pitch attitude to hold the nosewheel clear of the surface (position 2), and lift off in approximately the same attitude (position 3). After liftoff, lower the pitch attitude slightly (position 4), and accelerate to V_X or the recommended speed (position 5) while remaining in ground effect.

with the nosewheel slightly off the ground. If you fail to release some of the back pressure, the aircraft may rotate to an extremely nose-high attitude which increases drag significantly. In some airplanes, the tail skid may contact the takeoff surface if the pitch attitude is too high.

With the aircraft in the nose-high attitude throughout the takeoff roll, the wings will progressively relieve the landing gear of more and more of the airplane's weight. This reduces the retarding effect of the surface. If you maintain the proper attitude, the airplane will fly itself off the ground at an airspeed that is slower than the normal liftoff airspeed. In fact, because of ground effect, most aircraft will become airborne at a speed that is slower than the safe climb speed.

THE CLIMB

After liftoff, lower the nose slightly, but make sure you keep the wheels above the surface as you accelerate within ground effect. An airplane climbing out of ground effect requires a higher airspeed or an increased angle of attack to maintain the same amount of lift. This is due to the increase in induced drag or thrust required as ground effect decreases.

Decrease the pitch attitude slightly after liftoff to allow the airplane to accelerate to a safe climb speed.

If there are obstacles in your departure path, accelerate to V_X or the recommended speed and begin an obstacle clearance climb. When you are clear of all obstacles, accelerate to V_Y and retract the flaps as recommended by the airplane manufacturer.

SOFT-FIELD APPROACHES AND LANDINGS

The soft-field approach and landing procedure also is intended for a runway surface covered with snow, soft sand, mud, high grass, loose rocks, or rough, uneven terrain. Before you attempt a landing on this type of field, refer to the POH and note any pertinent recommendations. You should also make an effort to determine how soft or rough the landing surface is. For example, after heavy rain or snow, the surface may be too soft or hazardous for a safe landing. In addition, you should consider if a takeoff from that field is practical, or even possible, after you land. Of course, in an emergency, your choices are usually limited, and you may have to land on a very soft or rough surface.

DESCRIPTION

The objective of a soft-field landing is to support the weight of the aircraft with the wings during the landing roll as long as it is practical. In other words, you want to delay the weight transfer to the wheels during the roll-out until the aircraft is at the slowest possible speed. This minimizes drag and stresses that may be imposed on the landing gear by a soft or rough surface.

PROCEDURE

The soft-field approach is similar to a normal landing approach. Most likely, you will use the same final approach speed that you used for a short-field procedure. Full flaps are usually recommended and, in most cases, you will use a normal descent angle. There is no advantage to a steep approach unless obstacles restrict the final approach path.

After you are established on the final approach in the landing configuration, maintain precise control of your airspeed, descent rate, and ground track. Make smooth, timely adjustments, as required.

Although airspeed, pitch, and power control are important throughout the approach, they are critical during the landing flare and touchdown. Maintain some power throughout the flare to help arrest the rate of descent. You also must control drift. At touchdown, the longitudinal axis of the aircraft should be aligned with the landing area. [Figure 10-15]

During the flare before touchdown for a soft-field landing, use at least a minimal amount of power to control the rate of descent.

As airspeed decreases prior to touchdown, you will have to increase back pressure on the control wheel. The actual touchdown should occur in a nose-high attitude just above the stalling speed. The use of power during the flare provides increased airflow over the airplane's tail surfaces and helps you control the descent. The amount of power required will vary with different airplanes, landing weights, and density altitudes.

While the use of flaps during soft-field procedures decreases landing speeds and is normally recommended, there are certain disadvantages. On low-wing airplanes, the flaps may be damaged by mud, slush, or stones thrown up by the wheels. Generally, leaving the flaps down is preferable to trying to retract them during the landing roll. Raising the flaps is usually less critical than giving your full attention to controlling the aircraft.

On soft surfaces, deceleration is rapid after touchdown, so you may not need to use brakes. Braking at this time may even cause a nose-over

Figure 10-15. Begin the landing flare normally (position 1), transition through a level flight attitude (position 2), establish a nose-high landing attitude (position 3), and hold this attitude as speed decreases during touchdown (position 4). After touchdown, hold sufficient elevator back pressure to keep the nosewheel off the ground. Then, lower it gently while you still have sufficient elevator control effectiveness. Throughout the landing roll, maintain positive directional control.

tendency. In fact, to prevent bogging down or getting stuck, you may have to add power to keep the airplane moving. Taxi procedures generally require more power, and you should maintain back pressure on the control yoke to decrease weight on the nosewheel. When you are clear of the landing area, complete your after-landing check.

COMBINED PROCEDURES

For the sake of simplicity and ease of description, we have discussed only one takeoff or landing procedure at a time. However, in the course of your flying career, you are likely to encounter situations where more than one procedure is necessary. For example, if you are required to land in a crosswind on a soft field with obstacles at the end of the runway, you would need to use a combination of crosswind, soft-field, and short-field procedures. Under these circumstances, you will need to modify the procedures you have learned to fit the existing conditions. As you gain experience, you will develop the knowledge and the skills necessary to cope with a variety of conditions and be able to use the full performance capabilities of your aircraft.

CHECKLIST

After studying this section, you should have a basic understanding of:

- ✓ **Short-field takeoff and climb** — How to safely achieve maximum performance by using the appropriate procedures, speeds, and configurations.
- ✓ **Short-field approach and landing** — What the procedures and performance considerations are, including use of the appropriate airspeed, configuration, descent angle, and braking.
- ✓ **Soft-field takeoff and climb** — What the objectives and procedures are, including how to obtain maximum performance before liftoff and after liftoff while flying within ground effect.
- ✓ **Soft-field approach and landing** — What the procedures and performance considerations are, including use of the proper airspeed, configuration, and landing flare/touchdown technique.
- ✓ **Combined procedures** — How you may need to modify or combine different takeoff or landing procedures to fit a given situation, such as a short field with a soft surface.

SECTION D

LAZY EIGHTS AND PYLON EIGHTS

As a commercial pilot applicant, you will also be required to become proficient in advanced figure eight maneuvers. These maneuvers may include lazy eights, eights-around-pylons, and eights-on-pylons. Lazy eights are considered performance maneuvers, while eights-around- and eights-on-pylons are ground reference maneuvers.

LAZY EIGHTS

The lazy eight consists of a series of climbs, descents, turns, and various combinations of each. By practicing this training maneuver, you will continue to develop and improve your coordination, orientation, planning, division of attention, and ability to maintain precise control of the aircraft. Throughout the maneuver, airspeed, altitude, bank angle, and pitch attitude, as well as control pressures, are constantly changing. Because of these constant changes, you cannot fly the lazy eight mechanically or automatically. A high degree of piloting skill and a sound understanding of the associated performance factors are required.

DESCRIPTION

The lazy eight is essentially two consecutive 180° turns in opposite directions, with each turn including a climb and a descent. It is called a "lazy" eight because the longitudinal axis of the aircraft appears to scribe a flight pattern about the horizon that resembles a figure eight lying on its side. A good way to visualize the lazy eight is to break each 180° turn into segments. [Figure 10-16]

PROCEDURE

As with all of the training maneuvers, you should first check the area for other traffic. Since you will be changing directions and altitudes continually, be particularly careful to maintain vigilance throughout the maneuver. While scanning the area, look for visual reference points that you can use for orientation. You should also try to determine the direction of the wind.

Use an entry altitude that will allow you to complete the entire maneuver at least 1,500 feet above ground level. Start from a straight-and-level attitude at normal cruise power and at the recommended airspeed. A turn into the wind during the first part of the maneuver is best. This helps you maintain loops of equal size. A crosswind will make the loops

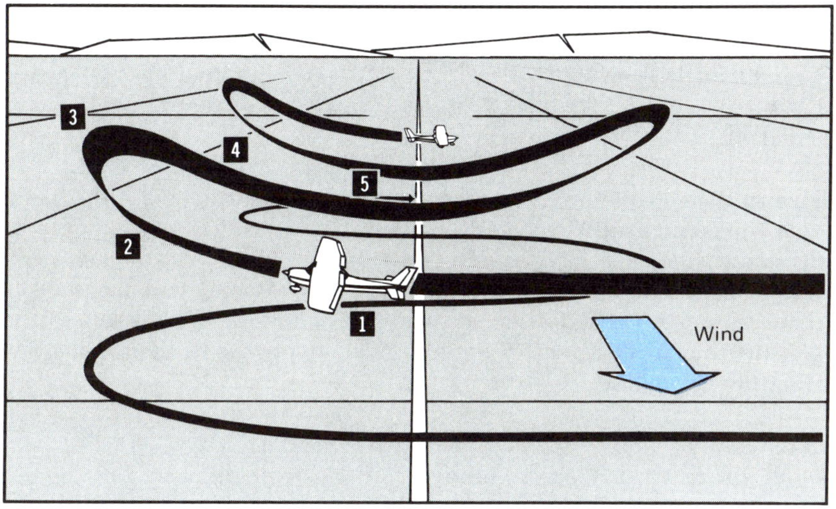

Figure 10-16. This three-dimensional view shows the complete maneuver, with the key points during the first 180° of turn identified by numbered callouts. Position 1 is the entry point, position 2 is the 45° point, position 3 is the 90° point, position 4 is the 135° point, and position 5 is the 180° point. You may want to refer back to this illustration during the accompanying discussion.

of the lazy eight cross the horizon at different points, and the longitudinal axis of the aircraft will draw an asymmetrical figure about the horizon. Making turns into the wind also helps you keep the aircraft within a given area. [Figure 10-17]

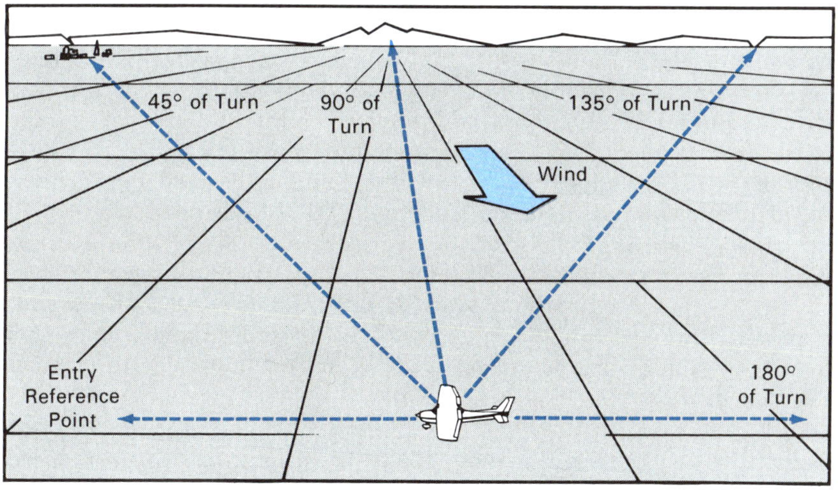

Figure 10-17. During the entry, align your flight path with an initial reference point. Begin the first climbing turn when you are abeam the 90° reference point. Additional reference points at 45°, 135°, and 180° are also useful. All of these references should be on the horizon, well beyond the maneuvering area.

When you start the gradual climbing turn in the direction of the 45° reference point, you must carefully plan and control your pitch attitude and bank angle. You should reach the 45° point with your steepest nose-high pitch attitude. At this point, your bank angle should be 15° and increasing.

Between the 45° and 90° reference points, continue to increase the bank angle while you gently decrease the pitch attitude. Since your airspeed is still decreasing, use right rudder pressure, as required, to counteract torque. You should pass through a level flight pitch attitude at the 90° point. Throughout the first 90° of turn, you increase the bank angle at a constant rate so you reach the maximum bank angle of approximately 30° at the 90° point.

As the nose of your airplane passes through the 90° point, slowly begin to roll out of the 30° bank and continue to lower the nose for the descending turn. You should reach the slowest airspeed during the maneuver at the 90° point. From this point through the 180° point, your airspeed gradually increases and your bank angle decreases.

When you reach the 135° reference point, your airplane should be at the steepest nose-low pitch attitude and your bank angle should be about 15° and decreasing. As your airspeed increases, gradually relax rudder pressure while you continue to slowly roll out of the turn and simultaneously raise the nose. You must time your rate of roll out and pitch change so you reach straight-and-level flight after 180° of turn. At the 180° point, you should be at the same altitude and airspeed that you used to enter the maneuver.

As you reach level flight after 180° of turn, immediately begin the climbing turn in the opposite direction. You should not fly the airplane straight and level, but only pass through the straight-and-level attitude at the 180° point of turn. Use the same procedure throughout the next 180° of turn in the opposite direction. Preplanning the events in each 45° segment helps you visualize and anticipate what to do next.

When properly flown, the peak of the loop above the horizon should come at the 90° point, and the lowest altitude should come as the aircraft is passing through straight-and-level flight. In addition, you must plan for the continual variation of bank angle and pitch attitude throughout the maneuver.

The symmetry of the loops during a lazy eight depends on your ability to control airspeed by adjustments of the pitch attitude.

One of the key factors in making symmetrical loops is proper airspeed control. Since the power is set before you begin the maneuver, you control airspeed by varying the pitch attitude. During the first 90° of turn, which is the climbing segment, the airspeed should decrease from the entry speed to slightly above the stall speed at the 90° point. This will occur only if you constantly adjust your pitch attitude throughout the maneuver. You should pass through a level flight pitch attitude at the 90° point, then gradually establish a nose-low pitch attitude that allows your aircraft to accelerate to the entry speed after 180° of turn.

If you use too steep a pitch attitude in the climbing turn, you may stall the airplane before it reaches the 90° point. The nose of the aircraft should pass through the 90° reference point at the minimum speed for the maneuver which is slightly above stall speed.

Too low a pitch attitude in the second portion of the turn results in an excessive dive. This causes the aircraft to exceed entry speed at the 180° point. The excessive nose-low pitch attitude also causes you to lose too much altitude. As a result, you enter the second half of the lazy eight at the wrong airspeed and altitude.

Since the maneuver may seem easier to perform with steeper banks, you may have a tendency to steepen the bank angle beyond normal. The steeper bank will cause more rapid changes in pitch and bank, with a resulting lack of precise control. In most training airplanes, you should perform the lazy eight as a slow, lazy maneuver with only 30° of bank at the steepest point.

The maximum bank angle during a lazy eight normally is 30°.

EIGHTS-AROUND-PYLONS

Although no longer a listed task for the commercial pilot practical test, eights-around-pylons require many of the piloting skills you developed while learning the basic turn maneuvers, as well as skills necessary for maneuvering by reference to ground objects. They improve your skills in planning, coordination, aircraft control, wind drift correction, and division of attention.

DESCRIPTION

The objective of this ground reference maneuver is to fly a figure eight ground track around two points, or pylons, on the ground. This requires two 360° turns, one in each direction, with a brief straight-and-level segment between the turns. Normally, you will have to vary the bank angle throughout the turns to maintain a constant turn radius. Since the amount of bank required depends on the distance between the pylons, as well as the wind conditions, choosing the correct pylons is important.

PROCEDURE

Begin eights-around-pylons with normal clearing procedures. Then, carefully select the pylons. They should be in a line perpendicular to the wind. In addition, the pylons should be easy to see and they should be located in an open area clear of hazardous obstructions. The distance

Make sure the distance between the pylons is compatible with your average groundspeed and bank angle during the maneuver.

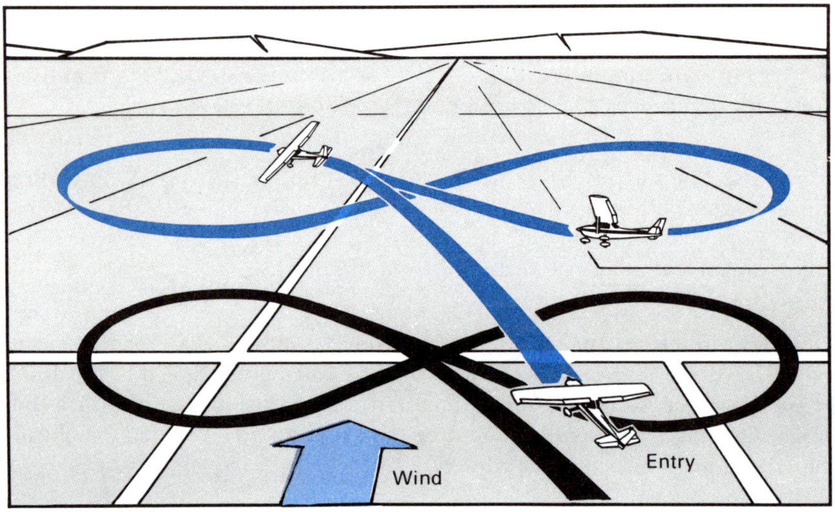

Figure 10-18. In this example, the pylons are road intersections. Enter between the pylons from a downwind position. Maintain the recommended entry airspeed and a constant altitude (600 to 1,000 feet AGL) throughout the maneuver. Your bank angle should not exceed approximately 30° to 40° at the steepest point.

between them must be appropriate for your anticipated groundspeed and bank angle. [Figure 10-18]

Vary your bank angle, as necessary, to maintain a constant radius of turn.

During turning segments of this maneuver, you must adjust the angle of bank to maintain a constant turn radius. Since the highest groundspeed occurs while you are flying downwind, the steepest bank angle is required just after the entry point. From this point, gradually decrease the bank to the shallowest angle while you are flying directly into the wind on the outer rim of the figure eight. During the brief, diagonal straight-and-level segment between the turns, maintain a heading that compensates for wind drift. Follow the same procedure for the turn around the second pylon. Be sure to maintain the entry altitude and airspeed throughout the maneuver.

EIGHTS-ON-PYLONS

Eights-on-pylons also involve flying a figure eight around two points, or pylons, on the ground. However, with this maneuver, you do not attempt to maintain a constant altitude and a uniform radius around the pylon. Instead, you fly the airplane at an altitude and airspeed that allows you to keep a reference point (usually near the wingtip) on the pylon. The reference point should appear to pivot about the pylon. The objective of eights-on-pylons is to refine your ability to control the airplane at traffic pattern altitude over a varied ground track while dividing your attention between instrument indications and visual cues outside the aircraft. [Figure 10-19]

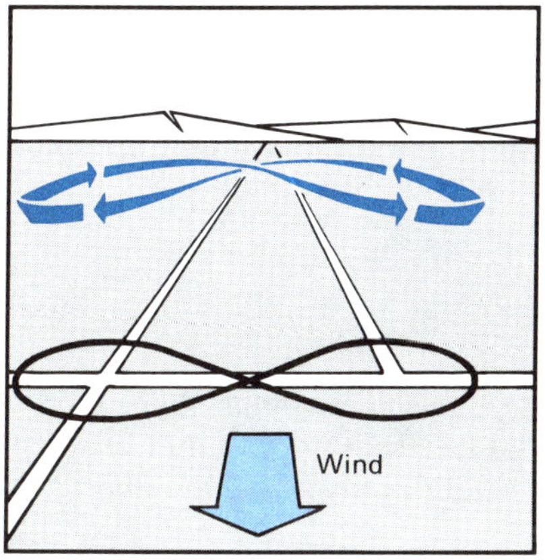

Figure 10-19. A complete maneuver consists of a turn in one direction around the first pylon, followed by a turn in the opposite direction around the second pylon. Notice that altitude decreases slightly during segments of the maneuver where the strongest headwind occurs.

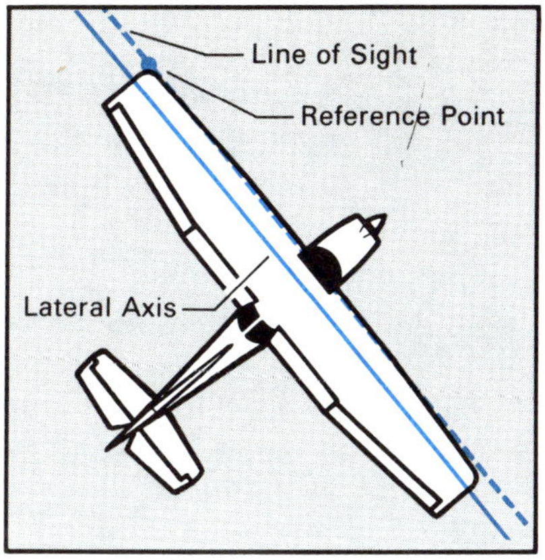

Figure 10-20. Choose a reference point on, or near, the wingtip, so your line of sight through the reference point is parallel to the lateral axis of the aircraft.

DESCRIPTION

Since you perform this maneuver at a relatively low altitude, select pylons that are in an open area and are not near hills or obstructions. Obstruction-induced turbulence, as well as updrafts and downdrafts caused by uneven terrain, increase the difficulty of the maneuver. Select pylons with approximately the same elevation to avoid the added burden of adjusting altitude for variations in terrain. The pylons should also be in a line which is perpendicular to the wind, and they should be spaced to provide three to five seconds of straight-and-level flight between the turns. [Figure 10-20]

Carefully select pylons that will be visible, clear of obstructions, and properly oriented in relation to the prevailing wind.

The reference point may vary considerably on different airplanes. It may be above the wingtip on a low-wing aircraft, below the wingtip on a high-wing airplane, or forward of the wingtip on a tapered-wing airplane. This reference point is also affected by your sitting height and position in the airplane.

Another key element in eights-on-pylons is altitude. When an aircraft is in turning flight, a specific altitude will make the aircraft appear to pivot about the pylon rather than simply turn around it. This is known as the **pivotal altitude**, and it is governed by the aircraft's groundspeed. As groundspeed increases, pivotal altitude increases, and vice versa. Pivotal altitude does not change as you vary the bank angle unless the bank is steep enough to affect your groundspeed.

Pivotal altitude increases as groundspeed increases, and it decreases as groundspeed decreases.

Pivotal Altitude = $TAS^2 \div 15$.

You can estimate pivotal altitude by using a simple formula: TAS (in m.p.h.)^2 ÷ 15 = Pivotal Altitude. As an example, if your true airspeed is 115 m.p.h. (100 knots), your estimated pivotal altitude is 882 feet (115^2 = 13225 ÷ 15 = 882). Since you will seldom know the exact TAS, groundspeed, and elevation, you will need to determine your actual pivotal altitude by experimenting.

PROCEDURE

As you should do for all training exercises, clear the area before you begin, and remain vigilant throughout the maneuver.

To begin eights-on-pylons, clear the area, select the pylons, and adjust the power for the recommended entry airspeed. Select an entry altitude that you estimate will be close to the pivotal altitude as you approach the pylons. [Figure 10-21]

Throughout the maneuver, your bank angle should be no more than 30° to 40° at the steepest point. During the banked segments, align your reference point and line of sight with the pylon. [Figure 10-22]

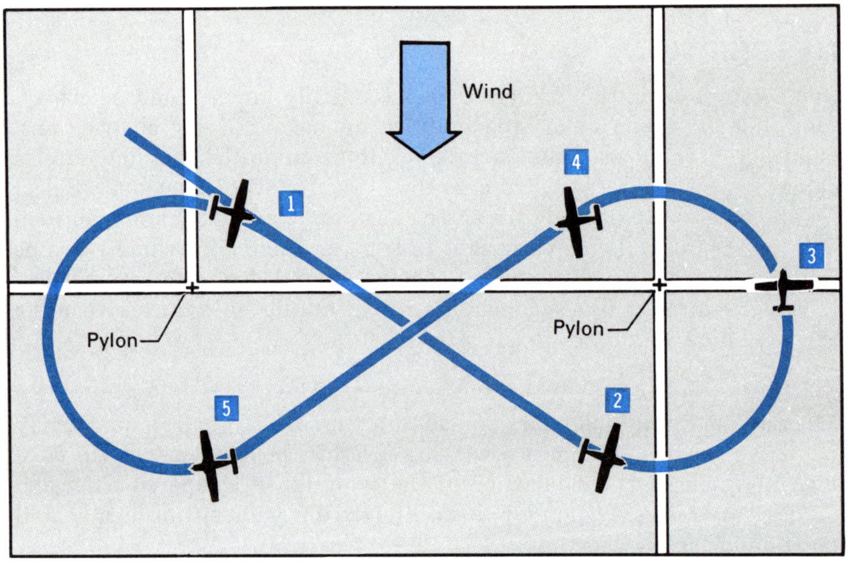

Figure 10-21. Enter diagonally between the pylons from downwind (position 1). Maintain straight-and-level flight until you are approximately abeam the first pylon (position 2). While heading downwind, you will have the highest groundspeed, therefore, the highest pivotal altitude. Your lowest groundspeed and pivotal altitude occur while flying directly into the wind (position 3). As you proceed around the first pylon, your groundspeed and pivotal altitude gradually increase as you return briefly to straight-and-level flight (position 4). Maintain the straight-and-level segment for three to five seconds and crab into the wind, as necessary, to correct for wind drift. When you are abeam the second pylon (position 5), begin a turn in the opposite direction.

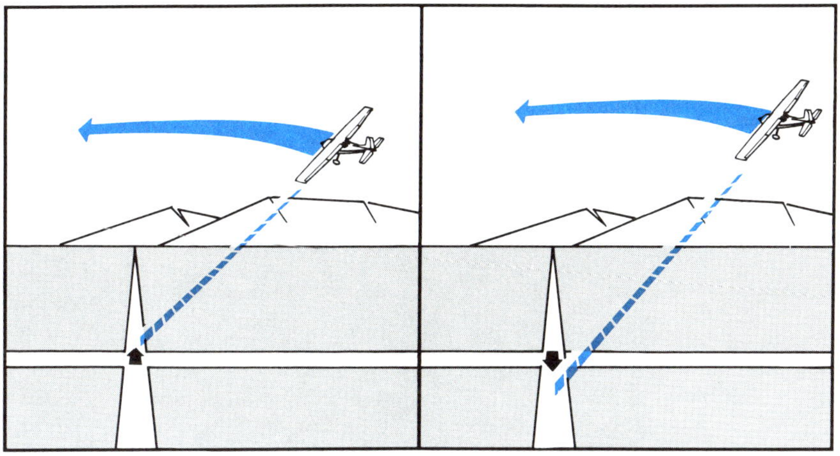

Figure 10-22. Movement of your line of sight reference point in relation to the pylon is the best indication that you need to adjust the altitude. If the reference point appears to move ahead of the pylon, as shown on the left, you should increase the altitude slightly. A climb has a double effect on pivotal altitude. First, the climb itself provides a correction and, second, the climb decreases groundspeed. This, in turn, lowers the pivotal altitude. If the reference point seems to move behind the pylon, as shown on the right, the aircraft is too high, and you must descend to the pivotal altitude. This also has a double corrective effect, since the process of descending to the pivotal altitude increases airspeed which, in turn, raises the pivotal altitude.

It is important that you use altitude changes, rather than rudder pressure, to hold the reference point on the pylon. Use of the rudder may yaw the wing back temporarily, but the end result will be a yawing, uncoordinated flight condition. You should maintain coordinated flight throughout the maneuver.

Do not attempt to keep your reference point on the pylon by uncoordinated use of the rudder.

CHECKLIST

After studying this section, you should have a basic understanding of:

✓ **Lazy eights** — What the main performance factors are and how to achieve maximum performance.

✓ **Entry and recovery procedures** — How to enter and recover from the lazy eight.

✓ **Eights-around-pylons** — How to describe the maneuver and what the basic requirements are.

✓ **Eights-on-pylons** — How to describe the maneuver, what pivotal altitude is, and how to adjust it to maintain the reference point on the pylons.

ABBREVIATIONS

A

AC — convective outlook
ADF — automatic direction finder
ADIZ — air defense identification zone
ADM — aeronautical decision making
AFSS — automated flight service station
AIM — Aeronautical Information Manual
AIRMET — airmen's meteorological information
ALS — approach light system
ALSF — ALS with sequenced flashers
ALTN — alternate
AM — amplitude modulation
AMA — area minimum altitude
AME — aviation medical examiner
APP — approach (also AP)
APP CON — approach control (also APC)
ARINC — Aeronautical Radio, Inc.
ARP — airport reference point
ARSR — air route surveillance radar
ARTCC — air route traffic control center
ARTS — automated radar terminal system
ASOS — Automated surface observation system
ASR — airport surveillance radar
ATC — Air Traffic Control
ATCRBS — ATC radar beacon system
ATIS — automatic terminal information service
AWOS — automated weather observation system
AWW — severe weather forecast alert

B

BC — back course
BCN — beacon
BRG — bearing

C

CAS — calibrated airspeed
CAT — clear air turbulence
CAT — category
CDI — course deviation indicator
CEIL — ceiling (also CIG)
C_L — coefficient of lift
C_{Lmax} — coefficient of lift maximum
CLNC — clearance
CLNC DEL — clearance delivery
COP — change-over point
CP — center of pressure
CRM — cockpit resource management
CRP — compulsory reporting point
CRS — course
CTAF — common traffic advisory frequency
CVFP — charted visual flight procedures
CWA — center weather advisories

D

DA — density altitude
DA(H) — decision altitude (height)
DEP — departure (also DP)
DEP CON — departure control (also DPC)
DH — decision height
DME — distance measuring equipment
DOD — Department of Defense
DOT — Department of Transportation
DUATS — direct user access terminal system

E

EAS — equivalent airspeed
EFAS — enroute flight advisory service (Flight Watch)
EFC — expect further clearance
EGT — exhaust gas temperature

F

FA — area forecast
FAC — final approach course
FAF — final approach fix
FD — wind and temperatures aloft forecast
FDC — Flight Data Center
FIR — Flight Information Region
FL — flight level
FLIP — Flight Information Publication

G

GCA — ground controlled approach
GPS — global positioning system
GS — glide slope (also groundspeed)
GSIP — glide slope intercept point

H

HAA — height above airport
HAT — height above touchdown
HIRL — high intensity runway lights
HIWAS — hazardous in-flight weather advisory service
HSI — horizontal situation indicator
HVOR — high altitude VOR

I

IAC — instrument approach chart
IAF — initial approach fix
IAP — instrument approach procedure
IAS — indicated airspeed
ICAO — International Civil Aviation Organization
IF — intermediate fix
IFR — instrument flight rules
ILS — instrument landing system
IM — inner marker
IMC — instrument meteorological conditions
INS — inertial navigation system
INT — intersection
IR — IFR military training route
ISA — International Standard Atmosphere
IVSI — instantaneous vertical speed indicator

K

KIAS — knots indicated airspeed
KTAS — knots true airspeed

L

LAA — local airport advisory
LCL — local
LDA — localizer-type directional aid
LDI — landing direction indicator
LDIN — lead-in light system
LIM — locator inner marker
LIRL — low intensity runway lights
LLWAS — low-level wind shear alert system
LMM — locator middle marker
LOC — localizer
LOM — locator outer marker
LORAN — long range navigation
LR — lead radial
LT — local time
LVOR — low altitude VOR

M

MAA — maximum authorized altitude
MAC — mean aerodynamic chord
MALS — medium intensity approach light system
MALSF — MALS with sequenced flashers
MALSR — MALS with RAIL
MAP — missed approach point
MAP — manifold absolute pressure
MCA — minimum crossing altitude
MDA(H) — minimum descent altitude (height)
MDH — minimum descent height
MEA — minimum enroute altitude
METAR — aviation routine weather report
MF — medium frequency
MHA — minimum holding altitude
MIRL — medium intensity runway lights
MKR — marker
MLS — microwave landing system
MM — middle marker
MOA — military operations area
MOCA — minimum obstruction clearance altitude
MRA — minimum reception altitude
MSA — minimum safe altitude
MSAW — minimum safe altitude warning
MTR — military training route
MVA — minimum vectoring altitude
MVFR — marginal VFR

N

NA — not authorized
NAR — North American Route (also refers to North Atlantic Route)
NASA — National Aeronautics and Space Administration
NAWAU — National Aviation Weather Advisory Unit
NDB — nondirectional beacon/radio beacon
NMC — National Meteorological Center
NOAA — National Oceanic and Atmospheric Administration
NoPT — no procedure turn
NORDO — no radio
NOS — National Ocean Service
NOTAM — Notice to Airmen
NSSFC — National Severe Storms Forecast Center
NTSB — National Transportation Safety Board
NWS — National Weather Service

O

OBS — omnibearing selector
ODALS — omnidirectional approach light system
OM — outer marker
OTS — out of service

P

PA — pressure altitude
PAPI — precision approach path indicator
PAR — precision approach radar
PCL — pilot controlled lighting
PIREP — pilot report
PPR — prior permission required
PT — procedure turn
PTS — practical test standard

R

R — radial
RA — radio altimeter
radar — radio detection and ranging
RADAT — freezing level data
RAIL — runway alignment indicator lights
RAPCON — radar approach control (USAF)
RB — relative bearing
RCAG — remote communications air/ground facility
RCLS — runway centerline light system
RCO — remote communications outlet
REIL — runway end identifier lights
RMI — radio magnetic indicator
RMK — remark
RNAV — area navigation
RRL — runway remaining lights
RRWDS — radar remote weather display system
RSC — runway surface condition
RTS — return to service
RVR — runway visual range
RVV — runway visibility value
rwy — runway

S

SALS — short approach light system
SAR — search and rescue
SCATANA — Security Control of Air Traffic and Air Navigation Aids
SD — radar weather report
SDF — simplified directional facility
SFL — sequenced flashing lights
SHF — super high frequency
SID — standard instrument departure
SIGMET — significant meteorological information
SPECI — Aviation selected special weather report
SQ — squawk
SSALS — simplified SALS
SSALSF — SSALS with SFL
SSALSR — SSALS with RAIL
SSV — standard service volume
STAR — standard terminal arrival route
STBY — standby
STC — sensitivity time control

T

TA — transition altitude
TACAN — tactical air navigation
TAF — terminal aerodrome forecast
TAS — true airspeed
TCAS — traffic alert and collision avoidance system
TCH — threshold crossing height
TDZ — touchdown zone
TDZ/CL — touchdown zone and runway centerline lighting
TDZE — touchdown zone elevation
TDZL — touchdown zone lights
TEC — tower enroute control
TEL-TWEB — telephone access to TWEB
TERPS — U.S. Standard for Terminal Instrument Procedures
TIBS — telephone information briefing service
TLv — transition level (also TL)
TPA — traffic pattern altitude
TRACON — terminal radar approach control
TVOR — terminal VOR
TWEB — transcribed weather broadcast

U

UA — pilot report (PIREP) message identifier
UFN — until further notice
UNICOM — aeronautical advisory station
UTC — Coordinated Universal Time (Zulu time)

V

V-speeds — performance speeds (see FAR Part 1.1)
VASI — visual approach slope indicator
VDP — visual descent point
VFR — visual flight rules
VIS — visibility
VMC — visual meteorological conditions
VOR — VHF omnidirectional radio range
VOR/DME — collocated VOR and DME navaids
VORTAC — collocated VOR and TACAN navaids
VOT — VOR test facility
VR — VFR military training route
VSI — vertical speed indicator
VVI — vertical velocity indicator

W

WA — AIRMET message identifier
WFO — Weather Forecast Office
WMSC — Weather Message Switching Center
WPT — waypoint
WS — SIGMET message identifier
WSFO — Weather Service Forecast Office
WSO — Weather Service Office
WST — convective SIGMET message identifier
WW — severe weather watch bulletin
WX — weather

Z

Z — Zulu time (UTC)

INDEX

I

INDEX

A

M

N

O

P

R

S

T

NOTAMs

N

NOTAMs

The NOTAMs section is designed to inform you of recent developments that could affect your training.

PRIVATE VERSUS COMMERCIAL OPERATIONS

Common carriage and private carriage are common law terms which are applied to aircraft operations and these terms determine which regulations an operator must follow. Generally, private carriage may be conducted under FAR Parts 125 or 91, Subpart F. In commercial operations, an operator becomes a common carrier when it "holds itself out" to the public, or to a segment of the public, as willing to furnish transportation to any person who wants it. There actually are four elements in defining a common carrier; (1) a holding out or a willingness to (2) transport persons or property from (3) place to place (4) and for compensation.

FAR Part 125 applies to the nonscheduled operation of U.S.-registered civil airplanes with a seating capacity of 20 or more passengers, or a maximum payload capacity of 6,000 pounds or more. Common carriage cannot be involved. Thus, no certificate holder may conduct any operation under Part 125 which results from "holding out" to the public to furnish transportation. Further, no one who holds an operating certificate under FAR Part 121 or Part 135 is eligible to operate under Part 125. Operations conducted under Part 125 inside the United States must still comply with, and operate under, FAR Part 91.

To serve as pilot in command or second in command under Part 125, you must have at least a commercial pilot certificate with an appropriate category, class, and type rating. You must also have an appropriate medical certificate and an instrument rating. To act as pilot in command, you must have at least 1200 hours of flight time as a pilot, including 500 hours of cross-country flight time, 100 hours of night time, including at least 10 night takeoffs and landings, and 75 hours of actual or simulated instrument flight time, at least 50 hours of which were in actual flight. In addition, as a pilot in command under Part 125, you must have made at least three takeoffs and landings in the type aircraft in which you are to serve within the preceding ninety days. These may be accomplished in an approved visual simulator.

BIG BOLD TYPE ON APPROACH CHARTS

As part of an ongoing process of improving chart clarity and readability, Jeppesen began adding bigger, bolder text to approach charts, starting March 10, 1995. This enhancement is a result of input from pilots and chart user groups. The intention of the big, bold type is to highlight the most important numbers and fixes/waypoints so they can be easily identified by pilots, especially during critical phases of a flight. In addition, Jeppesen is making selected changes to airport charts, SIDs, and STARs. At this time, radar approach charts are not being changed. The accompanying chart for the ILS Runway 15 approach at Butte, Montana, shows the new enhancements. You will also notice the colored contours to highlight rising terrain. [Figure N-1]

In the chart heading, bigger, bolder text is used to highlight communications frequencies, the primary navaid frequency and identifier, and the airport elevation (item 1). The Morse code identifier for the primary navaid has been removed from the heading, but remains on the plan view.

The chart plan view uses the bigger, bolder text to highlight the primary navaid frequency and identifier, and the inbound approach course in the navaid box (item 2). The inbound approach course is also shown in large type along the final approach path (item 3). Names of fixes which are part of the procedure are also enhanced (item 4). Secondary airports are subdued in a lighter color to differentiate them from the primary airport and reduce visual clutter (item 5). General information, including equipment requirements, procedure restrictions, and pilot controlled lighting, is located in a standardized plan view box (item 6).

In the profile view of the chart, big bold type is used for navaids and fixes (item 7), the TDZE, when charted, the runway touchdown zone elevation (item 8), the final approach course (item 9), minimum altitudes and glide slope interception altitudes (item 10), and the procedure turn limit (item 11). The missed approach text contains big bold type for altitudes, headings or courses, and navaid or fix identifiers (item 12). The airport elevation is omitted when the TDZE or runway threshold is charted. The landing minimums are also highlighted with big bold type, including all DA(H)s and MDA(H)s (item 13), and when applicable, Category II radio altimeter readings (RAs).

COLORED CONTOURS ADDED TO APPROACH CHARTS

The first Jeppesen approach charts with the new color contours were introduced in June 1994. The colors, which are graduated tints of brown, depict gradients of terrain elevation. For contrast, terrain high points and manmade structures, and their associated elevations, will continue to be shown in black on approach charts. Jeppesen also plans to apply the new color specifications to area charts and other types of charts which currently show terrain contours or area minimum altitudes (AMAs). The new colored charts will gradually be phased in during the next couple of years.

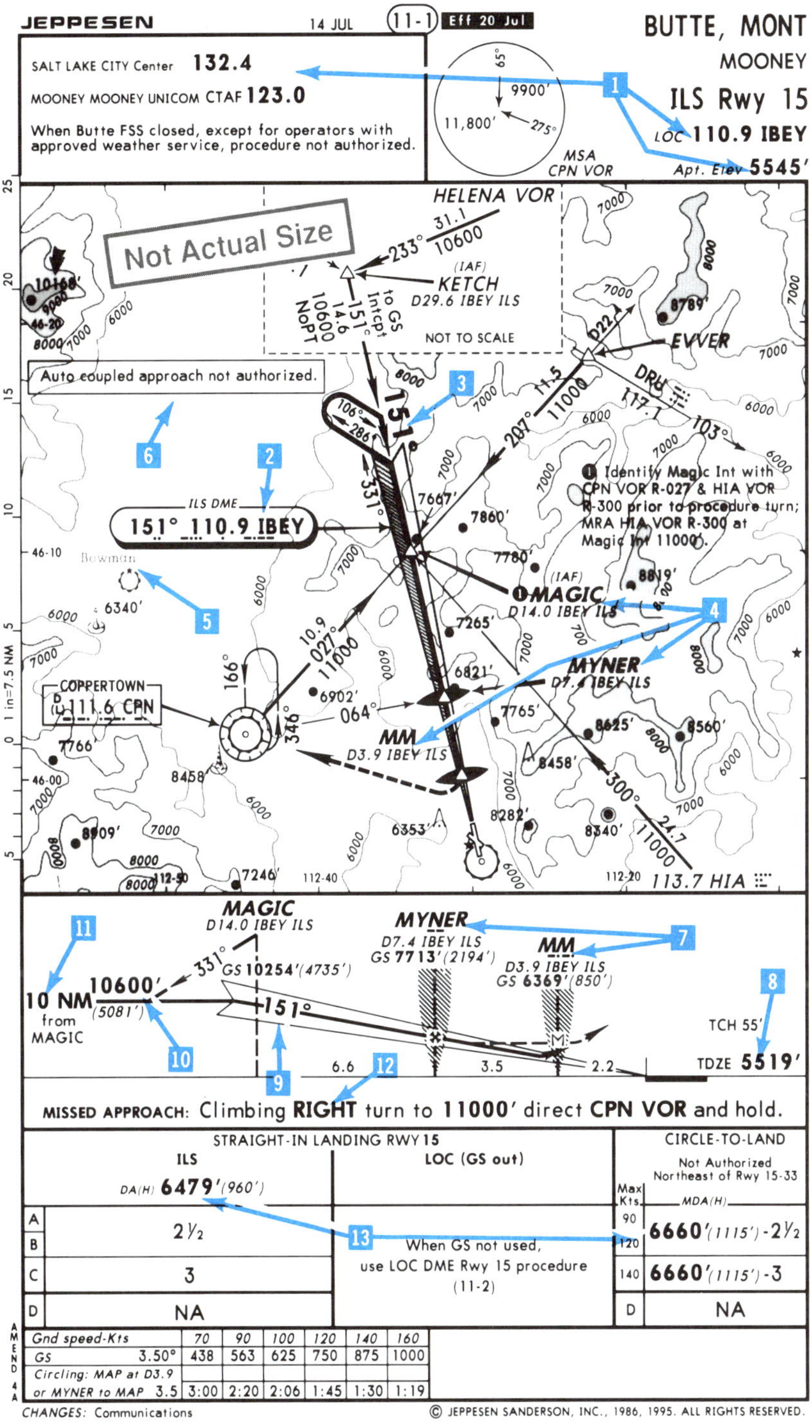

Figure N-1. This approach chart shows the bigger, bolder type which enhances readability. Also shown are the shaded color terrain contours.

JEPPESEN TESTS APPROACH BRIEFING STRIP

Before flying an instrument approach, a thorough review of the procedure must be completed. The FAA recommends a pre-approach briefing as a standard flight deck procedure, especially for airline, corporate, and other commercial flight crews. Also, in recent years, crew resource management (CRM) concepts have become more widely accepted and used. Flight safety is enhanced when all flight crew members are involved and familiar with the approach procedure.

As part of the ongoing effort to improve chart format and readability, Jeppesen has worked closely with the Air Transport Association (ATA) and other interested organizations to review chart-related issues. With the increasing use of new aircraft, advanced avionics, and electronic flight management systems, it appears that new adaptations of the standard chart format may be necessary. A primary objective of revising the chart format is to improve aviation safety by reducing chart clutter and crew workload associated with using the charts in the low altitude regime.

As a result, Jeppesen and the ATA have developed a new format which is being applied to selected approach charts on a trial basis. These charts are issued to a limited number of airlines and other operators who have agreed to test them operationally. A major feature is the addition of a briefing strip at the top of the chart. [Figure N-2]

The information in this strip (item 1) was arranged to be consistent with pre-approach briefing actions. It begins with the primary navaid frequency and identifier, the final approach course, and the glide slope crossing altitude at the outer marker, or for non-precision approaches, the minimum crossing altitude at the FAF. Next is the decision altitude (height) or minimum descent altitude (height) for a straight-in landing. Under the title is the four-letter database identifier for the airport, useful for equipment such as GPS and other RNAV systems. The next line on the briefing strip includes the airport and runway or touchdown zone elevations, all notes applicable to the approach, a depiction of approach lighting, and the missed approach procedure. The third and fourth lines contain the communications frequencies, generally in the order in which they are used.

The big, bold type discussed earlier highlights the most important information on the plan view, profile view, and landing minimums sections of the chart. The minimum safe altitude (MSA) circle is moved from the heading of the chart and displayed on the plan view (item 2). Added to the profile view is symbology for initial pilot actions for missed approach, indicating "up" and "out" (item 3). Complete missed approach instructions are contained in the briefing strip. Sidestep landing minimums, when depicted, are shaded with a screen to differentiate them from normal straight-in minimums (item 4).

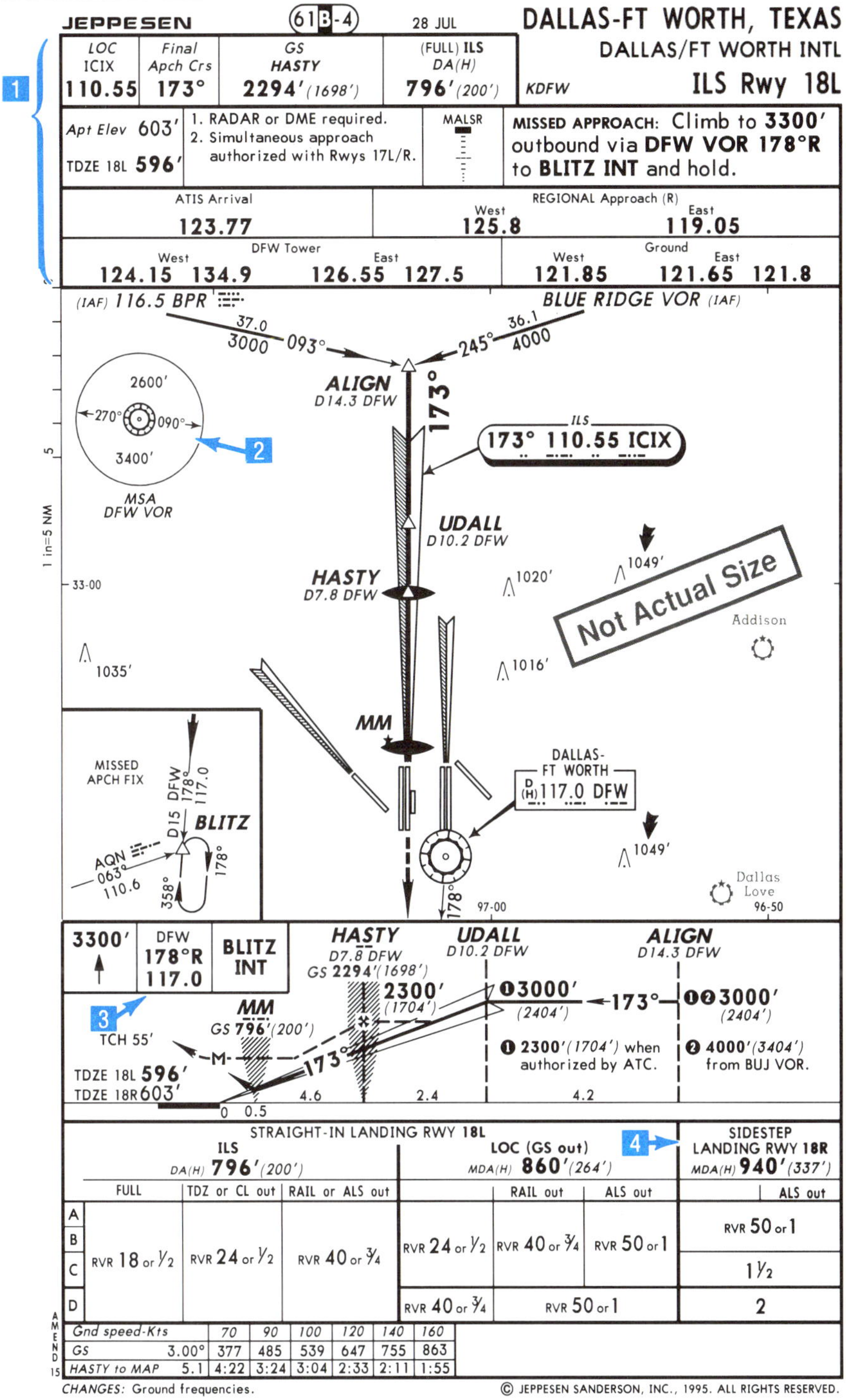

Figure N-2. This modified Jeppesen approach chart is intended to provide an improved depiction of key flight information that also meets requirements for a pre-approach briefing. The sequence of elements is based on modern human factor and CRM considerations.

SUPPORT MATERIALS

Look for these support materials to complement your Instrument/Commercial Manual:

- **Instrument/Commercial Exercise Book** — Self-test questions and answers to reinforce your textbook study
- **Instrument/Commercial Syllabus** — A curriculum guide for your ground and flight training
- **FAR/AIM Manual** — Includes Instrument and Commercial FAR Study Lists and Self-Test Exercises, plus an order form for free one-time update
- **TechStar Electronic Flight Computer** — A versatile, easy-to-use computer with many special features like weight shift capability
- **ATC Clears** — Unique cassette provides practice in copying instrument clearances, using standard clearance shorthand
- **Holding Pattern Computer** — Easy-to-follow pictorial presentations to help you visualize standard and nonstandard holding patterns
- **In-Flight Training Service** — A special low-cost, 4-month subscription to Jepp charts for instrument students
- **Professional Pilot Logbook** — Convenient format designed for active pilots; holds up to 10 years of data
- **Instrument Training Hood** — Adjustable headband for comfortable fit; designed so it can be worn with eyeglasses
- **Softside Flight Case** — An attractive, durable flight case to carry your charts, logbook, flight computer, and other flying accessories

These items are among the wide variety of Jeppesen materials that are available through your authorized Jeppesen Dealer. If there is no Jeppesen Dealer in your area, you can contact us directly:

Jeppesen Sanderson
Sanderson Training Systems Division
55 Inverness Drive East
Englewood, CO 80112-5498

Help yourself become a better pilot and lower the cost of training too!

The Jeppesen FS-200
PC-Based Instrument Flight Simulator

Now you can practice flying instrument approaches in the convenience of your own home without having to invest thousands of dollars in flight time just to develop basic IFR skills. Capable of running on your home PC, you can save time and money by practicing at home and going to the airport better prepared!

The Jeppesen FS-200 Instrument Flight Simulator is a complete system to help you learn and practice your IFR procedures. This system allows you to make all flight control inputs through physical controls, never using your mouse or keyboard during flight operations. Available with two hardware options, each includes the following standard components:

- Photo-realistic software
- Jeppesen NavData North American database
- FS-200 Pilot's Operating Handbook
- Jeppesen U.S. Low Flight Planning Chart
- One-Year Subscription Service, including one NavData database update, two Instrument Skill Enhancement Modules, software revisions occurring within the revision year and toll-free technical support.

The Jeppesen FS-200 Basic Package (includes all listed features plus Basic hardware control console)

$799.95

The Jeppesen FS-200 Advanced Package (includes all listed features plus Advanced hardware console)

$1,495.00

Optional ATC Packages

$49.95

To place your order or for additional information call us today

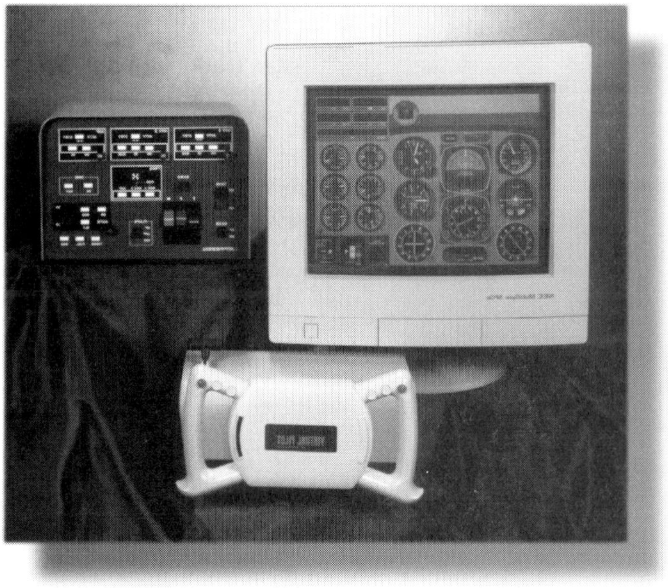

Computer System Requirements: IBM or compatible PC using a 486 or higher microprocessor, 8MB system RAM, memory, 5MB hard disk space, 1MB video RAM, color SVGA card and monitor, and a Soundblaster-compatible card for sound features.

Visit Your Jeppesen Dealer or Call 1-800-621-JEPP

Prices subject to change

Airway Manual Express

You Buy Where You Fly

This instrument chart service delivers small, conveniently segmented into easy-to-use geographic coverages. Complete new contents are issued every 56 days, eliminating time spent on filing revisions (Mid-Cycle 28-day updates are available).

You only purchase the coverage(s) you need. Airway Manual Express is available by subscription and single-issue retail purchase.

Area	Geographic Coverage	Single Issue	Annual Subscription (6 Issues)
Southeast	**GA,NC,SC**	**$19.95**	**$86.95**
Mid-South	**AL,KY,MS,TN**	**$19.95**	**$86.95**
Florida	**FL**	**$17.75**	**$76.95**
Mid-Atlantic	**DE,DC,MD, VA,WV**	**$17.75**	**$76.95**
East	**NJ,NY,PA**	**$19.95**	**$86.95**
Northeast	**CT,NH,ME, MA,RI,VT**	**$13.95**	**$61.45**
East Central	**IN,OH**	**$17.75**	**$76.95**
Southwest	**AZ,CO,NM, NV,UT**	**$17.75**	**$76.95**
California	**CA**	**$19.95**	**$86.95**
Alaska	**AK**	**$17.75**	**$76.95**
Central	**CO,KS,MO,NE**	**$19.95**	**$86.95**
No. Central	**IA,MN,WI**	**$19.95**	**$86.95**
Great Lakes	**IL,WI**	**$19.95**	**$86.95**
So. Central	**AR,LA,OK**	**$19.95**	**$86.95**
Texas	**TX**	**$20.95**	**$95.75**
Michigan	**MI**	**$13.95**	**$61.45**
Northwest	**WA,OR,ID**	**$13.95**	**$61.45**
West Central	**MT,ND, SD,WY**	**$17.75**	**$76.95**

Airway Manual Service Binders

Functional. Durable. Protects and lasts. Jeppesen's leather binders are made of top-grain cowhide and include steel hinges and safety locks to secure the rings. • Our sturdy Duraflo binder has the look and feel of our leather binders. Our flow-mold process laminates and binds sturdy, ethylene acetate with a clear, tough surlyn outer surface.

LB-2 Leather, 2-inch Rings, Stay-Open Bar. Item Number AM621122 $51.00
LB-1 Leather, 1-inch Rings, No Stay-Open Bar. Item Number AM621125 $46.45
FB-2 Duraflo, 2-inch Rings, Stay-Open Bar. Item Number AM621138 $32.25
FB-1 Duraflo, 1-inch Rings, No Stay-Open Bar. Item Number AM621139 $27.95
PB-2 Vinyl, 2-inch Rings, No Stay-Open Bar. Item Number AM621126 $22.75
PB-1.5 Vinyl, 1 1/2-inch Rings, No Stay-Open Bar. Item Number AM621131 $11.25

Visit Your Jeppesen Dealer or Call 1-800-621-JEPP

Prices subject to change